D1126958

# COMPUTER-AIDED ANALYSIS AND DESIGN OF ELECTROMAGNETIC DEVICES

# COMPUTER-AIDED ANALYSIS AND DESIGN OF ELECTROMAGNETIC DEVICES

S. Ratnajeevan H. Hoole
Associate Professor of Engineering
Harvey Mudd College
Claremont, California

Elsevier
New York • Amsterdam • London

Elsevier Science Publishing Co., Inc.
655 Avenue of the Americas, New York, New York 10010

Sole distributors outside the United States and Canada:
Elsevier Science Publishers B.V.
P.O. Box 211, 1000 AE Amsterdam, the Netherlands



Library of Congress Cataloging-in-Publication Data

Hoole, S. Ratnajeevan H.
Computer-aided analysis and design of electromagnetic devices/S. Ratnajeevan H. Hoole.
p. cm.
Bibliography: p.
Includes index.
ISBN 0-444-01327-X
1. Electric machinery—Design and construction—Data processing. 2. Electromagnetic fields—Data processing. I. Title.
TK2331.H66 1989
621.31'042—dc 19 88-15406
CIP

Current printing (last digit):
10 9 8 7 6 5 4 3 2 1

Manufactured in the United States of America

*This book is dedicated*
*to my parents*
*the late Rev. Richard H.R. Hoole and*
*Jeevamany Hoole*
*for all that they have given me*

*my wife*
*Dushyanthi*
*for her encouragement and numerous sacrifices*

*and my two little daughters*
*Mariyahl Mahilmany and Elisapeththu Elilini*
*who lost much fun as a result of this undertaking.*

# CONTENTS

# PREFACE

Computational electromagnetics has come of age, and sophisticated software packages are now available for the field analysis of large problems in a matter of minutes as opposed to much longer times by classical methods. As a result, the computer-assisted analysis of electromagnetic devices can no longer be viewed as an esoteric subject to be confined to graduate courses and must of necessity be moved to the undergraduate level so that the new generation of engineers is amply prepared for modern industry. Our ability today to solve all manner of field problems in the most general geometric shapes, with arbitrary source and material distributions, makes it now natural to teach the solution of field problems through numerical methods, rather than the somewhat more restricted classical schemes.

This book is an attempt, a successful one it is hoped, to present field analysis in simple terms so that it may be used for a senior undergraduate or first-level graduate course. Alternatively, the book may be used in conjunction with classical electromagnetics courses. Except for sections 2.5, 2.6, 4.2, and 5.5 and chapters 8 and 10, the content of the book is suitable for presentation in an undergraduate class. Algorithms have been given in pseudo-code for their heuristic value and so as to serve as a guide to students in writing their software for design purposes. As a teacher, I have found this a useful compromise between the spoon-feeding that results when computer listings are given and abandoning the student in the wilderness, as it were, by asking him (or her) to write a program merely on the basis of a lecture. My debt to my graduate students, especially to Konrad Weeber, who helped check these codes, is gratefully acknowledged. The ease with which my students used these codes to write their own software augurs well for the teaching of electromagnetics through numerical schemes.

The references given at the end of the book are merely to indicate a starting point for further reading and are not meant to be exhaustive or to ignore much good work by others not in the list. No explicit exercises are given, because the programming of the theory and the reworking of the problems worked in this book constitute sufficient homework. Matrix solution is introduced early in chapter 2 so as to allow the student to work in the now-common personal-computer environment without any reliance on library routines.

My thanks to Professor Martin Kaplan, Associate Head, Electrical and Computer Engineering Department, Drexel University, who first put to me the idea of writing this book. My thanks also to Dr. Shankara Reddy for his intelligent criticism of the text.

# 1 COMPUTATIONAL ELECTROMAGNETICS

## 1.1 NUMERICAL METHODS IN ELECTROMAGNETIC DEVICES

### 1.1.1 Numerical as against Classical Methods

Electromagnetic devices such as transistors, electrical machines, and waveguides have their behavior governed by the electromagnetic fields which flow in them; the fields, in turn, obey Maxwell's equations (Maxwell 1904; Panofsky and Phillips 1969). Thus, in order to be able to predict performance characteristics, it is necessary in the design of these devices to solve the Maxwell equations governing the fields. The differential (or alternative integral) form of Maxwell's equations has traditionally made electromagnetics a heavily mathematically oriented discipline.

Before the advent of the computer, recourse to elaborate mathematics had to be made to solve the equations of electromagnetics, using such solution concepts as series expansions, separation of variables, Legendre polynomials, Bessel functions, Schwarz–Christoffel transformations, Laplace transforms, and the like. By these methods, the solution of the electromagnetic field inside a complex device under design is a lengthy process, often taking days to solve a trivial problem. Indeed, it is frequently the case that no closed form solution is possible without making drastic simplifying assumptions on device geometry, current and charge distributions, and so on. As a consequence of these assumptions, the ensuing solution is not completely reliable and defeats the purpose of the analysis—which is to design accurately. Human error is another source of inaccuracy when these methods are used. In fact, a slight mistake is common and difficult to detect especially in the early stages of a several page derivation, and can be very frustrating.

Happily, however, with the advent of the digital computer and subsequent advances in computing power and storage devices (available at ever-decreasing costs), it is now possible to use simple numerical approximation schemes to solve large-scale problems (large in terms of equation complexity as well as device size) in a matter of a few minutes. Although these schemes are known as *approximation* methods, the term is in a sense misleading, since it is possible to increase the accuracy as much as desired, at some additional computing cost. True, these digital, approximate solutions may be less accurate than the closed form solutions from classical analysis for simple problem shapes such as circles and rectangles (which rarely exist in the real design environment). However, in the real world, which involves complex geometries and electromagnetic configurations, numerical schemes yield far more accurate solutions than are possible by classical analysis, in view of the latter methods' dependence on simplifying assumptions.

The importance of numerical techniques is now slowly being acknowledged in the light of available computing resources, and software is being introduced into the classroom as a teaching aid (Hoburg and Davis 1983; Davis and Hoburg 1985; Hoole and Hoole 1986). One may even go so far as to hazard the assertion that numerical schemes are generally superior to, and are therefore preferable to, classical methods; the exception being found in those few problems possessing neat closed form solutions. This is not to say that classical schemes are not useful. Indeed, they are often the foundation of numerical methods. In fact, their use in helping us to understand the behavior of fields, such as the cutoff frequency in a rectangular waveguide, cannot be overemphasized. When it comes to large problems involving complex geometries, however, closed form solutions are just not possible and only numerical schemes will do the job. With the sophisticated software packages now available for field computation, large problems may be solved in minutes as opposed to much longer times (if at all possible) taken by classical methods.

### 1.1.2 History of Computational Electromagnetics

While the theory of approximations has long been well known to mathematicians (Mikhlin and Smolitsky 1967), most numerical work in electromagnetics initially involved evaluating closed form expressions from classical analysis for specific device values (Binns and Lawrenson 1963). The Schwarz (1869) and Christoffel (1870) transformations were once popular for the solution of elaborate problems. Here we would use complex transforms to map the shape of the device into a simpler form, solve the field in the simplified device, and take the inverse transform for the real answer. A simple example of the Schwarz–Christoffel transform is that of two "parallel" L-shaped plates at different potentials, shaped like a pipe (such as occur in a pipe carrying some flow, which may be magnetic flux). The transform would convert the L-shaped object to a parallel plate capacitor whose solution in closed form is well known. This solution may be transformed back to its original shape. However, because of human limitations, only simple problems could be attempted by these methods.

In the 1920s and 1930s advances were made in graphical methods for the solution of field problems (Johnson and Green 1927; Stevenson and Park 1927). Although somewhat approximate, these methods allowed large, elaborate problems to be solved for the first time. Bewley's work in the 1930s for devising easy field plotting schemes is of particular note. He subsequently elaborated upon his methods in a book (Bewley 1948).

Most of the period up to the 1960s merely witnessed the use of lumped parameter circuit models of electromagnetic systems, as opposed to continuum models, a *continuum* being loosely defined as continuous space in which every location is associated with a value of an electromagnetic field quantity such as flux density, potential, etc. (Duffin 1959). These models were based on drastic assumptions and were developed for the analysis of electric machines (Kron 1959; Park 1929; Draper 1959; Adkins 1962; Ellison 1969). Lumped parameter models, although developed from fundamental laws on the behavior of magnetic fields, were derived with a view to obtaining terminal (or input/output) characteristics of devices and not for getting detailed information on what happens inside a device. As such, the computation involved is very simple and typically deals with six or fewer unknowns governed by linear equations. Another advantage here is that the lumped constants of the model may often be experimentally derived, so that for a given device, the model would be found quite accurate over the usual operating range of parameters (such as current, speed, etc.) in an electric machine.

In the early 1940s, analog methods were also being considered (Karplus 1958), wherein we would seek the solution to an analogous system that we knew how to solve in the laboratory. One such example is using a conducting sheet of paper to model a field problem physically as a current flow problem through the conducting sheet. It is easily shown that this problem, as well as those in two-dimensional electrostatics and magnetostatics, is governed by the Poisson equation. Once the analog problem is physically set up, using a probe and voltmeter, we trace over the paper and identify equipotential lines and compute all quantities of interest therefrom. Cherry (1949) used the analogy between a waveguide and a stretched elastic membrane to get the eigenfunctions of a waveguide at a resonant frequency by measuring the amplitude of oscillation of a membrane. Another analog method consists of the correspondence between the finite difference approximations of field equations and current equations in a resistive network. Liebmann's approach is of particular interest in this area. He used the finite difference approximations of field equations to set up resistive networks physically, to simulate magnetostatic devices (1949) and waveguides and resonant cavities (1952). He got his answers through measurement. The drawbacks to this method are, as much as measurement errors, lack of generality and the cost and time involved in performing the experiments.

In the finite difference method based on differential equations, the solution region is subdivided into a rectangular mesh (an approximation) and the values of a scalar potential field are sought at all the grid points (Southwell 1940; Binns and Lawrenson 1963). Engineers, with intuition derived through experience in circuit theory (Laithwaite 1967; Silvester 1967; Carpenter 1968, 1975; Carpenter and Djurovic 1975), were able to use network analog models and generalize the finite difference method to inhomogeneous problems with graded meshes in such a way as to be easily understood. Initially, solutions were obtained for small problems by hand relaxation, which in turn gave way to bigger problems on the digital computer in the 1960s. The disadvantage of this method is that curved boundaries and varying sources cannot be modeled using simple, general data structures, so that general purpose software is difficult to write for complex problems. Moreover, open boundaries have to be modeled by a close boundary with artificial boundary conditions. Recent years have witnessed the development of very general finite difference schemes that overcome some of these deficiencies (Jensen 1972; Chung 1981; Tseng 1984). These elegant variants, however, require complex programming efforts and special data structures.

Some of the earlier numerical work was also developed from variational methods (Gould 1957), based on the now-famous Euler's brachistochrone problem. This problem gives us that particular shape of a curve between two points in space, along which a mass may slide down without friction in the shortest time. A variational scheme may commonly be described as *differential* in that we replace the condition of the satisfaction of a differential equation governing an unknown function by the equivalent requirement that an integral function of the unknown function shall be at a minimum. Indeed, integral variational principles are also used where we use a functional that satisfies an integral relationship at its minimum (McDonald, Friedman, and Wexler 1974).

Kornhauser and Stakgold (1952) have used differential variational schemes to analyze a waveguide, and Sheingold (1953) used one to study an iris. Their schemes are remarkably similar to the techniques of modern finite element analysis. They relied upon a Rayleigh–Ritz scheme, which assumed the solution to be a sum of coordinate functions, with their weights to be determined by variational calculus (Van Bladel 1964). These variational schemes differed from present-day finite elements only in that the solution domain was not split into subdomains over each of which a different trial function may be assumed. As a consequence of their trial solution being applicable to the whole domain, the choice of coordinate functions was critical to the accuracy of the solution. Thus, their methods are of use only in a few problems for which a good estimate of the solution may be made.

With the development of the computer, statistical methods involving long solution times were also employed for the solution of field problems. The Monte Carlo simulation technique (Ehrlich 1959) is used to find the potential at one point within a device that has been divided into a mesh. We start at the point where we want the solution and "walk" through the nodes until we reach a boundary. The solution at the starting node is expressed as a statistical formula in terms of the charge densities at the nodes through which the walk is performed. Because this is a statistical formula, its accuracy increases with the number of walks we perform. Thus, for convergence of the solution several walks must be made. What we finally have is the solution at just one node. The validity of the method for general gradient (Neumann) boundary conditions has not been thoroughly investigated. It has also been found that the time taken to get the potential at one node is much greater than that for identifying the potential everywhere in the mesh by relaxation (Binns and Lawrenson 1963). As a result, the Monte Carlo method in electromagnetics has fallen into general disuse. However, the method is still extremely useful in other disciplines such as statistical chemistry where no alternative techniques are available.

Real numerical modeling of the continuum may be broadly divided into integral methods (Silvester 1968; Harrington 1968; Mayergoyz 1983), differential methods (Forsythe and Wasow 1960), and variational methods (Gould 1957). The variational methods are really based on the differential or integral form of the equation to be solved. Integral schemes for materially homogeneous problems have long been known and used. The equation for integration naturally follows from the potential due to a unit point charge being given at a distance $r$ by $(4\pi\varepsilon r)^{-1}$. Thus, any charge cluster of given density $\rho$ may be subdivided into small volumes $dR$ so that the charge in that volume, $\rho dR$, is effectively a point charge causing a potential $(4\pi\varepsilon r)^{-1}\,\rho dR$ at a point at distance $r$. On the basis of the superposition principle (Ferraro 1970), we may integrate such effects at any point and get the potential there. When material inhomogeneities are involved, this scheme cannot be used. To overcome this in electrostatics, we may resort to dipole moment techniques whereby we eliminate an inhomogeneous region and replace the effects coming from inside the eliminated region by a surface charge (Edwards and Van Bladel 1961). In magnetics, Carpenter (1967) has introduced the analogous idea of a magnetic shell (or magnetic pole) density.

We shall see how these simple concepts, derived from an understanding of the physics of fields, fall under the more general boundary elements schemes that are derived mathematically from the governing equations (Brebbia 1978). This integral method, involving the idea of the equivalent dipole layer, has assumed a sophisticated form under the boundary integral method (Harrington 1968; Jeng and Wexler 1977). Here we solve for the surface charge by which an inhomogeneous region is replaced; to this end we subdivide the surface of the region being eliminated into elements (Lindholm 1980, 1981), postulate a trial function for the unknown charge over each element, and use some criterion to determine the best values of these trial functions. The merits of the method (Salon 1985) are few equations for solution (in view of the reduction in dimensionality by 1, such as from a volume to a surface) and ease in treating open boundary value problems such as are posed by a radiating antenna propagating out into space. From the solution of the surface source (charges or their magnetic analogs) or any other given space charge distribution of current or charge, it is possible to obtain by integration the solution at a few points of interest in space. This approach, however, has some shortcomings: (1) matrix inversion is required when we are given the voltage (or vector potential) distribution as opposed to charge (or current) distribution; (2) the matrix equation is not assured to be diagonally dominant, so that some efficient techniques such as the renumbered preconditioned conjugate gradients technique (Hoole, Cendes, and Hoole 1986) cannot be used for solution; (3) the matrix is fully populated and unsymmetric so that storage demands are heavy on the computing environment; and (4) the work involved increases with the number of points of interest and when material inhomogeneities are involved. It is in three-dimensional problems (especially magnetic field problems) that the boundary integral method strongly challenges the finite element method.

It was not until the late 1960s that the finite element method was first applied in electromagnetics (Winslow 1967). The finite element method is a general method for the solution of differential equations. Here, we subdivide the solution region into subdomains, called *elements*, and postulate a trial function (with free parameters) over each of the elements. It is commonly found convenient to have interpolation nodes on the element and to define the trial function in terms of the unknown values of the unknown variable of the differential equation at the node. As a result, the nodal values become the free parameters. The finite element method essentially consists of finding the values of the free parameters with respect to some optimality criterion such as minimum error, energy extremum, functional orthogonality, etc., which we shall leave to later sections. This method, now widely accepted as one of the most powerful numerical schemes available, is one that engineers may take true pride in, for having founded intuitively, leaving for later the rigorous justifications of the method. Indeed, there was a day when engineers used the method on the grounds simply that it worked, and were decried by purists for practicing an inexact art.[1] Today, these pioneering engineers stand vindicated.

## 1.2 THE ELECTROMAGNETIC FIELD EQUATIONS

### 1.2.1 Maxwell's Equations

It is beyond the scope of this book to deal with the detailed statement of the basic laws of electromagnetics (such as those laws

[1] Personal communication by Dr. M. V. K. Chari, Manager, Electromagnetics Program, R&D Center, General Electric Company, Schenectady, New York.

we owe to Coulomb, Lenz, Ampere, and others). These laws are elaborately laid out elsewhere in the basic literature. However, we are concerned here with the solution of the Maxwell equations (Maxwell 1904) for design and analysis purposes; therefore, it is relevant and appropriate to derive them briefly. While the original laws of electromagnetics were generalized into integral forms, all these integral equations have differential counterparts.

Ampere's law gives us the magnetic field strength $H$ at a distance $r$ from an infinitely long wire carrying a current $I$, as $I/2\pi r$. When generalized, this becomes Ampere's circuital theorem, which states that the total MMF (magnetomotive force) drop along any closed loop $l$ is the totality of currents (conduction and displacement) crossing the loop surface $S$:

$$\int \mathbf{H} \,.\, \mathbf{dl} = \iint \left\{ \mathbf{J} + \frac{\partial \mathbf{D}}{\partial t} \right\} dS, \tag{1.2.1}$$

where the conduction current density $\mathbf{J}$ is related to the electric field $\mathbf{E}$ by the constitutive relationship

$$\mathbf{J} = \sigma \mathbf{E}, \tag{1.2.2}$$

and $\sigma$ is the material conductivity. Using Stokes's theorem (A14), according to which, for any vector $\mathbf{H}$

$$\int \mathbf{H} \,.\, \mathbf{dl} = \iint \nabla \times \mathbf{H} \,.\, \mathbf{dS}, \tag{1.2.3}$$

and comparing eqns. (1.2.1) and (1.2.3), we obtain the differential form of eqn. (1.2.1):

$$\nabla \times \mathbf{H} = \mathbf{J} + \frac{\partial \mathbf{D}}{\partial t}. \tag{1.2.4}$$

The corresponding Maxwell law for the electrical field $\mathbf{E}$ comes from Faraday's law that the voltage induced in a loop $l$ of surface area $S$ is the rate of change of flux crossing the surface. Generalizing with Lenz's sign convention for the direction of $\mathbf{E}$, this Maxwell law in integral form is:

$$\int \mathbf{E} \,.\, \mathbf{dl} = -\iint \frac{\partial \mathbf{B}}{\partial t} \,.\, \mathbf{dS}. \tag{1.2.5}$$

Again through the use of Stokes's theorem (A14) on the line integral, this takes the differential form of interest to us:

$$\nabla \times \mathbf{E} = -\frac{\partial \mathbf{B}}{\partial t}. \tag{1.2.6}$$

Maxwell's law for the magnetic flux density **B** related to **H** by the constitutive relationship

$$\mathbf{B} = \mu \mathbf{H} \tag{1.2.7}$$

where $\mu$ is the permeability, comes from the law that no magnetic poles exist in isolation. In integral form, this is the same as saying that the totality of flux flowing out of a closed surface is zero:

$$\iint \mathbf{B} \, . \, d\mathbf{S} = 0. \tag{1.2.8}$$

Using Gauss's theorem (A13) to convert a surface integral to a volume integral

$$\iint \mathbf{B} \, . \, d\mathbf{S} = \iiint \nabla \, . \, \mathbf{B} \, dV \tag{1.2.9}$$

and comparing with eqn. (1.2.9), we get the point form of (1.2.8):

$$\nabla \, . \, \mathbf{B} = 0. \tag{1.2.10}$$

In electrostatics we deal with the electric flux density **D** which is related to electric field intensity **E** by the constitutive law

$$\mathbf{D} = \varepsilon \mathbf{E}, \tag{1.2.11}$$

where $\varepsilon$ is material permittivity. Unlike in magnetics, however, since charges (the source of electric flux) do exist, the law corresponding to eqn. (1.2.8) is:

$$\iint \mathbf{D} \, . \, d\mathbf{S} = \iiint \rho \, dV. \tag{1.2.12}$$

This states that the total electric flux coming out of a closed surface $S$ is the sum of the charges enclosed. The law will be readily recognized as a generalization of Coulomb's law, giving the force **E** on a unit charge, placed at a distance $r$ from a charge $q$, as $q/4\pi\varepsilon r$. Using Gauss's theorem (A13) again, we obtain the differential form of the integral law (1.2.12):

$$\nabla \, . \, \mathbf{D} = \rho. \tag{1.2.13}$$

In this book, therefore, we are concerned with the solution of Maxwell's equations, given in integral form by eqns. (1.2.1), (1.2.5), (1.2.8), and (1.2.12), and in alternative differential form by eqns. (1.2.4), (1.2.6), (1.2.10), and (1.2.13), which in turn are subject to the constitutive relationships given in eqns. (1.2.2), (1.2.7), and (1.2.11), coming from material behavior.

### 1.2.2 Low and High Frequency Electromagnetics

The subject of computational electromagnetics may be divided into low and high frequency electromagnetics. It is a natural dichotomy in view of the governing equations and the specialized nature of the division between low and high frequency in electromagnetics. It may generally be said that electrical engineers presently interested in field analysis are those dealing with antennae, waveguides and resonant cavities, electrical machines, magnetic recording heads, transmission lines, lightning, and electron devices. Barring a few exceptions, these seven groups of engineers rarely interact, despite having so much in common. Those who do cross these boundaries are usually developers of numerical methodology and associated software.

Typical low frequency devices are electrical machines, electron devices, transmission lines, and magnetic recording heads, while those that typify high frequency devices are waveguides, resonant cavities, and radiating devices such as antennae. In terms of equations, the difference between these systems is that the displacement current in low frequency devices is negligible, while in high frequency devices it is not. Quite often, the conduction current in high frequency devices is negligible or zero, in view of the zero conductivity of the media through which waves are transmitted. As a result the method of analysis for these devices differs considerably, so that the study of the two kinds of systems may be considered different, despite the common thread of numerical methodology that runs through them. Indeed, few people routinely work simultaneously with both types of devices.

For purposes of low frequency applications the displacement term is small, and therefore eqn. (1.2.4) reduces to

$$\nabla \times \mathbf{H} = \mathbf{J}. \tag{1.2.14}$$

At high frequencies, in contrast, although **D** is small its rate of change in time is high. In high frequency problems in general, eqn. (1.2.4) is directly applicable. However, in many problems in lossless media (such as the transmission of waves through free space or waveguides), since the medium has no conductivity $\sigma$, no conduction current may exist in keeping with eqn. (1.2.2).

### 1.2.3 Scalar and Vector Potentials

In the preceding section we have identified the equations governing a device. Once they are solved within a device of interest, the designer will be able to extract and predict the characteristics of the device. However, the equations we are interested in solving are explicitly in terms of vector fields (**D** or **E** in electricity and **B** or **H** in magnetism), and there are three inconveniences associated with this. First, these vector fields undergo a discontinuity at material interfaces and are therefore multivalued along such boundaries. It may be shown (Panofsky and Phillips 1969; Carter 1961) that the discontinuities at boundaries (free of surface charge and current densities) are such that the field intensities are tangentially continuous, so that

$$H_{t1} = H_{t2}, \tag{1.2.15}$$

$$E_{t1} = E_{t2}, \tag{1.2.16}$$

and the flux densities are normally continuous, giving us:

$$B_{n1} = B_{n2}, \tag{1.2.17}$$

$$D_{n1} = D_{n2}. \tag{1.2.18}$$

Eqns. (1.2.15) and (1.2.16) are shown by applying eqns. (1.2.1) and (1.2.5), respectively, to a thin rectangular loop running along the surface. Eqns. (1.2.17) and (1.2.18) are proved by applying eqns. (1.2.8) and (1.2.12), respectively, to a cylindrical pillbox of short height with the lids on either side of the surface as shown in Figure 1.2.1. The second inconvenience of solving directly for the field vector results from our having to solve coupled equations. For

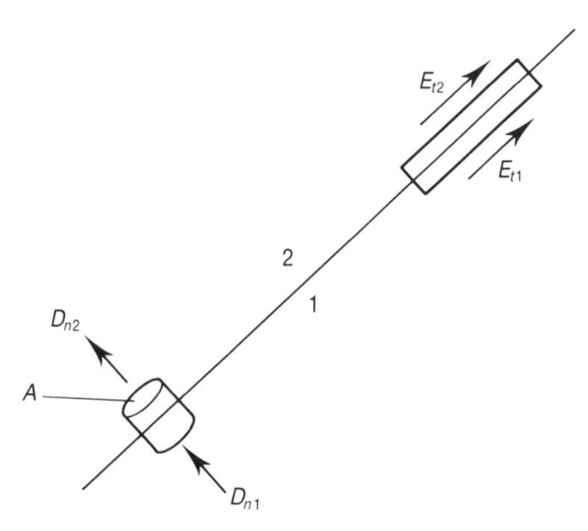


**Figure 1.2.1.** Loop and cylinder for proving continuity conditions on fields.

example, in a simple direct current (DC) magnetic field problem, we would solve eqns. (1.2.3) and (1.2.10) for a given current density **J**. In solving coupled equations by numerical approximation, the system of equations is often overdetermined (meaning that there are more equations than there are unknowns) on account of the approximations. As a result, neither equation is exactly satisfied and the answer tries to satisfy both in some optimal manner. Therefore, we resort to some technique such as using the Lagrange multiplier (Chari et al. 1985a,b) to determine the weight that each equation is to have. Again, here, unless the weight is judiciously chosen we may have to deal with matrices with a zero on the diagonal and attendant problems in matrix solution. The third and most important inconvenience of solving directly for the field vector arises from the very nature of vectors. At each point in space, we need to solve for more than one component (two in two dimensions and three in three dimensions), so that not only will our final matrix equation be large, but its bandwidth will also be correspondingly larger.

We have identified three disadvantages of solving directly for the field vectors. In static electricity, all three of the shortcomings may be avoided by making the electric scalar potential the principal variable. Under static conditions, eqn. (1.2.6) reduces to:

$$\nabla \times \mathbf{E} = 0, \tag{1.2.19}$$

which in comparison with the well-known vector identity (A4):

$$\nabla \times \nabla\phi = 0, \tag{1.2.20}$$

and with sign convention for the direction of field intensity with respect to the direction of drop in electric scalar potential $\phi$ (or voltage as some prefer to say), allows us to define

$$\mathbf{E} = -\nabla\phi = -\mathbf{u}_x\frac{\partial\phi}{\partial x} - \mathbf{u}_y\frac{\partial\phi}{\partial y} - \mathbf{u}_z\frac{\partial\phi}{\partial z}. \tag{1.2.21}$$

Putting this into eqn. (1.2.13), using eqn. (1.2.11), gives us the Poisson equation for electricity:

$$-\varepsilon\nabla^2\phi = \rho. \tag{1.2.22}$$

Eqn. (1.2.22) clearly overcomes all three disadvantages of solving directly for the electric field intensity. First, although **E** may be discontinuous at interfaces, $\phi$ is always continuous, changing its gradient at interfaces to accommodate the step discontinuity in **E**. Second, eqn. (1.2.22) incorporates both eqns. (1.2.6) and (1.2.13), so that no question of weight attaches to it. Third, whether in two or three dimensions, we have replaced a vector unknown by a scalar and thereby have reduced the computational effort. These advantages, however, have been purchased at some cost to accuracy, as we shall see later in section 10.9. This loss of accuracy results from our having to differentiate the potential according to eqn. (1.2.21) to get the electric field; the process of differentiation causes the error. Nonetheless, any such loss of accuracy may be overcome by using better approximations, as we shall see, and need not unduly concern us. It may also be shown that energy-based computations, such as those for capacitance, which may be directly related to potential without any differentiation, are very accurate by the finite element method for the electric potential.

In magnetics, by switching to the vector potential, the first and second of these advantages—namely, replacing a discontinuous unknown by a continuous one and two equations by one—may always be realized. The third advantage of replacing a vector unknown by a scalar is obtained only in two-dimensional problems. The magnetic vector potential **A** is derived by comparing the equation for the nondivergence of magnetic flux (1.2.10) with the vector identity (A5):

$$\nabla . \nabla \times \mathbf{A} = 0, \tag{1.2.23}$$

so that by definition of the vector potential,

$$\begin{aligned} \mathbf{B} = \nabla \times \mathbf{A} = \mathbf{u}_x\left[\frac{\partial \mathbf{A}_y}{\partial z} - \frac{\partial \mathbf{A}_z}{\partial y}\right] + \mathbf{u}_y\left[\frac{\partial \mathbf{A}_z}{\partial x} - \frac{\partial \mathbf{A}_x}{\partial z}\right] \\ + \mathbf{u}_z\left[\frac{\partial \mathbf{A}_x}{\partial y} - \frac{\partial \mathbf{A}_y}{\partial x}\right], \end{aligned} \tag{1.2.24}$$

where the expansion of $\nabla \times \mathbf{A}$ has been done using the identities (A1) and (A3) of Appendix A. Putting this into eqn. (1.2.4) and using eqn. (1.2.7), we get:

$$\mu^{-1}\nabla \times \nabla \times \mathbf{A} = \mathbf{J}. \tag{1.2.25}$$

Now, as shown in Appendix C, for the vector **A** to be unique and therefore determinable, its divergence in addition to its curl has to be defined. We conveniently take, to use common terminology, as gauge:

$$\nabla . \mathbf{A} = 0. \tag{1.2.26}$$

We shall see later that this is a special case of the Lorentz gauge, used when the displacement current is significant. Thus, in general the equation pair (1.2.25–1.2.26) must be solved. Putting eqn. (1.2.26) into the vector identity (A9),

$$\nabla \times \nabla \times \mathbf{A} = \nabla(\nabla \cdot \mathbf{A}) - \nabla^2\mathbf{A} = -\nabla^2\mathbf{A}, \qquad (1.2.27)$$

we get from eqn. (1.2.26):

$$-\mu^{-1}\nabla^2\mathbf{A} = \mathbf{J}. \qquad (1.2.28)$$

Thus, we see that although we have a continuous variable governed by one equation, we still must solve for three components. A great simplification arises in two-dimensional problems where no changes occur in one direction, say the $z$ direction, so that

$$\frac{\partial}{\partial z} \equiv 0. \qquad (1.2.29)$$

In this case, all the magnetic flux will be on the $xy$ plane. From the equation for the curl of $\mathbf{B}$, eqn. (1.2.4), using (1.2.29) and $B_z = 0$, it will be seen that

$$\mathbf{J} = \mathbf{u}_z J, \qquad (1.2.30)$$

where $\mathbf{u}$ stands for a unit vector in the direction indicated by the subscript. Since eqn. (1.2.28) represents three Poisson equations for the three components of $\mathbf{A}$, the vector $\mathbf{A}$ will have only one component:

$$\mathbf{A} = \mathbf{u}_z A. \qquad (1.2.31)$$

This value of vector potential will automatically have zero divergence in view of eqn. (1.2.29). By either putting these single component vector expressions into eqn. (1.2.29) or putting the vector potential into eqn. (1.2.24) and the resulting expression for flux density into eqn. (1.2.4), and, thereafter, equating the magnitudes, we obtain the Poisson equation for the magnitude $A$ of the vector potential $\mathbf{A}$:

$$-\nu\nabla^2 A = J, \qquad (1.2.32)$$

where $\nu$ is the reluctivity or $\mu^{-1}$.

Thus, it is seen that in two-dimensional magnetostatics at least, under which many design tasks are fitted, using the vector potential does replace a vector unknown by a scalar, giving us a scalar boundary value problem.

## 1.2.4 Some Useful Boundary Conditions

It has been shown in Appendix B that to solve Poisson equations of the type of eqn. (1.2.22) or eqn. (1.2.32), either $\phi$ or its normal derivative must be specified. That is, when we solve these differential equations we need a closed system with a Dirichlet or Neumann boundary surrounding it. Indeed, every system is closed, in that space, however large, may be bounded by a surface. We may then say that this surface does not feel the effects of what is inside in view of its being so far away from the charges or currents in the middle where all the action is. That is, at the surface so far away from the physical device being designed, no electric or magnetic field may be seen. We may, therefore, say that this our surface is at a constant potential, say $\phi = 0$, which we may take as our reference. This is consistent with the definition of the potential $\phi \to 0$ as $r \to \infty$. Symmetry conditions, however, often allow us to solve within a smaller section of the device with Neumann conditions (meaning the normal gradient of the potential is specified) or Dirichlet conditions (meaning the potential is specified) along the symmetry lines. Now, as shown in Figure 1.2.2, if we define a coordinate system with the $x$ direction along the tangent and the $y$ direction along the normal to the surface, we see that since no change in potential occurs along the tangent, $\partial\phi/\partial x$ is zero. Therefore, according to eqn. (1.2.21) for electric field systems, which expands in two dimensions as

$$\mathbf{E} = -\mathbf{u}_x \frac{\partial\phi}{\partial x} - \mathbf{u}_y \frac{\partial\phi}{\partial y}, \tag{1.2.33}$$

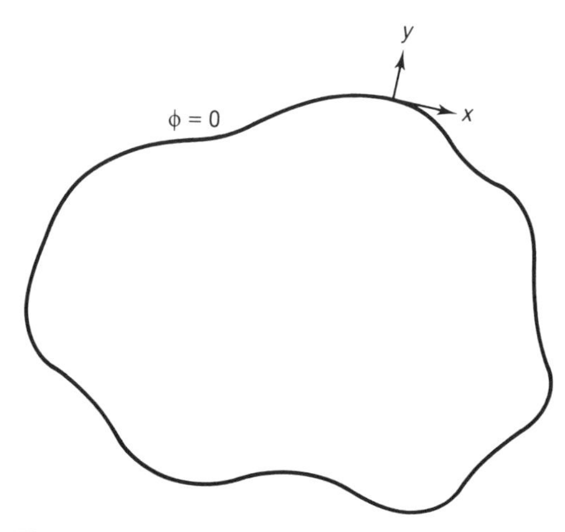


**Figure 1.2.2.** Local coordinate system at equipotential.

the $x$ component of flux is zero, and this means that the electric flux flows normal to the surface. Similarly, according to eqn. (1.2.24) for magnetic field systems, in two-dimensions with eqns. (1.2.29), (1.2.30), and (1.2.31) operating, the magnetic field reduces to

$$\mathbf{B} = \mathbf{u}_x \frac{\partial A}{\partial y} - \mathbf{u}_y \frac{\partial A}{\partial x}. \tag{1.2.34}$$

Therefore, if $\partial A/\partial x$ is zero, then the normal component is zero, meaning that magnetic flux flows along the bounding surface. It is a point of great importance. We shall repeatedly recall later that in electric field problems, the flux flows normal to equipotentials, and in magnetic field problems, along equipotentials.

The magnetic field in parts without current may also be modeled by the magnetic scalar potential $\Omega$, such that the field intensity is its gradient. In this case, the analogy with the electric scalar potential may be used for solution. However, at the faraway boundary, since the fields rotate around the currents in the middle, the magnetic field is tangential so that, in the absence of the normal component, the Neumann condition $\partial\Omega/\partial n = 0$ will apply. One is cautioned against arguing that, since the boundary is far away and no magnetic field is present there, the potential is zero there. This is indeed true in theory at infinity where both $\Omega = 0$ and $\partial\Omega/\partial n = 0$. However, in practical analysis, the boundary is at a finite distance. The condition $\Omega = 0$ kills the tangential component of the magnetic field, whereas the condition $\partial\Omega/\partial n = 0$ suppresses the normal component. A close examination will show that the latter condition does not distort the field significantly at a close-by boundary, while the former does indeed distort the solution by making the flux leave the solution region. A caution along similar lines may be offered in dealing with the other potentials $A$ and $\phi$.

In solving problems, it is not always possible to solve over an infinitely large area, and this so-called boundary at infinity must be placed at a finite distance, unless special techniques such as ballooning (Silvester et al 1977) or integral methods are used. Effectively, imposing a boundary at a finite distance will yield slightly higher field densities than obtain, because we are making the potential drop to zero in a finite distance rather than over infinity. The higher gradient imposed on the potential is what causes a higher field, as seen from eqns. (1.2.33) and (1.2.34). Moreover, if we had a solution region bounded by a box, as in the case of an electric field, at the vertical edge of a corner, the equipotential says that the $y$ component of flux is zero and at the horizontal edge the $x$ component is forced to vanish. Therefore, in effect no electric field can exist at the corner and we have not only made the field drop faster, but we have also forced it to vanish. For this reason, whenever possible, we should try to use close-to-circular shapes for the boundary.

One way of economizing on our solution is to use symmetry to reduce our solution region. Symmetry may result in a Neumann boundary or a Dirichlet boundary. Consider the two infinitely long current-carrying conductors shown in Figure 1.2.3A, both with

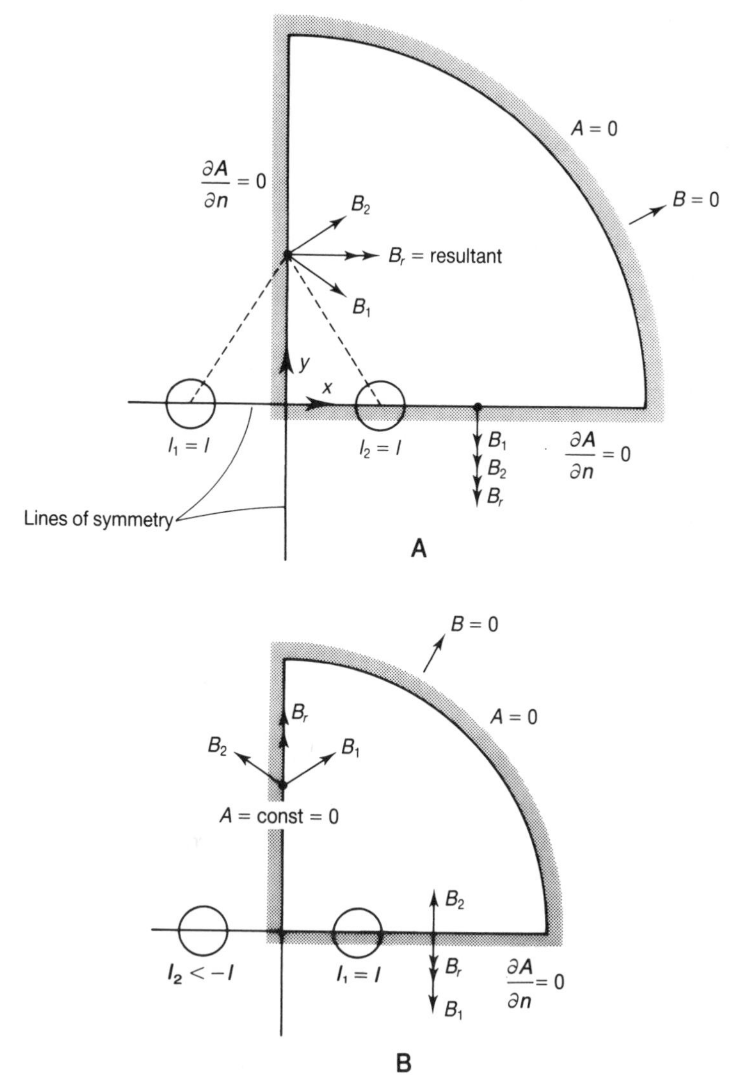


**Figure 1.2.3.** A two-conductor magnetic field problem. **A.** Same currents. **B.** Opposite currents.

currents in the same direction. This is clearly a two-dimensional problem, since no changes occur along the lines. For this problem note that two lines of symmetry exist. Along the vertical line, the magnetic fields of the two conductors, whose directions are determined by the corkscrew rule, have their $y$ components canceling each other so that the resultant is normal to the boundary. Now, since an equipotential is along the direction of magnetic flux, the vertical boundary has zero normal gradient of vector potential and so is a Neumann boundary. Along the horizontal line of symmetry magnetic flux is always normal, so that again it is a Neumann boundary. Thus, we need to solve the problem within the shaded boundary of Figure 1.2.3A, completely enclosed by Neumann and Dirichlet boundaries, to get our solution. Now supposing one of the currents is reversed as shown in Figure 1.2.3B while the horizontal boundary remains Neumann, the vertical boundary becomes a flux line so that it has to be at a constant potential. The "faraway" boundary also is at constant potential, and both of these boundaries meet, so that these sections must be at the potential of 0. The problem we solve is shown in the shaded area of Figure 1.2.3A. If we had been interested in solving the electric field about the transmission line, when both conductors are of the same sign (as shown in Figure 1.2.4), although the electric field will be along the vertical and horizontal lines of symmetry, because the equipotentials are normal to electric fields, the boundary conditions would be the same as in the corresponding magnetic field problem. Again, when the current in one of the conductors is reversed, the boundary conditions would correspond to those of the go and return transmission lines when solving for the vector potential as in Figure 1.2.3B.

An additional condition may be imposed on the transmission line problem of Figure 1.2.4, where we are analyzing the electric potential. This condition applies to any perfect conductor (conductivity $\sigma \to \infty$) in which no potential difference may exist; if it did, then according to eqn. (1.2.21) an electric field $\mathbf{E}$ should exist. Therefore, according to eqn. (1.2.14), an infinite current density $\mathbf{J}$ should be present—a total impossibility. Therefore, unlike in the magnetic field problem, we may leave out the insides of the conductor and make the surface of the conductor a Dirichlet boundary with the potential there being the voltage of transmission $V$.

Like the perfect conductor in electric field problems, the perfect conductor in magnetic field problems is unsaturated steel with permeability $\mu \to \infty$. Consider the rotor of a relatively long synchro-

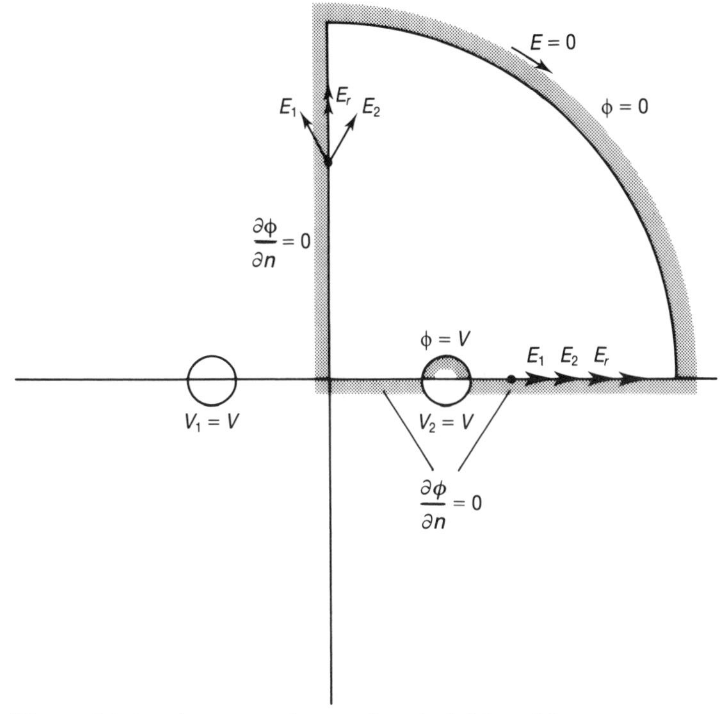


**Figure 1.2.4.** A two-conductor electric field problem.

nous machine on an open circuit shown in Figure 1.2.5A, encased in a steel stator in which the teeth and slots of the machine are not shown. Since the machine is long in comparison with its other dimensions, we may employ a two-dimensional model, except near the end regions. The rotor consists of two salient poles magnetized by a current coil wound around the rotor. Normally, even if we used symmetry to solve a quarter of the problem, we would attempt the configuration of Figure 1.2.5A, with a faraway boundary along which the vector potential $A = 0$. The vertical line of symmetry, being a flux line, is a Dirichlet boundary along which also $A = 0$ because it meets the faraway boundary at $A = 0$. The horizontal line of symmetry is a Neumann boundary with zero normal gradient, because flux meets the boundary normally. We have seen, however, that modeling distant boundaries leads to inaccuracy and more work, and we may wish to avoid it if possible. In this problem we can avoid solving outside in space by using a Neumann boundary condition at the surface of highly permeable

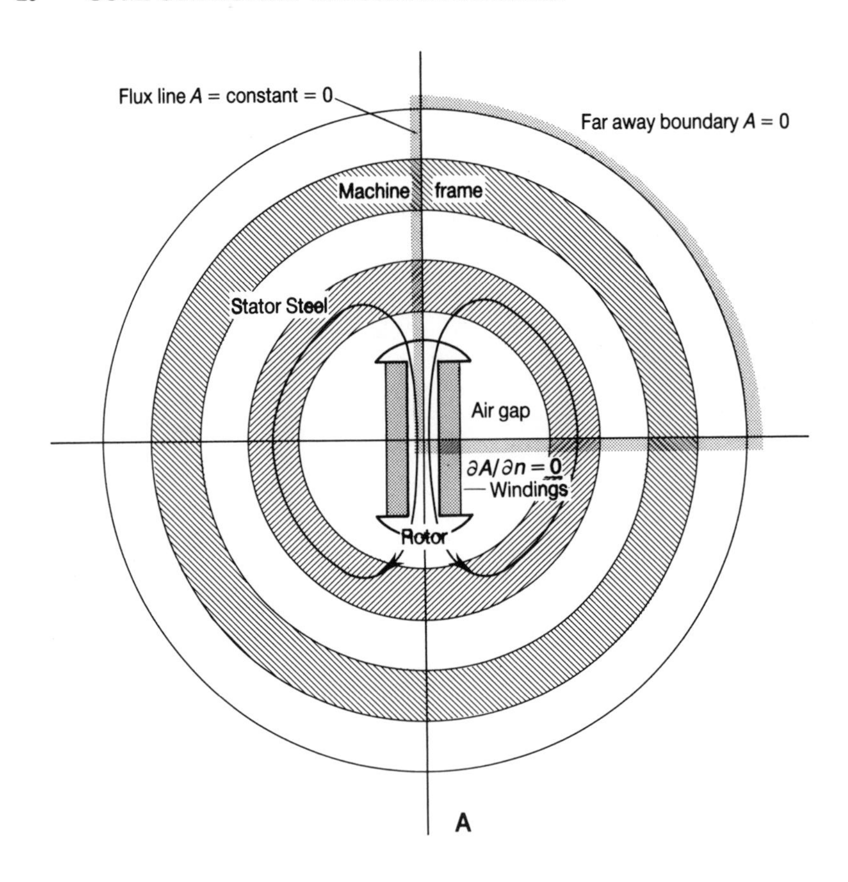


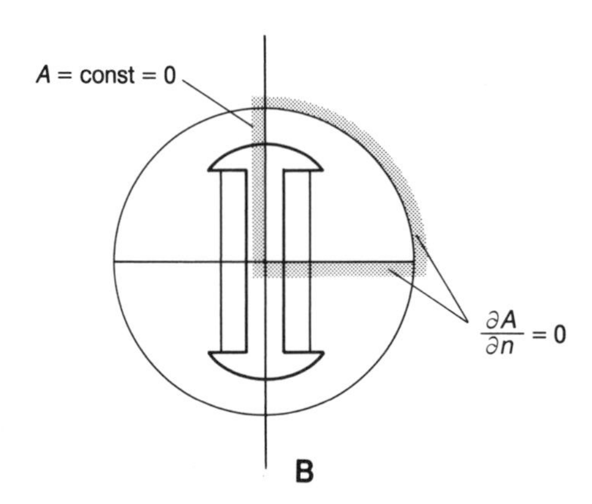


**Figure 1.2.5.** Boundary conditions on a synchronous machine. **A.** The full problem. **B.** The reduced problem, using high steel permeability.

materials. If the machine is on no load, the flux will be due only to the rotor coils and not to the back MMF from the load currents. The magnetic field in the stator is therefore low, and it may be assumed to be unsaturated with infinite permeability. Since $\mu$ is high, according to eqn. (1.2.7) **H** is negligible (**B** at its highest is usually about 2.5 Tesla and is never known to have exceeded 11 Tesla). Now, according to eqn. (1.2.15) the tangential component of **H** outside the steel in the air between the rotor and the stator is the same as it is in the steel—which is nothing. Thus, any flux coming into highly permeable steel must meet it normally. Using symmetry, therefore, this problem may have only one quarter solved, without our having to use a faraway boundary at all, as shown in Figure 1.2.5B. Be it noted that if we were not using symmetry, since the entire boundary will be Neumann, we would need to establish the reference potential by specifying $A$ at least at one point—a good choice would be where the vertical line of symmetry meets the stator. But in this case, since we are using symmetry, the vertical boundary is Dirichlet and we may take $A$ to be zero all along that boundary.

Another boundary condition useful in modeling alternating current (AC) magnetic field systems allows us to eliminate perfect conductors from our solution region. We have already seen from eqn. (1.2.14) that no electric field can exist in a perfect conductor, so that by converting eqn. (1.2.6) to phasor form,

$$\nabla \times \mathbf{E} = -j\omega\mathbf{B}, \tag{1.2.35}$$

we see that since **E** is zero no AC magnetic flux density **B** can exist in a good conductor. Now, by the continuity eqn. (1.2.17), the normal component of **B** outside the conductor also has to be zero, so that the conductor surface is a flux line. We may eliminate the conductor from the solution region by a Dirichlet boundary along the surface. This physical property of conductors is often used to shield the environment from electric machines that carry heavy currents, which may possibly induce parasitic currents in the surroundings. A good copper sheet is wrapped around the machine so that flux cannot penetrate through the sheet. We shall see more of this effect under the eddy current problem of section 5.2.

## 1.3 AXISYMMETRIC PROBLEMS

Axisymmetric configurations are defined simply by what the term *axisymmetry* connotes. These have an axis about which the configuration is symmetric. As a simple illustration, if we view the cross section of an infinitely long, cylindrical conductor and rotate it about its axis, we will see no change. When dealing with such configurations, it is possible to achieve a reduction in the dimension of the problem by one. Indeed, for the cylindrical conductor, axisymmetry achieves a reduction of one and translational symmetry down the length offers another reduction, so that the problems becomes one-dimensional.

If we convert our analysis of problems to cylindrical coordinates $r, \theta, z$ from the rectilinear coordinates $x, y, z$ that we have used up to now, in several problems we will encounter axi- (or cylindrical) symmetry that reduces a three-dimensional problem in $r, \theta, z$ to one that is two-dimensional in $r$ and $z$. That is, no changes will be seen if we hold $r$ and $z$ constant and move from a point allowing only $\theta$ to change:

$$\frac{\partial}{\partial \theta} = 0. \tag{1.3.1}$$

This time, the field quantities are functions of $r$ and $z$ in place of $x$ and $y$. A typical electric field problem with axisymmetry is the transmission line insulator of Figure 1.3.1A. A magnetic field

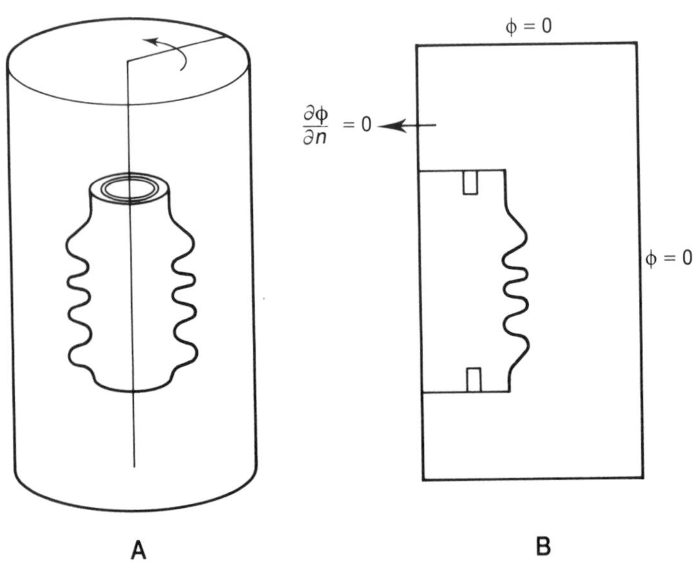


**Figure 1.3.1.** The insulator string. **A.** In three dimensions. **B.** In axisymmetric two dimensions.

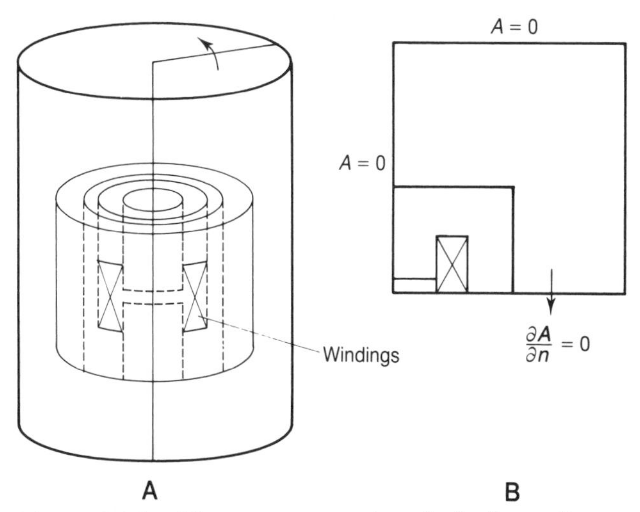


**Figure 1.3.2.** The pot-core reactor. **A.** In three dimensions. **B.** In axisymmetric two dimensions.

problem is the pot-core reactor of Figure 1.3.2A. In both cases, if we rotate the plane shown in the figures about the axis of symmetry, no changes will be seen. Let us first pose the related field problem in three dimensions. We shall construct a faraway cylindrical boundary $S$ about the symmetry axis and close the problem, if no other suitable boundary is defined. As shown in Appendix C, either the normal or tangential component of the field, whether electric or magnetic, ought to be specified. By symmetry, moreover, the specified value should be the same around the cylinder.

To see what the governing equation is, first let us take up the $\nabla$ and $\nabla\times$ operators in cylindrical coordinates. They are, as developed by Ferraro (1970) using $dr$, $rd\theta$, and $dz$ for the line elements in place of $dx$, $dy$, and $dz$,

$$\begin{aligned}\nabla\phi &= \mathbf{u}_r\frac{\partial\phi}{\partial r} + \mathbf{u}_\theta r^{-1}\frac{\partial\phi}{\partial\theta} + \mathbf{u}_z\frac{\partial\phi}{\partial z} \\ &= \mathbf{u}_r\frac{\partial\phi}{\partial r} + \mathbf{u}_z\frac{\partial\phi}{\partial z},\end{aligned} \tag{1.3.2}$$

$$\begin{aligned}\nabla\,.\,\mathbf{A} &= \frac{\partial A_r}{\partial r} + \mathbf{u}_\theta r^{-1}\frac{\partial A_\theta}{\partial\theta} + \frac{\partial A_z}{\partial z} \\ &= \frac{\partial A_r}{\partial r} + \frac{\partial A_z}{\partial z},\end{aligned} \tag{1.3.3}$$

$$\begin{aligned}\nabla \cdot \nabla &= r^{-1}\frac{\partial}{\partial r}\left[r\frac{\partial \phi}{\partial r}\right] + r^{-2}\frac{\partial^2}{\partial \theta^2}\phi + \frac{\partial^2}{\partial z^2}\phi \\ &= r^{-1}\frac{\partial}{\partial r}\left[r\frac{\partial \phi}{\partial r}\right] + \frac{\partial^2}{\partial z^2}\phi,\end{aligned} \tag{1.3.4}$$

and

$$\begin{aligned}\nabla \times \mathbf{A} &= \mathbf{u}_r\left[r^{-1}\frac{\partial A_z}{\partial \theta} - \frac{\partial A_\theta}{\partial z}\right] + \mathbf{u}_\theta\left[\frac{\partial A_r}{\partial z} - \frac{\partial A_z}{\partial r}\right] \\ &\quad + \mathbf{u}_z r^{-1}\left[\frac{\partial r A_\theta}{\partial r} - \frac{\partial A_r}{\partial \theta}\right] \\ &= -\mathbf{u}_r\frac{\partial A_\theta}{\partial z} + \mathbf{u}_\theta\left[\frac{\partial A_r}{\partial z} - \frac{\partial A_z}{\partial r}\right] + \mathbf{u}_z r^{-1}\frac{\partial r A_\theta}{\partial r},\end{aligned} \tag{1.3.5}$$

where we have used the full expressions for the special two-dimensional case.

Therefore, for electric field problems, we have, corresponding to eqns. (1.2.33) and (1.2.22):

$$\mathbf{E} = -\mathbf{u}_r\frac{\partial \phi}{\partial r} - \mathbf{u}_z\frac{\partial \phi}{\partial z}, \tag{1.3.6}$$

$$-\varepsilon\nabla^2\phi = -\varepsilon\left\{r^{-1}\frac{\partial}{\partial r}\left[r\frac{\partial \phi}{\partial r}\right] + \frac{\partial^2}{\partial z^2}\phi\right\} = \rho. \tag{1.3.7}$$

Eqn. (1.3.6) indicates that an electric field is always zero in the $\theta$ direction in an axisymmetric problem. It further tells us that if the normal (or $r$) component on the curved part of the boundary $S$ is zero, then $\partial\phi/\partial r$ is zero. On the curved part of the cylindrical boundary, the direction $r$ is normal to $S$, making it a Neumann boundary. Along the lids of the cylinder, $r$ is along the surface, meaning that the potential does not change along the surface, so that it will be a Dirichlet boundary. If, on the other hand, the tangential component (the $\theta$ component, which is already zero, and the $z$ component) is zero, then $\partial\phi/\partial z$ is zero; this makes the lids of the cylinder Neumann boundaries and the curved surface of the cylinder a Dirichlet boundary. Along the axis of symmetry, we have two charged conductors, so that the electric field will be directed along the symmetry line, making it a Neumann boundary. The boundary value problem arrived at by these considerations is shown in Figure 1.3.1B.

In axisymmetric magnetic field problems, the coils go symmetrically round the axis, and therefore the current density $\mathbf{J}$ has only one component $\mathbf{u}_\theta J$, dropping the superfluous subscript $\theta$ on $J$ in the absence of the two components. As a result, the vector potential also has only one component $\mathbf{u}_\theta A$, so that from eqns. (1.2.24) and (1.3.5),

$$\mathbf{B} = -\mathbf{u}_r \frac{\partial A}{\partial z} + \mathbf{u}_z r^{-1} \frac{\partial rA}{\partial r}. \tag{1.3.8}$$

Putting this into eqn. (1.2.14) using eqn. (1.3.5),

$$\nabla \times \mu^{-1}\mathbf{B} = \mathbf{u}_\theta \mu^{-1}\left[\frac{\partial B_r}{\partial z} - \frac{\partial B_z}{\partial r}\right] = \mathbf{u}_\theta J, \tag{1.3.9}$$

we have the equation governing the vector potential:

$$-\mu^{-1}\left[\frac{\partial^2}{\partial z^2} A + \frac{\partial}{\partial r} r^{-1} \frac{\partial rA}{\partial r}\right] = J. \tag{1.3.10}$$

Let us suppose that the current flows in the circular coil so as to make flux flow upwards in the center of the coil—the considerations not being altered if the direction is reversed. It may be readily gathered from Figure 1.3.2A that the coils will push flux through the middle of the coil so that the flux will flow up along the axis of symmetry, diverge away from the axis as it flows towards the top lid, and return down outside the device and reenter the coil from the bottom. Thus, the flow of flux is along the axis of symmetry, the upper and lower lids, and the circular walls. Naturally, it will be normal to the horizontal plane of symmetry that cuts the device into two parts. From eqn. (1.3.8), therefore, since the $z$ component of flux is zero on the lids, the term $rA$ cannot change with respect to $r$; but since $r$ changes along the lid, the only possibility is for $A$ to be constant on the lids. Similarly, along the axis of symmetry and along the curved walls of the cylinder, the $r$ component of flux density is zero, making $A$ a constant with respect to $z$, so that $A$ is constant along the axis of symmetry and the cylindrical walls. And, since these parts meet the lids at constant potential, they must all be at the same constant potential, which may be taken as the reference potential 0. Along the plane of symmetry that cuts the device in two, since the $r$ component of flux density is zero (from eqn. (1.3.8)), $\partial A/\partial z$ has to be zero, making it a Neumann boundary. These considerations give us the configuration of Figure 1.3.2B for solution.

## 1.4 ANALOGY WITH OTHER DISCIPLINES

This text deals with the computer-assisted analysis of electromagnetic devices. However, it is readily recognized that the numerical

**Table 1.4.1.** Unity of Disciplines in Poisson's Equation

| | |
|---|---|
| 1. Magnetics | $A$ = Magnetic vector potential<br>$\mu$ = permeability<br>$\mathbf{J}$ = current density<br>$\mathbf{B}$ = flux density = $\nabla \times A\mathbf{u}_z$<br>$\mu^{-1}\nabla^2 A = -J$ |
| OR | $\Omega$ = magnetic scalar potential<br>$\mu$ = permeability<br>$\mathbf{J}$ = current density = 0<br>$\mathbf{B}$ = flux density = $-\nabla\Omega$<br>$\mu\nabla^2\Omega = 0$ |
| 2. Electrostatics | $\phi$ = electric potential<br>$\varepsilon$ = permittivity<br>$\rho$ = electric charge density<br>$\mathbf{D}$ = flux density = $-\varepsilon\nabla\phi$<br>$\varepsilon\nabla^2\phi = -\rho$ |
| 3. Fluid flow | $p$ = velocity potential<br>$\rho$ = density<br>$q$ = mass production<br>$\mathbf{V}$ = velocity = $-\nabla p$<br>$\rho\nabla^2 p = -q$ |
| OR | $f$ = stream function<br>$\rho$ = density<br>$q$ = mass production = 0<br>$\mathbf{V}$ = velocity = $\nabla \times f\mathbf{u}_z$<br>$\rho\nabla^2 f = 0$ |
| 4. Heat flow | $T$ = temperature<br>$k$ = conductivity<br>$q$ = heat source density<br>$\mathbf{V}$ = conduction velocity = $-k\nabla T$<br>$k\nabla^2 T = -q$ |
| 5. Ground water flow | $\phi$ = piezometric head<br>$k$ = permeability<br>$q$ = recharge/pumping<br>$\mathbf{V}$ = velocity = $-k\nabla\phi$<br>$k\nabla^2\phi = -q$ |
| 6. Torsion (2-D) | $\phi$ = stress function<br>$G$ = Young's modulus<br>$\theta$ = angle of twist/length<br>$\boldsymbol{\tau}$ = shear stress = $\nabla \times \phi\mathbf{u}_z$<br>$G^{-1}\nabla^2\phi = -2\theta$ |
| 7. Elastic membranes | $u$ = transverse deflection<br>$T$ = tension in the membrane<br>$F$ = transversely distributed load<br>$T\nabla^2 u = F$ |

techniques that underlie the analysis are so general that they are equally applicable to other disciplines.

It is envisaged that two classes of engineers may want to extend the principles expounded here to other fields. First, electrical engineers reading this text will at some point be called upon to solve problems from other disciplines, especially when dealing with coupled problems. For example, an electrical engineer designing a transistor may also need to perform a thermal analysis to determine the electrical properties required in the electrical analysis. Second, engineers from other disciplines may likewise need to analyze electromagnetic fields that may interact with their systems. It is equally possible for a nonelectrical engineer to desire to learn the state of the numerical art in electrical engineering for possible use in his or her work.

For these reasons, it is useful to give a table that shows the analogies between the applicable disciplines. The most important commonness of principle between field analysis in electrical engineering and other branches of engineering lies in the Poisson equation. This is shown in Table 1.4.1, taken from Hoole, Hoole, Jayakumaran, and Hoole (1988). Thus wherever an electrical equation is being discussed in the text, by substitution of the corresponding variables of the other discipline, the theory may be applied there.

# 2 ELECTRIC FIELDS: ANALYSIS BY INTEGRAL METHODS

## 2.1 INTEGRAL METHODS

### 2.1.1 On Integral Methods

Integral methods of solving Maxwell's equations make use of the integral forms of the Maxwell equations and their corollaries—eqns. (1.2.1), (1.2.5), (1.2.8), and (1.2.12). Integral methods express the action at a distance of electric and magnetic sources. In essence, these methods permit us to integrate, by superposition, the total effect at a field point in space of all the causal quantities lying at the source points of space. The sources in electric and magnetic field problems are charge and current, respectively. In uniform media, for determining the behavior of electric field systems from known charge distributions and magnetic field systems from known current distributions, integral methods are ideal and involve computations bordering on the trivial.

However, when material inhomogeneities are present or it is the voltage distribution that is specified, as in most practical electric field problems, the method is often required to draw on the full power of the computing environment and may be considered impracticable for large problems. On the other hand, when it comes to open boundary problems, i.e., those problems encompassing all space as the domain with zero field at the boundary at infinity as the boundary condition—integral methods are found to handle such problems with relative ease, compared with differential methods relying upon the differential form of Maxwell's equations.

### 2.1.2 Electric Fields from Known Charge Distributions

The analysis of an electric field system from a known charge distribution, as a direct field problem, is indeed rare in practice; for, in most practical systems, it is the voltage that is specified on certain parts of a device and not the charge, and it is the applied voltage differentials that determine the charges that are induced. For example, in a two conductor transmission system making up the go and return paths across which a potential difference is applied, although one conductor will contain positive charges and the other negative charges, these two opposite charges will be attracted to each other through Coulomb forces and therefore will move to the inner side of the circuit loop as shown in Figure 2.1.1. In this system, it will be the transmission voltages that will be known and not the charge distributions on the lines.

Be that as it may, as an indirect problem, the computation of electric fields from a known charge distribution is an important one, because even when it is the voltage distribution that is specified, we usually first solve for the charge distribution and then, from the now known charge distribution, we may need to compute the electric field at the various points at which it may be desired.

To determine the electric field from a known charge distribution, we shall have recourse to the integral form of Maxwell's equation:

$$\iint \mathbf{D} \,.\, \mathbf{dS} = \iiint \rho \, dV. \tag{1.2.12}$$

Now consider a sphere of radius $r$, surrounding a pointlike charge $q$, occupying an infinitesimal volume $\delta V$ at the center. When we apply eqn. (1.2.12) to this system, by spherical symmetry, the flux density $\mathbf{D}$ will be a normally directed constant all over the surface so that $\mathbf{D} \,.\, \mathbf{dS}$ is $D\,dS$ and $D$ will come out of the surface integral sign, giving us $4\pi r^2$, the surface area of the sphere, for the surface integral of $dS$. On the other hand, the volume integral will be

**Figure 2.1.1.** Charge realignment in a transmission line.

$q(\delta V)^{-1}\,\delta V = q$. This, therefore, results in:

$$\mathbf{D} = \mathbf{u}_r \frac{q}{4\pi r^2}, \tag{2.1.1}$$

and from eqn. (1.2.11),

$$\mathbf{E} = \mathbf{u}_r \frac{q}{4\pi \varepsilon r^2}, \tag{2.1.2}$$

which we may recognize as a form of Coulomb's law, stating the force on a unit charge at a distance $r$ from a charge $q$. This is not surprising, since it is from this law that the Maxwell law given in eqn. (1.2.12) was generalized. To find the potential $\phi$ at the same point, if we now employ the equivalent of eqn. (1.2.21) in spherical coordinates and use the fact that **E** only has a component in the radial direction, we have

$$E_r = -[\nabla\phi]_r = -\frac{\partial \phi}{\partial r} = \frac{q}{4\pi\varepsilon r^2}. \tag{2.1.3}$$

Integrating this and using as reference the boundary condition that $\phi \to 0$ as $r \to \infty$ (implying that the effect of the charge is not felt far away), we get

$$\phi = \frac{q}{4\pi\varepsilon r}. \tag{2.1.4}$$

If we generalize eqns. (2.1.2) and (2.1.4) for a charge distribution $\rho$ using the superposition theorem, then from an elemental volume $dV$, the "point" charge $q$ is $\rho\, dV$ and we have

$$\mathbf{E} = \frac{1}{4\pi\varepsilon} \iiint \frac{\mathbf{r}\rho\, dV}{r^3}, \tag{2.1.5}$$

where the unit vector $\mathbf{u}_r$ has been written as $\mathbf{r}/r$, and

$$\phi = \frac{1}{4\pi\varepsilon} \iiint \rho \frac{dV}{r}. \tag{2.1.6}$$

In both the above equations, it is to be noted that we are computing the field at just one point (the field point) and to do so the integration is over all regions containing a charge density $\rho$. Thus, in the process of integration, the position vector **r**, pointing from the location of a pointlike charge $\rho\, dV$ at the source point to the field point, always has the tip of the arrow of the vector **r** at the field point, while the beginning of the arrow moves over all the locations where charges may lie. Therefore for a known charge distribution $\rho$, using eqns (2.1.5) and (2.1.6), **E** and $\phi$ may be computed at any desired point. For example, if we are computing the potential $\phi$, we would divide the source distribution into small elemental sizes, convert the integral into a summation form, and sum the effects.

For two-dimensional problems with symmetry in the $z$ direction, say, the corresponding equations are slightly different. To demonstrate the difference between two- and three-dimensional systems through example, a rectangle in three dimensions is a thin rectangular plate without thickness, while in two dimensions, since we cannot have changes in the $z$ direction, the object must be a rectangle at any cross section down the $z$ axis, so that it will be an object that is rectangular in cross section and infinitely long running down the $z$ axis, as shown in Figure 2.1.2. In two dimensions, similarly, a point charge is in fact a line charge, extending from $-\infty$ to $+\infty$ down the $z$ axis. Therefore, in two dimensions, we shall apply eqn. (1.2.12) to a cylindrical surface of unit length with the line charge along the axis, as shown in Figure 2.1.3. By symmetry, the electric field must be along the radial direction, so that $\mathbf{D}\,.\,\mathbf{dS}$ will be zero along the lids of the cylinder—where **D** and **dS** are normal to each other. As a result, we obtain

$$\mathbf{E} = \mathbf{u}_r \frac{q}{2\pi\varepsilon r}, \tag{2.1.7}$$

where $q$ is the charge per meter length of line. Generalizing this, we get

$$\mathbf{E} = \frac{1}{2\pi\varepsilon} \iint \rho \mathbf{r} \frac{dR}{r^2}, \tag{2.1.8}$$

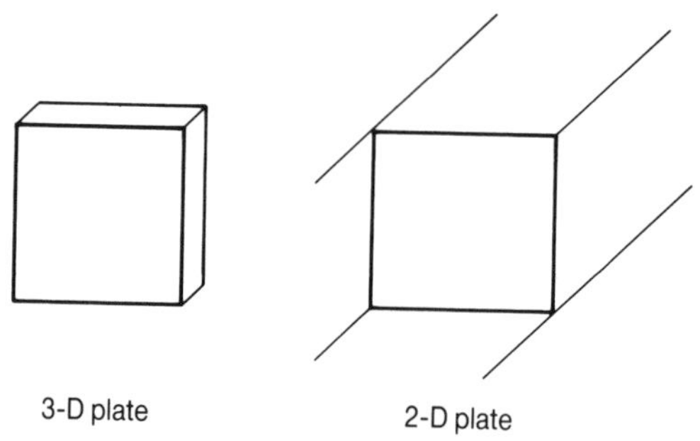


**Figure 2.1.2.** The rectangle in three and two dimensions.

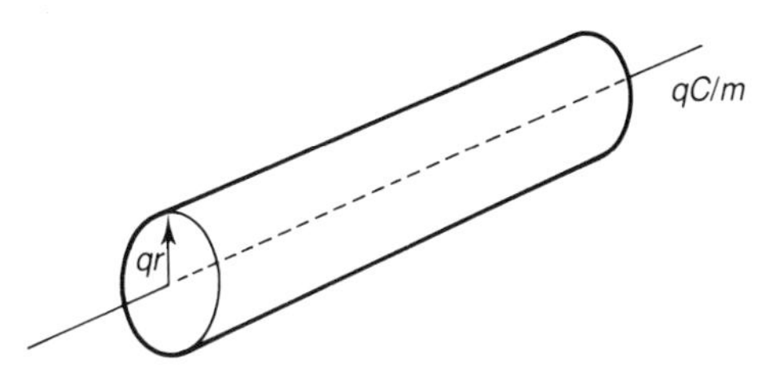


**Figure 2.1.3.** Gaussian surface for a point charge in two dimensions.

and

$$\phi = -\frac{1}{2\pi\varepsilon}\iint \rho lnr \, dR, \tag{2.1.9}$$

where the integrals are surface integrals.

Taking the three-dimensional thin-plate situation again, with obvious extensions to two dimensions as has been shown, we may divide the source region containing the charges into, say, $n$ rectangular parts of area $\Delta_i$, in each of which $q_i$ may be assumed to be at a constant density $\rho_i = q_i/\Delta_i$ (the approximation being more valid for smaller areas $\Delta$). Applying eqn. (2.1.6) we have the potential at a point $j$ as

$$\phi_i = \sum_{i=1}^{n} \frac{q_i}{4\pi\varepsilon\Delta_i}\int_\Delta \frac{1}{r_j}\, dR = \sum_{i=1}^{n} A_{ji}q_i, \tag{2.1.10}$$

where

$$A_{ji} = \frac{1}{4\pi\varepsilon\Delta_i}\int_\Delta \frac{1}{r_j}\, dR, \tag{2.1.11}$$

as shown in Figure 2.1.4, and $r_j$ is the distance between the source point $dR$ and the field point $j$ and the integrals are over the elemental area $\Delta_i$. Now, for elemental areas that are small in relation to the distance $r_j$ from the field point, we may justifiably assume that $r_j$ does not vary in the process of integrating over the area and assume a constant value $r_{ij}$. Consequently, at a point separated from an elemental area $\Delta_i$ by a distance $r_{ij}$, we have:

$$A_{ji} = \frac{1}{4\pi\varepsilon r_{ij}}, \tag{2.1.12}$$

which follows naturally from Coulomb's law. The distance $r_{ij}$ between the field point $(x_j, y_j)$ and the representative source point

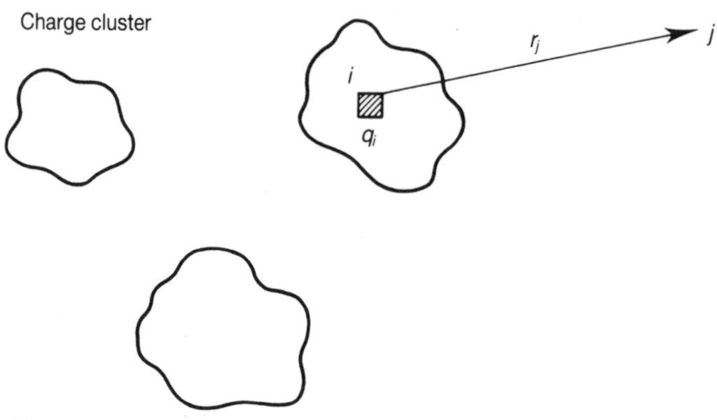


**Figure 2.1.4.** Integral solution of effect of space charge.

at the middle of the elemental piece $(x_i, y_i)$ may be computed from $\sqrt{[(x_i - x_j)^2 + (y_i - y_j)^2]}$. Eqn. (2.1.12) may be used at any point $j$ in space that does not, itself, contain a charge, since the assumption that $\Delta_i$ is small in relation to the distance $r_{ij}$, on the basis of which eqn. (2.1.12) was derived, is not valid when the source and field points are the same. Moreover, even if we tried using eqn. (2.1.12), when computing the effect of the charge at that point caused by the charge on itself, $r_{ij}$ is $r_{jj}$, which is zero, and, therefore, eqn. (2.1.12) cannot be used to compute $A_{jj}$. For the self-caused term, an approximate value to use is the geometric mean self-distance of the element $\Delta_i$ from what may be called its middle. For an $a \times b$ rectangle, for example,

$$
\begin{aligned}
r_{\text{self}} &= \sqrt{\int_{-a/2}^{a/2} \int_{-b/2}^{b/2} [x^2 + y^2] \frac{dx\,dy}{ab}} = \sqrt{\{\tfrac{1}{12}[a^2 + b^2]\}} \\
&= 0.2887\sqrt{[a^2 + b^2]}.
\end{aligned}
\tag{2.1.13}
$$

In the case of an $a \times a$ square subdivision, the self-distance becomes $0.40825a$. An exact value is obtained by working out the actual integral in eqn. (2.1.11). Referring to Figure 2.1.5, for an $a \times b$ rectangle we obtain

$$
\begin{aligned}
A_{ji} &= \frac{1}{4\pi\varepsilon ab} \int_{-a/2}^{a/2} \int_{-b/2}^{b/2} \frac{1}{\sqrt{[x^2 + y^2]}}\, dx\,dy \\
&= \frac{1}{4\pi\varepsilon ab}\left[2a\ln\frac{\sqrt{a^2 + b^2} + b}{a} + 2b\ln\frac{\sqrt{a^2 + b^2} + a}{b}\right]
\end{aligned}
\tag{2.1.14}
$$

after much complicated manipulation, as has been demonstrated in Appendix E. In several, if not most, practical problems, we may not

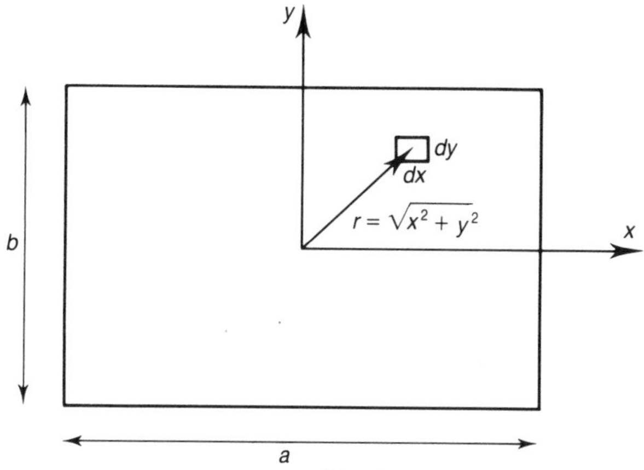


**Figure 2.1.5.** Meaning of $\iint r^{-1}\,dR$.

always be able to have subdivisions into rectangles, because devices come in all manner of shapes, and, therefore, an analytical expression such as eqn. (2.1.14) may be nearly impossible for several of the shapes we may possibly come across. It profits us well, therefore, to investigate as an exercise how the integral may be performed numerically. To this end, consider the small $a \times b$ area shown in Figure 2.1.6, over which we wish to evaluate the integral $r^{-1}\,dR$. Divide this piece into $n^2$ small pieces, where $n$ is any even integer greater than 2. That $n$ ought to be even is required, so that the distance $r$ should not be zero for any of the smaller bits into which the plate is divided. Now performing the integral, we have the Function Sum_InvR presented in Alg. 2.1.1.

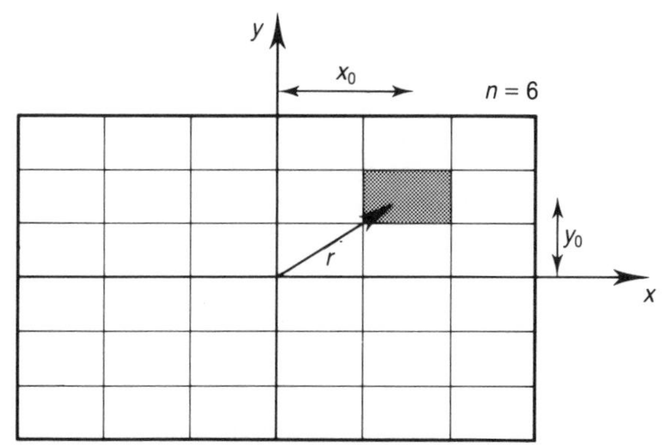


**Figure 2.1.6.** Numerical scheme for integrating $r^{-1}$.

**Algorithm 2.1.1. Integrating $r^{-1}$ over a Rectangle**

```
Function Sum_InvR(a, b, n)
{Function: To integrate r^-1 over a rectangle
Output: Sum_InvR, the integral
Inputs: a and b, the width and length of the rectangle, and n,
the number of subdivisions of each side}
Begin
  x0 ← a/2
  y0 ← b/2
  δx ← a/n
  δy ← b/n
  x ← δx/2
  Integral ← 0.0
  For i ← 1 To n Do
    y ← δy/2
    For j ← 1 To n Do
      Dist ← √{(x0 − x)² + (y0 − y)²}
      Integral ← Integral + 1/Dist
      y ← y + δy
    x ← x + δx
  Sum_InvR ← Integral*δx*δy{same area δxδy for each
 element}
 End
```

Obviously, Alg. 2.1.1 will be inaccurate if $n$ is 2 and highly accurate and expensive for very large $n$. Therefore, to study the numerical convergence of a procedure such as the above with increasingly fine subdivisions into smaller bits and, at the same time, to use it efficiently, we need to make a function of it, say, Sum_InvR with $a$, $b$, and $n$ as input and the resulting integral from the algorithm above as output. Thereafter, we may write another function, say, Integral_InvR and call it repeatedly, increasing $n$ in steps of two until convergence has occurred. This is shown in Alg. 2.1.2, in which convergence is flagged when changes in the answer are less than 0.1%. This percentage may be modified, depending on the level of accuracy required. If this algorithm is implemented for a plate that is $1\,\mathrm{m} \times 1\,\mathrm{m}$, the answer will be found to be 3.5254, as given by the analytical expression.

When Alg. 2.1.2 is implemented on the computer, it is found that convergence is very slow. Particularly, as $n$ increases, the work increases as $n^2$. Clearly, therefore, numerical schemes for computing $A_{jj}$ are unacceptable on grounds of cost and time; on the other

**Algorithm 2.1.2. Integrating $r^{-1}$ over a Rectangle and Checking for Convergence**

```
Function Integral_InvR(a, b)
{Function: Numerically integrates r^-1 over an a × b rectangle
and makes sure that the answer is accurate by checking for
convergence. Requires the function Sum_InvR defined in Alg.
2.1.1}
Begin
  Old ← Sum_InvR(a, b, 2)
  New ← Sum_InvR(a, b, 4)
  i ← 4
  While Abs[(Old–New)/New] > 0.001 Do
    i ← i + 2
    Old ← New
    New ← Sum_InvR(a, b, i)
  Integral_InvR ← New
End
```

hand, closed-form evaluations are not possible. Therefore, we need to seek an intermediate scheme, such as an approximation on the integral. The self-distance approximation of eqn. (2.1.13) for a $1 \times 1$ square gives us an integral of $\Delta_i/r_{\text{self}} = 2.4492$, compared with the exact value of 3.5254, an error of 30.52%. This again proves unacceptable and we need to seek a better approximation. To this end, consider the exact integral over the rectangle in eqn. (2.1.14) when $a$ is the same as $b$:

$$A_{jj} = \frac{1}{4\pi\varepsilon a^2} 4a ln(\sqrt{2} + 1) = \frac{1}{4\pi\varepsilon} \frac{3.52549}{\sqrt{\Delta_j}}. \tag{2.1.15}$$

Similarly, for a circular division into subareas of radius $R$, employing cylindrical coordinates, we have

$$\begin{aligned} A_{jj} &= \frac{1}{4\pi\varepsilon\Delta_i} \iint \frac{1}{r}\, dR = \frac{1}{4\pi\varepsilon\pi R^2} \int_{r=0}^{R} \int_{\theta=0}^{2\pi} \frac{1}{r} r \, d\theta \, dr \\ &= \frac{1}{4\pi\varepsilon} \frac{2\pi R}{\pi R^2} = \frac{1}{4\pi\varepsilon} \frac{2\sqrt{\pi}}{\sqrt{\pi R^2}} \\ &= \frac{1}{4\pi\varepsilon} \frac{3.5449}{\sqrt{\Delta_j}}. \end{aligned} \tag{2.1.16}$$

Comparing with eqn. (2.1.15), we note that although we have changed from a rectangular element to a circular one, $A_{jj}$ remains virtually unchanged in form. Therefore, as suggested by Harrington (1968), we may in general say with a reasonable degree of accuracy that

$$A_{jj} = \frac{1}{4\pi\varepsilon}\frac{3.53}{\sqrt{\Delta_j}}. \tag{2.1.17}$$

This equation may then be used to evaluate the self-contributory term of pieces of any shape.

### 2.1.3 Fields from a Known Voltage Distribution in Homogeneous Space

More often than not, practical problems have the voltage distribution specified, in place of the charge distribution. This is so because of the common presence of almost perfect conductors, in which no voltage gradient may be present—in case an infinite current might be caused. As a consequence, any charges that are induced (or placed) on the conductors must of necessity move and rearrange themselves so as to preserve the constant voltage distribution along the conductors. Thus, the problem will be to solve the electric field everywhere, given the voltages at which some parts of a device, such as a capacitor or a transmission line, are maintained.

In such a problem, the charges as described earlier will redistribute themselves to maintain the required voltage profiles, and, therefore, we need first to determine the charge distribution. Thereafter, using the procedure of the previous section, we may determine the voltage and electric field distributions everywhere from the now known charge distribution, by generalizing eqn. (2.1.6) in three dimensions to volume charges $\rho$ and surface charges $\sigma$:

$$\phi = \frac{1}{4\pi\varepsilon}\iiint\frac{\rho\, dV}{r} + \frac{1}{4\pi\varepsilon}\iint\frac{\sigma\, dS}{r}. \tag{2.1.18}$$

Alternatively, in two dimensions eqn. (2.1.9) may be generalized to

$$\phi = -\frac{1}{2\pi\varepsilon}\iint \rho \ln_e r\, dR - \frac{1}{2\pi\varepsilon}\int \sigma \ln_e r\, dS. \tag{2.1.19}$$

To implement this, as before, we discretize the surfaces of specified voltages into, say, $n$ convenient small areas $\Delta_i$, each at a voltage $\phi_i$ and containing an as yet undetermined charge $q_i$. This voltage $\phi_i$ at the element $i$ is on account of the totality of effects from the $n$ charged elements. Therefore, for each elemental subdivision $i$, we would have one such eqn. (2.1.10) relating the $n$ charges $q_j$, $j = 1, \ldots, n$. Whereas in the earlier problem the $q_i$'s were known and we merely had to perform the summation given in eqn. (2.1.10) to obtain the voltages, this time it is the voltages that are known, and therefore we need to invert the $n \times n$ matrix (eqn. (2.1.10)). The resulting matrix equation will be full because each charge is related to every other charge in eqn. (2.1.10), and symmetric because $r_{ij}$, the distance between elemental pieces $i$ and $j$, is the same as $r_{ji}$. This matrix equation may be solved, and then from the charge distribution the voltage and electric field distribution everywhere may be found by a direct application of eqn. (2.1.10).

## 2.2 ANALYSIS OF A HIGH VOLTAGE DC LINE: FROM KNOWN CHARGES

We have already seen in the preamble to section 2.1.2 that few problems have known charge distributions from which the voltage distribution is required to be found. However, even when it is the voltage that is specified on conductors, after solving for the charge distribution to compute the voltage at a particular point in space, we need to know how to do it from the now known charge distribution. Two problems in which the charge distribution is specified are (1) a single infinitely long transmission line suspended from a tall transmission tower, and (2) a charged sphere far removed from other objects, as may happen in a high-voltage testing laboratory. The requirement of isolation is to avoid a redistribution of charges. But it is inane even to attempt to solve these two problems numerically when they possess neat closed-form solutions, easily obtainable by the application of eqn. (1.2.12) to closed surfaces surrounding the bodies. For the transmission line with $qC/m$, using a meter-long coaxial cylinder, it may be shown that the electric field is $q/2\pi\varepsilon r$ outside and $qr/2\pi a^2$ inside, where $a$ is the radius of the line and a uniform density of charge over the cross section of the line has been assumed. Similarly, for the sphere, using a concentric sphere as the Gaussian surface, it may be shown that the electric field is $q/4\pi\varepsilon r^2$ outside and $qr/4\pi\varepsilon a^3$ inside.

To demonstrate and verify the numerical scheme we may need to use for a more general charge distribution, we shall take up here the solution of an isolated, infinitely long DC transmission line. As shown in Figure 2.2.1A, let us subdivide the radius $a$ into $m$ sections and the 360° traversing around the line into $n$ sections to give us elemental pieces of the line, sufficiently small to be considered point charges. If we wish to consider the effect of an elemental piece of area $\Delta$ located at polar coordinates $r, \theta$ at a field point $x_j, y_j$, the distance $r_{ij}$ between the source and field points is $\sqrt{\{[x_j - r\cos\theta]^2 + [y_j - r\sin\theta]^2\}}$. The charge on an elemental area is $q\Delta/\pi a^2$, since the charge density is $q/\pi a^2$. For the elemental piece shown in Figure 2.2.1B, the area $\Delta$ is $\frac{1}{2}[r_i\delta\theta + r_0\delta\theta][r_0 - r_i] = \frac{1}{2}\delta\theta[r_0^2 - r_i^2]$ from the rule for the area of a trapezium. So, applying eqn. (2.1.9), we have Alg. 2.2.1, which finds the field at a point

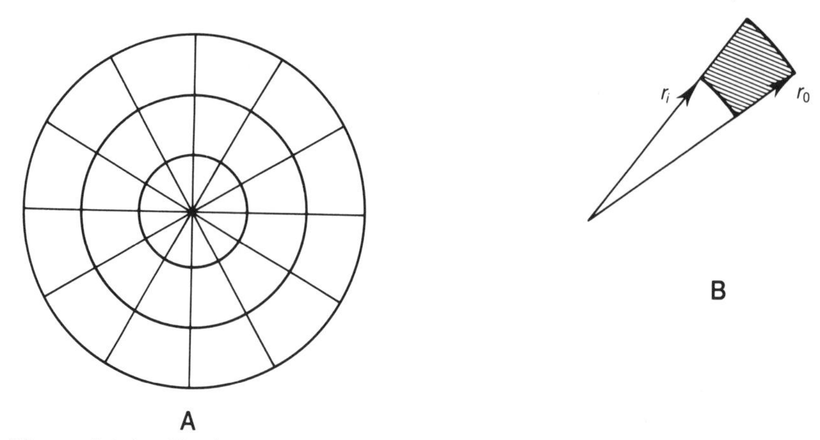


**Figure 2.2.1.** Finding the potential caused by a cylindrical transmission line.

## Algorithm 2.2.1. Field Caused by a Long Cylindrical Wire

**Program CircularWire**
{Function: Computes the electric potential $\phi$ caused at a field point $(x_j, y_j)$ by a wire of circular cross section and radius $a$, containing a uniform charge distribution $q$. The subdivisions $n$ and $m$ of Figure 2.2.1 are assumed}
**Begin**
  $\delta\theta \leftarrow 2\pi/n$
  $\delta r \leftarrow a/m$
  $r_j \leftarrow 0.0$
  $\phi \leftarrow 0.0$
  $\rho \leftarrow q/\pi a^2$
  Factor $\leftarrow -\rho/2\pi\varepsilon$
  **For** $i \leftarrow 1$ **To** $m$ **Do**
    $r_0 \leftarrow r_i + \delta r$
    $r \leftarrow [r_i + r_0]/2$
    $\theta \leftarrow \delta\theta/2$
    $\Delta \leftarrow \frac{1}{2}\delta\theta[r_0^2 - r_i^2]$
    **For** $j \leftarrow 1$ **To** $n$ **Do**
      Dist $\leftarrow \sqrt{\{[x_j - r\cos\theta]^2 + [y_j - r\sin\theta]^2\}}$
      $\phi \leftarrow \phi$ − Factor*Δ***ln**$_e$ Dist
      $\theta \leftarrow \theta + \delta\theta$
    $r_i \leftarrow r_0$
    $r_0 \leftarrow r_i + \delta r$
  **Print** $\phi$
**End**

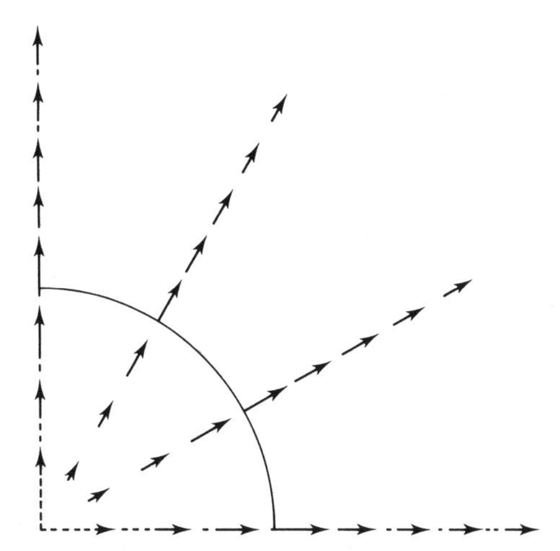

**Figure 2.2.2.** Electric field about a cylindrical DC line.

$(x_j, y_j)$ by summing the effects of all the small pieces at the field point (assumed to be outside the conductor).

Using eqn. 2.1.5, the electric field strength **E** may alternatively be computed. Figure 2.2.2 gives a plot of the electric field **E** about a cylindrical conductor, taken from Hoole (1982).

It is left as an exercise to the reader to implement this procedure and verify the algorithm by comparison with the exact algebraic values. It will be noted that the accuracy of the numerical solution can be increased by increasing $n$ and $m$, by which we shall be making the elemental pieces more and more like the point sources for which eqn. (2.1.9) was derived.

A problem will arise if the field point $(x_j, y_j)$ happens to be at the middle of one of the elements. The distance between the field and source points will then be zero, and the ensuing division by zero will result in an arithmetic overflow and consequent crash of the program. When the field point is in the conductor, the contribution from the element containing the field point should be computed using eqn. (2.1.17).

## 2.3 MATRIX SOLUTION BY GAUSSIAN ELIMINATION

Before proceeding to the next section on solving an electric field system from a known voltage distribution, it is in order to mention something of matrix solution schemes. This is because in applying the conditions for having a specified voltage distribution, we shall encounter a matrix equation of the form of eqn. (2.1.10) giving the required relationships between the unknown charges on our elemental subdivision of the surfaces. The matrix is symmetric, but fully populated. Once we form the matrix equation, say,

$$\mathbf{A}\mathbf{x} = \mathbf{B}, \tag{2.3.1}$$

from our approximations, we need to solve the equation to obtain explicitly the solution **x**. For the moment we shall content ourselves with just getting the solution, without devoting much thought to the speed of solution and storage requirements (or efficiency), which topic we shall take up in detail in a later chapter.

Matrix solution methods fall into two broad categories—*iterative* methods where we progressively improve the solution until an acceptable accuracy is reached, and *direct* methods where the solution is obtained, once and for all, to the precision of the computing machine employed. Broadly speaking, direct methods consume memory, and iterative schemes may not have guaranteed convergence, so that solution times can be high (Jennings 1977).

In this section we shall briefly describe the Gaussian elimination scheme for direct matrix solution—one of the most general methods of matrix solution available (Stewart 1973). Suffice it to say for now that the implementation presented here does not exploit any possible symmetry or sparsity of the coefficient matrix **A**. In the algorithm presented here, we only use the fact that the matrix is diagonally dominant, so that in the process of elimination, we shall never encounter a zero along the diagonal. In general matrix solution by Gaussian elimination, however, zeros may occur on the diagonal, in which case we may resort to the swapping of some rows.

In solving eqn. (2.3.1) by Gaussian elimination, we make the diagonal term of row 1 unity by dividing the whole equation by the diagonal coefficient of row 1, and then subtract an appropriate multiple of that row from every subsequent row, to make all entries of column 1 that lie below row 1 vanish. Then we proceed to row 2, make the diagonal 1, and make all entries of columns 2 that lie below row 2 vanish. Proceeding in this manner we will arrive at a matrix equation with all terms below the diagonal zero. The last row of the equation now only involves the last unknown $x_n$ of the unknown column vector, so that it may be found. The penultimate row involving $x_{n-1}$ and $x_n$ may then be used to find $x_{n-1}$. Proceeding thus, all variables down to $x_1$ may be found. This process is known as back substitution. This procedure is algorithmically described in Alg. 2.3.1. We shall appreciate the algo-

**Algorithm 2.3.1. Gaussian Elimination**

```
Procedure Gauss (x, A, B, n)
{Function: Solves the n × n matrix equation Ax = B by Gauss-
ian elimination
Output: x = the solution, an n vector
Inputs: A = a positive definite n × n matrix; B = the right-
hand side of the equation; n = the size of the matrix}
Begin
  {Begin: Make the matrix upper triangular}
  For i ← 1 To n − 1 Do
    {Begin: scale row i}
    For j ← i + 1 To n Do
      A[i, j] ← A[i, j]/A[i, i]
    B[i] ← B[i]/A[i, i]
    A[i, i] ← 1.0
    {End: Scale row i}
    {Begin: Make column i below row i zero
    That is, subtract A[j, i] times row i from row j for j > i}
    For j ← i + 1 To n Do
      B[j] ← B[j] − A[j, i]*B[i]
      For k ← i + 1 To n Do
        A[j, k] ← A[j, k] − A[j, i]*A[i, k]
      A[j, i] ← 0.0
    {End: Making column i below row i zero}
  {End: Making the matrix upper triangular}
  {Begin: Back substitution}
  x[n] ← B[n]/A[n, n]
  For i ← 1 To n − 1 Do
    j ← n − i
    x[j] ← B[j]
    For k ← j + 1 To n Do
      x[j] ← x[j] − A[j, k]*x[k]
End
```

rithm more fully through the handworked example of the next section.

## 2.4 ELECTRIC FIELDS IN THE PRESENCE OF CONDUCTORS

### 2.4.1 Analysis of a Charged Plate in Space

In this section, we shall take up two examples in which the voltage distribution is specified along certain parts of a homogeneous medium and we are required to find the general resulting voltage distribution in space. Physically this corresponds to finding the distribution following the application of an electric voltage to a conductor. We shall first decide on a simple problem that permits hand solution, so that the procedures whereby matrix eqn. (2.1.10) is built up and solved are clearly understood.

Consider a thin, perfectly conducting plate 8 m × 0.1 m isolated in space and charged to a specified voltage of 100 V. We wish to determine the capacitance of the plate and the voltage it causes at another point in space. To solve this problem, we need to find the charge distribution down the length of the conductor—assuming, of course, that no changes occur along the narrow width. To this end, we shall divide the length of the conductor into eight pieces as shown in Figure 2.4.1, remembering that a finer subdivision will yield more accurate results. By the symmetry of the configuration, we may say that the charge on the first element will be the same as on the last; on the second, the same as on the seventh, and so on, so that only four charges $q_1, \ldots, q_4$ are to be determined. Now according to eqn. (2.1.10), the potential at the first element, due to the charges on the eight elements, is, using eqns. (2.1.12) and (2.1.14) (or eqn. (2.1.17) in place of the latter),

$$4\pi\varepsilon_0\phi_1 = \left\{2{*}1.0 ln \frac{\sqrt{(1^2 + 0.1^2)} + 0.1}{1.0} + 2{*}0.1 ln \frac{\sqrt{(1^2 + 0.1^2)} + 1.0}{0.1}\right\} q_1$$

$$+ \frac{q_2}{1.0} + \frac{q_3}{2.0} + \frac{q_4}{3.0} + \frac{q_4}{4.0} + \frac{q_3}{5.0} + \frac{q_2}{6.0} + \frac{q_1}{7.0} \tag{2.4.1}$$

$$= 8.136q_1 + 1.167q_2 + 0.7q_3 + 0.583q_4 = 100{*}4\pi\varepsilon_0,$$

where, the medium being air, $\varepsilon_0$, the permittivity of free space, has been used. Writing out similar equations for the other three pieces,

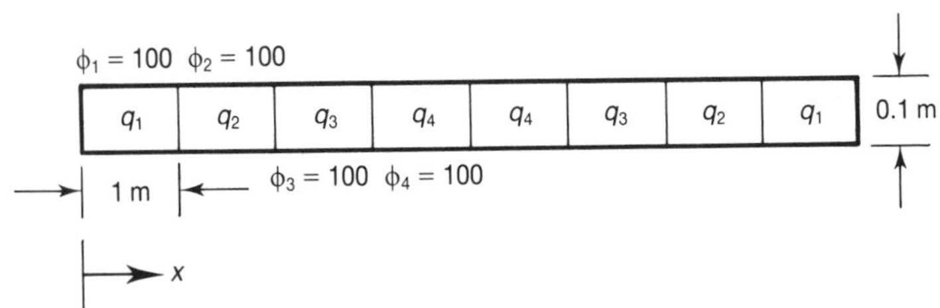


**Figure 2.4.1.** Division of a thin isolated plate.

we have the matrix equation:

$$\begin{bmatrix} 8.136 & 1.167 & 0.7 & 0.583 \\ 1.167 & 8.193 & 1.25 & 0.833 \\ 0.7 & 1.25 & 8.326 & 1.5 \\ 0.583 & 0.833 & 1.5 & 0.993 \end{bmatrix} \begin{bmatrix} q_1 \\ q_2 \\ q_3 \\ q_4 \end{bmatrix} = \begin{bmatrix} 100.0 \\ 100.0 \\ 100.0 \\ 100.0 \end{bmatrix} 4\pi\varepsilon_0. \quad (2.4.2)$$

Attention is drawn to the matrix being symmetric and full. Now, to solve this by Gaussian elimination (described in the previous section), we first make the diagonal term of the first row unity by dividing throughout by 8.136:

$$\begin{bmatrix} 1.0 & 0.14344 & 0.08604 & 0.07166 \\ 1.167 & 8.193 & 1.25 & 0.833 \\ 0.7 & 1.25 & 8.326 & 1.5 \\ 0.583 & 0.833 & 1.5 & 8.993 \end{bmatrix} \begin{bmatrix} q_1 \\ q_2 \\ q_3 \\ q_4 \end{bmatrix} = \begin{bmatrix} 12.2911 \\ 100.0 \\ 100.0 \\ 100.0 \end{bmatrix} 4\pi\varepsilon_0. \quad (2.4.3)$$

Now we subtract 1.167 times the first equation from the second to make location (2, 1) zero; 0.7 times the first equation from the third equation to make location (3, 1) zero; and similarly for the last row:

$$\begin{bmatrix} 1.0 & 0.14344 & 0.08604 & 0.07166 \\ 0.0 & 8.02561 & 1.14959 & 0.74938 \\ 0.0 & 1.14959 & 8.26577 & 1.44984 \\ 0.0 & 0.74938 & 1.44984 & 8.95122 \end{bmatrix} \begin{bmatrix} q_1 \\ q_2 \\ q_3 \\ q_4 \end{bmatrix} = \begin{bmatrix} 12.2911 \\ 85.6563 \\ 91.3963 \\ 92.8343 \end{bmatrix} 4\pi\varepsilon_0. \quad (2.4.4)$$

Now proceeding to use row 2, we eliminate the terms at (3, 2) and (4, 2) by first dividing throughout row 2 by 8.02561 and then subtracting 1.14959 times row 2 from row 3 and similarly 0.74938 times row 2 from row 4:

$$\begin{bmatrix} 1.0 & 0.14344 & 0.08604 & 0.07166 \\ 0.0 & 1.0 & 0.14324 & 0.09337 \\ 0.0 & 0.0 & 8.10111 & 1.3425 \\ 0.0 & 0.0 & 1.3425 & 8.88125 \end{bmatrix} \begin{bmatrix} q_1 \\ q_2 \\ q_3 \\ q_4 \end{bmatrix} = \begin{bmatrix} 12.2911 \\ 10.6729 \\ 79.1268 \\ 84.8363 \end{bmatrix} 4\pi\varepsilon_0. \quad (2.4.5)$$

The same is done using row 3:

$$\begin{bmatrix} 1.0 & 0.14344 & 0.08604 & 0.07166 \\ 0.0 & 1.0 & 0.14324 & 0.09337 \\ 0.0 & 0.0 & 1.0 & 0.16572 \\ 0.0 & 0.0 & 0.0 & 8.65878 \end{bmatrix} \begin{bmatrix} q_1 \\ q_2 \\ q_3 \\ q_4 \end{bmatrix} = \begin{bmatrix} 12.2911 \\ 10.6729 \\ 9.76741 \\ 71.7263 \end{bmatrix} 4\pi\varepsilon_0. \quad (2.4.6)$$

Now that the matrix equation has been upper triangulated, we are ready for back substitution. From the last equation,

$$q_4 = \frac{71.7263}{8.65878} = 8.28334. \tag{2.4.7a}$$

And, from the third, second, and first equations of eqn. (2.4.6), respectively,

$$q_3 = 9.76741 - 0.16572q_4 = 8.39471, \tag{2.4.7b}$$

$$q_2 = 10.6729 - 0.09337q_4 - 0.14324q_3 = 8.69697, \tag{2.4.7c}$$

$$q_1 = 12.2911 - 0.07166q_4 - 0.08604q_3 - 0.14344q_2 = 9.72777, \tag{2.4.7d}$$

where the voltages have been scaled by a factor of $4\pi\varepsilon_0$, for convenience. This solution for a finer subdivision is shown plotted in Figure 2.4.2. Be it noted that the charge distribution is essentially flat in the middle of the conductor and tends to rise at the ends. And, since our answers improve with a finer subdivision of the charge distribution, the reader is encouraged to computerize the

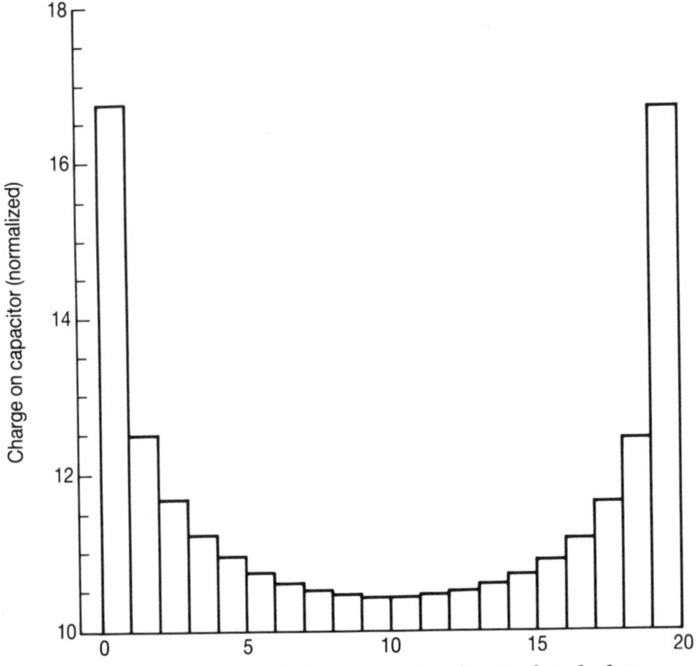


**Figure 2.4.2.** Solution of charge on the thin isolated plate.

above algorithm and solve this problem using smaller elements closer towards the ends of the conductor, so that the rise in the charge there may be modeled better. It will be found that the charges rise even more at the tips, so as to satisfy the requirement of high fields near sharp parts.

From this solution, we may compute the capacitance $C$:

$$C = \frac{q}{V} = \frac{q_1 + q_2 + q_3 + q_4}{100}. \tag{2.4.8}$$

Similarly, the potential may be computed at any given point using eqn. (2.1.10) and the now known charges.

Attention is also drawn to the fact that the working of this example corresponds to applying the boundary integral eqn. (2.5.21), described in section 2.5, to the closed surface $S$ of Figure 2.4.3A as the outer periphery moves to infinity—as a result, the outer boundary will have surface charge $g$ and $\phi$ zero, so that the only contribution to the integral will come from the rest of the boundary. Moreover, over the rest of the boundary, the integral of $\phi r^{-3}\mathbf{r} \cdot \mathbf{u}_n$ will reduce to nothing, in view of the same path being traversed twice in opposite directions since the direction of the normal reverses sign. That leaves us only with the integral $r^{-1}g$. If we take $g$ to be $g_1$ on one side of the curve and $g_2$ on the other, then by applying Gauss's theorem to the small pillbox shown in Figure 2.4.3A, it may be shown that

$$g_2 = \sigma + g_1, \tag{2.4.9}$$

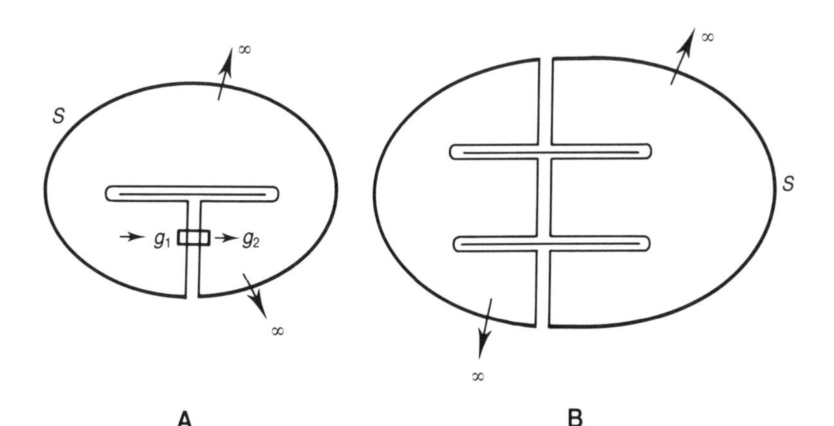


**Figure 2.4.3.** Relationship to boundary integral method. **A**. The isolated plate solution. **B**. The parallel plate capacitor solution.

where $\sigma$ is the charge between the two colinear and opposite paths. $\sigma$ exists only along the plate. Thus, as the integral is evaluated, we have

$$\phi = -\frac{1}{4\pi\varepsilon}\iint r^{-1}g_1\,dS + \frac{1}{4\pi\varepsilon}\iint r^{-1}[\sigma + g_1]\,dS$$
$$= \frac{1}{4\pi\varepsilon}\iint r^{-1}\,\sigma\,dS, \tag{2.4.10}$$

which is what we have used.

### 2.4.2 Analysis of a Capacitor with Fringing: From Known Voltages

As another example of solving for the electric field from known voltages, let us consider a parallel plate capacitor, across the two 1 m × 1 m square plates across which a 100 V difference is applied. Conventionally we would neglect fringing and assume that the electric field flows perpendicularly to the plates, whereas it actually tends to curve out close to the edges—because when the flow of electric flux is through a larger area, the resistance to flow is diminished. The purpose of this exercise is to determine quantitatively the error inherent in neglecting fringing, by comparing the numerical results with the analytical result obtained by ignoring the fringing effect.

First let us neglect the fringing effect so that all flux is normal to the plates. Now consider a small cylindrical pillbox with a lid of area $\delta A$ embedded in one of the plates, and let us apply eqn. (1.2.12) to the cylindrical surface shown in Figure 2.4.4. The only place where we shall have a $\mathbf{D}\,.\,\mathbf{dS}$ term is the outside cap, the curved surface of the cylinder being parallel to the flux. The charge enclosed in the cylinder is $q\delta A/A$, where $A$ is the total area of the plate. This would give us $q/A$ for $\mathbf{D}$ directed from plate to plate, say the $y$ direction. Now using eqns. (1.2.11) and (1.2.33),

$$E_y = -\frac{\partial\phi}{\partial y} = \frac{q}{\varepsilon A}, \tag{2.4.9}$$

integrating which,

$$\phi = -\frac{qy}{\varepsilon A} + c \tag{2.4.10}$$

where $c$ is an arbitrary constant. Now applying the plate potentials

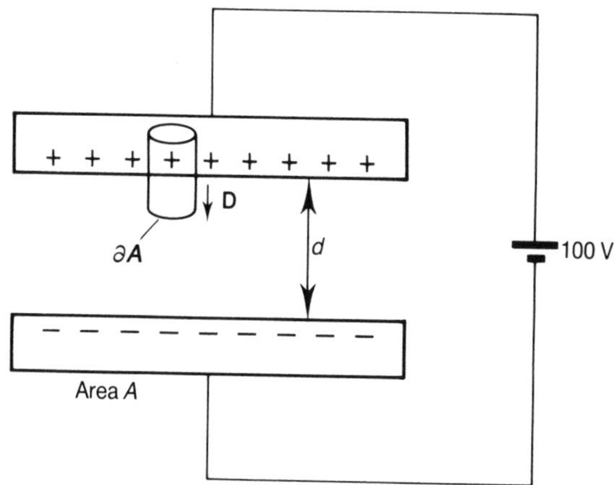


**Figure 2.4.4.** The parallel plate capacitor.

as boundary conditions;

$$V = \frac{qd}{\varepsilon A}, \tag{2.4.11}$$

where $d$ is the separation between the plates. We have the well-known expression for capacitance:

$$C = \frac{q}{V} = \frac{\varepsilon A}{d}. \tag{2.4.12}$$

Now to check the validity of this expression, we may proceed to solve this problem numerically without neglecting fringing, as demonstrated in the preceding section. Each plate may be divided into four zones as seen in Figure 2.4.5. By symmetry, the charge on an elemental piece in one zone will be the same as that on the corresponding elements in the other three zones on the same plate, and a mirror image of that on the four corresponding elements on

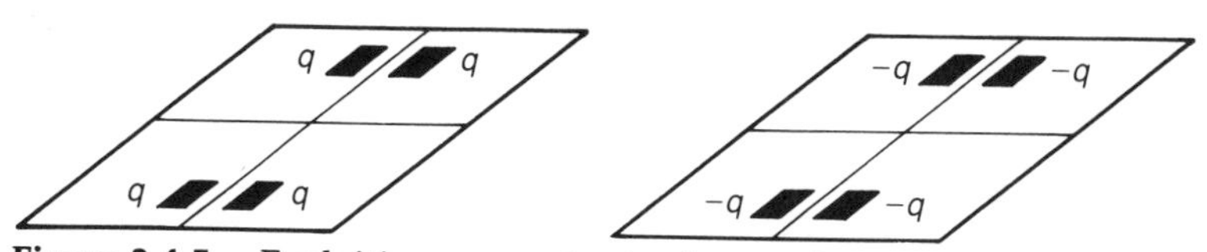


**Figure 2.4.5.** Exploiting symmetry in the parallel plate capacitor.

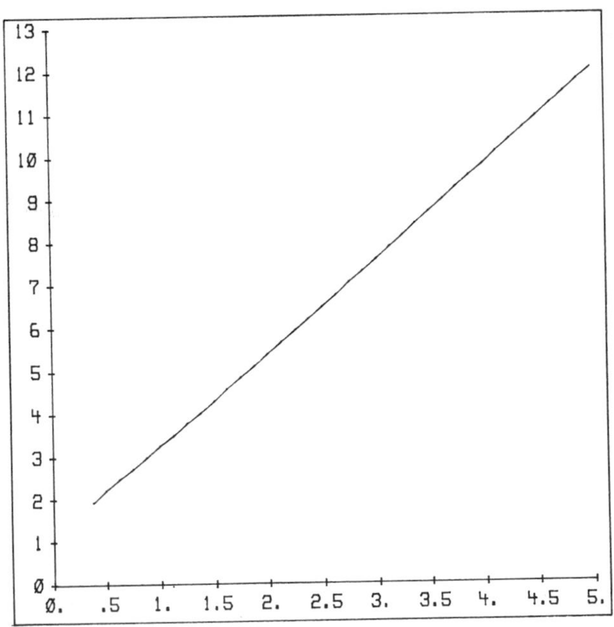


**Figure 2.4.6.** Capacitance and plate separation.

the other plate. Using this, both plates may be discretized so as to produce the same discretization in the four quarters and only an eighth of the problem need be solved.

It is left as an exercise to the reader to solve this problem for different values of plate separation $d$ from 2 m–0.1 m and to compare the values of capacitance with eqn. (2.4.12) (the answer obtained neglecting fringing). The results one should get are shown in Figure 2.4.6. A note of caution is called for on using small values of the plate separation $d$. Care should be taken that the element size is much larger than $d$; otherwise, the self-induced voltage term on an element might be comparable to the induced term from the corresponding term on the opposite plate. Special integration schemes ought to be employed to compute the latter.

Just as it was shown at the end of section 2.4.1 that the evaluation of the charges around a single flat plate there corresponds to applying the boundary integral method to the closed surface of Figure 2.4.3A, it may be shown that this problem and its working correspond to the application of the boundary integral technique to the surface of Figure 2.4.3B.

# 2.5 FIELDS FROM A KNOWN VOLTAGE DISTRIBUTION IN INHOMOGENEOUS SPACE

## 2.5.1 The Boundary Integral Method

In the presence of dielectrics in an electric field system, the electric fields will bend at the material interfaces so as to preserve the continuity of the normal component of flux density and the tangential component of the electric field as shown in eqns. (1.2.16) and (1.2.18). As a result, the equations in section 2.1, which were derived for homogeneous media, are no longer valid and we must seek new procedures for evaluating the electric fields.

In this section we will learn about the *boundary integral* method, one of the most advanced numerical techniques used for the solution of field problems. In Appendix B, it is shown that a Poissonian field such as the electric field $\phi$ is uniquely determined in a closed region $V$ with boundary $S$ when either $\phi$ or its normal gradient is prescribed on $S$. The boundary integral method expresses the electric field at a point in a region $V$ bounded by $S$, as a sum of the effects of interior charge and the surface potential and charge distribution; that is, the contributions from the exterior charges are reexpressed in terms of the boundary potentials and an equivalent surface charge layer. In eqn. (2.1.18) and its implementation in section 2.1.3, we divided the surfaces containing charges into elements over each of which the unknown charges were assumed to be unknown constants. This was a forerunner of the boundary integral method. Because of the division of surfaces into elements the method is also known as the *Boundary element method* (Brebbia 1978). As we shall see later in chapter 8, this method falls under the wide, encompassing umbrella of the finite element method.

To solve for the surface potentials and charges, we need to first derive the relationship that exists between them. Before doing that, however, we ought to consider some vector identities. For a field point $(x_f, y_f, z_f)$ and a source point $(x_s, y_s, z_s)$ the vector $\mathbf{r}$ of the above equations is

$$\mathbf{r} = \mathbf{u}_x(x_f - x_s) + \mathbf{u}_y(y_f - y_s) + \mathbf{u}_z(z_f - z_s), \tag{2.5.1}$$

$$r = [(x_f - x_s)^2 + (y_f - y_s)^2 + (z_f - z_s)^2]^{1/2}. \tag{2.5.2}$$

When the operator $\nabla$ operates on the field point, as in finding the electric field from the gradient of the electric potential, we must differentiate with respect to $x_f$, $y_f$, $z_f$ in the operator $\nabla$ of eqn. (A1). Similarly, when $\nabla$ operates on the sources as in taking the divergence of electric flux density $\mathbf{D}$ to get the charges $\rho$, the differentiations will be with respect to $x_s$, $y_s$, $z_s$.

In what is to come, we need to evaluate $\nabla^2 r^{-1}$ at the field points. To this end, consider this problem in two parts; first when $r \neq 0$ and then $r = 0$ (or when we are trying to evaluate the effect of a source at its own location). Taking the first case and using eqn. (A1),

$$\begin{aligned}\nabla r^{-1} &= \mathbf{u}_x \frac{\partial r^{-1}}{\partial r}\frac{\partial r}{\partial x_f} + \mathbf{u}_y \frac{\partial r^{-1}}{\partial r}\frac{\partial r}{\partial y_f} + \mathbf{u}_z \frac{\partial r^{-1}}{\partial r}\frac{\partial r}{\partial z_f} \\ &= \frac{\mathbf{u}_x(-r^{-2})(x_f - x_s) + \mathbf{u}_y(-r^{-2})(y_f - y_s) + \mathbf{u}_z(-r^{-2})(z_f - z_s)}{[(x_f - x_s)^2 + (y_f - y_s)^2 + (z_f - z_s)^2]^{1/2}} \\ &= -r^{-3}\mathbf{r},\end{aligned} \tag{2.5.3}$$

using eqn. (2.5.2) for $r$. Similarly,

$$\nabla r^{-3} = -3r^{-5}\mathbf{r} \tag{2.5.4}$$

where $r \neq 0$. Also from eqns. (2.5.1) and (A1),

$$\nabla \cdot \mathbf{r} = \frac{\partial(x_f - x_s)}{\partial x_f} + \frac{\partial(y_f - y_s)}{\partial y_f} + \frac{\partial(z_f - z_s)}{\partial z_f} = 3. \tag{2.5.5}$$

Therefore, from eqns. (A2) and (A6),

$$\begin{aligned}\nabla^2 r^{-1} &= \nabla \cdot \nabla r^{-1} = -\nabla \cdot [r^{-2}\mathbf{r}] \\ &= -r^{-3}\nabla \cdot \mathbf{r} - \mathbf{r} \cdot \nabla r^{-3} \\ &= -3r^{-3} - \mathbf{r} \cdot [-3r^{-5}\mathbf{r}] \qquad \text{(using eqn. (A11))} \\ &= 0,\end{aligned} \tag{2.5.6}$$

using eqns. (2.5.4) and (2.5.5) and the fact that $\mathbf{r} \cdot \mathbf{r}$ is $r^2$. Also, we shall later have recourse to the identity

$$\begin{aligned}\nabla ln_e r &= \mathbf{u}_x \frac{\partial}{\partial r} ln_e r \frac{\partial r}{\partial x_f} + \mathbf{u}_y \frac{\partial ln_e r}{\partial r}\frac{\partial r}{\partial y_f} + \mathbf{u}_z \frac{\partial ln_e r}{\partial r}\frac{\partial r}{\partial z_f} \\ &= \frac{[\mathbf{u}_x(r^{-1})(x_f - x_s) + \mathbf{u}_y(r^{-1})(y_f - y_s)y + \mathbf{u}_z(r^{-1})(z_f - z_s)]}{[(x_f - x_s)^2 + (y_f - y_s)^2 + (z_f - z_s)^2]^{1/2}} \\ &= r^{-2}\mathbf{r}.\end{aligned} \tag{2.5.7}$$

Now taking up the evaluation of $\nabla^2 r^{-2}$ at the field point where $r$ vanishes, we resort to an alternative procedure in order to avoid the singularity. Integrating this over a closed volume $V$ (containing the point $r = 0$ and bounded by the surface $S$ on which $r \neq 0$) and applying Gauss's theorem (A13) we obtain

$$\iiint \nabla^2 r^{-1}\, dV = \iiint \nabla \,.\, \nabla r^{-1}\, dV = \iiint \nabla \,.\, [-r^{-3}\mathbf{r}]\, dV$$
$$= -\iint r^{-3}\,\mathbf{r}\,.\,\mathbf{dS} = -\iint d\Omega = -4\pi \qquad (2.5.8)$$

by the definition of a solid angle $\Omega$ subtended at an interior point by a closed surface. The result, eqn. (2.5.8), may not be clear to those readers not versatile with the concept of a solid angle. To interpret it in simpler terms, consider the situation in two dimensions, as shown in Figure 2.5.1. In two dimensions, the potential would have been an integral of $ln_e r$ in place of $r^{-1}$, and therefore in place of eqn. (2.5.8) we would have

$$\iint \nabla^2 ln_e r\, dR = \iint \nabla \,.\, \nabla ln_e r\, dR = \iint \nabla \,.\, r^{-2}\mathbf{r}\, dR \text{ (from eqn. (2.5.7))}$$
$$= \int r^{-2}\mathbf{r}\,.\,\mathbf{dS} = \int r^{-2}\,.\,r\,.\,r\, d\theta = \int d\theta = 2\pi \qquad (2.5.9)$$

using (A13). To explain, the vector $\mathbf{dS}$ has the arc length $dS$ as magnitude and is directed along the outward normal. $\mathbf{r}\,.\,\mathbf{dS}$ is therefore $r\,dS\,\cos\alpha = r\,dC$, where $dC = dS\cos\alpha$ is the distance traversed along the boundary perpendicular to the vector $\mathbf{r}$ as shown and defined in Figure 2.5.1A. We know also that $dC = 2r\sin\frac{1}{2}d\theta = r\,d\theta$ in the limit, where $d\theta$ is the angle subtended by

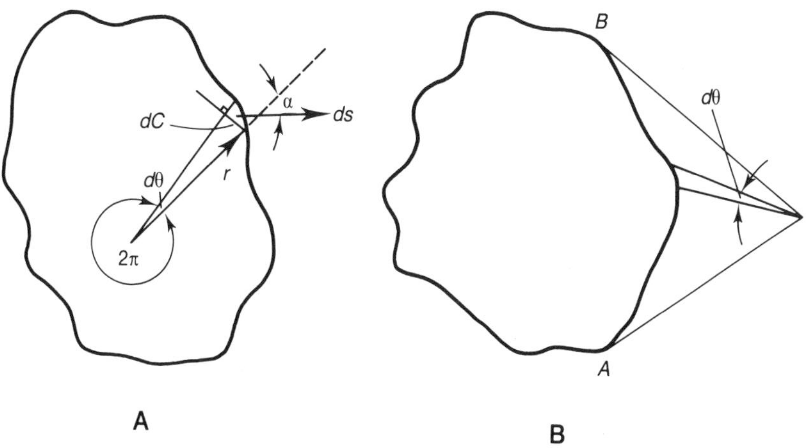


**Figure 2.5.1.** Definition of subtended angle.

$dC$ at the field point. Also, the integral reduces to that of $d\theta$. When the field point is inside the contour $S$, the answer clearly works out to $2\pi$. However, when the field point is outside, as demonstrated in Figure 2.5.1B, the answer is zero because the integral of $d\theta$ traversing from $A$ to $B$ is nullified when going from $B$ to $A$. This confirms what has already been shown. Correspondingly in three dimensions, the solid angle subtended by a closed surface at a point inside is $4\pi$. Therefore, we may summarize results (2.5.6) and (2.5.7) thus:

$$\iiint \nabla^2 r^{-1}\, dV = -4\pi \text{ if the field point is in } V \qquad (2.5.10a)$$

$$= 0 \text{ otherwise} \qquad (2.5.10b)$$

$$= -4\pi\delta(\mathbf{r}), \qquad (2.5.10)$$

where $\delta(\mathbf{r})$ is the Dirac delta function, which is 1 for $\mathbf{r} = 0$, and 0 otherwise. And in two dimensions, from eqn. (2.5.9):

$$\iint \nabla^2 ln_e r\, dV = 2\pi \text{ if the field point is in } V \qquad (2.5.11a)$$

$$= 0 \text{ otherwise} \qquad (2.5.11b)$$

$$= 2\pi\delta(\mathbf{r}). \qquad (2.5.11)$$

Albeit with less rigor, the same result (eqn. (2.5.10)) might equally well have been derived from physical principles (meaning Coulomb's law). We have already shown that the electric potential $\phi$ satisfies

$$-\varepsilon\nabla^2\phi = \rho, \qquad (1.2.22)$$

and has a uniform space solution:

$$\phi = \frac{1}{4\pi\varepsilon}\iiint r^{-1}\rho\, dV. \qquad (2.1.6)$$

Now consider a unit point charge occupying a small volume $\delta V$ so that $\rho$ there is $1/\delta V$. Substituting eqn. (2.1.6) in eqn. (1.2.22) and noticing that while $\nabla$ operates on the field points, the integration over $V$ involves the source points, there obtains

$$-\varepsilon\nabla^2\phi = -\frac{\varepsilon}{4\pi\varepsilon}\iiint \nabla^2 r^{-1}\rho\, dV = 1 \text{ if } r \text{ is at the point charge} \qquad (2.5.12)$$

$$= 0 \text{ otherwise,}$$

which gives us the desired result. One more result, Green's second theorem, is required before we are ready to derive the boundary integral equations. This states that for two functions $f$ and $g$,

$$\iiint [f\nabla^2 g - g\nabla^2 f]\, dV = \iint [f\nabla g - g\nabla f] \cdot \mathbf{dS}. \qquad (2.5.13)$$

To prove this, setting $\phi = f$ and $\mathbf{A} = \nabla g$ in the identity (A11):

$$f\nabla^2 g + \nabla f \cdot \nabla g = \nabla \cdot f\nabla g. \qquad (2.5.14)$$

Integrating this over the volume $V$ and applying Gauss's theorem (A13) to the right-hand side term, we obtain

$$\iiint [f\nabla^2 g + \nabla f \cdot \nabla g]\, dV = \iint f\nabla g \cdot \mathbf{dS}. \qquad (2.5.15)$$

Similarly with $\phi = g$ and $A = \nabla f$ in eqn. (A11):

$$\iiint [g\nabla^2 f + \nabla g \cdot \nabla f]\, dV = \iint g\nabla f \cdot \mathbf{dS}. \qquad (2.5.16)$$

That which is to be proved, eqn. (2.5.13), follows upon subtracting eqn. (2.5.16) from eqn. (2.5.15).

Now we are ready for the boundary integral equations. To obtain them, consider a region $V$ bounded by $S$ and containing a charge distribution $\rho$. Now set $r^{-1}$ in place of $f$ and $\phi$ in place of $g$ in eqn. (2.5.13):

$$\iiint [r^{-1}\nabla^2\phi - \phi\nabla^2 r^{-1}]\, dV = \iint [r^{-1}\nabla\phi - \phi\nabla r^{-1}]\, \mathbf{dS}. \quad (2.5.17)$$

Within the volume $V$, according to eqn. (1.2.22), $\nabla^2\phi$ is $-\rho/\varepsilon$. The second integral may be split into two parts, the first over a small infinitesimal sphere surrounding the field point and the second over all other parts of $V$ (which reduces to zero in view of eqn. (2.5.10)). The first part of the integral becomes $-4\pi\phi(r)$, also because of eqn. (2.5.10). Therefore, writing $\mathbf{dS}$ in terms of the unit vector in the normal direction:

$$\mathbf{dS} = \mathbf{u}_n\, dS, \qquad (2.5.18)$$

$$4\pi\phi(r) = \iiint \frac{\rho}{\varepsilon r}\, dV + \iint [r^{-1}\mathbf{u}_n \cdot \nabla\phi - \phi\mathbf{u}_n \cdot \nabla r^{-1}]\, dS, \quad (2.5.19)$$

which is an equation relating $\phi$ at a field point to $\phi$ and its normal gradient $\mathbf{u}_n \cdot \nabla\phi$ along the surface. Now we shall define $g$, the normal gradient of the electric potential:

$$\mathbf{u}_n \cdot \nabla\phi = \frac{\partial\phi}{\partial n} = g, \qquad (2.5.20)$$

consistent with the definitions (A1) and (A2). Observe that eqn. (2.5.3) may be used for $\nabla r^{-1}$ only when $r \neq 0$. We have in general:

$$4\pi\phi(r) = \iiint \frac{\rho}{\varepsilon r}\, dV + \iint [r^{-1}g + \phi\nabla r^{-1} \cdot \mathbf{u}_n]\, dS. \quad (2.5.21)$$

To solve for the electric potential caused by a set of localized charge distributions, let us surround all the charges by a surface $S$. Now let us discretize the surface $S$ into $n$ small surface elements and the charge clouds into $m$ small volume elements as shown in Figure 2.5.2, over each of which $g$ and $\phi$ are taken to vary according to a trial function, conveniently a constant over the entire element (see Brebbia 1978b for higher orders). Observe that a slight inconsistency arises from this assumption of constant $\phi$ and $g$ on each element. For example, on a Neumann boundary where the normal gradient of the potential $g$ is zero, if we let $\phi$ be constant on the element, then differentiating $\phi$ along the boundary, the tangential gradient of the potential will be seen to be zero. Thus we would have set the entire field to zero on the Neumann boundary. At a Dirichlet boundary, however, since it is the gradient $g$ that we compute, no such inconsistency will be inherent. After obtaining

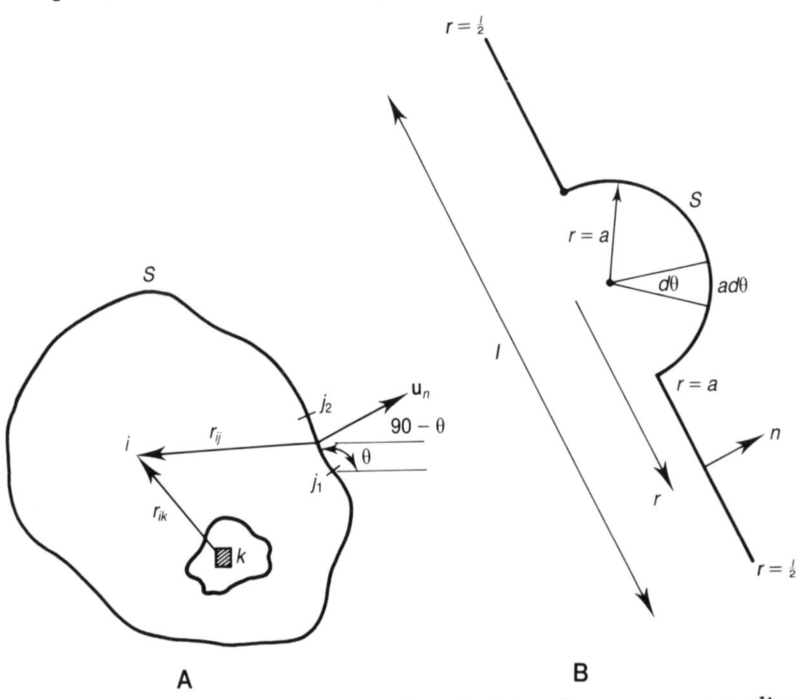


**Figure 2.5.2.** The boundary integral method in a homogeneous medium. **A.** Definitions. **B.** Self-contributions. **C.** Relationship between $ds$ and $d\theta$.

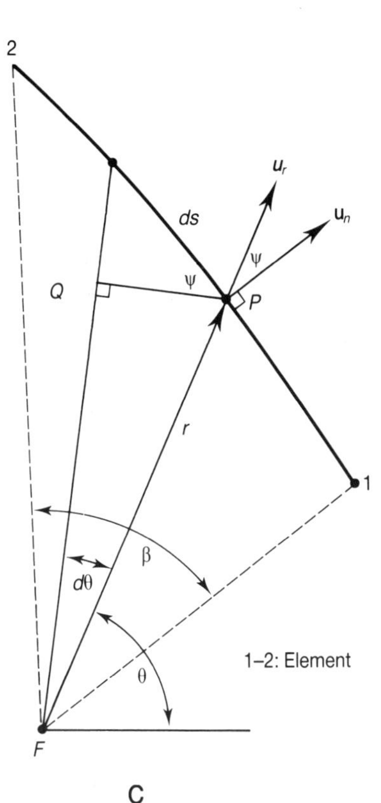


**Figure 2.5.2.** (*continued*)

the boundary element solution with zeroth-order elements, if we reconstruct the field inside the boundary by using eqn. (2.5.21), the interior solution will be smooth whereas close to the boundaries, especially Neumann boundaries, the solution will not look too good. Ideally, therefore, we must use an approximation on $\phi$ that is of one higher order than that on $g$. Observe further that for consistency $\phi$ need not necessarily be of one higher order than $g$ in $x$ and $y$, since $\phi$ along the boundary specifies the gradient in a direction orthogonal to that in which $g$ is obtained by differentiation. Then we have for the potential on element $i$:

$$\begin{aligned} 4\pi\phi_i &= \sum_{k=1}^{m} \rho_k \varepsilon^{-1} \iiint r_{ik}^{-1}\, dV + \sum_{j=1}^{n} g_i \iint_j r_{ij}^{-1}\, dS \\ &\quad + \sum_{j=1}^{n} \phi_j \iint_j \nabla_j \mathbf{r}_{ij}^{-1} \cdot \mathbf{u}_{nj}\, dS \\ &= \sum_{k=1}^{m} t_{ik}\rho_k + \sum_{j=1}^{n} p_{ij} g_j + \sum_{j=1}^{n} q_{ij}\phi_j, \end{aligned} \qquad (2.5.22)$$

where the coefficients

$$p_{ij} = \varepsilon^{-1} \iint_j r_{ij}^{-1}\, dS, \tag{2.5.23}$$

$$q_{ij} = \iint_j \nabla r_{ij}^{-1} \cdot \mathbf{u}_{nj}\, dS, \tag{2.5.24}$$

$$t_{ij} = \iiint_k r_{ik}^{-1}\, dV \tag{2.5.25}$$

may be evaluated, taking special care when $i = j$, as described by Salon (1982). The suffixes by the integrals indicate the element over which the integration is to be performed.

To demonstrate, let us consider the two-dimensional situation, which promotes greater clarity and is more heuristic. Here it is seen from eqn. (2.1.9) that $-ln_e r/2\pi$ takes the place of $1/4\pi r$ in eqn. (2.1.6). And corresponding to eqns. (2.5.17), (2.5.19), and (2.5.22) we have:

$$\iint [-ln_e r \nabla^2 \phi + \phi \nabla^2 ln_e r]\, dR = \int [-ln_e r \nabla \phi + \phi \nabla ln_e r] \cdot \mathbf{dS}, \tag{2.5.26}$$

$$2\pi\phi = -\iint \frac{\rho}{\varepsilon} ln_e r\, dR - \int [ln_e r \mathbf{u}_n \cdot \nabla \phi - \phi \mathbf{u}_n \cdot \nabla ln_e r]\, dS, \tag{2.5.27}$$

$$2\pi\phi_i = -\sum_{k=1}^{m} t_{ik}\rho_k - \sum_{j=1}^{n} p_{ij} g_j + \sum_{j=1}^{n} q_{ij}\phi_j, \tag{2.5.28}$$

where the new values of the coefficients $t$, $p$, and $q$ may be gathered by comparing eqns. (2.5.27) and (2.5.28). When $i \neq j$, the computations may be reduced to the trivial by assuming that $r_{ij}$ is constant over the element $j$. We then have

$$p_{ij} = ln_e r_{ij} S_j, \tag{2.5.29}$$

where $S_j$ is the length $\sqrt{[(x_{j2} - x_{j1})^2 + (y_{j2} - y_{j1})^2]}$ of the element $j$. To evaluate $q_{ij}$, first note that from eqn. (2.5.7), $\nabla ln_e r$ is $r^{-2}\mathbf{r}$. We now write from Figure 2.5.2A:

$$\mathbf{r}_{ij} = \mathbf{u}_x(x_i - x_j) + \mathbf{u}_y(y_i - y_j), \tag{2.5.30}$$

$$\mathbf{u}_n = \mathbf{u}_x \sin\theta - \mathbf{u}_y \cos\theta, \tag{2.5.31}$$

where the coordinates $(x_i, y_i)$ and $(x_j, y_j)$ are taken at the middle of the elements so that they are representative points on the elements $i$ and $j$ and

$$\cos\theta = \frac{x_{j2} - x_{j1}}{\sqrt{[(x_{j2} - x_{j1})^2 + (y_{j2} - y_{j1})^2]}}, \tag{2.5.32}$$

and points $j1$ and $j2$ are the tips of the element $j$. Therefore $q_{ij}$ becomes, upon taking the dot products of the vectors $\mathbf{r}_{ij}$ and $\mathbf{u}_n$,

$$q_{ij} = [(x_{j2} - x_{j1})^2 + (y_{j2} - y_{j1})^2]^{-1} \times [\sin\theta(x_{j2} - x_{j1}) - \cos\theta\,(y_{j2} - y_{j1})]S_j. \tag{2.5.33}$$

Similarly,

$$t_{ik} = ln_e r_{ik} R_k/\varepsilon. \tag{2.5.34}$$

On the other hand, when we are evaluating the self terms for $i = j$, special care is required to avoid the singularity at $r = 0$ (Schneider and Salon 1980). This is done by modifying the boundary $S$ at the field point $r = 0$ as shown in Figure 2.5.2B. The integral over the element of length $l$ is replaced by one over the semicircle of radius $a$ and two integrals from $r = a$ to $r = \frac{1}{2}l$. The latter two integrals are equal, since, although the surface $S$ is a contour, we must keep in mind that $r$ is the magnitude of the radius vector and has no sign, and that we are dealing with a surface integral in which the surface element $dS$ (in this case equal to $dr$ in magnitude) points out in the normal direction. That is, although we replace the integral with respect to $dS$ by another with respect to $dr$, the integral does not become a contour integral, which if it were, $dr$ would be along the tangent. For this part of the integral, the direction vector $\mathbf{r}$ is along the tangent to the element. The actual integral we want is obtained by letting $a$ tend to 0 in the limit. As we let $a$ vanish, we get the actual boundary without the semicircle. Observe that when we move along the boundary of the semicircle, $r$ is the constant $a$, the normal direction is the radial direction, and, in cylindrical coordinates, $dS$ is $a\,d\theta$ with $\theta$ going from 0 to $\pi$ for the half-circle. Thus we have

$$\begin{aligned} p_{ii} &= \lim_{a\to 0}\left[\int_{\text{semicircle}} ln_e r\, dS + 2\int_{r=a}^{l/2} ln_e r\, dS\right] \\ &= \lim_{a\to 0}\left[\int_{\theta=0}^{\pi} ln_e a\, a\, d\theta + 2\int_{r=a}^{l/2} ln_e r\, dr\right] \\ &= \lim_{a\to 0}\left[a ln_e a\int_{\theta=0}^{\pi} d\theta + 2\int_{r=a}^{l/2} (dr ln_e r - r\, d ln_e r)\right] \end{aligned} \tag{2.5.35a}$$

by partial integration

$$= \lim_{a\to 0} [a ln_e a[\theta]_0^\pi + 2[r ln_e r - r]_a^{l/2}]$$

$$= \lim_{a\to 0} [a\pi ln_e a + 2[\tfrac{1}{2} l ln_e \tfrac{1}{2} l - a ln_e a - \tfrac{1}{2} l + a]]$$

$$= l ln_e \tfrac{1}{2} l - l.$$

$$q_{ii} = \lim_{a\to 0} \left[ \int_{\text{semicircle}} \mathbf{u}_n . \nabla ln_e r \, dS + 2 \int_{r=a}^{l/2} \mathbf{u}_n . \nabla ln_e r \, dS \right]$$

$$= \lim_{a\to 0} \left[ \int_{\text{semicircle}} \mathbf{u}_n . \mathbf{u}_r \frac{\partial}{\partial r} ln_e r \, dS + 2 \int_{r=a}^{l/2} \mathbf{u}_n . \mathbf{u}_r \frac{\partial}{\partial r} ln_e r \, dS \right]$$

$$= \lim_{a\to 0} \left[ \int_{\theta=0}^{\pi} \mathbf{u}_r . \mathbf{u}_r \left[\frac{1}{r}\right]_{r=a} a \, d\theta + 2 \int_{r=a}^{l/2} \mathbf{u}_n . \mathbf{u}_r \frac{1}{r} dr \right]$$

$$= \pi. \tag{2.5.35b}$$

In these derivations we have used the fact that $a ln_e a$ vanishes as $a \to 0$. That it does indeed vanish is seen by applying l'Hopital's rule: the limit of $a ln_e a = ln_e a / a^{-1}$ is the same as the limit of $d ln_e a / d a^{-1} = a^{-1}/-a^{-2} = -a$. In evaluating $q_{ii}$ we have substituted the operator $\mathbf{u}_n . \nabla$ for $\partial/\partial n$ where $n$ is the normal direction.

In evaluating the non-self elements, that is for $i \neq j$, we have assumed $r$ to be a constant. This assumption is not satisfactory when the elements $i$ and $j$ are close to each other. In this event, we may use Gaussian quadrature to perform the integrations, more of which is explained in section 8.6 (Silvester 1970). For now, a simple explanation will suffice. Under this procedure we assume that the function $f$ being integrated over the element has a certain polynomial order in distance. Assuming higher orders will give us greater accuracy. Let us assume, for example, that the function $f$ varies as a quadratic function of distance. If $\zeta$ is the normalized distance from the left end of the element, then $f$ is a quadratic in $\zeta$: $f = a + b\zeta + c\zeta^2$. By our definition of $\zeta$, $dS$ is $L d\zeta$ where $L$ is the length of the element on the boundary. Thus our integral is $\int f \, dS = \int L f \, d\zeta$ with $\zeta$ ranging from 0 to 1. Integrating we have $[a\zeta + \tfrac{1}{2} b\zeta^2 + \tfrac{1}{3} c\zeta^3]L$. Imposing the limits, the integral is shown to be $[a + \tfrac{1}{2} b + \tfrac{1}{3} c]L$. To know the integral then, we must know $a$, $b$, and $c$. Evaluating $f$ at three points will give us these constants $a$, $b$, and $c$. Taking these three points as the left end (point 1, $\zeta = 0$, $f_1 = a$), the middle (point 2, $\zeta = 0.5$, $f_2 = a + 0.5b + 0.25c$), and the right end (point 3, $\zeta = 1$, $f_3 = a + b + c$) and solving for $a$, $b$, and $c$ gives us $a = f_1$, $b = -3f_1 + 4f_2 - f_3$, and $c = 2f_1 - 4f_2 + 2f_3$. Substituting these values in the integral, we have $\int f \, dS = (1/6)[f_1 + 4f_2 + f_3]L$. For the second order variation assumed, 1/6, 4/6, and 1/6 are known as the quadrature weights of the three nodes at which $f$ was evaluated. These weights are independent of $f$ and are usually found in tables (Silvester 1970). If the quadrature weights are known, then the values of the integrand $f$ computed at these positions are multiplied by the weights and summed to give the integral.

In the evaluation of $\int (\partial ln_e r/\partial n)\, dS$, as shown in Figure 2.5.2C, a convenient formula exists that equates the integral to the angle $\beta$ subtended by the element 1-to-2 at the field point $F$. The integrand, as already shown by expanding in cylindrical coordinates, is $\mathbf{u}_n \cdot \nabla ln_e r = \mathbf{u}_n \cdot \mathbf{u}_r(\partial/\partial r) ln_e r = \mathbf{u}_n \cdot \mathbf{u}_r/r = \cos\psi/r$ where $\psi$ is the angle that the direction vector $r$ makes with the normal to the boundary. Now evaluating the side $PQ$ of Figure 2.5.2C from the two abutting triangles, we have $r\,d\theta = dS\cos\psi$. Using this in the integral: $\int \cos\psi/r\, dS = \int r\, d\theta/2 = \int d\theta = \beta$, proving what we sought to show. Using this exact result is naturally to be commended in preference to the approximate quadrature scheme. When this is applied to the self-term $q_{ii}$, eqn. (2.5.35b) results immediately. No self-term $t_{ik}$ exists because $i$ is on a surface element and $k$ is in the volume charge distribution. Thus, from eqn. (2.5.8) we have (with all the singularities eliminated) the matrix equation:

$$\pi\phi_i = -\sum_{k=1}^{m} t_{ik}\rho_k - \sum_{\substack{j=1\\ j\neq i}}^{n} p_{ij}g_j + \sum_{\substack{j=1\\ j\neq i}}^{n} q_{ij}\phi_j, \tag{2.5.36}$$

which may be assembled and solved for $\phi$ and $g$ along the boundary. Since the solution of the Poisson equation requires the specification of $\phi$ or its normal gradient along the boundary as shown in Appendix B, only one of the quantities $\phi$ and $g$ will be unknown on each element.

The expression eqn. (2.5.19) is valid only when the region enclosed by $V$ is homogeneous, and, as a result, is not of much practical value when dielectrics of different permittivities are present. To extend it to inhomogeneous cases, consider a dielectric body in a larger region as shown in Figure 2.5.3. This may be easily

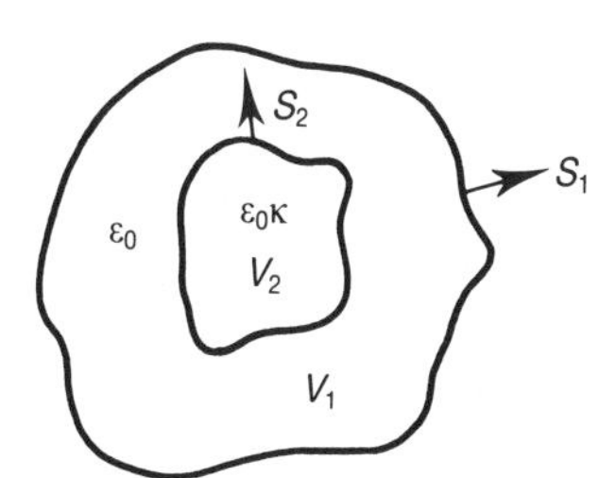


**Figure 2.5.3.** The boundary integral method in an inhomogeneous medium.

extended to several such bodies. Consider the two homogeneous regions $V_1$ and $V_2$ bounded, respectively, by $S_1 + S_2$ and $S_2$, where $\phi$ or its normal gradient is specified on the exterior boundary $S_1$. Also, for the sake of simplicity, let us ignore any charge distribution $\rho$, its presence being easily accommodated by adding an extra term. Writing out eqn. (2.5.19) for the two homogeneous regions $V_1$ and $V_2$ with $\rho = 0$:

$$4\pi\phi(r) = \iint_1 [r^{-1}g - \phi\nabla r^{-1} \cdot \mathbf{u}_n]\, dS_1 - \iint_1 [r^{-1}g - \phi\nabla r^{-1} \cdot \mathbf{u}_n]\, dS_2, \tag{2.5.37}$$

$$4\pi\phi(r) = \iint_2 [r^{-1}g - \phi\nabla r^{-1} \cdot \mathbf{u}_n]\, dS_2, \tag{2.5.38}$$

where the field point $r$ belongs to $V_1$ and $V_2$, respectively. The subscripts by the integral signs indicate the region, 1 or 2, in which the integrands are to be evaluated. Observe that the sign of the second integral is reversed in eqn. (2.5.37) to allow for the fact that the normal to $S_2$ from inside $V_1$ is pointing inwards, whereas by definition it should point out of the bounding surface. Neither equation may be solved independently, because along $S_2$, neither of the quantities $\phi$ and $g$ is known as required for uniqueness of $\phi$.

To solve this problem, we need to use the appropriate boundary conditions at the interface $S_2$ between $V_1$ and $V_2$. Let $\varepsilon_0\kappa$ be the permittivity of the dielectric occupying $V_2$, where $\kappa$ is the dielectric constant or relative permittivity of the region, and $\varepsilon_0$ is the permittivity of the region $V_1$. By continuity, we must have

$$\phi_1 = \phi_2, \tag{2.5.39}$$

and by eqn. (1.2.18) dictating the continuity of the normal component of flux density $\mathbf{D} = -\varepsilon\nabla\phi$,

$$-\varepsilon_0\nabla\phi_1 \cdot \mathbf{u}_{n1} = -\varepsilon_0\kappa\nabla\phi_2 \cdot \mathbf{u}_{n2}, \tag{2.5.40}$$

which reduces to

$$g_1 = -\kappa g_2 \tag{2.5.41}$$

in view of the definition (2.5.20). The negative sign accounts for the normals from two sides of a surface, $n_1$ and $n_2$, being oppositely directed. We have seen that to solve the boundary integral equations, $\phi$ or $g$ ought to be given on every element. We have also seen that this information is missing along the elements of the

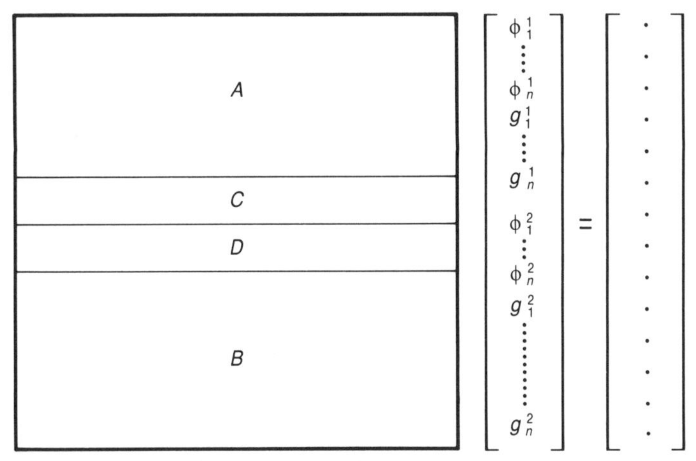


**Figure 2.5.4.** Matrix assembly for Figure 2.5.3.

surface $S_2$. Thus, we are an equation per element short in the set of equations comprising eqn. (2.5.37) and an equation per element short in the set comprising eqn. (2.5.38). What we need, therefore, is two equations per element on $S_2$ to solve the two sets (2.5.37) and (2.5.38). These two equations are given by eqns. (2.5.39) and (2.5.41). A demonstration of solving a field problem by this method is found in section 2.6.2. It is pointed out that when the four equations (2.5.37), (2.5.38), (2.5.39), and (2.5.41) are put together (as shown in Figure 2.5.4 for a two region problem in keeping with the suggestion of Trowbridge (1980)) the resulting matrix equation will have no symmetry or any other property that would make the solution easy. In Figure 2.5.4, block A stands for eqn. (2.5.37) for region 1, block B is eqn. (2.5.38) for region 2, and blocks C and D, respectively, give the boundary conditions eqns. (2.5.39) and (2.5.41) connecting the variables from region 1 to those of region 2. However, despite the form of the matrix equation, the advantage of the boundary integral method is the reduction in dimensionality to elements along a two-dimensional surface for three-dimensional problems and along a line for two-dimensional problems.

## 2.5.2 Electric Fields Impinging on Dielectrics

We have seen that the boundary integral method may be used to solve electric fields by posing the problem in terms of the unknown potential $\phi$ and its normal gradient $g$ along the exterior surface bounding the region and the interfaces along material inhomogeneities.

An alternative integral formulation has been given by Phillips (1934) to solve the problem of the presence of a dielectric in an electric field in terms of the surface potential only. Not only does this achieve a reduction in dimensionality, but it also reduces the number of unknown variables along the surfaces.

This approach, too, like the boundary integral method of the previous section, is based on the result of eqn. (2.5.19) derived from Green's theorem (2.5.13). So as to be general, let $V_1$ be an open region, with the surface $S_1$ going to infinity in eqn. (2.5.37); that is, $g$ and $\phi$ become zero on $S_1$ and so the integrals over $S_1$ vanish:

$$4\pi\phi(r) = -\iint_1 [r^{-1}g - \phi\nabla r^{-1} \cdot \mathbf{u}_n]\, dS_2. \tag{2.5.42}$$

Let us now define $\phi_i$, the impressed potential. That is, $\phi_i$ is the potential in $V_1$, if no dielectric were present. We shall solve for $\phi$ on the surface $S_2$ in terms of $\phi_i$. Therefore, in this method we may sometimes be required to carry out a two part solution, first for $\phi_i$ when no dielectric is present and then for $\phi$. Also observe that $\phi_i$ is harmonic in view of its being the potential that would exist in uniform space in the absence of the dielectric. The adjective harmonic describes a function $\phi$ that obeys Laplace's equation:

$$\nabla^2\phi = 0. \tag{2.5.43}$$

To solve this problem involving an inhomogeneity, now apply eqn. (2.5.17) to the region $V_1$ with the field point outside $V_1$; i.e., in $V_2$:

$$0 = -\iint_1 [r^{-1}g - \phi\nabla r^{-1} \cdot \mathbf{u}_n]\, dS_2 \tag{2.5.44}$$

Here the volume integral of $\phi\nabla^2 r^{-1}$ has vanished outside $V_1$ in keeping with eqn. (2.5.10b). Doing the same thing to the equation for $V_2$ with the field point outside $V_2$:

$$0 = \iint_2 [r^{-1}g - \phi\nabla r^{-1} \cdot \mathbf{u}_n]\, dS_2. \tag{2.5.45}$$

The four equations (2.5.38), (2.5.42), (2.5.44), and (2.5.45) apply to any harmonic potential as defined in eqn. (2.5.43). The potential $\phi_1$ in region $V_1$ is also harmonic since it is the final potential in a uniform medium. Therefore, $\phi = \phi_1 - \phi_i$ is also harmonic. Applying eqn. (2.5.42) applicable to harmonic functions in $V_1$, we have for

$\phi = \phi_1 - \phi_i$ with $r$ in $V_1$:

$$4\pi(\phi_1 - \phi_i) = -\iint [r^{-1}(g_1 - g_i) - (\phi_1 - \phi_i)\nabla r^{-1} . \mathbf{u}_n]\, dS_2. \quad (2.5.46)$$

Similarly, with the field point again in $V_1$, with $\phi = \kappa\phi_2 - \phi_i$ in eqn. (2.5.48):

$$\begin{aligned} 0 &= \iint [r^{-1}(\kappa g_2 - g_i) - (\kappa\phi_2 - \phi_i)\nabla r^{-1} . \mathbf{u}_n]\, dS_2 \\ &= \iint [r^{-1}(g_1 - g_i) - (\kappa\phi_1 - \phi_i)\nabla r^{-1} . \mathbf{u}_n]\, dS_2 \end{aligned} \quad (2.5.47)$$

by the interface conditions (2.1.58) and (2.1.60). Adding eqns. (2.5.46) and (2.5.47):

$$4\pi\phi_1 - 4\pi\phi_i = -\iint (\kappa - 1)\phi_1\nabla r^{-1} . \mathbf{u}_n\, dS_2, \quad (2.5.48)$$

so that, rearranging, we have the potential in $V_1$:

$$\phi_i = \phi_1 + \frac{1}{4\pi}\iint (\kappa - 1)\phi_1\nabla r^{-1} . \mathbf{u}_n\, dS_2. \quad (2.5.49)$$

Similarly, integrating the surface quantities for field points inside the surface $V_2$, we have with $\phi = \phi_2 - \kappa^{-1}\phi_i$ in eqn. (2.5.38):

$$\begin{aligned} 4\pi(\phi_2 - \kappa^{-1}\phi_i) = \iint [r^{-1}&(g_2 - \kappa^{-1}g_i) \\ &- (\phi_2 - \kappa^{-1}\phi_i)\nabla r^{-1} . \mathbf{u}_n]\, dS_2. \end{aligned} \quad (2.5.50)$$

And, applying eqn. (2.5.44) to the harmonic function $\phi_2 - \phi_i$ inside $V_2$:

$$\begin{aligned} 0 &= -\iint [r^{-1}(g_1 - g_i) - (\phi_1 - \phi_i)\nabla r^{-1} . \mathbf{u}_n]\, dS_2 \\ &= -\iint [r^{-1}(\kappa g_2 - g_i) - (\phi_2 - \phi_i)\nabla r^{-1} . \mathbf{u}_n]\, dS_2 \end{aligned} \quad (2.5.51)$$

by the interface conditions (2.5.39) and (2.5.41). Now adding eqn. (2.5.50) to $\kappa^{-1}$ times eqn. (2.5.51), we eliminate $g$ and obtain

$$4\pi(\phi_2 - \kappa^{-1}\phi_i) = -\iint (1 - \kappa^{-1})\phi_2\nabla r^{-1} . \mathbf{u}_n\, dS_2. \quad (2.5.52)$$

Rearranging:

$$\phi_i = \kappa\phi_2 + \frac{1}{4\pi}\iint (\kappa - 1)\phi_2\nabla r^{-1} . \mathbf{u}_n\, dS_2. \quad (2.5.53)$$

In eqn. (2.5.49) we have the impressed potential in terms of the exterior surface potential; and in eqn. (2.5.53) we have it in terms of the interior surface potentials. As the field points approach the surface from both sides, the rapid changes in potential are on account of the dipolelike layer induced on the surface. In the limit (van Bladel 1964, pp. 73–75), the surface potential $\phi$ settles to the average given by eqns. (2.5.49) and (2.5.53):

$$\phi_i = \tfrac{1}{2}(\kappa + 1)\phi + \frac{1}{4\pi}\iint (\kappa - 1)\phi \nabla r^{-1} \cdot \mathbf{u}_n \, dS_2. \qquad (2.5.54)$$

The singularity in the integrals may be avoided by taking a hemisphere on the surface, as we did before for the line integral. Now, dividing the surface into elements, on each of which the potential $\phi$ is assumed to be known or an unknown constant, this equation may be easily built up for the unknown constant values of potential and solved.

In two-dimensional problems, as may be observed from eqn. (2.5.17), the integrand of eqn. (2.5.54) becomes $-(1/2\pi)\nabla ln_e r \cdot \mathbf{dS}$ in place of $(1/4\pi)\nabla r^{-1} \cdot \mathbf{dS}$ so that we have

$$\phi_i = \tfrac{1}{2}(\kappa + 1)\phi + \frac{1}{2\pi}\int (\kappa - 1)\phi \nabla ln_e r \cdot \mathbf{u}_n \, dS_2. \qquad (2.5.55)$$

## 2.5.3. Some Notes on Higher-Order Boundary Elements

We have seen in section 2.5.1 the details of the boundary element method and how to compute the coefficient matrix for elements of order zero. It was also observed that at Neumann boundaries, the use of zeroth-order elements would cause problems since, in effect, we set the field to zero at the boundary.

Higher-order elements are a straightforward way of enhancing accuracy. Indeed, when we go from zeroth order to first order, the matrix size for the problem does not change, but significant gains in accuracy are made. For instance, in using zeroth-order potential $\phi$, we would compute $\phi$ at the middle of an element. But with first order, we would compute at the ends of the element and postulate that the potential varies linearly from end to end. Observe now that, for a closed mesh, there are as many element middles as there are ends. Since accuracy is increased without any more work (the most significant part of computation being matrix inversion), one must take the trouble to use at least first-order elements. It is stressed that the matrix size does go up when we go from first order to second order, where we must explicitly compute the potential at the ends and the middle. Also, in three-dimensional computation, work increases even when we change from order zero to order one, since there are more vertices than middles, whether we use triangular or quadrilateral elements to discretize the surface.

The terms that we seek to evaluate are of the form $\int \phi(\partial ln_e r\, dS/\partial n)$ and $\int g ln_e r\, dS$. Previously, in section 2.5.1, we assumed that both the potential and its normal gradient $g$ were constant on an element. This simplified matters for us significantly since, as a result, we were able to pull them out of the integral sign. We then dealt with performing the integrals $\int (\partial ln_e r\, dS/\partial n)$ and $\int ln_e r\, dS$.

Let us see how the integrals turn out for order one, since the results may be easily extended to other higher orders in like manner. $\phi$ now varies on an element 12 of Figure 2.5.5A, as

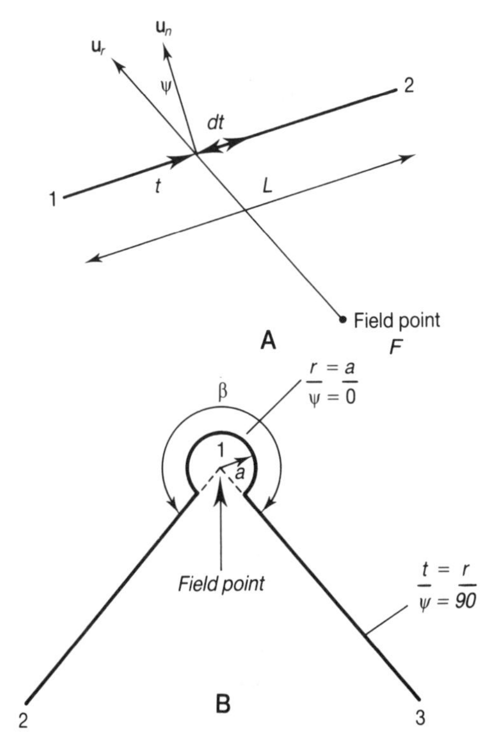


**Figure 2.5.5.** Coefficients for high-order elements. **A.** Off-diagonal terms. **B.** Diagonal terms.

$\phi(\zeta) = \phi_1 + \zeta(\phi_2 - \phi_1)$ where $\zeta = t/L$ is the distance of the point at which $\phi$ is expressed, scaled by the length $L$ from 1 to 2. It is clear then that the integral $\int \phi(\partial ln_e r\, dS/\partial n)$ may be split into two parts proportional to $\int(\partial ln_e r\, dS/\partial n)$ and $\int \zeta(\partial ln_e r\, dS/\partial n)$. The first term may be handled as before and, as proved earlier, is the angle subtended at the field point $F$ by the side 12. For the second term, using $\partial ln_e r/\partial n = \cos\psi/r$, proved earlier, we may evaluate this integral using Gaussian quadrature formulae. However, when we are working out the diagonal coefficient at node 1, say, we must divide the contour into two parts as shown in Figure 2.5.5B. This time the self-term will come in evaluating at the vertex between two elements, and the circle will not be half a circle but a more general arc. We observe that along the lines beyond the partial circle to nodes 2 and 3, $t = r$, $\psi = 90°$, so that the integral collapses. Over the arc, $\psi = 0$ since the radius is along the normal. Therefore the term $\int(\partial ln_e r\, \partial n/dS) = \int(1/r)r\, d\theta = \beta$, while the term $\int \zeta(\partial ln_e r\, \partial n/dS)$ over the circle $r = a$ is $\int Lr(1/r)r\, d\theta \to 0$ as $a \to 0$. Observe that the exterior angle is taken for $\beta$ in keeping with the definition of Figure 2.5.2C where $\theta$ grows counterclockwise.

Similarly, if we allow $g$ to be of order one, we must evaluate integrals of the form $\int ln_e r\, dS$ and $\int \zeta ln_e r\, dS$. The first term is as dealt with earlier in section 2.5.1. The second term for off-diagonal elements is dealt with using the quadrature formulae. The diagonal term takes the form, at node 1,

$$\begin{aligned}\lim_{a\to 0}&\left\{\int_{\text{Arc}} \zeta ln_e r\, dS + \int_{r=a}^{L_1} \zeta ln_e r\, dS + \int_{r=a}^{L_2} \zeta ln_e r\, dS\right\}\\ &= \lim_{a\to 0}\left\{\int_{\text{Arc}} a ln_e a\, a\, d\theta + \int_{r=a}^{L_1} rL_1 ln_e r\, dr + \int_{r=a}^{L_2} rL_2 ln_e r\, dr\right\}.\end{aligned} \tag{2.5.56}$$

$a^2 ln_e a$ must vanish as $a$ tends to zero, since we have already shown in section 2.5.1 that $a ln_e a$ vanishes. The integrals $r ln_e r\, dr$ may be rewritten as $\frac{1}{2} ln_e r\, dr^2$ and integrated by parts to give $\frac{1}{2}r^2 ln_e r - \int \frac{1}{2} r dr = \frac{1}{2}r^2 ln_e r - \frac{1}{4}r^2$. These two terms assume values at the upper limits $r = L_1$ and $L_2$. Their values at the lower limit $r = a$ vanish as $a$ tends to zero. This may be proved by l'Hopital's rule, as used in section 2.5.1.

It is hoped that these brief notes on the procedures for the evaluation of integrals give enough information for one to understand the general methodology for boundary element matrix coefficient computation and start solving serious problems. The work by Lindholm (1981), Lean and Wexler (1985), and Kalaichelvan and Lavers (1987) may be consulted by the more avid reader.

## 2.6 ELECTRIC FIELDS IN THE PRESENCE OF DIELECTRICS: EXAMPLES

### 2.6.1 A Cylindrical Dielectric in a Uniform Field

We have derived eqns. (2.5.54) and (2.5.55) for the solution of the electric field in the presence of a dielectric discontinuity in an impressed field. To demonstrate the method, we shall take up here the solution of a long cylindrical cavity in an infinite dielectric immersed in a constant (or uniform) electric field. It is pointed out, however, that this being a numerical technique, the cavity cross section may be of any shape. A cylindrical cavity is chosen here for the simple reason that, by symmetry, no changes will occur normal to a plane that contains the impressed field and slices the cylinder into two halves. Therefore, this problem may be tackled in two dimensions if one is principally interested in the field in the middle. This is indeed often the case because, according to Gauss's theorem (A13) applied to any closed surface $S$ within the cavity, since no charge is contained within the surface,

$$\iint \mathbf{D} \cdot d\mathbf{S} = 0. \tag{2.6.1}$$

This is possible only if $\mathbf{D}$ is a constant inside, and, therefore, it suffices to determine it anywhere in the cavity. Therefore, we determine this constant field at the middle of the cavity and thereby reduce the dimension of the problem. Observe that by adjusting the value of the dielectric constant $\kappa$, the same problem becomes that of a dielectric in a uniform electric field.

Eqn. (2.5.55) allows us to solve for the potentials on the dielectric surface, provided the potential in the absence of the dielectric is known. As shown in Figure 2.6.1, let the impressed electric field be a constant $E_0$ in the $x$ direction. Then by eqn. (1.2.21),

$$-\frac{\partial \phi_i}{\partial x} = E_0, \tag{2.6.2}$$

solving which,

$$\phi_i = -E_0 x, \tag{2.6.3}$$

except for a constant of integration, which may be ignored in view of our freedom to choose a reference potential. Knowing the

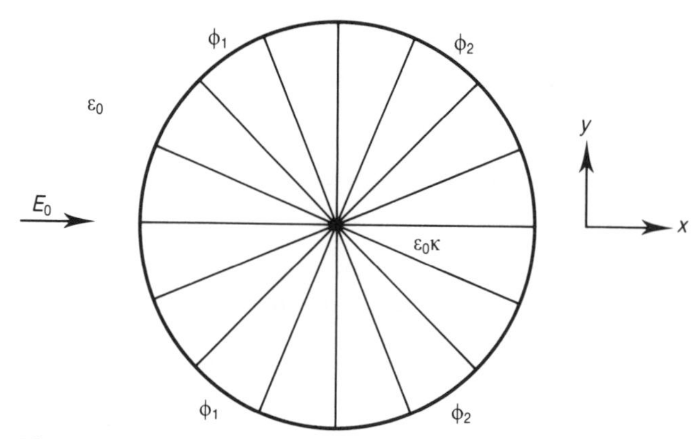


**Figure 2.6.1.** Discretization for cavity in a dielectric.

impressed potential everywhere, we may now symmetrically discretize the surface of the cylinder into, say, 16 small, straight lines. By symmetry, the potential will be unknown on only 8 of these elements. For two elements, $p$ and $q$, at coordinates $(x_p, y_p)$ and $(x_q, y_q)$, the expression (2.5.55) becomes:

$$\phi_i(x_p, y_p) = \tfrac{1}{2}(\kappa + 1)\phi(x_p, y_p) + \frac{1}{2\pi}\sum_{q=1}^{16}(\kappa - 1)\phi(x_q, y_q)\nabla ln_e r_{pq} \cdot \mathbf{u}_{nq}\, dS_2, \tag{2.6.4}$$

where the working out of the vectors $r_{pq}$ and the unit vector $\mathbf{u}_{nq}$ normal to the element $q$ and the dot product of these vectors is already detailed in eqns. (2.5.29)–(2.5.36). With an equation such as eqn. (2.6.4) for every surface element $p$, the surface potentials $\phi_p$ may be solved and thence $\phi$ anywhere.

## 2.6.2 A Dielectric Filled Parallel-Plate Capacitor

Another example of using an integral method for the solution of electric fields in the presence of dielectrics is the same capacitor problem taken up before in section 2.4.2, but now filled with a dielectric. Observe that we have already solved this problem in the absence of the dielectric there; that is, from the charge distribution computed in section 2.4.2, we may compute the impressed electric field $\phi_i$ in the absence of dielectrics required for the application of eqn. (2.5.55) to the surface shown in Figure 2.6.2. Now the surface

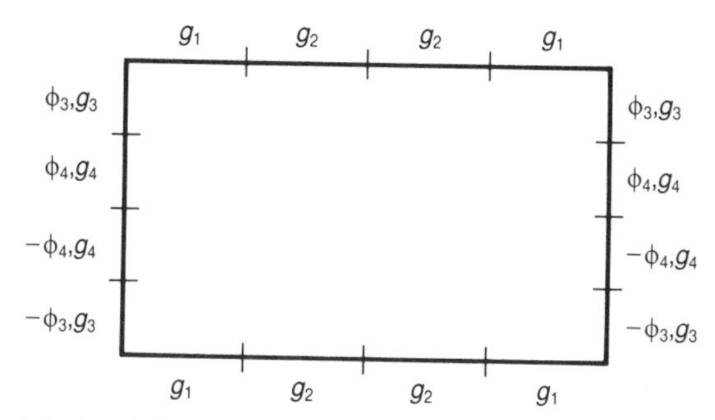


**Figure 2.6.2.** Discretization for dielectric filled parallel plate capacitor.

may be divided into elements over each of which a constant potential is assumed, and these constants may be computed. The potentials at the plates need not have expressions written out for them, since they are already known.

As an alternative, we may use the boundary integral method of eqns. (2.5.37) and (2.5.38) to the regions interior and exterior to the dielectric and match them using the boundary conditions. Notice that only the normal gradient needs to be solved at the plates.

# 3 ELECTRIC FIELDS: ANALYSIS BY DIFFERENTIAL METHODS

## 3.1 DIFFERENTIAL METHODS

### 3.1.1 On Differential Methods

The integral methods we have seen in chapter 2 express the total effect of all the causal quantities at a particular point; that is, they express action at a distance. On the other hand, the differential methods concentrate on the differential forms of Maxwell's equations and so express the relationship of a variable to its values at neighboring points, both of which may be caused by a causal quantity far away. Differential methods therefore express the laws in terms of local relationships.

### 3.1.2 Finite Difference Solution

To solve the Poisson equation

$$-\varepsilon\nabla^2\phi = \rho, \tag{1.2.22}$$

one of the older and still commonly, albeit decreasingly, used approximation methods is the finite difference method. It is a method simple in conception and easy of implementation. The method is known as a *differential* method, unlike the previously treated integral method, because we locally approximate the differential equation being solved, using approximations for the differential operators. We specifically solve for the scalar field governed by the Poisson equation at a few selected points that are the nodes of a rectangular grid placed so as to completely encompass the solution region. Each boundary of the device must run along a line of the grid, and, as we have shown in Appendix B, the boundary of a region of solution needs the Neumann or Dirichlet condition to be specified at every point for uniqueness.

In the finite difference method, the region of solution is divided into a rectangular grid and we explicitly solve for the potential at the grid points, using linear algebraic equations relating the potential at a node to those at the neighboring points by approximating the Poisson equation at that node. Thus, at every node we will have one equation associated with it. To approximate the second derivative of the potential $\phi$ in the Poisson equation with respect to $x$, consider the node at position $(x, y)$ in the $h \times k$ rectangular grid shown in Figure 3.1.1. We write for the first derivatives:

$$\frac{\partial}{\partial x}\phi(x + \tfrac{1}{2}h, y) = \frac{\phi(x + h, y) - \phi(x, y)}{h}, \tag{3.1.1}$$

$$\frac{\partial}{\partial x}\phi(x - \tfrac{1}{2}h, y) = \frac{\phi(x, y) - \phi(x - h, y)}{h}, \tag{3.1.2}$$

so that we have:

$$\begin{aligned}\frac{\partial^2}{\partial x^2}\phi(x, y) &= \frac{\frac{\partial}{\partial x}\phi(x + \frac{1}{2}h, y) - \frac{\partial}{\partial x}\phi(x - \frac{1}{2}h, y)}{h} \\ &= \frac{\phi(x - h, y) + \phi(x + h, y) - 2\phi(x, y)}{h^2},\end{aligned} \tag{3.1.3}$$

and similarly:

$$\frac{\partial^2}{\partial y^2}\phi(x, y) = \frac{\phi(x, y - k) + \phi(x, y + k) - 2\phi(x, y)}{k^2}. \tag{3.1.4}$$

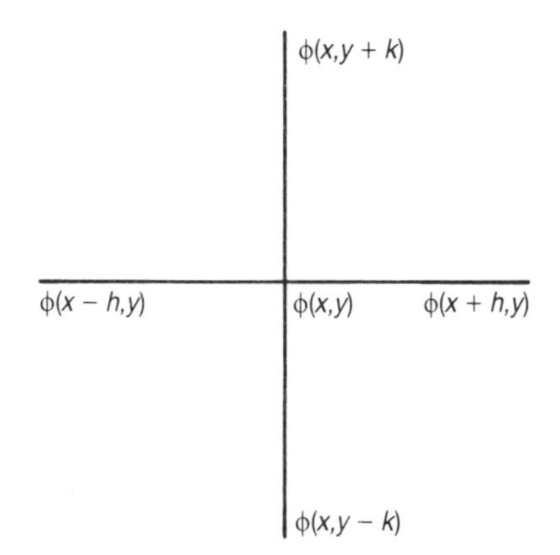


**Figure 3.1.1.** An interior point of a finite difference mesh.

However, eqns. (3.1.3) and (3.1.4) do not tell us the degree of error inherent in them. To quantify the error in the difference equations, a more elaborate derivation of the finite difference approximations, based on Taylor's series approximations, is required. Consider the Taylor's series expansion for the unknown scalar $\phi$ at a node of the $h \times k$ rectangular grid, of coordinates $(x + h, y)$, as shown in Figure 3.1.1:

$$\begin{aligned}\phi(x + h, y) = \phi(x, y) &+ \frac{h}{1!}\frac{\partial}{\partial x}\phi(x, y) + \frac{h^2}{2!}\frac{\partial^2}{\partial x^2}\phi \\ &+ \frac{h^3}{3!}\frac{\partial^3}{\partial x^3}\phi + \frac{h^4}{4!}\frac{\partial^4}{\partial x^4}\phi + O(h^5),\end{aligned} \tag{3.1.5}$$

where $O(h^5)$ stands for terms of order $h^5$. Similarly,

$$\begin{aligned}\phi(x - h, y) = \phi(x, y) &- \frac{h}{1!}\frac{\partial}{\partial x}\phi(x, y) + \frac{h^2}{2!}\frac{\partial^2}{\partial x^2}\phi \\ &- \frac{h^3}{3!}\frac{\partial^3}{\partial x^3}\phi + \frac{h^4}{4!}\frac{\partial^4}{\partial x^4}\phi + O(h^5).\end{aligned} \tag{3.1.6}$$

Summing the above two equations, we obtain the finite difference approximation to the second partial derivative with respect to $x$ of the unknown potential $\phi$:

$$\frac{\partial^2}{\partial x^2}\phi(x, y) = \frac{\phi(x - h, y) + \phi(x + h, y) - 2\phi(x, y)}{h^2} + O(h^2). \tag{3.1.7}$$

It may similarly be shown that

$$\frac{\partial^2}{\partial y^2}\phi(x, y) = \frac{\phi(x, y - k) + \phi(x, y + k) - 2\phi(x, y)}{k^2} + O(k^2). \tag{3.1.8}$$

Thus, we see that the error in the approximation is of order 2 in $h$ and $k$, meaning that, for example, if $h$ is 0.1, the terms that we neglect would be multiples of 0.01, making the approximations highly accurate. Substituting eqns. (3.1.7) and (3.1.8) in the Poisson eqn. (1.2.22), we have the complete finite difference approximation:

$$\begin{aligned}&\frac{\phi(x + h, y) + \phi(x - h, y) - 2\phi(x, y)}{h^2} \\ &\quad + \frac{\phi(x, y + k) + \phi(x, y - k) - 2\phi(x, y)}{k^2} = -\frac{\rho(x, y)}{\varepsilon}.\end{aligned} \tag{3.1.9}$$

Thus, eqn. (3.1.9) represents an equation for every interior point of the region of solution surrounded by four nodes. However, to solve

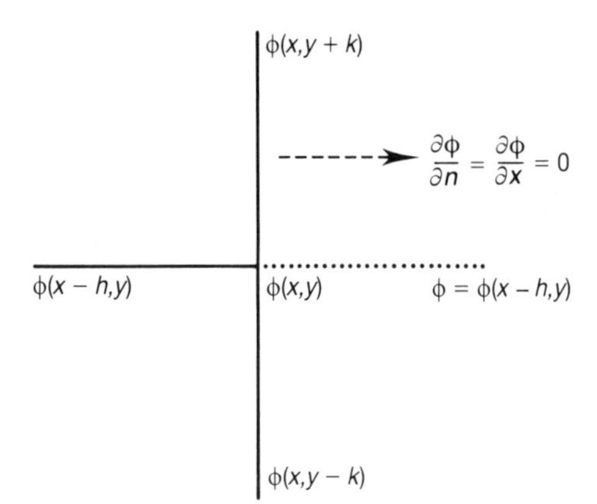


**Figure 3.1.2.** A boundary node of a finite difference mesh.

for the field at every node, we need an equation for every node. Nodes on Dirichlet sections of the boundary need no equations since $\phi$ is specified there and therefore known. At Neumann boundaries, however, the situation for the example of a vertical boundary on the right is as depicted in Figure 3.1.2. Here we have a node at which an equation is required, but without a node at $(x + h, y)$ to use in eqn. (3.1.9). Now, at the hypothetical node at that point, outside the region of solution, should the potential be $\phi$, then by the Neumann condition,

$$\frac{\partial}{\partial n}\phi = \frac{\partial}{\partial x}\phi = \frac{\phi - \phi(x-h,y)}{2h} = 0. \qquad (3.1.10)$$

From this we may say that $\phi$ is $\phi(x - h, y)$, thus allowing us to write, in place of eqn. (3.1.9),

$$\frac{2\phi(x-h,y) - 2\phi(x,y)}{h^2} + \frac{\phi(x,y+k) + \phi(x,y-k) - 2\phi(x,y)}{k^2} = -\frac{\rho(x,y)}{\varepsilon}. \qquad (3.1.11)$$

Similar expressions may be obtained for points along vertical boundaries on the left as well as for horizontal ones below and above. Thus it is that we arrive at an equation for every point, all of which, when solved together, will yield a unique solution.

Once the potential $\phi$ is obtained at the nodes, we know that

$$\frac{\partial}{\partial x}\phi = \frac{\phi(x+h,y) - \phi(x-h,y)}{2h}, \qquad (3.1.12)$$

$$\frac{\partial}{\partial y}\phi = \frac{\phi(x,y+k) - \phi(x,y-k)}{2k}, \qquad (3.1.13)$$

so that the field densities may be obtained by $\nabla\phi$ or $\nabla \times \phi\mathbf{u}_z$ as given by eqns. (1.2.33) and (1.2.34). Moreover, the energy of the system, usually a multiple of $(\nabla\phi)^2$, may also be easily computed.

### 3.1.3 Finite Difference Network Models

For the solution of practical engineering problems, the finite difference approximation derived in eqn. (3.1.9) clearly must be generalized to incorporate inhomogeneities (or heterogeneities) and nonuniform meshes (necessary for the modeling of curved boundaries). A useful concept, which allows us to perform this generalization without getting involved in elaborate mathematics, is that of the analogy between the finite difference approximation and electrical network models (Liebmann 1949, 1952; Carpenter 1968; Laithwaite 1967; Silvester 1967).

As seen in the simple analogous electrical network of Figure 3.1.3A, eqn. (3.1.9) expresses the continuity of currents at node 0 with $\phi$ being the potential at the nodes, $-\rho$ being the flow out of that node, and $h^2/\varepsilon$ and $k^2/\varepsilon$ being the impedances, respectively, of horizontal and vertical branches of the electrical network. To see this, we rearrange eqn. (3.1.9) and obtain

$$\frac{\phi_1 - \phi_0}{h^2/\varepsilon} + \frac{\phi_2 - \phi_0}{k^2/\varepsilon} + \frac{\phi_3 - \phi_0}{h^2/\varepsilon} + \frac{\phi_4 - \phi_0}{k^2/\varepsilon} = -\rho_0. \quad (3.1.14)$$

Thus, our finite difference approximation for the whole region of solution is the analogue of an electrical network covering that region, with impedance between nodes a distance $h$ apart, in a region of material $\varepsilon$, being given by

$$Z = h^2/\varepsilon. \quad (3.1.15)$$

This has its natural extension to nonuniform meshes, since only the distance $h$ will vary between connected nodes. To extend the above equation to inhomogeneous media, supposing the impedance $Z$ lies along a material discontinuity, then, since an impedance $Z$ is the same as two parallel impedances of value $2Z$, we may consider each of the two branches $2Z$ (as shown in Figure 3.1.3B) to be contributed by the two regions on either side of the branch. Therefore, where we have a material $\varepsilon_1$ on one side of the branch and $\varepsilon_2$ on the other, by extension, we would have $Z_1 = 2h^2/\varepsilon_1$ and $Z_2 = 2h^2/\varepsilon_2$ in parallel, so that the total impedance along the

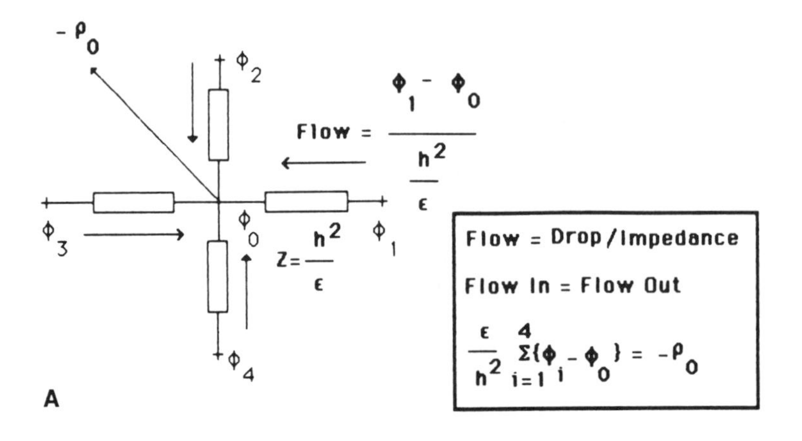


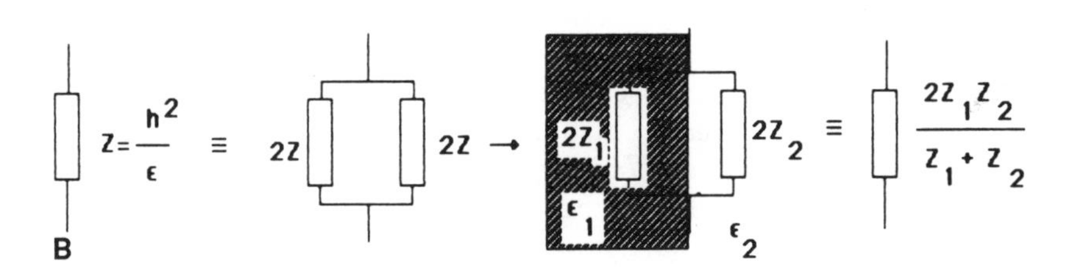


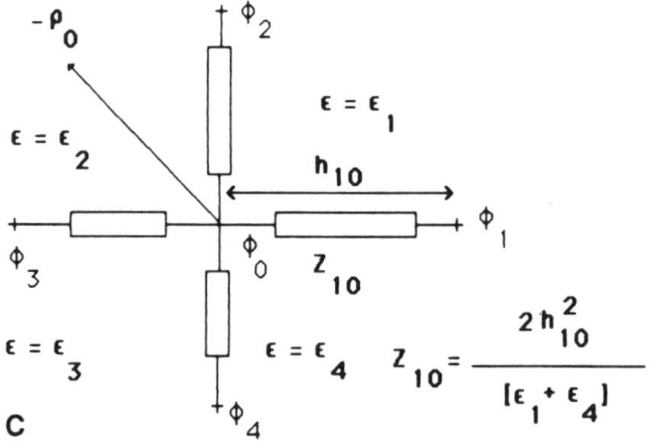


**Figure 3.1.3.** Finite difference network models. **A.** Circuit analog of Poisson equation for homogeneous media. **B.** Impedance at material discontinuity. **C.** Generalized nodal network.

discontinuity is

$$Z_{12} = \frac{Z_1 Z_2}{Z_1 + Z_2} = \frac{2h^2}{\varepsilon_1 + \varepsilon_2}. \tag{3.1.16}$$

The generalized equation for the node 0 of Figure 3.1.3C is therefore

$$\frac{\phi_1 - \phi_0}{2h_{10}^2/(\varepsilon_1 + \varepsilon_4)} + \frac{\phi_2 - \phi_0}{2h_{20}^2/(\varepsilon_2 + \varepsilon_3)} + \frac{\phi_3 - \phi_0}{2h_{30}^2/(\varepsilon_2 + \varepsilon_3)} + \frac{\phi_4 - \phi_4}{2h_{40}^2/(\varepsilon_3 + \varepsilon_4)} = -\frac{\rho_0}{4} - \frac{\rho_0}{4} - \frac{\rho_0}{4} - \frac{\rho_0}{4}. \tag{3.1.17}$$

Therefore, the above equation may be used to construct the impedances across inhomogeneous surfaces and to regard the $-\rho$ terms as being contributed by four quarters surrounding a node. Thus, for a node at the corner of a square conductor with charge density $\rho$, the flow will be $-\rho/4$, and along the edge of a conductor, $-\rho/2$.

## 3.2 LIEBMANN'S ITERATION

A point to note in the foregoing finite difference technique is that the matrix equation is a sparse one, because on an equation corresponding to a particular node, at most only five coefficients appear. However, because of the presence of Neumann boundaries, the matrix is not a symmetric one, so that new, efficient solution techniques such as the Cholesky decomposition scheme and the Preconditioned and Renumbered Conjugate Gradients Algorithm (to be described in chapter 6) cannot be employed.

A very great simplification results when we use an $h \times h$ square mesh. For then, eqn. (3.1.9) simplifies to

$$\phi(x + h, y) + \phi(x, y + h) + \phi(x - h, y) + \phi(x, y - h) - 4\phi(x, y) = -\frac{h^2 \rho(x, y)}{\varepsilon}. \tag{3.2.1}$$

This fact allows us to employ a simple data structure defining the four nodes $n$, $s$, $w$, $e$ surrounding each node $i$ (on the north, south, west, and east) and the source $\rho$ at $i$. Thus, the data for each node will comprise four integers and a real number. The equation thereat is

$$\phi(i) = \frac{\phi(e) + \phi(n) + \phi(w) + \phi(s) + [h^2 \rho(i)/\varepsilon]}{4}. \tag{3.2.2}$$

Rather than get involved in assembling the matrix equation, the simplest procedure is to use Liebmann's iterations (Liebmann 1949; 1952), also called the Successive Over Relaxation (SOR) or Gauss–Seidel method of solution, for the $n \times n$ matrix equation

$$\mathbf{Ax} = \mathbf{B}. \tag{3.2.3}$$

In this algorithm an initial guess is made of the solution, and in each iteration we go through every row and compute the variable corresponding to that row, assuming of course that all the other variables are known. To algorithmically describe this:

$$x[i] := \frac{\mathbf{B}[i] - \sum_{k \neq i; k=1}^{n} A[i,k]^{*}x[k]}{A[i,i]} \qquad i = 1, \ldots, n \tag{3.2.4}$$

**Algorithm 3.2.1. Liebmann's (or SOR) Matrix Solution Scheme**

```
Procedure Liebmann(x, A, B, n, Precis)
{Function: Solves the matrix equation Ax = B
Output: x = the solution, an n vector
Inputs: A = the n × n matrix of coefficients; B = the right-
hand side, an n vector; n = the size of the equation; Precis =
the percentage precision to which convergence is required}
Begin
  Repeat
    MaxX ← 0
    MaxChange ← 0
    For i ← 1 To n Do
      Old ← x[i]
      Sum ← 0
      For k ← 1 To n Do
        Sum ← Sum + A[i, k]*x[k]
      x[i] ← (B[i] − Sum + A[i, i]*x[i])/A[i, i]
      Change ← x[i] − Old
      x[i] ← Old + 1.25* Change {Acceleration}
      AbsChange ← Abs(Change)
      AbsX ← Abs(x[i])
      If AbsChange > MaxChange
        Then MaxChange ← AbsChange
      If AbsX > MaxX
        Then MaxX ← AbsX
    Error ← 100*MaxChange/MaxX
  Until (Error < Precis)
End
```

must be implemented until the elements $x[i]$ imperceptibly change. This is described in Alg. 3.2.1. Note that the checking for $k \neq i$ required in eqn. (3.2.4) is avoided by adding $A[i,i]*x[k]$ to the quantity Sum and then taking it out. Acceleration is also used, with a factor of 1.25, exploiting the consistent convergence of the approximate solution from one side of the true solution. That is, if $x[i]$ moves, say, from 100 in an iteration to 140 in the next, then the true solution is taken to require a movement in the same direction from 100 to 140 and beyond. Acceleration therefore nudges the answer further along the way by multiplying the change by the acceleration factor 1.25, so that the answer is modified to 100 + 1.25*40 = 150. Schemes exist for computing the ideal acceleration

**Algorithm 3.2.2. Liebmann's Finite Difference Scheme**

```
Program Finite_Difference
{Function: Solves a finite difference problem with uniform
h × h mesh in a uniform medium of material value ε
Output: x = the solution, an n vector
Inputs: n = the number of unknown nodes numbered first;
Precis = the percentage precision to which convergence is
required; a data file, giving the nodes e, w, n, and s
surrounding a node and ρ thereat}
Begin
  Repeat
    MaxX ← 0
    MaxChange ← 0
    For i ← 1 To n Do
      Old ← x[i]
      Read nodes e, n, w, and s surrounding node i and ρ at i
      x[i] ← (x[e] + x[n] + x[w] + x[s] + h²ρ/ε)/4
      Change ← x[i] − Old
      x[i] ← Old + 1.25*Change {Acceleration}
      AbsChange ← Abs(Change)
      AbsX ← Abs(x[i])
      If AbsChange > MaxChange
        Then MaxChange ← AbsChange
      If AbsX > MaxX
        Then MaxX ← AbsX
    Error ← 100*MaxChange/MaxX
  Until (Error < Precis)
End
```

factor that lies in the range 1 to 2 and may be employed (Zienkiewicz and Lohner 1985). A factor that is too large will push the approximate solution beyond the true solution, and therefore a safe value of 1.25 is used here.

It is easily shown that in a uniform finite difference mesh, since the diagonal terms are 4 and the off-diagonal terms are $-1$, Alg. 3.2.1 reduces to Alg. 3.2.2, where no storage needs to be allocated for a matrix. Moreover, a lot of time is saved by avoiding the several unnecessary multiplications $A[i,k]^*x[k]$ inherent to Alg. 3.2.1, even when the coefficient $A[i,k]$ is zero.

## 3.3 SOME EXAMPLE FLOW PROBLEMS

### 3.3.1 Flow of Fluid through a Pipe

As an example of demonstrating the application of the finite difference method, consider the incompressible flow of water through a pipe, as shown in Figure 3.3.1A. Here a narrow pipe carries water and meets a larger pipe of double the radius. We need to determine the pattern of flow within. Although this is a three-dimensional problem, along the vertical plane of symmetry which cuts the pipes into two halves lengthwise the problem is two-dimensional, and therefore an inkling of the processes within the pipes may be obtained by analyzing the flow along the plane, close and perpendicular to which no changes occur. It is pointed out that this two-dimensional analysis is approximate. It is not generally true that the solution is two-dimensional on the plane of geometric and electromagnetic symmetry of a three-dimensional object. What is true is that $\partial/\partial z \equiv 0$ on the plane. To make the analysis truly two-dimensional $\partial^2/\partial z^2$ must also be zero, and this is not necessarily true. For instance, consider two blocks of dielectric of rectangular cross section on the $xy$ plane and large length in the $z$ direction, facing flux coming from afar in the $x$ direction. The middle of the length of each block is at $z = 0$. The large length may be used to say that at the midsection $z = 0$, both $\partial/\partial z$ and $\partial^2/\partial z^2$ are zero, so that the problem is two-dimensional. However, if, say, one block were much smaller in length, although the $xy$ plane is a midplane of symmetry with $\partial/\partial z = 0$, the flux would develop a $z$-component close to the plane $z = 0$ and the two-dimensional model is no longer valid.

It is stressed therefore that this exercise in two-dimensional

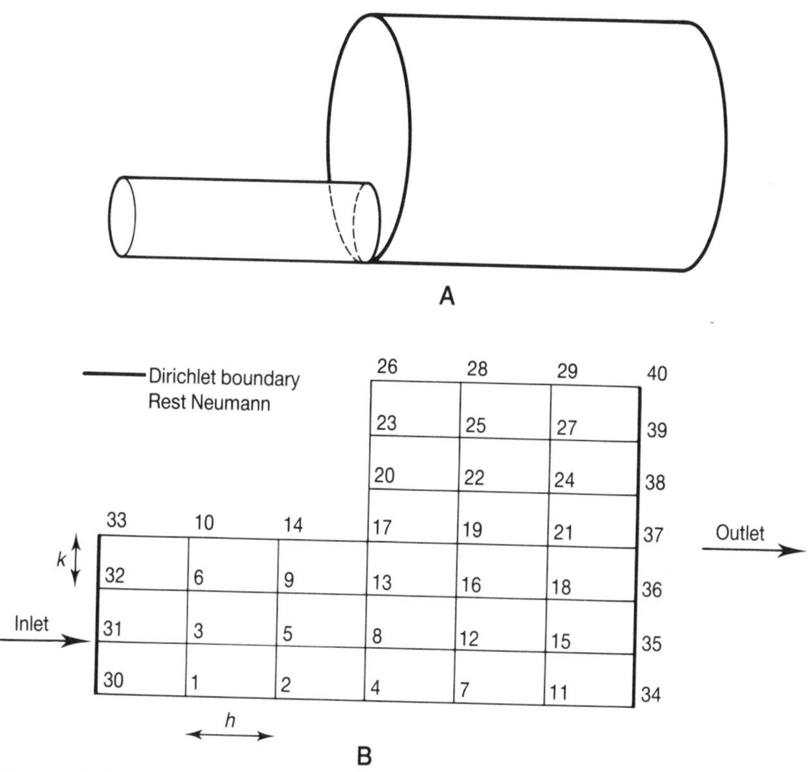


**Figure 3.3.1.** The pipe flow problem. **A.** The pipe junction. **B.** Finite difference mesh for pipe.

analysis is only to obtain an estimate of the field. The two-dimensional analysis would be exact, however, if the water were guided through pipes of different rectangular cross sections with wide dimensions in the $z$ direction.

The vector field in this problem is the velocity of flow, which is proportional to the gradient of the pressure $p$ driving the fluid, so that

$$V = -k\nabla p, \tag{3.3.1}$$

where the negative sign denotes flow from higher to lower pressure regions. Conservation of mass dictates that

$$\nabla \cdot \rho \mathbf{V} = 0, \tag{3.3.2}$$

where $\rho$ is density, so that for the homogeneous medium

$$\nabla^2 p = 0. \tag{3.3.3}$$

As for boundary conditions, provided we take the inlet and outlet sufficiently far away from the junction, these two boundaries will behave as if they were in the midst of an infinitely long pipe and will therefore have the same pressure along their cross section. These may then be made Dirichlet boundaries, with the inlet and outlet respectively at, say, 100 and 0 $Nm^{-2}$. Since the rest of the boundary represents pipe wall through which water may not exit, an appropriate boundary condition is that the velocity of flow, normal to these walls, is zero; i.e., they are Neumann boundaries with $\partial p/\partial n = 0$. Thus, we have the whole region of the pipe surrounded by Neumann or Dirichlet boundaries as needed for uniqueness.

To solve for the pressure everywhere within, we divide the solution region, as shown in Figure 3.3.1B, into a square mesh. The Neumann boundaries have been extended as indicated in section 3.1.1. The data for this problem are given in Figure 3.3.2. The list of nodes, giving, in order, the nodes to the right, above, to the left,

a. h = k = (Smaller Pipe Radius)/3
b. NU = Number of Unknowns = 29
c. NN = Number of Nodes = 40
d. press(30) = press(31) = ... = press(33) = 100.0
e. press(36) = press(37) = ... = press(40) = 0.0

f.

| Node NO. | e | n | w | s | p |
|---|---|---|---|---|---|
| 1 | 2 | 3 | 30 | 3 | 0.0 |
| 2 | 4 | 5 | 1 | 5 | 0.0 |
| 3 | 5 | 6 | 31 | 1 | 0.0 |
| 4 | 7 | 8 | 2 | 8 | 0.0 |
| 5 | 8 | 9 | 3 | 2 | 0.0 |
| 6 | 9 | 10 | 32 | 3 | 0.0 |
| 7 | 11 | 12 | 4 | 12 | 0.0 |
| 8 | 12 | 13 | 5 | 4 | 0.0 |
| 9 | 13 | 14 | 6 | 5 | 0.0 |
| 10 | 14 | 6 | 33 | 6 | 0.0 |
| \| | \| | \| | \| | \| | \| |
| \| | \| | \| | \| | \| | \| |
| \| | \| | \| | \| | \| | \| |
| 29 | 40 | 27 | 28 | 27 | 0.0 |

**Figure 3.3.2.** Data for pipe flow problem in Figure 3.3.1B.

and below each node and the Poissonian source term, is required for the iterations. These data are usually entered and written on file.

Since we know that the answers must lie between 0 and 100, it is usually convenient to take the unknowns press($i$) = 50 for $i$ = 1,29, as the starting solution. Thereafter, we implement Alg. 3.2.2, described in the previous section for finite difference meshes. The resulting solution was plotted using the algorithm to be described in sections 3.4 and 9.2 and is presented in Figure 3.3.3. Alternatively, computer libraries may be used to plot the equipotential contours.

The following are to be noted in this example:

1. A finer grid will yield more accurate answers, since according to eqns. (3.1.7) and (3.1.8) the error is of order $h^2$. This also means more nodes and a correspondingly larger computing effort. A refined mesh and the corresponding finer equipotential plots are seen in Figure 3.3.4.
2. Shifting the two Dirichlet boundaries away from the pipe junction makes the boundary conditions more accurate. This will require more grid spaces and, therefore, more nodes and computation again.

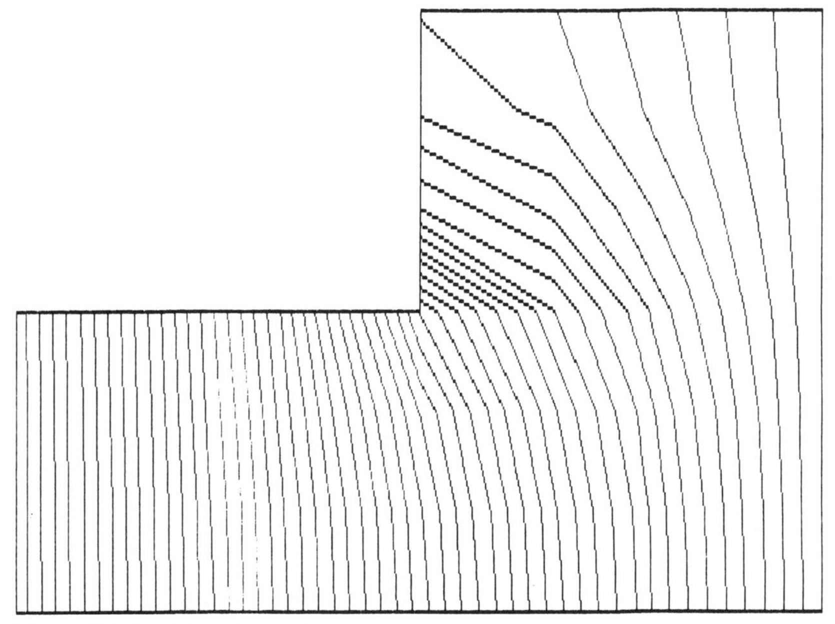

**Figure 3.3.3.** Plot of solution.

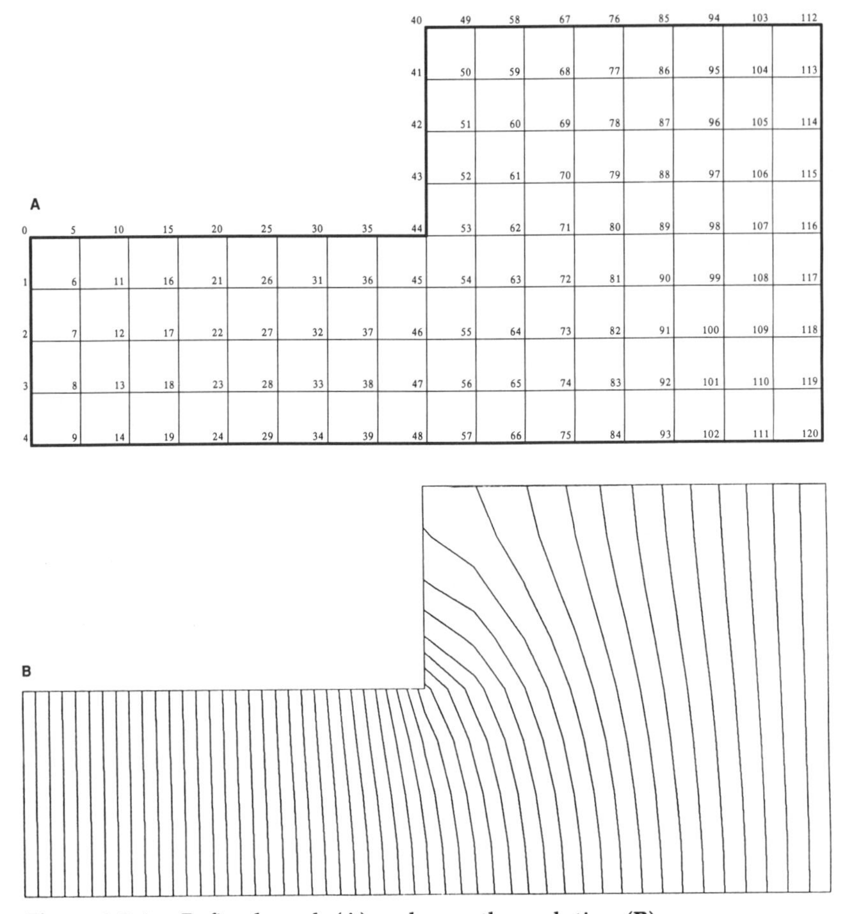


**Figure 3.3.4.** Refined mesh (**A**) and smoother solution (**B**).

3. The nodes have been numbered so that the unknown ones come earlier. This obviates the need to keep track of which nodes are known and which are not.
4. The nodes have also been numbered, starting close to the boundary of the device, with the new numbers radiating from there; i.e., node 1 is surrounded by nodes 2 and 3, then 2 by 4 and 5, and so on. This makes the matrix banded and thereby improves the convergence properties of the Liebmann iterations.
5. At each iteration, the change in the value of the solution is accelerated by a factor of 1.25 so as to speed convergence.

6. For the square mesh chosen, grid size $h$ is not used anywhere in the computations since $\rho$ is zero. However, it will be required for computing the velocities through differentiation according to eqns. (3.1.12) and (3.1.13).

### 3.3.2 Flow of Current through a Conductor

A problem of considerable practical importance in electrical engineering is that of determining the current distribution in a conductor. This may, for example, be the distribution under the earth when a potential is applied to an earthing rod; here the distribution may be required for safety studies in determining the distance from the rod within which it is not safe to tread or to determine the earthing resistance of the rod. Figure 3.3.5 gives the electric potential distribution resulting from such a study, where the nonconducting foundation of a building redistributes the voltages.

Alternatively, the current flow may be required to determine the heating in a conductor so that the optimal load of the conductor

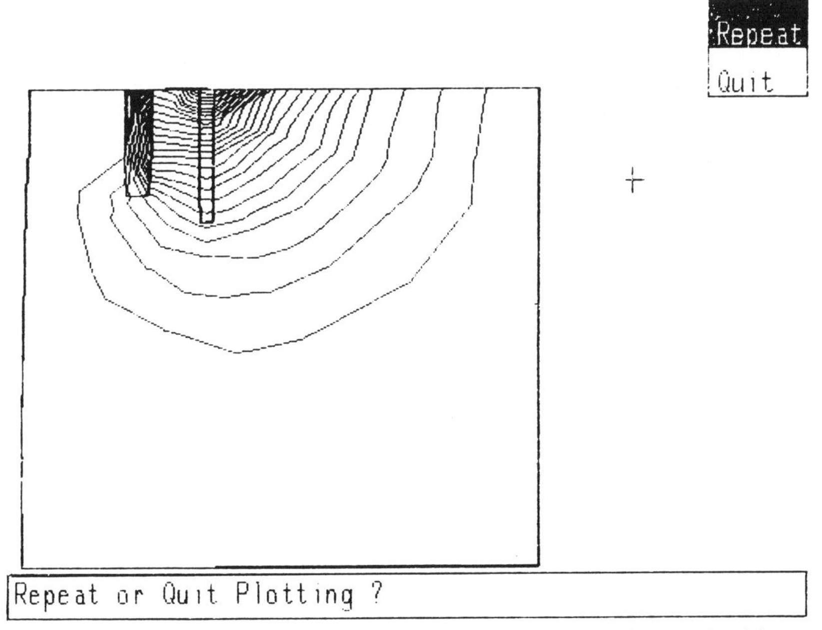


**Figure 3.3.5.** Potential distribution about earthing rod close to a building.

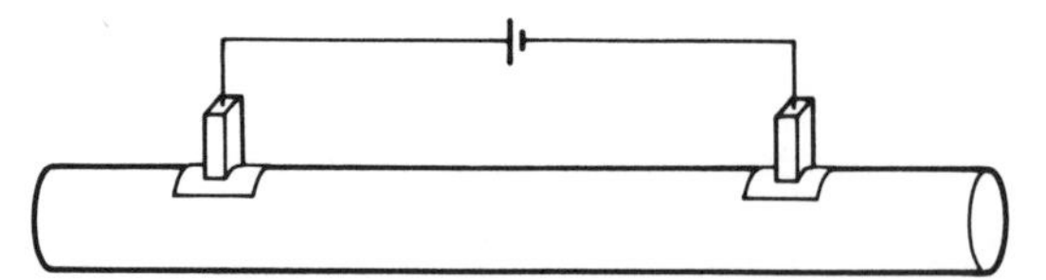

**Figure 3.3.6.** A current-carrying conductor.

may be determined. Figure 3.3.6 shows a cylindrical conductor that has opposite electric potentials applied at symmetric locations from the two ends. Let us attempt to determine the voltage distribution in the conductor. In general for a current flow problem such as this, we have for the current density **J** and electric field **E** related by the conductivity $\sigma$:

$$\mathbf{J} = \sigma\mathbf{E}, \tag{1.2.2}$$

$$\mathbf{E} = -\nabla\phi, \tag{1.2.21}$$

so that

$$\mathbf{J} = -\sigma\nabla\phi. \tag{3.3.4}$$

And, by the continuity of current across any surface $S$,

$$\iiint \mathbf{J} \cdot \mathbf{dS} = 0, \tag{3.3.5}$$

which upon the application of Gauss's theorem (A13) gives us

$$\nabla \cdot \mathbf{J} = 0. \tag{3.3.6}$$

Combining eqns. (3.3.4) and (3.3.6), we obtain

$$-\sigma\nabla^2\phi = 0. \tag{3.3.7}$$

We have just seen how to solve this equation in two dimensions by the finite difference method. Observe that for the problem at hand, a plane through the points where the potential is applied would have no changes normal to it by symmetry. So, constructing our coordinate system with this plane containing the $x$ and $y$ axes, we reduce this approximately to a two-dimensional problem (at least on this plane; please see the discussion in section 3.3.1 on this approximation). And since no current may exit the conductor except through the electrodes, the entire boundary, save the portions in contact with the electrode, must have no current density component normal to the wall; by eqn. (3.3.4), therefore, these parts have the Neumann condition $\partial\phi/\partial n$ zero. Figure 3.3.6, therefore, defines the boundary value problem to be solved. If the applied potentials are equal and opposite, by electrical and geometric symmetry, we may solve within a half of this configuration. It is stressed that even if the electrical symmetry does not exist, we may impose it by adding a suitable constant to the potentials so as to result in symmetry. Observe that if $\phi$ is a solution to the Poisson eqn. (1.2.22), then $\phi + c$ is also a solution:

$$-\varepsilon\nabla^2[\phi + c] = -\varepsilon\nabla^2\phi = \rho. \tag{3.3.8}$$

So, for example, if the applied potentials are 10 and 100, then $c = -(10 + 100)/2 = -55$. That is, we subtract 55 from the boundary conditions at the electrodes to make them have the symmetric values $-45$ and 45, impose the condition $\phi = 0$ at the symmetric plane, solve the reduced problem at a considerably lower cost, and then add 55 to our solution to get the actual physical potential with respect to the actual reference being used by the power supply. This process is the same as applying $\phi = 55$ at the line of symmetry and $\phi = 10$ at the low-voltage electrode, as shown in Figure 3.3.7A.

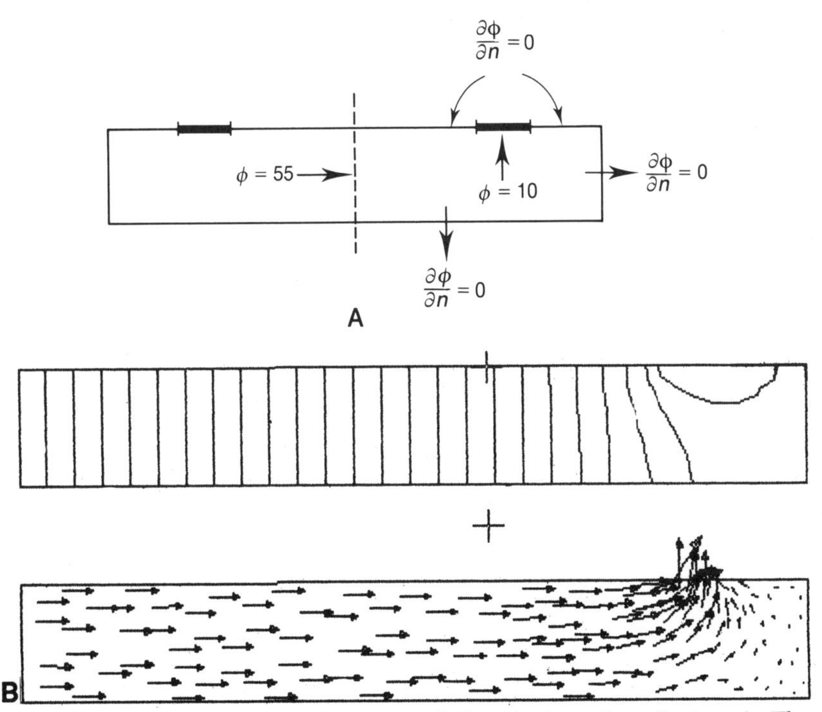


**Figure 3.3.7.** Reduced two-dimensional description of conductor. **A.** The boundary value problem. **B.** The solution.

The solution for this half problem is shown in Figure 3.3.7. Observe also how the uniformity of the solution region makes this problem equally suitable for treatment by the boundary integral method of section 2.5.1.

## 3.4 PLOTTING RESULTS

We have seen so far how to obtain the numerical value of the electric potential field at any point in space by the integral method, and at finite difference grid points by the finite difference method. Unfortunately, what the computer returns is a string of numbers representing the potential at the various points at which it was sought; these numbers are not immediately meaningful, so that we would not know from the numbers whether our program is "bug" free—and even if it were, an intuitive feeling for the behavior of the fields does not follow from the numerical output.

The best way to check the validity of our results and at the same time get a pictorial representation of the behavior of the device is to draw equipotential plots. To this end, the potential at grid points of a rectangular mesh is required. This is automatically given by the finite difference solution, and, when dealing with an integral solution, we may arbitrarily construct a rectangular grid and compute the potentials at the grid points.

The plotting is done by first identifying the rectangle above and to the right of each node. The procedure for such identification is to pick, from the data of Figure 3.3.2, the node to the right of that node, then the one above it, and the one to the left of that. For example, to get the rectangle for node 5 of Figure 3.3.1B, we first pick node 8 of the fifth row of data, then the second entry 13 of the eighth row of data, and the third entry 9 of the 13th row of data. Simple checks for repetition of nodes ensure that we do not consider rectangles beyond Neumann boundaries. To get rectangles to the right of Dirichlet boundaries, it is often convenient to enter data for Dirichlet nodes also, giving the nodes surrounding them. Such data, although not used in the computation of the fields, will prove useful under postprocessing. Thus, for node 30, for example, we would enter the data 1, 31, 0, 0, the last two zeros being dummy numbers.

To come to plotting the equipotentials, we first need to establish how many equipotential contours we wish to draw (50 is a good number for the typical screen) without overcrowding or the contours being so few as to give little information as a result of no lines being present in large portions of the device. Once this number is chosen, we search through all the nodes and establish the variation in potential. This, in turn, establishes the contours to be plotted. For example, for our problem of Figure 3.3.1A, the potential varies from 0 to 100, so that with 51 contours to be plotted, the equipotentials to be plotted will be 0, 2, 4, 6, . . . , 100.

For purposes of plotting, we search through the mesh and identify the rectangles within which the equipotential contours need to be drawn. We draw them by simple graphics commands such as Move$(x, y)$ to move the graphics cursor to point $(x, y)$, and Line$(x, y)$ to draw a line from where the cursor is to the point $(x, y)$. The procedure for drawing contours is best explained by the example of Figure 3.4.1. Here we have identified an $h \times k$ rectangle $ABCD$ within which contours need to be plotted. The coordinates of the four vertices are known (vertex $A$ at $x_0, y_0$), and the finite difference solution algorithm has returned the solution, say, 91, 97, 105, and 98 at the four corners as shown. Supposing we are plotting contours of potential $0, 2, 4, \ldots, 100$; then side $AB$ would intersect the contours 92, 94, and 96. Contour 92 would be a distance $h(92\text{–}91)/(97\text{–}91)$, or $h/6$ from corner $A$. This contour may go to side $BC$, $CD$, or $DA$. By examination, we identify side $DA$ as the only side containing potential 92 at a distance $k(92\text{–}91)/(98\text{–}91) = k/7$ from $A$, and by the sequential commands (available on most graphics systems) Move$(x_0 + h/6, y_0)$ and Line$(x_0, y_0 + k/7)$, draw that contour. After dealing similarly with the other two contours on side $AB$, we proceed to side $BC$, identify the contours to be plotted, and search only on sides $CD$ and $DA$ to see if they arrive there. We do not search on side $AB$, since plots from there to side $BC$ would

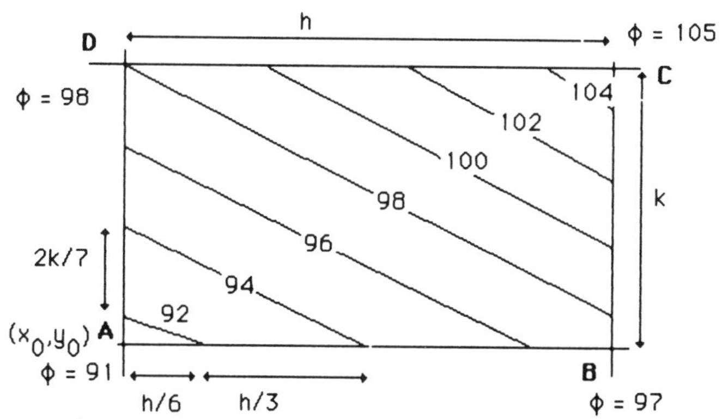


**Figure 3.4.1.** Plotting equipotentials in a typical rectangle.

have been already drawn. Likewise, when dealing with side *CD*, we draw only those leading to side *DA*. This algorithm is formalized in section 9.2.

Physically, in a scalar potential field where the field vector is given by a gradient, the vectors run normal to the equipotentials. On the other hand, a vector field given by a single component vector potential runs along the contours. In either case, where the lines are closely clustered the vector field is higher, since the potential changes faster.

In plotting the equipotentials, it is also important to identify the location of the device in relation to the equipotential lines—that is, to superimpose a drawing of the device upon the potential contours. When dealing with a finite difference mesh, this may easily be done by searching through the data structures for the nodes. Suppose, for example, that for node $i$, the surrounding nodes are $e$, $n$, $w$, and $s$. Then if, say, the line from node $i$ to node $e$ is along the boundary, then the line $i \rightarrow e$ must be a Dirichlet boundary, or failing that, a Neumann boundary, for the problem to be properly posed. In either event, this may be verified from the data. If it is a Dirichlet boundary, then both node numbers $i$ and $e$ must be greater than the number of unknowns *NU*, and if so, we may draw a line from $i$ to $e$ using the Move and Line commands. On the other hand, if it is a Neumann boundary, we will find $n$ and $s$ to have the same number (or, if the alternative data structure described above is employed, one of them will be zero). If this is found to be true, again we shall draw the line. Proceeding in like manner through all the nodes, the complete geometry may be drawn.

It is left as an exercise to the reader to plot the equipotentials of Figure 3.3.3 for the mesh of Figure 3.3.1B, and Figure 3.3.4 for the pipe problem and the other problems we have worked in this chapter.

## 3.5 SOME EXAMPLES OF ELECTRIC FIELD PROBLEMS

### 3.5.1 The Analysis of a Three-Phase Cable

To demonstrate the finite difference method through further examples, consider the three-phase cable system shown in Figure 3.5.1A. So as to be able to model this problem using finite

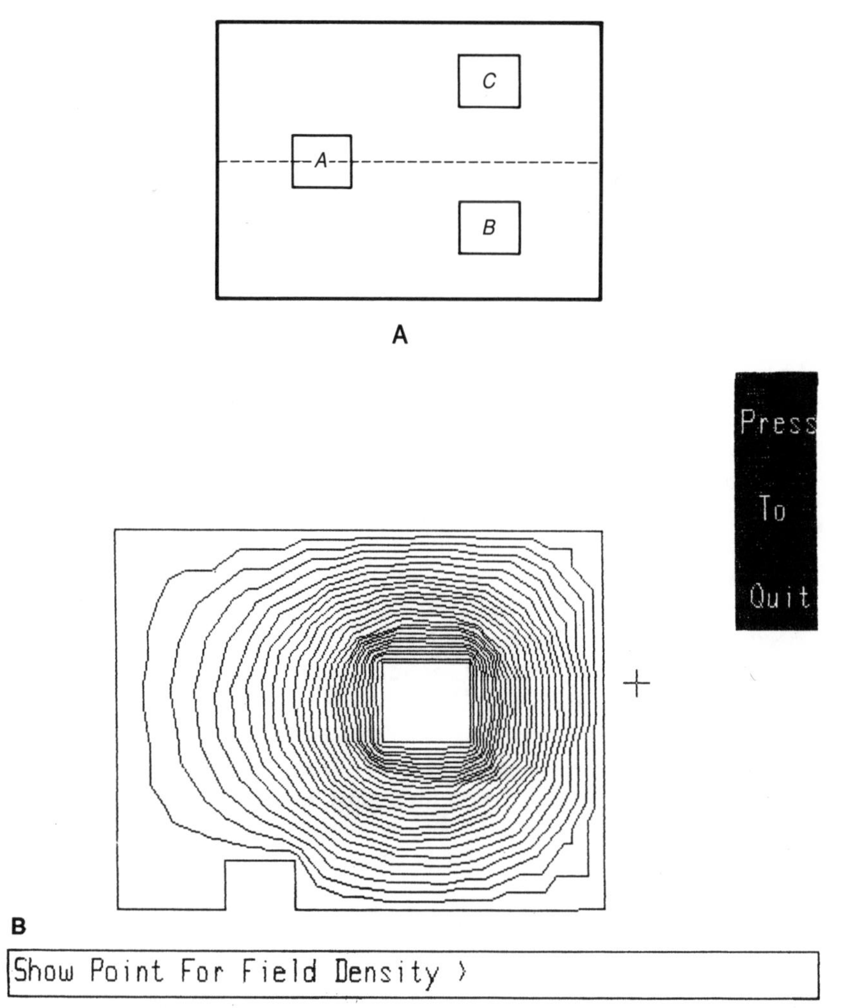


**Figure 3.5.1.** The three-conductor cable. **A**. The boundary value problem. **B**. The solution.

differences, a rectangular geometry is assumed. Each interior conductor is carrying a phase of the power supply—with voltages $100 \cos \omega t$, $100 \cos(\omega t + 2\pi/3)$, and $100 \cos(\omega t + 4\pi/3)V$, respectively, on lines $A$, $B$, and $C$. The earthed exterior aluminum sheath carries the current back from the load.

Let the electric field at the instant in time $t$ such that $\omega t = \pi/2$ be determined. The conductors are at voltages $100 \cos 90 = 0$, $100 \cos 210 = -86.66$, and $100 \cos 270 = 86.66$ V. The sheath, being earthed, is always at zero potential. Examination of the electric field at a point on the line of symmetry will show that it is normal to the line (with vectors equal in magnitude running towards conductor $B$ and away from $A$). As a result, equipotential lines run along the line of symmetry. Since this line passes through the conductor $A$ and the sheath, both at potential zero, the line also must be at potential zero. Thus, the upper or lower half of the problem may be solved. The solution is shown in Figure 3.5.1B.

## 3.5.2 The Analysis of an Axisymmetric Electron Lens

The fields around electron lenses have attracted interest with the increased use of these lenses for directing molten metals such as gold onto electronic circuit boards (Orloff and Swanson 1979). An axisymmetric lens arrangement is shown in Figure 3.5.2A. The high-voltage electrode at $A$ contains the metal, which melts upon the passage of high currents. The molten drops flow as a beam along the electric field. The secondary electrodes $B$, $C$, and $D$ at specified applied voltages are meant to produce a field so as to direct the field (and therefore the metal) onto the point $E$ where the circuit boards may be placed.

It is immediately realized that the positions of the intermediate electrodes and their voltages are of immense industrial interest. Repeated analyses with different positions and voltages will yield an optimal design. This is an axisymmetric problem, and the considerations behind eqn. (1.3.7) of section 1.3 tell us that we need to solve:

$$-\varepsilon\left[\frac{\partial^2\phi}{\partial r^2} + r^{-1}\frac{\partial\phi}{\partial r} + \frac{\partial^2\phi}{\partial z^2}\right] = 0, \tag{3.5.1}$$

since no charge clouds are present (until the beam begins to flow). The solution region is shown in Figure 3.5.2B, and the boundary conditions are as shown (see section 1.3). Because of the presence of the additional term $r^{-1}\,\partial\phi/\partial r$, at node 0 with nodes 1, 2, 3, and 4, respectively, to its east, north, west, and south, we have at all interior nodes:

$$\begin{aligned}-\varepsilon\{h^{-2}[\phi_1 + \phi_3 - 2\phi_0] + 0.5r_0h^{-1}[\phi_1 - \phi_3]\\ + k^{-2}[\phi_2 + \phi_4 - 2\phi_0]\} = 0.\end{aligned} \tag{3.5.2}$$

And at Neumann boundaries along the $z$ direction, the extra term in the middle merely vanishes. At Neumann boundaries along the $r$ direction, of course, no modification to the equation is required.

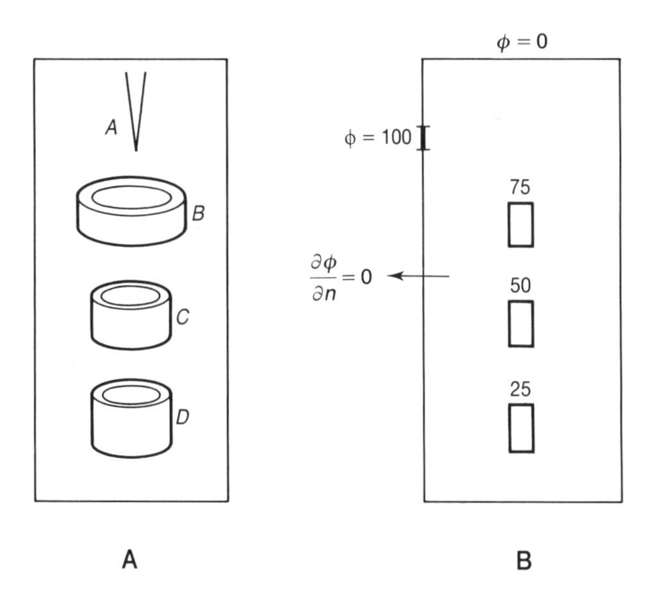

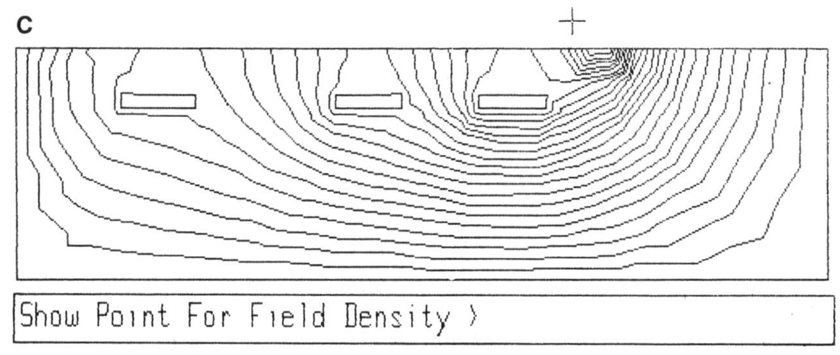


**Figure 3.5.2.** The electron lens. **A**. The device. **B**. The boundary value problem. **C**. The solution.

The permittivity may be dropped from the equation because of the homogeneous nature of the domain. The equipotential contours from the solution of the problem for the electrode tip at 100 V and the focusing electrodes at 75, 50, and 25 V are shown in Figure 3.5.2C.

For inhomogeneous regions, while the second derivatives may be extended using the network concepts already detailed, the midterm may be written as an average of the derivative to the right and left of node 0, $0.5\{\varepsilon_1 r_0 h^{-1}[\phi_1 - \phi_0] + \varepsilon_2 r_0 h^{-1}[\phi_0 - \phi_3]\}$. It is left to the reader to ponder on developing this further.

## 3.6 SOURCE-BASED NONLINEARITY: ELECTRON DEVICES

### 3.6.1 Nonlinear Problems

A linear problem may be stated to be one in which the unknown appears in the first power in the governing equation—in other words, when sources such as charges are doubled, the fields will also be doubled, or, equivalently, the effect of two sources is the sum of the effects of the sources when they act severally. On the other hand, in a nonlinear equation, the power of the unknown will not be 1 in the governing equation, so that a doubling or superposition of the causes will not necessarily produce a doubling or superposition of the effects. Taking as example our Poisson equation governing the electric field,

$$-\varepsilon \nabla^2 \phi = \rho, \tag{1.2.22}$$

it may be made nonlinear in two ways. First, when the material has properties depending on the field, the material value $\varepsilon$ is a function of $\phi$, as a result of which the left-hand side has $\phi$ appearing as a nonunit power, depending on the kind of function $\varepsilon$ is of $\phi$. This situation occurs principally in magnetics. Secondly, when the source $\rho$ is a complex non-first-order function of $\phi$, the equation again becomes nonlinear. This is a common situation with transistors and other electron devices.

The numerical solution of nonlinear problems is usually accomplished by some kind of linearization of the governing equation and by iteratively improving upon an initial guess of the solution. The linearization involves some inherent approximation, as a result of which the exact answer is never obtained at the first iteration. Thus, it is the procedure behind the linearization that determines the efficiency of the scheme. Two common methods exist for the linearization of nonlinear equations and are described below.

### 3.6.2 The Chord Method

In the chord method of linearizing the nonlinear equation to be solved, we assume some value of the unknown $\phi$ in the governing equation from the guess (or previous iteration), leaving out one linear term in the governing equation to be determined. Thus, eqn. (1.2.22) is linearized as:

$$-\varepsilon(\phi^{i-1})\nabla^2\phi^i = \rho(\phi^{i-1}). \tag{3.6.1}$$

So, by computing the values of $\varepsilon$ and $\rho$ from the values $\phi^{i-1}$ from the previous step in the iterative process, that value $\phi^i$ that gives the best fit of the governing equation is computed. This will need to be repeated until $\phi$ changes no more.

Needless to say, that convergence depends on how close the initial guess is to the actual solution. A point of note is that answers tend to fluctuate about the true value so that a retardation factor (typically 0.1) is called for, to force convergence from one side of the solution. To demonstrate, consider the nonlinear equation:

$$\phi^2 = 2, \tag{3.6.2}$$

whose solution we know to be 1.4142. A chord linearization will make this

$$\phi^{i-1}\phi^i = 2, \tag{3.6.3}$$

where $\phi^{i-1}$ is the answer from the previous iteration and the unknown $\phi^i$, to be determined from the current iteration, appears in the power 1. A good starting guess is

$$\phi^0 = 1, \tag{3.6.4}$$

giving

$$\phi^1 = \frac{2}{\phi^0} = 2, \tag{3.6.5}$$

$$\phi^2 = \frac{2}{\phi^1} = 1. \tag{3.6.6}$$

Clearly, the answer is nonconvergent. If we apply a retardation factor of 0.1 after computing $\phi^1$ in eqn. (3.6.5), we have:

$$\phi^1 = \phi^0 + 0.1(\phi^1 - \phi^0) = 1 + 0.1(2 - 1) = 1.1, \quad (3.6.7)$$

$$\phi^2 = \frac{2}{\phi^1} = \frac{2}{1.1} = 1.818, \quad (3.6.8)$$

$$\phi^2 = \phi^1 + 0.1(\phi^2 - \phi^1) = 1.1 + 0.1(1.818 - 1.1) = 1.1718. \quad (3.6.9)$$

After several such iterations, we will arrive at the correct answer. The method is therefore seen to be highly inefficient, although simple in concept.

### 3.6.3 The Newton Method

The Newton method for solving nonlinear equations is about the most powerful method available and assumes that the first (Frechet) derivative of the governing operator equation, which is also an operator, exists (Rall 1969). Convergence depends on whether the starting solution is within the "radius" of convergence and, if it is, takes place within 10 iterations or so. In the unusual event of our start having been bad, we may then easily afford to try another starting solution, since computation is not costly.

In the Newton scheme we solve for the error $\Delta\phi$ in the current approximation to the unknown $\phi$. Linearization is achieved by dropping higher order terms of $\Delta\phi$ on the assumption that $\Delta\phi$ is small—hence the existence of a radius of convergence. To state the procedure formally, let

$$\mathscr{R}(\phi) = 0 \quad (3.6.10)$$

be the nonlinear equation we are trying to solve. If $\phi^{i-1}$ is an approximate solution, with error $\Delta\phi$, then $\phi = \phi^{i-1} + \Delta\phi$. The trick in applying the method requires our somehow writing:

$$\begin{aligned} \mathscr{R}(\phi^{i-1} + \Delta\phi) &= \mathscr{R}(\phi^{i-1}) + \Delta\mathscr{R}(\phi^{i-1}, \Delta\phi) \\ &= \mathscr{R}(\phi^{i-1}) + \mathscr{F}(\phi^{i-1})\,\Delta\phi = 0, \end{aligned} \quad (3.6.11)$$

where $\Delta\mathscr{R}$ is a second operator operating on $\phi$ and $\Delta\phi$, but is linear in $\Delta\phi$. The fraction $\Delta\mathscr{R}/\Delta\phi$ as $\Delta\phi \to 0$ is known as the Frechet derivative $\mathscr{F}$ of the operator $\mathscr{R}$ (Rall 1969). Clearly, therefore, the scheme depends on our finding $\Delta\mathscr{R}$ and writing it in the form $\mathscr{F}(\phi)\,\Delta\phi$ so that it is linearized in $\Delta\phi$. Once this is done, since $\phi^{i-1}$ is known, $\Delta\phi$ may be found from eqn. (3.6.11), which is linear in $\Delta\phi$, and thence the new approximation $\phi^i$ is computed. To see the meaning of all this, let us return to our example of finding $\sqrt{2}$. The operator form of this equation, using the residual, is:

$$\mathscr{R}(\phi) = \phi^2 - 2 = 0, \tag{3.6.12}$$

and to linearize it, we use the Taylor expansion

$$\begin{aligned}\mathscr{R}(\phi^{i-1} + \Delta\phi) &= (\phi^{i-1} + \Delta\phi)^2 - 2 \\ &\simeq (\phi^{i-1})^2 + 2\phi^{i-1}\,\Delta\phi - 2 = 0,\end{aligned} \tag{3.6.13}$$

or

$$\Delta\phi = \frac{2 - (\phi^{i-1})^2}{2\phi^{i-1}}, \tag{3.6.14}$$

where it is seen that

$$\Delta\mathscr{R}(\phi^{i-1}, \Delta\phi) = 2\phi^{i-1}\,\Delta\phi, \tag{3.6.15}$$

which, for scalar operators $\mathscr{R}$ in general, is $\{\partial\mathscr{R}(\phi^{i-1})/\partial\phi\}\,\Delta\phi$. Applying eqn. (3.6.14) starting with $\phi^0 = 1$, we have

$$\phi^1 = \phi^0 + \Delta\phi = 1 + \frac{2 - 1^2}{2 \times 1} = 1.5, \tag{3.6.16}$$

$$\phi^2 = \phi^1 + \Delta\phi = 1.5 + \frac{2 - 1.5^2}{2 \times 1.5} = 1.4167. \tag{3.6.17}$$

Proceeding thus, an acceptable answer will be obtained in a few steps. The rapid convergence property of the scheme is readily apparent.

### 3.6.4 The Semiconductor Equations

In this section, the nonlinear equations governing the electric field in semiconductor devices will be described so that their solution may be demonstrated. The reader not interested in this development, but who wishes to know only the numerical solution of the equations, may simply assume eqn. (3.6.30) to be the equation that is to be solved, and proceed to section 3.6.5.

In a semiconductor device, the governing equation for the electrostatic potential $\phi$ is the Poisson equation (Sze 1981; Streetman 1980)

$$-\varepsilon\nabla^2\phi = \rho = -q[n - p - C], \tag{3.6.18}$$

where $q$ is the electric charge, $n$ and $p$ are the electron and hole carrier densities, and $C = N_D^+ - N_A^-$ is the electrically active net impurity concentration. These terms are themselves exponential functions of the potential, so that the equations have a source-based nonlinearity. Eqn. (3.6.18) is subject to the continuity of the electron and hole currents $\mathbf{J}_n$ and $\mathbf{J}_p$, over time $t$:

$$-\frac{1}{q}\nabla\cdot\mathbf{J}_n - G + R + \frac{\partial n}{\partial t} = 0, \tag{3.6.19a}$$

$$\frac{1}{q}\nabla\cdot\mathbf{J}_p - G + R + \frac{\partial p}{\partial t} = 0, \tag{3.6.19b}$$

where

$$\mathbf{J}_n = -q\mu_n n\nabla\phi + qD_n\nabla n, \tag{3.6.20a}$$

$$\mathbf{J}_p = -q\mu_p p\nabla\phi - qD_p\nabla p. \tag{3.6.20b}$$

$G$ allows for generation, such as impact ionization or carrier generation through external radiation, and $R$ accounts for recombination processes. In eqns. (3.6.20a–b), $\mu_n$ and $\mu_p$ are the electron and hole mobilities corresponding to the diffusion coefficients $D_n$ and $D_p$, which in turn depend on the electric field. For a nondegenerate system, these are related by the Einstein equation (Sze 1981; Seeger 1982):

$$D_i = \mu_i\beta, \qquad i = n, p, \tag{3.6.21}$$

where:

$$\beta = q/kT. \tag{3.6.22}$$

$k$ is Boltzmann's constant and $T$ is the carrier temperature. Under general nonequilibrium conditions, with quasi-Fermi levels (or imrefs) $q\phi_n$ and $q\phi_p$, where $\phi_n$ and $\phi_p$ are the corresponding quasi-Fermi potentials (Sze 1981; Streetman 1980; Seeger 1982), we have the Boltzmann approximation

$$n = n_i e^{\beta(\phi-\phi_n)}, \tag{3.6.23}$$

$$p = n_i e^{\beta(\phi_p-\phi)}, \tag{3.6.24}$$

where $n_i$ is the intrinsic density given by

$$n_i^2 = N_c N_v e^{-E_g/kT} \tag{3.6.25}$$

where $N_c$ and $N_v$ are the effective densities of state at the conduction and valence band edges and $E_g$ is the energy gap of the material. Of course, for nondegenerate materials, the two quasi-Fermi levels are equal and the same as the Fermi level. Thus, with the help of eqns. (3.6.21–25), eqns. (3.6.20a–b) reduce to

$$\mathbf{J}_n = -q\mu_n n \nabla \phi_n, \tag{3.6.26}$$

$$\mathbf{J}_p = -q\mu_p p \nabla \phi_p. \tag{3.6.27}$$

Therefore, for the analysis of electron devices under steady-state (non-time-varying) conditions, the Poisson eqn. (3.6.17), using eqns. (3.6.23) and (3.6.24), becomes

$$-\varepsilon\nabla^2\phi = \rho = -q[n_i e^{\beta(\phi-\phi_n)} - n_i e^{\beta(\phi_p-\phi)} - C], \tag{3.6.28}$$

subject to the continuity equations (3.6.19a–b):

$$\nabla \cdot \mathbf{J}_n = -\nabla \cdot \{q\mu_n n \nabla \phi_n\} = q[R - G], \tag{3.6.29a}$$

$$\nabla \cdot \mathbf{J}_p = -\nabla \cdot \{q\mu_p p \nabla \phi_p\} = q[G - R]. \tag{3.6.29b}$$

In the numerical analysis (Heimer 1973; Barnes and Lomax 1974, 1977; Adachi et al. 1979; Bank et al 1983; Fichtner et al. 1983) of the electric field in devices governed by eqns. (3.6.28–29), the procedure commonly followed (Gummel 1964) is to assume particular values for $\phi_n$ and $\phi_p$ and solve eqn. (3.6.28) for $\phi$ (Adachi et al. 1979). Thereafter, using the previously assumed value of $\phi_p$ and the newly computed value of $\phi$, we would solve eqn. (3.6.19a) for $\phi_n$; then, using the improved values of $\phi$ and $\phi_n$, we would compute $\phi_p$. This cycle is repeated using the improved values, until convergence of the three quantities $\phi$, $\phi_n$, and $\phi_p$ is achieved. These equations are also frequently scaled into dimensionless form (Fichtner et al. 1983; de Mari 1968).

Here we are principally concerned with the solution of eqn. (3.6.28) for given values of $\phi_n$ and $\phi_p$, because it involves the term $\nabla^2$ and represents the principal computational effort. For this case, it reduces with obvious notation to the form

$$-\varepsilon\nabla^2\phi = q[C - n_0 e^{\beta\phi} + p_0 e^{-\beta\phi}], \tag{3.6.30}$$

where the unknown $\phi$ must be solved for, with the Neumann boundary condition

$$\varepsilon\nabla\phi \cdot \mathbf{u}_n = Q, \tag{3.6.31}$$

where $\mathbf{u}_n$ is a unit vector along the normal to the boundary and $Q$ is the surface charge, or the Dirichlet boundary condition

$$\phi = \phi_0. \tag{3.6.32}$$

Also, at interfaces within the region of solution, it is possible to have line charge accumulations; that is, we may have $\rho$ in eqn. (3.6.18) as a line density as opposed to a surface density. The efficiency with which we solve eqn. (3.6.30) is critical, because being nonlinear, it must be done iteratively and poses iterations within iterations between eqns. (3.6.28) and (3.6.29).

### 3.6.5 Newton's Linearization of Semiconductor Equations

We wish to solve the nonlinear eqn. (3.6.30) for the unknown variable $\phi$. Newton's scheme provides us with an exponentially convergent set of linear equations whose solution will lead us to the solution of the nonlinear equation we seek. For a given inaccurate solution $\phi$, the residual operator is

$$\mathscr{R}[\phi] = -\varepsilon\nabla^2\phi - q[C - n_0e^{\beta\phi} + p_0e^{-\beta\phi}] = 0. \qquad (3.6.33)$$

As described in section 3.6.3, in Newton's method, if the error in $\phi$ is $\Delta\phi$, then the exact solution $\phi + \Delta\phi$ will make the residual, eqn. (3.6.33), vanish, so that, from eqn. (3.6.11), as an approximation,

$$\mathscr{R}[\phi + \Delta\phi] = \mathscr{R}[\phi] + \mathscr{F}[\phi]\,\Delta\phi = 0, \qquad (3.6.34)$$

where $\mathscr{F}[\phi]$ is the Frechet derivative of $\mathscr{R}$. Eqn. (3.6.34), linear in $\Delta\phi$, may be used repeatedly to improve upon $\phi$. The Frechet derivative, when it exists, is defined in the limit as $\Delta\phi \to 0$, by

$$\mathscr{F} = \frac{\mathscr{R}[\phi + \Delta\phi] - \mathscr{R}[\phi]}{\Delta\phi}. \qquad (3.6.35)$$

The determination of the Frechet term is much less trivial than when dealing with scalar equations such as that which obtains when finding $\sqrt{2}$, but the procedure is similar. Consider the term

$$\begin{aligned}
\mathscr{F}\Delta\phi &= \mathscr{R}[\phi + \Delta\phi] - \mathscr{R}[\phi] \\
&= \{-\varepsilon\nabla^2(\phi + \Delta\phi) - q[C - n_0e^{\beta(\phi+\Delta\phi)} + p_0e^{-\beta(\phi+\Delta\phi)}]\} \\
&\quad - \{-\varepsilon\nabla^2\phi - q[C - n_0e^{\beta\phi} + p_0e^{-\beta\phi}]\} \\
&\cong \{-\varepsilon(\nabla^2\phi + \nabla^2\Delta\phi) - q[C - n_0(e^{\beta\phi} + \beta e^{\beta\phi}\Delta\phi) \qquad (3.6.36) \\
&\quad + p_0(e^{-\beta\phi} - \beta e^{-\beta\phi}\Delta\phi)]\} \\
&\quad - \{-\varepsilon\nabla^2\phi - q[C - n_0e^{\beta\phi} + p_0e^{-\beta\phi}]\} \\
&= -\varepsilon\nabla^2\Delta\phi + qn_0\beta e^{\beta\phi}\,\Delta\phi + qp_0\beta e^{-\beta\phi}\,\Delta\phi,
\end{aligned}$$

where the expressions arise from the general Taylor expansion $f(\phi + \Delta\phi) = f(\phi) + \Delta\phi\, df(\phi)/d\phi$, neglecting $(\Delta\phi)^2$ and higher order terms. Therefore, the linearization of eqn. (3.6.33) by the Newton scheme eqn. (3.6.34) results in an expression for estimated error $\Delta\phi$ in $\phi$:

$$-\varepsilon\nabla^2 \Delta\phi + q\beta[n_0 e^{\beta\phi} + p_0 e^{-\beta\phi}]\Delta\phi = \varepsilon\nabla^2\phi + q[C - n_0 e^{\beta\phi} + p_0 e^{-\beta\phi}]. \tag{3.6.37}$$

Eqn. (3.6.37) is a linear Poisson equation in $\Delta\phi$, which may be repeatedly solved by either the finite difference scheme or the finite element method to be described in a later chapter, until convergence takes place. This equation may be, to keep the discussion simple and with obvious notation, expressed as

$$-\varepsilon\nabla^2 \Delta\phi + f[\phi]\, \Delta\phi = \varepsilon\nabla^2\phi + g[\phi], \tag{3.6.38}$$

where $f$ and $g$ are known functions that take a long time to evaluate digitally. It is the evaluation of these terms that makes the computation of electromagnetic fields within electron devices on the computer so time consuming.

## 3.7 THE ANALYSIS OF A TRANSISTOR

We shall have a demonstration of the procedures of the previous section, by using them to analyze the MOSFET device of Figure 3.7.1 at thermal equilibrium, taken from Mayergoyz and Korman (1987). The problem may be analyzed in two dimensions at the plane midway down the length, as shown in Figure 3.7.2.

The potential is governed by eqn. (3.6.30). $\varepsilon$ becomes $\varepsilon_s$, the permittivity of silicon. The concentration $C$ in the $n$ regions is $10e^{18}\,\mathrm{cm}^{-3}$ and in the $p$ regions is $10e^{-16}\,\mathrm{cm}^{-3}$; $n_0$ and $p_0$ are both $1.45e^{10}\,\mathrm{cm}^{-3}$; and $\beta^{-1}$ is 0.026 V. The applicable boundary conditions are as follows. Along the ohmic contacts, the condition is Dirichlet:

$$\phi = \beta^{-1} ln_e \frac{C/2 + \sqrt{[C^2/4 + n_0^2]}}{n_0}. \tag{3.7.1}$$

Along the artificial boundaries at the sides, the Neumann condition

$$\frac{\partial\phi}{\partial n} = 0 \tag{3.7.2}$$

applies. Along the dioxide–semiconductor interface, the condition is more complex and is given by

$$\phi + \frac{\delta\varepsilon_s}{\varepsilon_d}\frac{\partial\phi}{\partial n} = V_g + \frac{\delta}{\varepsilon_d} Q(\phi), \tag{3.7.3}$$

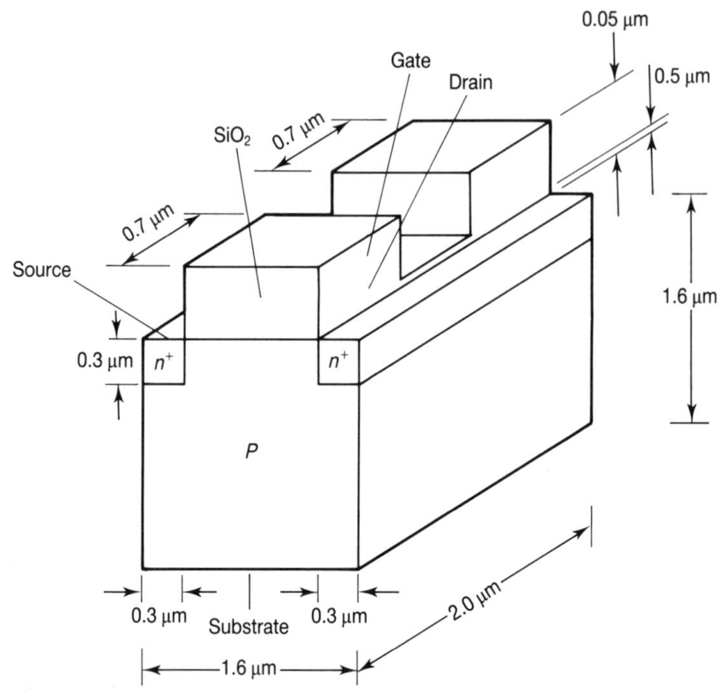


**Figure 3.7.1.** A MOSFET device.

where $\varepsilon_d$ is the permittivity within the dioxide; $V_g$ is the gate potential; $\delta$ is the thickness of the dioxide (0.05 μm); and $Q$ is the surface density of trapped electric charges at the interface, which nonlinearly depends on the potential. In this problem $Q$ is zero. The above boundary condition is derived from the requirement that the difference in the normal flux densities inside and outside the dioxide should be the surface charge $Q$, a condition that follows from applying Gauss's theorem (A13) to a small cylindrical pillbox with lids on either side of the interface. Inside the dioxide, the normal component of flux density is

$$D_n = \frac{\varepsilon_d[V_g - \phi]}{\delta}, \tag{3.7.3}$$

and outside it is:

$$D_n = \varepsilon_s \frac{\partial \phi}{\partial n}. \tag{3.7.4}$$

The described problem was solved using an approximately 70 × 40 finite difference mesh, and the equipotential contours from the resulting solution are seen plotted in Figure 3.7.3.

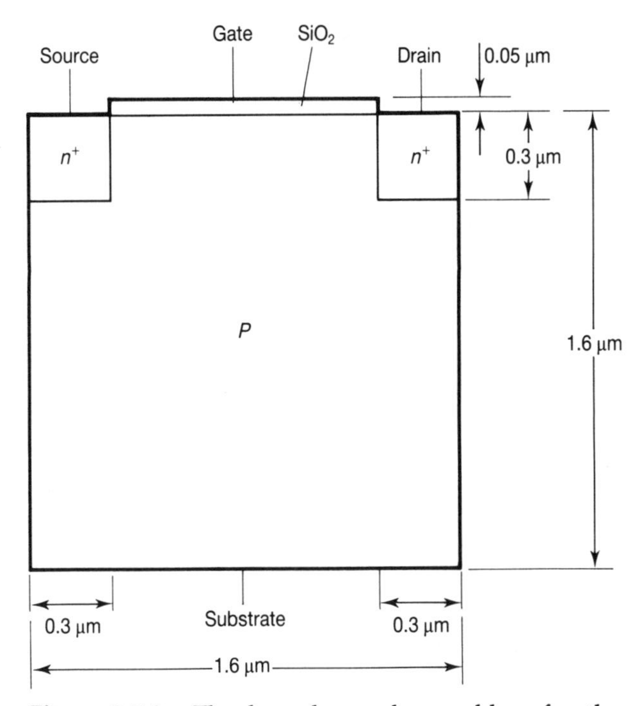


**Figure 3.7.2.** The boundary value problem for the MOSFET device.

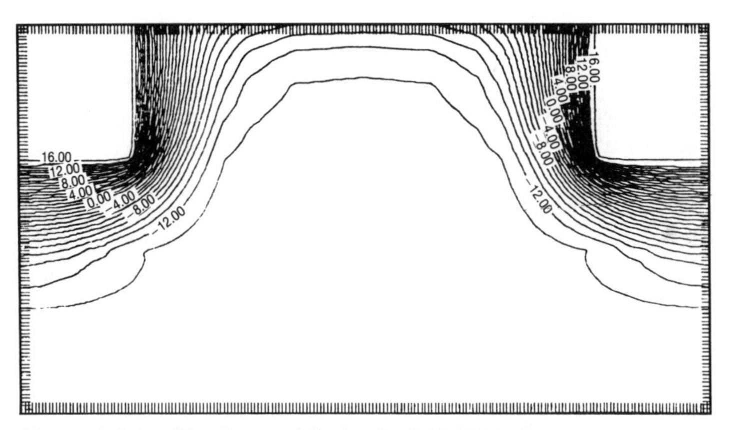


**Figure 3.7.3.** Equipotentials in the MOSFET device.

## 3.8 LIMITATIONS OF THE FINITE DIFFERENCE APPROXIMATION

What we have derived in eqn. (3.2.9) is the finite difference approximation for homogeneous media. So long as the region $(x - h, x + h) \times (y - k, y + k)$ lies entirely within a region of unchanging $\varepsilon$ and our mesh is uniformly $h \times k$, our equation is valid. What has been described is the usual form of the finite difference method. As already pointed out, special forms of the method are available that overcome many of the weaknesses described below (Jensen 1972; Chung 1981; Tseng 1984). These special forms often require elaborate algorithms. In reading on, therefore, it must be kept in mind that the discussion pertains to finite differences as the topic is commonly understood. The limitations of this method, and how they may be overcome, are summarized below.

**Inhomogeneous Media**

In practical engineering problems, we nearly always run into inhomogeneities in medium, as for example when we have a different permeability, permittivity, or conductivity, respectively, in magnetic field, electric field, and current flow problems. Such situations are handled by the circuit models of section 3.1.3, but the same accuracy of order $h^2$ will not be had.

**Open Boundaries**

Again, in an open-boundary problem, such as a collection of charges in space, solving the problem would involve, in reality, the Dirichlet condition that the potential (voltage) be zero at infinity. A commonly employed and crude but useful artifice is to employ a boundary at a far but limited distance from the charges and say that the potential there is zero. The effect of this is to make the potential fall (or rise) to zero faster and thereby result in smaller (or larger) potentials and larger vector fields than actually obtain. An impossible alternative is to cover a very large region by a finite difference mesh; this results in an impracticable number of nodes at which the potential is to be determined and will bog down the computing environment for unacceptably large periods. A useful way of getting around this is to use a graded mesh, as shown in Figure 3.8.1A. Here, we use a uniform mesh for the region containing the charges, just as we would when using an artificially close boundary. Thereafter, instead of imposing a zero potential, we grade the mesh, so that the next vertical line to the right and left is a distance

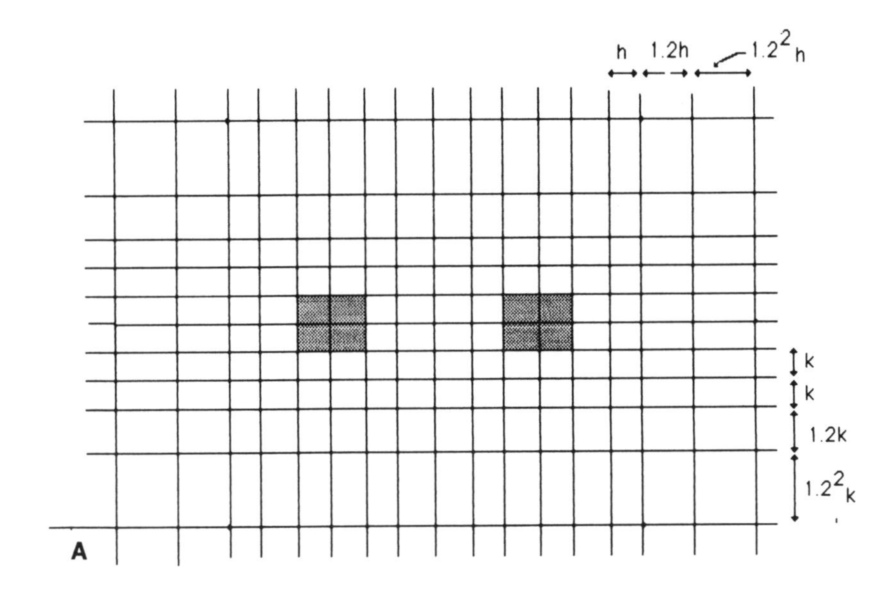


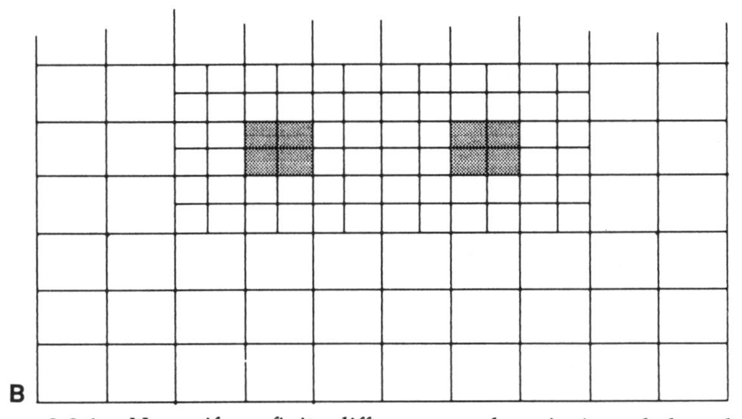


**Figure 3.8.1.** Nonuniform finite difference meshes. **A**. A graded mesh for open-boundary problems. **B**. Matched meshes of different sizes.

$1.2h$ away and, similarly, those above and below are $1.2k$ away. Thereafter, we would use lines $1.2^2h$ and $1.2^2k$ away. The 10th such line would be shifted by a factor of $1.2^{10}$, which, for all practical purposes, is at infinity. Thus, by introducing an additional few hundred nodes, we would have covered all of space. Again, the accuracy would be lowered from order $h^2$.

**Small Parts**

In analyzing a device of uniform dimensions, we may suddenly come across a part of relatively small dimensions, such as teeth in an electrical machine or a thin base of a frying pan being designed for good thermal characteristics. Here a fine mesh may be required for the small part, but to use the same mesh throughout the device

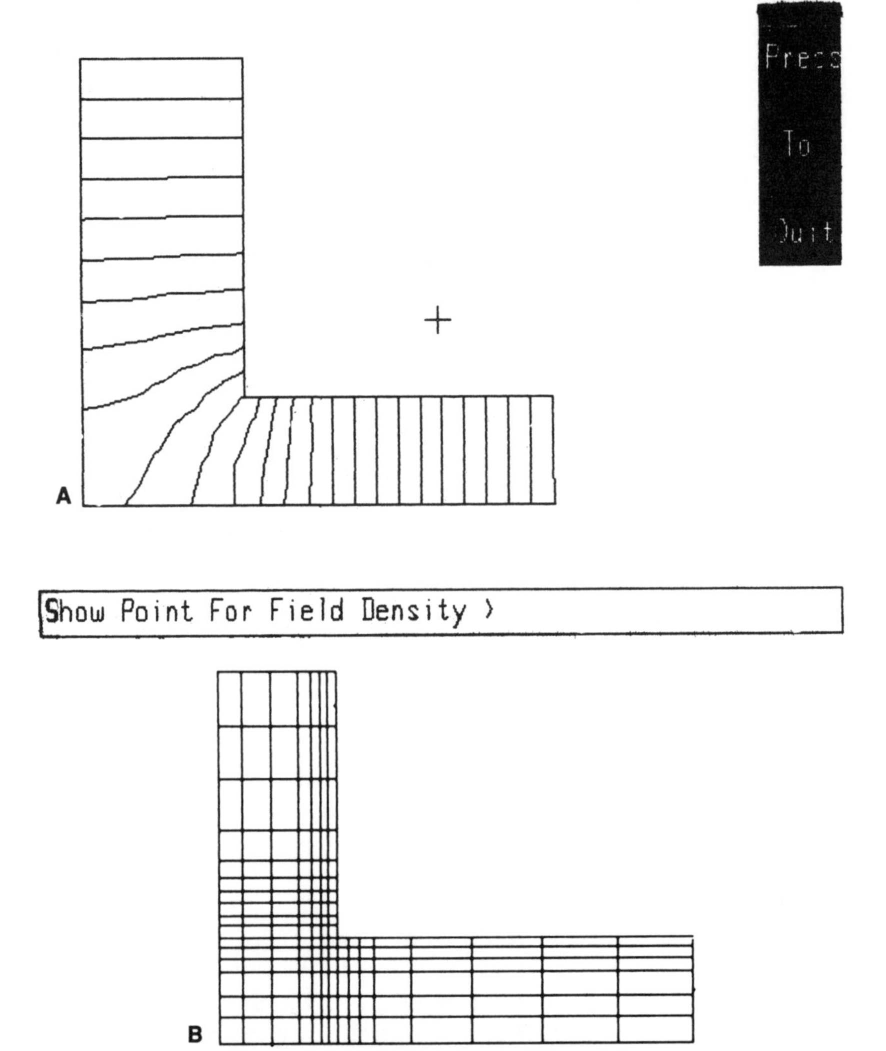


**Figure 3.8.2.** A nonuniform mesh for L-shaped capacitor. **A**. Flow at pipe corner. **B**. Nonuniform mesh for L-shaped capacitor.

may be expensive since it would result in an unnecessarily large number of nodes for solution. To overcome this we may use a graded mesh, as in Figure 3.8.1A, to arrive at a larger mesh size, which is then used for the other regions; or, less accurately and more conveniently, we may match two meshes of different sizes, as shown in Figure 3.8.1B.

**Regions of High Gradients**

Another use of nonuniform meshes is in regions of high gradients. Where the field varies less, our finite difference approximation is more accurate, since we have assumed the series expansion of the scalar field to be at most a cubic polynomial, as seen in eqns. (3.1.8) and (3.1.9). Thus to fit greater variation, since we do not have higher terms in our approximations, we must reduce $h$ to justify the validity of dropping terms of order $h^4$ and higher. Therefore, in problems like the one shown in Figure 3.8.2A, where we have a scalar potential problem for the electric field between two charged plates of an L-shaped capacitor with the potentials on the two plates known, we would have a high gradient in potential near the inner corner. (Note that, equivalently, this is a stream function problem describing the flow through a pipe with the stream function defined on the two plates.) As before we may use a fine uniform mesh everywhere, but this would not be required away from the corner, where the field does not vary much and the fine mesh is therefore wasteful. On the other hand, we may use a fine mesh that later gives way to a coarser one, away from the corner, as shown in Figure 3.8.2B.

# 4 MAGNETIC FIELD SYSTEMS

## 4.1 INTEGRAL METHODS: THE MAGNETIC FIELD IN FREE SPACE

### 4.1.1 The Governing Equations: The Biot–Savart Law

We have just been through integral and differential techniques for solving electric fields. Despite so much being in common to electric and magnetic fields, they are yet very different. The fundamental difference arises from the fact that the electrostatic field is irrotational, as seen from eqn. (1.2.6), whereas the magnetic field is nondivergent, as seen in eqn. (1.2.10). As a consequence, electric fields are easily described by a scalar potential $\phi$, while magnetic fields are more naturally described by a vector potential **A**, which increases the dimension of the problem. As we shall see, much of the research effort behind the numerical techniques for magnetics has to do with reducing the dimension of the problem so as to make it more tractable.

In this chapter, as we did with electric fields, we shall examine some integral and differential methods available for the solution of magnetic fields. In the course of doing so, we shall resort to several of the mathematical results previously derived in chapters 2 and 3.

We have already seen in chapter 1 that the magnetic field may be described by the magnetic vector potential **A**, which in a uniform medium is governed by

$$-\mu^{-1}\nabla^2\mathbf{A} = \mathbf{J}. \tag{1.2.28}$$

Separating the three components of the vector **A**:

$$-\mu^{-1}\nabla^2 A_i = J_i, \tag{4.1.1}$$

where the subscript $i$ refers in turn to the $x$, $y$, and $z$ components. In two dimensions with no changes in the $z$ direction, we have already seen that the components $A_x$ and $A_y$ do not exist, and that eqn. (4.1.1) corresponds to eqn. (1.2.32).

Now, to solve eqn. (4.1.1) we may use the analogy of equations from the area of electrostatic fields, which has already been considered. We have shown that the electric potential field $\phi$ is governed by

$$-\varepsilon\nabla^2\phi = \rho, \tag{1.2.22}$$

and that the solution in homogeneous space is

$$\phi = \frac{1}{4\pi\varepsilon}\iiint \frac{\rho\, dV}{r}, \tag{2.1.6}$$

where we are integrating the effects of pointlike charges $\rho\, dV$ at a field point at a distance $r$. By the pure numerical analogy $\phi \to A_i$, $\varepsilon \to \mu^{-1}$, and $\rho \to J_i$ inherent in eqns. (1.2.22) and (4.1.1), therefore, we ought to have the solution:

$$A_i = \frac{\mu}{4\pi}\iiint \frac{J_i\, dV}{r}. \tag{4.1.2}$$

Assembling together the equations for the three components,

$$\mathbf{A} = \mathbf{u}_x A_x + \mathbf{u}_y A_y + \mathbf{u}_z A_z = \frac{\mu}{4\pi}\iiint r^{-1}\mathbf{J}\, dV. \tag{4.1.3}$$

Similarly, in analogy with eqn. (2.1.9) for two-dimensional systems,

$$\mathbf{A} = -\mathbf{u}_z \frac{\mu}{2\pi}\iint J_z ln_e r\, dR, \tag{4.1.4}$$

the two components in the $x$ and $y$ directions being absent in view of there being no current components in these two directions. Thus, whenever we are given a current distribution **J** in space of homogeneous permeability, we may integrate the effects of the elemental currents at a field point. It should be noted, however, that we may not use analogy to compute the magnetic field **B** from eqn. (2.1.5) for the electric field. This is simply because **E**, which is computed from $-\nabla\phi$, is not analogous to **B**, which is found from

$\nabla \times \mathbf{A}$:

$$\mathbf{B} = \nabla \times \mathbf{A} = \nabla \times \frac{\mu}{4\pi} \iiint \mathbf{J} \frac{dV}{r} = \frac{\mu}{4\pi} \iiint \nabla \times (r^{-1}\mathbf{J})\, dV$$

$$= \frac{\mu}{4\pi} \iiint (r^{-1}\nabla \times \mathbf{J} - \mathbf{J} \times \nabla r^{-1})\, dV \tag{4.1.5}$$

$$= \frac{\mu}{4\pi} \iiint \frac{\mathbf{J} \times \mathbf{r}}{r^3}\, dV,$$

where we have used identity (A12) to expand $\nabla \times (r^{-1}\mathbf{J})$ and the fact that $\nabla \times \mathbf{J}$ is zero. The latter arises from the differentiations, required by the operator $\nabla\times$, being performed by moving the field point while holding the source point fixed; for **B** is governed by the rate of change of **A** where we want **B** and not by any changes in the locations of the cause **J**. The term $\nabla r^{-1}$ is already evaluated in eqn. (2.5.3) as $-r^{-3}\mathbf{r}$. Eqn. (4.1.5) is known as the *Biot–Savart law*.

The Biot–Savart law of eqn. (4.1.5) may be easily applied numerically. The following example, although analytical, serves to demonstrate its utility and gives us a useful result for two-dimensional problems. Consider the infinitely long wire of Figure 4.1.1. It is of small cross section $S$, carries a current $I$, and lies along the $z$ axis. Let us try to compute the magnetic field at a point $P$, lying at a distance $d$ from the wire; we know from Ampere's law that the answer is $\mu I/2\pi d$; we will try to reduce eqn. (4.1.5) to this result. Now take a source consisting of an elemental length $dz$ of the

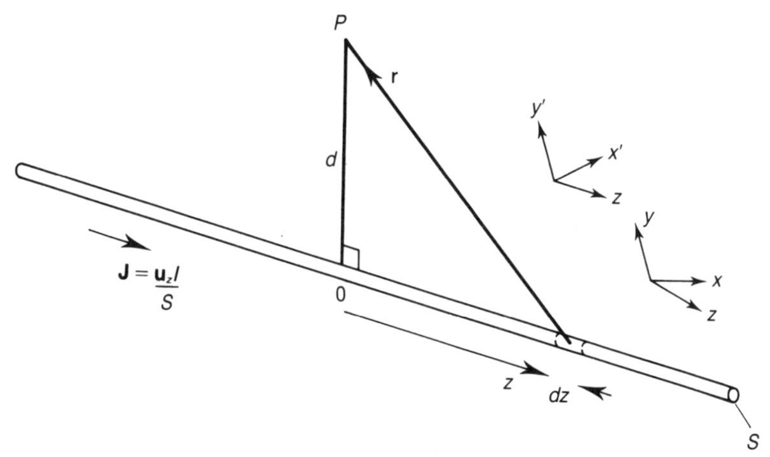


**Figure 4.1.1.** An infinitely long conductor.

wire, at a distance $z$ from the origin of a temporary coordinate system $O$ at the foot of the projection from $P$ to the wire. Let $OP$ be the $y'$ axis. The current density $\mathbf{J}$ has only one component and is $\mathbf{u}_z I/S$. The vector $\mathbf{r}$ is $d\mathbf{u}'_y - z\mathbf{u}_z$. Therefore, applying eqn. (4.1.5):

$$\begin{aligned}
\mathbf{B} &= \frac{\mu \iiint I\mathbf{u}_z \times (d\mathbf{u}'_y - z\mathbf{u}_z)\, dx'\, dy'\, dz}{4\pi S(d^2 + z^2)} \\
&= \frac{\mu}{4\pi S} \iint_{X\text{-sect}} dx'\, dy' \int_{-\infty}^{\infty} \frac{-I\, d\mathbf{u}'_x}{(z^2 + d^2)}\, dz \\
&= -\mathbf{u}'_x \frac{\mu I}{4\pi} \left( \frac{z}{d(z^2 + d^2)^{-1/2}} \right)_{-\infty}^{\infty} \\
&= \frac{-\mathbf{u}'_x \mu I}{2\pi d},
\end{aligned} \tag{4.1.6}$$

as expected. It is to be noted that the cross product $I\mathbf{u}_z \times (d\mathbf{u}'_y - z\mathbf{u}_z)$ has been evaluated as $-I\, d\mathbf{u}'_x$ using the identity (A3).

Now, in eqn. (4.1.6) we have a result for the magnetic field caused by a thin, long wire. Therefore for two-dimensional systems, if the current sources are divided into small, infinitely long wires, their effects may be summed to obtain the flux density at a point. However, the way the $x$ axis has been defined (namely the direction perpendicular to both the wire and the line from the wire to the field point), the $x$ direction will vary for each filamental wire. Referring to Figure 4.1.2, the direction $\mathbf{u}'_x$ referred to in eqn. (4.1.7) may be defined in terms of the new, absolute two-dimensional directions $x$ and $y$:

$$\begin{aligned}
\mathbf{u}'_x &= -u_x \sin\theta + \mathbf{u}_y \cos\theta = \frac{-\mathbf{u}_x y}{\sqrt{[x^2 + y^2]}} + \frac{\mathbf{u}_y x}{\sqrt{[x^2 + y^2]}} \\
&= r^{-1}[-y\mathbf{u}_x + x\mathbf{u}_y].
\end{aligned} \tag{4.1.7}$$

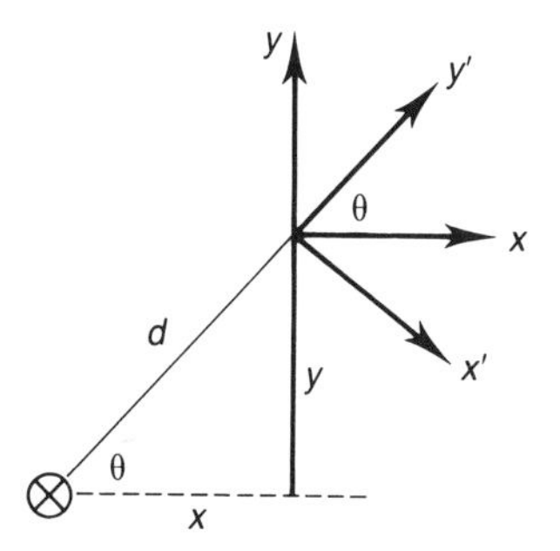


**Figure 4.1.2.** Mapping of coordinate systems.

Putting this into eqn. (4.1.6), using $r$ in place of $d$ and $J_z\,dx\,dy$ in place of the filamental current $I$ in the $z$ direction, we have:

$$\mathbf{B} = \frac{\mu}{2\pi}\iint r^{-2}[y\mathbf{u}_x - x\mathbf{u}_y]J_z\,dx\,dy. \tag{4.1.8}$$

Observe how the same result, eqn. (4.1.8), could have been obtained directly from eqn. (4.1.4) with the aid of eqn. (1.2.24) for two-dimensional magnetic fields:

$$\begin{aligned}\mathbf{B} = \nabla \times \mathbf{A} &= \nabla \times \mathbf{u}_z\frac{\mu}{2\pi}\iint J_z ln_e r\,dR \\ &= -\mathbf{u}_z \times \frac{\mu}{2\pi}\iint J_z \nabla ln_e r\,dR \quad \text{using (A11)}, \\ &= -\mathbf{u}_z \times \frac{\mu}{2\pi}\iint \frac{J_z\mathbf{r}}{r^2}\,dR \\ &= \frac{\mu}{2\pi}\iint J_z\frac{-\mathbf{u}_z x(x\mathbf{u}_x + y\mathbf{u}_y)}{r^2}\,dR \\ &= \frac{\mu}{2\pi}\iint r^{-2}[y\mathbf{u}_x - x\mathbf{u}_y]J_z\,dR,\end{aligned} \tag{4.1.9}$$

using (A3) for the cross product. In this derivation, the term $\nabla ln_e r$ has been expanded as $r^{-2}\mathbf{r}$ using eqn. (2.5.7). Thus, by dividing the conductor into several small elements over each of which $J_z$ may be modeled as a constant, the integrals may be rewritten as a summation of integrals over the elements and the flux density $\mathbf{B}$ evaluated at any desired point that, it is emphasized, is not itself a source point. When the field point is a source point, special care has to be taken to avoid the singularities that arise on account of $r$ going to zero.

### 4.1.2 The Magnetic Field of a Long, Rectangular Bus-bar

As an example of using the Biot–Savart law, consider a long, rectangular bus-bar carrying a known DC current $I$. This is clearly a two-dimensional problem, since the conductor has the same cross section anywhere along the $z$ axis, and therefore the applicable equation is (4.1.4) with only one component of vector potential, in the absence of the other two current components. Be it also observed that bus-bars (conductors used for carrying large power) being made of copper or aluminum, the permeability within the bus is the same as that of the surrounding air; that is, the geometry is of homogeneous permeability and the results of the preceding section are applicable. The problem then becomes very similar to that worked in section 2.2 and the attendant algorithm is 2.2.1. The difference is that the circular conductor there is now rectangular, so that the elemental subdivisions will be different and in the rectangular coordinate system in place of the cylindrical system previously used.

As a further example, consider a circuit consisting of a rectangular conductor, of size $a \times b$, carrying the current forward and a second and parallel rectangular conductor bringing the current back. This time the summation must be over both conductors, with one having a current density $J_z$ of $I/ab$ and the other its negative. The solution of this system may be plotted by placing the conductor system in a rectangular mesh encompassing the regions in which we want plots and computing the vector potential at all the nodes of the mesh, except those that are interior where the term $ln_e r$ will cause problems on account of $r$ going down to zero. The latter fact arises from the contradiction of an infinitely long conductor, which disallows us from postulating a reference potential at infinity that will contain the tips of the conductor if we move along the $z$ axis, but free space if we move along the $x$ or $y$ direction.

Now we have a rectangular mesh at the nodes of which we have computed the values of the vector potential. The equipotential plots may then be plotted by hand using the same algorithm we used for plotting with finite differences in section 3.4. A computerized version of this procedure is described in section 9.2.3. The results of the plot from an Apple Macintosh computer are presented in Figure 4.1.3.

Should we desire to obtain the magnetic field directly, rather than through the vector potential (which then has to be differentiated numerically), we may employ eqn. (4.1.9). Here all the parts containing the source $J_z$ would be divided into, say, $n$ smaller regions in each of which we shall assume $J_z$ to be a constant $J_k$. This gives us:

$$
\begin{aligned}
\mathbf{B}_i &= \frac{\mu}{2\pi} \iint_R r^{-2}[y\mathbf{u}_x - x\mathbf{u}_y]J_z \, dx \, dy \\
&= \frac{\mu}{2\pi} \sum_k \left\{ \mathbf{u}_x J_k \iint_\Delta r^{-2} y \, dx \, dy - \mathbf{u}_y J_k \iint_\Delta r^{-2} x \, dx \, dy \right\} \qquad (4.1.10) \\
&= \frac{\mu}{2\pi} \sum_k \{ \mathbf{u}_x p_{ik} J_k - \mathbf{u}_y q_{ik} J_k \},
\end{aligned}
$$

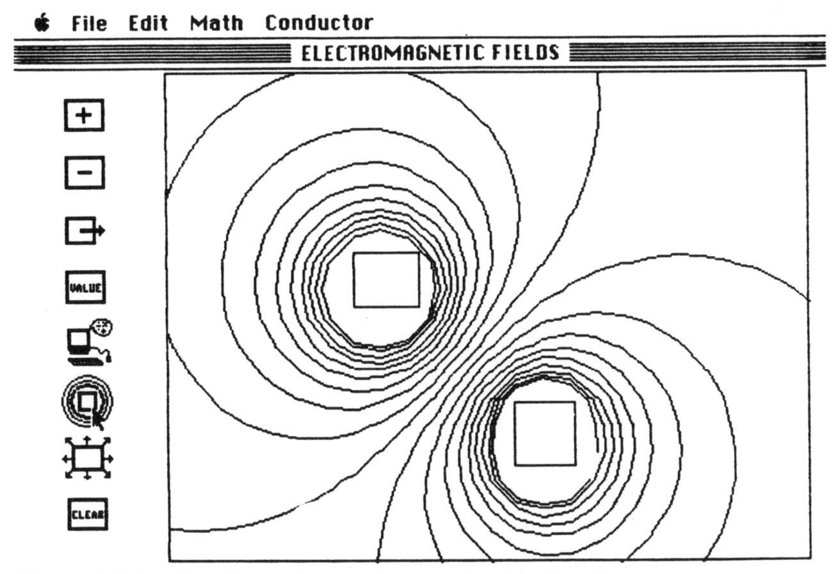


**Figure 4.1.3.** Flux lines about a two-conductor circuit.

where the integral over $R$ has been replaced by the sum of the integrals over the subdomains $\Delta_k$ and $p_{ik}$ and $q_{ik}$ are the computable coefficients:

$$p_{ik} = \iint r^{-2} y \, dx \, dy = y_{ik} r_{ik}^{-2} \Delta_k, \tag{4.1.11}$$

$$q_{ik} = \iint r^{-2} x \, dx \, dy = x_{ik} r_{ik}^{-2} \Delta_k, \tag{4.1.12}$$

where $\Delta_k$ is the area of an element and the above expressions are valid only when $i \neq k$ when no singularity will obtain. However, when we wish to evaluate the magnetic field at a point containing a current source $J_i$, we have $i = k$, so that the term $r_{ik}$ will vanish, giving us an integrable singularity. This problem may be overcome by dividing the elements into smaller ones. For example, for a rectangular element, Figure 4.1.4 gives a subdivision into 28 smaller rectangles over which the integrals may be reduced to summations such as:

$$p_{ii} = \sum_{j=1}^{28} y_{ij} r_{ij}^{-2} \Delta_j, \tag{4.1.13}$$

where $\Delta_j$ is the area of each subelement of the element $i$. Observe

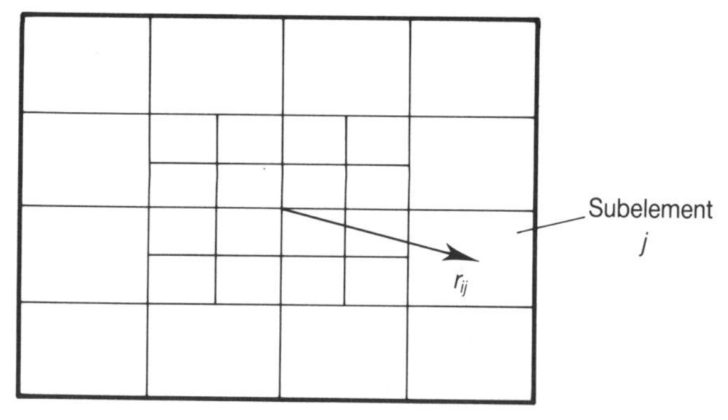


**Figure 4.1.4.** Integrating a function with a singularity.

the finer division close to the middle of the element so as to model the larger contributions of the term $r^{-2}$ as $r$ becomes smaller.

### 4.1.3 The Magnetic Field of a Coil

The Biot–Savart law (eqn. 4.1.5) also allows us to compute easily the magnetic field caused by a coil at a field point in free space. If the wire of the coil is sufficiently thin, let us define a local coordinate system $p, q, t$ within an elemental piece of the wire carrying a current $I$, where $t$ is along the direction of the wire in which the current flows and $p$ and $q$ are normal to it. **J** will then be along $t$; performing the integrations over the cross sections:

$$\mathbf{B} = \frac{\mu}{4\pi}\iiint \frac{I\mathbf{u}_t \times \mathbf{r}}{Sr^3}\, dp\, dq\, dt = \frac{\mu I}{4\pi}\int \frac{\mathbf{u}_t \times \mathbf{r}}{r^3}\, dt. \tag{4.1.14}$$

To compute the field, then, we divide the coil into $n$ small straight pieces $i$, each of length $t_i$, and reduce eqn. (4.1.14) to:

$$\mathbf{B} = \frac{\mu I}{4\pi}\sum_{i=1}^{n} t_i r^{-3}\mathbf{u}_{ti} \times \mathbf{r}, \tag{4.1.15}$$

where we have taken $r$ to be a constant over the element—this being valid for elements that are small compared to the distance of the field point $r$. To evaluate this numerically, let the piece $i$ go from $(x_{i1}, y_{i1}, z_{i1})$ to $(x_{i2}, y_{i2}, z_{i2})$. Taking the midpoint along the elemental wire $[\frac{1}{2}(x_{i1} + x_{i2}), \frac{1}{2}(y_{i1} + y_{i2}), \frac{1}{2}(z_{i1} + z_{i2})]$ as a representative source point $(x_s, y_s, z_s)$, we have for the field point $(x_f, y_f, z_f)$ at which we wish to compute the flux density:

$$\mathbf{r} = \mathbf{u}_x[x_f - \tfrac{1}{2}(x_{i1} + x_{i2})] + \mathbf{u}_y[y_f - \tfrac{1}{2}(y_{i1} + y_{i2})] + \mathbf{u}_z[z_f - \tfrac{1}{2}(z_{i1} + z_{i2})], \tag{4.1.16}$$

$$t_i = [(x_{i2} - x_{i1})^2 + (y_{i2} - y_{i1})^2 + (z_{i2} - z_{i1})^2], \tag{4.1.17}$$

$$\mathbf{u}_t = t_i^{-1}[\mathbf{u}_x(x_{i2} - x_{i1}) + \mathbf{u}_y(y_{i2} - y_{i1}) + \mathbf{u}_z(z_{i2} - z_{i1})]. \tag{4.1.18}$$

Thus, $\mathbf{u}_t \times \mathbf{r}$ could be evaluated from the identity (A3) and a general program may be written with the nodal coordinates of the elements defining the loop of wire and the current as input. With this information, for any field point the magnetic field thereat may be calculated.

Once such a program is written, to verify its proper operation, we may take as a test problem the simple configuration of a circular wire of radius $a$ and compute the field at a distance $z$ along the axis of a coil. To determine analytically the field at a point along the axis, consider the contribution under the integral sign of eqn. (4.1.14) by a small element $\mathbf{u}_t\, dt$ at a point a distance $z$ along the axis. Drawing the loop perpendicularly to the plane of the paper as in Figure 4.1.5, let the current be directed such that it goes into the paper at the top and comes out at the bottom. Before proceeding further, for any two vectors $\mathbf{A}$ and $\mathbf{B}$ an angle $\theta$ apart, constructing

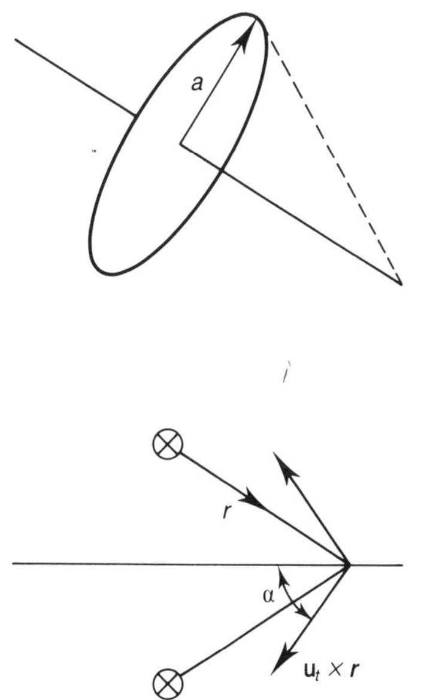


**Figure 4.1.5.** Magnetic field along the axis of a circular loop.

a coordinate system such that both vectors are on the $xy$ plane:

$$\mathbf{A} = \mathbf{u}_x A, \tag{4.1.19}$$

$$\mathbf{B} = \mathbf{u}_X B \cos\theta + \mathbf{u}_y B \sin\theta, \tag{4.1.20}$$

and applying the definition of the curl (A3), we have

$$\mathbf{A} \times \mathbf{B} = \mathbf{u}_z AB \sin\theta. \tag{4.1.21}$$

Since $\mathbf{u}_t$ is into the paper and of unit magnitude and $\mathbf{r}$ is along the paper in Figure 4.1.5, they are perpendicular to each other ($\theta = 90$) and we have that $\mathbf{u}_t \times \mathbf{r}$ has magnitude $r$ and is directed as shown in Figure 4.1.5. Similarly, the piece opposite the one being considered has $\mathbf{u}_t$ reversed in such a direction (out of the paper) that its contribution is as shown in Figure 4.1.5. The resultant then has a component only along the $z$ axis, the normal components cancelling each other. The combined contribution of these two symmetric (diametrically opposite) elements is therefore:

$$d\mathbf{B} = -\mathbf{u}_z \frac{\mu I}{4\pi} 2\cos\alpha \frac{r\,dt}{r^3} = -\mathbf{u}_z \frac{\mu I}{2\pi} \frac{a}{r} r^{-2}\,dt$$
$$= -\mathbf{u}_z \frac{\mu I}{2\pi} a r^{-3}\,dt, \tag{4.1.22a}$$

where we have used $a/r$ for $\cos\theta$. Integrating this over a half of the circle (since we have already accounted for the contribution from the other symmetric half) and noting that $r$ remains constant at $\sqrt{(a^2 + z^2)}$, we have:

$$\mathbf{B} = -\mathbf{u}_z \frac{\mu I}{2\pi} \frac{\pi a^2}{(a^2 + z^2)^{3/2}} = -\mathbf{u}_z \frac{\mu I a^2}{2(a^2 + z^2)^{3/2}}, \tag{4.1.22b}$$

using $\pi a$ for the integral $dt$ over a half-circle.

### 4.1.4 Inductance Computation: Neumann's Formula

The inductance is a physical quantity relating to coils that depends only on geometry and material property. It relates the energy $W$ stored in a coil to the current in it:

$$W = \tfrac{1}{2} L I^2. \tag{4.1.23}$$

Alternatively, by writing for a coil with current $I$ and voltage $V$ at a time $t$:

$$W = \tfrac{1}{2} L I^2 = \int_0^I L I\,dI = \int_0^t V I\,dt, \tag{4.1.24}$$

we have the more familiar definition

$$V = L\frac{dI}{dt}, \tag{4.1.25}$$

of inductance as relating the induced voltage to the rate of change of current in a coil. A formula attributed to Neumann exists that permits easy numeric evaluation of inductance.

The stored energy in a magnetic field is given by

$$W = \frac{1}{2}\iiint \mathbf{H} \,.\, \mathbf{B}\, dV. \tag{4.1.26}$$

(A rigorous proof of this is given by Panofsky and Phillips (1969), pp. 170–173.) To determine the inductance, we must relate this expression to eqn. (4.1.24). To this end, using eqn. (1.2.24):

$$W = \frac{1}{2}\iiint \mathbf{H} \,.\, (\nabla \times \mathbf{A})\, dV = \frac{1}{2}\iiint [\nabla \,.\, (\mathbf{H} \times \mathbf{A}) + \mathbf{A} \,.\, (\nabla \times \mathbf{H})]\, dV$$

using (A8),

$$= \frac{1}{2}\iint (\mathbf{H} \times \mathbf{A}) \,.\, d\mathbf{S} + \frac{1}{2}\iiint \mathbf{A} \,.\, \mathbf{J}\, dV \tag{4.1.27}$$

from (A14) and (1.2.14),

$$= \frac{1}{2}\iiint \mathbf{A} \,.\, \mathbf{J}\, dV,$$

where for all space the surface integral may be knocked off; as seen in eqn. (4.1.3), **A** varies as $r^{-1}$ and **H** (which is derived from **A** by differentiation) as $r^{-2}$, so that the integrand varies as $r^{-3}$. $dS$, however, varies as $r^2$, and, as a result, the multiple disappears for large $r$. This would not be correct for high-frequency fields involving radiation and the displacement current where, it may be shown, **H** varies as $r^{-1}$.

To derive Neumann's formula, let us take two coils 1 and 2, respectively, carrying currents $I_1$ and $I_2$. Consider a current element $I_1\, dt_1$ on coil 1. The magnetic vector potential $A_{12}$ caused there by the currents in coil 2 is derivable from the vector potential as given by eqn. (4.1.3):

$$A_{12} = \frac{\mu}{4\pi}\iiint_2 r^{-1}\mathbf{J}\, dV = \frac{\mu I_2}{4\pi}\int_2 r^{-1}\mathbf{u}_{t2}\, dt_2, \tag{4.1.28a}$$

where the integral over the wire cross section has been performed as in eqn. (4.1.14) and the suffix 2 by the integral sign indicates that the integral is over the coil 2.

Therefore, the energy stored in coil 1 as a result of the current in coil 2, according to eqn. (4.1.27), is

$$W_{12} = \frac{1}{2}\iiint_1 \mathbf{A}_{12} \cdot \mathbf{J}_1 \, dV = \frac{1}{2}\iiint_1 \mathbf{A}_{12} \cdot \mathbf{J}_1 \, dV = \frac{1}{2} I_1 \int_1 \mathbf{A}_{12} \cdot \mathbf{u}_{t1} \, dt_1$$
$$= \frac{\mu}{8\pi} I_1 I_2 \int_1 \int_2 \frac{\mathbf{u}_{t1} \cdot \mathbf{u}_{t2}}{r} dt_1 \, dt_2. \qquad (4.1.29)$$

Thence,

$$L_{12} = \frac{W_{12}}{I_1 I_2} = \frac{\mu}{8\pi} I_1 I_2 \int_1 \int_2 \frac{\mathbf{dt}_1 \cdot \mathbf{dt}_2}{r}, \qquad (4.1.28b)$$

which is Neumann's formula, in which the unit vectors along the tangents to the curves have been combined with the lengths $dt_1$ and $dt_2$. To implement this numerically (the only way possible in several practical cases), we divide, as in the previous section, the loops 1 and 2 into elemental pieces of length $t_i$, numbering $n$ and $m$, say, and reduce the integrations to summations:

$$L_{12} = \frac{W_{12}}{I_1 I_2} = \frac{\mu}{8\pi} I_1 I_2 \sum_{i1=1}^{n} \sum_{i2=1}^{m} \frac{\mathbf{t}_{i1} \cdot \mathbf{t}_{i2}}{r_{i1,i2}}. \qquad (4.1.30)$$

The means to finding the vectors **t** are in eqns. (4.1.16), (4.1.17), and (4.1.18). The dot product is as in eqn. (A2). Techniques requiring special treatment are called for in using this expression for self-inductance, when loops 1 and 2 are the same because $r_{i1,i2}$ goes to zero at times (Silvester 1968).

# 4.2 INTEGRAL METHODS IN INHOMOGENEOUS MEDIA

## 4.2.1 The Magnetization Vector in Permeable Structures

The integral equations above have been derived and are seen to be easy to use. However, they are valid only for homogeneous media, because they integrate the effects of the current densities **J** but do not account for the magnetization of commonly encountered permeable materials such as steel. For example, Figure 4.2.1A gives the actual field about a conductor near a highly permeable steel plate, whereas the field that would result from the application of the above integral equations is shown in Figure 4.2.1B. This is because the magnetic field is actually sourced by the currents and the

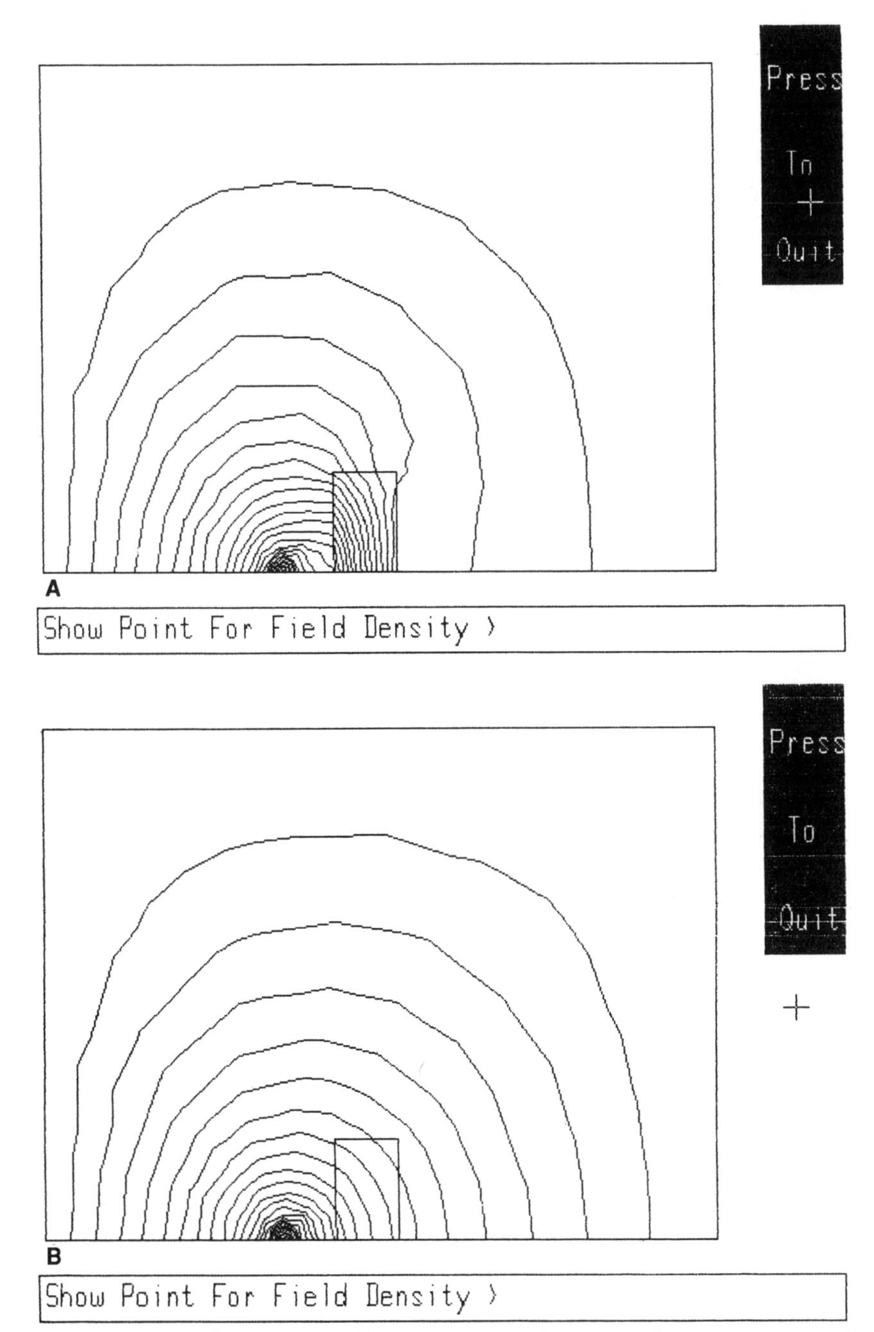


**Figure 4.2.1.** The magnetic field about a conductor. **A**. In the presence of steel. **B**. Without steel.

magnetization of the permeable parts that tend to offer a path of lower reluctance to the magnetic flux, thus pulling the flux into the steel. This magnetization **M** is defined so that it does not exist in regions of unit relative permeability, and it is proportional to the applied magnetic field strength:

$$\mathbf{M} = (\mu_r - 1)\mathbf{H} = \chi\mathbf{H}, \tag{4.2.1}$$

where $\mu_r$ is the relative permeability and $\chi$ is the magnetic susceptibility.

In this section we shall discover a means of solving directly for the vector magnetic field strength **H** in the presence of permeable structures, by integral methods. To this end, let us split the total field strength at a point into two parts: $\mathbf{H}_c$, the free space field strength that would exist if there were no permeable parts, and $\mathbf{H}_m$, the magnetization component that arises as a result of the presence of these parts;

$$\mathbf{H} = \mathbf{H}_c + \mathbf{H}_m. \tag{4.2.2}$$

$\mathbf{H}_c$, being the free space field strength, may be evaluated using eqn. (4.1.8) in two dimensions and eqn. (4.1.5) in three dimensions, and, as we have seen in eqn. (1.2.14), it satisfies

$$\nabla \times \mathbf{H}_c = \mathbf{J}. \tag{4.2.3}$$

This is the magnetic field we would experience if the permeable structure were absent. Therefore, the solution of this from eqns. (4.1.5) and (4.1.8) is

$$\mathbf{H}_c = \mu_0^{-1}\mathbf{B}_c = \frac{1}{4\pi}\iiint \left(\frac{\mathbf{J} \times \mathbf{r}}{r^3}\right) dV \qquad \text{in general,} \tag{4.2.4a}$$

$$= \frac{1}{2\pi}\iint r^{-2}[y\mathbf{u}_x - x\mathbf{u}_y]J_z\, dx\, dy \qquad \text{in 2-D.} \tag{4.2.4b}$$

Using eqn. (4.2.3) in eqn. (1.2.14),

$$\nabla \times \mathbf{H}_m = \nabla \times \mathbf{H} - \nabla \times \mathbf{H}_c = J - J = 0, \tag{4.2.5}$$

so that according to the vector identity (A5):

$$\mathbf{H}_m = -\nabla\Omega, \tag{4.2.6}$$

where $\Omega$ is the magnetic scalar potential. At this point, one approach to solving for the magnetic field (Simkin and Trowbridge

1979; Simkin 1981) is to use eqn. (1.2.10):

$$\nabla \cdot \mathbf{B} = \nabla \cdot \mu\mathbf{H} = \nabla \cdot \mu[\mathbf{H}_c + \mathbf{H}_m] = \nabla \cdot \mu[\mathbf{H}_c - \nabla\Omega] = 0, \quad (4.2.7)$$

giving us

$$\nabla \cdot \mu\nabla\Omega = \nabla \cdot \mu\mathbf{H}_c, \quad (4.2.8)$$

and as expected the applied field $\mathbf{H}_c$ acts as a source to the magnetization field caused by the gradient of $\Omega$. Since $\mathbf{H}_c$ is known, by solving the above Poisson equation for $\Omega$ we may find the magnetization field strength $\mathbf{H}_m$ and then the total field strength. The solution of the Poisson equation we have seen may be attempted by the finite difference method of section 3.1, or, as will be shown later in chapter 7, by the finite element method.

In this section, we are concerned with solving for the magnetic field strength by integral schemes. To proceed in this direction, taking up eqn. (1.2.10) again:

$$\nabla \cdot \mathbf{B} = \mu_0\nabla \cdot \mu_r\mathbf{H} = \mu_0\nabla \cdot [\mathbf{M} + \mathbf{H}] = \mu_0\nabla \cdot [\mathbf{M} + \mathbf{H}_c - \nabla\Omega] = 0, \quad (4.2.9)$$

where eqn. (4.2.1) has been used to substitute $\mathbf{M} + \mathbf{H}$ for $\mu_r\mathbf{H}$. But $\mathbf{H}_c$ being a free space solution in homogeneous space, it is the free space flux density scaled by $\mu_0$ and has zero divergence. Therefore:

$$-\nabla^2\Omega = -\nabla \cdot \mathbf{M}, \quad (4.2.10)$$

which is a Poisson equation for $\Omega$ sourced by $\nabla \cdot \mathbf{M}$, which we do not know as yet. Significantly, the equation has no material terms and therefore may be likened to a problem in homogeneous space. Consequently the free space solution of the Poisson equation may be used. In this, resorting again to the numerical analogy with electric charge systems is useful.

We will now generalize the Poisson equation (1.2.22) to a closed region $V$ with volume charge density $\rho$, bounded by a surface $S_1 + S_2 + C_1 + C_2 + \sum$ with surface charge density $\sigma$ as shown in Figure 4.2.2. The surface $\sum$ tends to infinity, $C_1$ and $C_2$ are colinear and opposite running curves with nothing in between, and $S_1$ and $S_2$ are also opposing curves, but they encase the surface charge distributions within the volume. Ignoring the integrations over the remote surface $\sum$ where $\phi$ and its derivative are zero, the previously derived solution eqn. (2.5.19) for three-dimensional systems

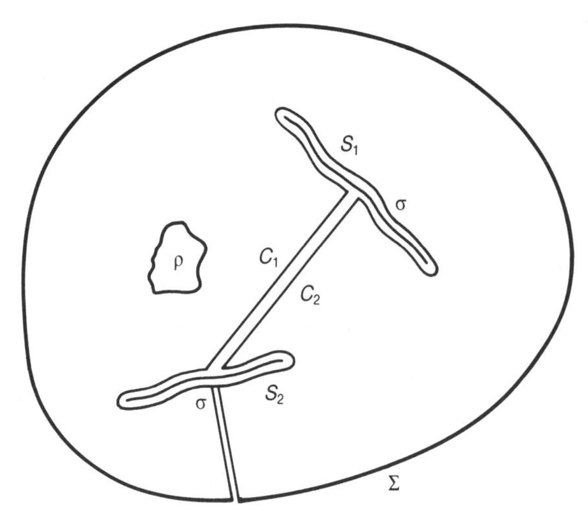


**Figure 4.2.2.** A general Poissonian configuration.

generalizes to

$$\begin{aligned}\phi &= \frac{1}{4\pi\varepsilon}\iiint r^{-1}\rho\, dV + \frac{1}{4\pi}\iint_{S_1+S_2}[r^{-1}\mathbf{u}_n \,.\, \nabla\phi - \phi\mathbf{u}_n \,.\, \nabla r^{-1}]\, dS \\ &\quad + \frac{1}{4\pi}\iint_{C_1+C_2}[r^{-1}\mathbf{u}_n \,.\, \nabla\phi - \phi\mathbf{u}_n \,.\, \nabla r^{-1}]\, dS \\ &= \frac{1}{4\pi\varepsilon}\iiint r^{-1}\rho\, dV + \frac{1}{4\pi}\iint_{S_1+S_2}[r^{-1}\,\mathbf{u}_n \,.\, \nabla\phi]\, dS \\ &= \frac{1}{4\pi\varepsilon}\iiint r^{-1}\rho\, dV + \frac{1}{4\pi}\iint_{S_1+S_2}[r^{-1}\mathbf{u}_n \,.\, \nabla\phi]\, dS,\end{aligned} \tag{4.2.11}$$

where since $\phi$ and $\nabla\phi \,.\, \mathbf{u}_n$ are continuous across $C$ and $\phi$ is continuous across $S$ according to eqns. (2.5.39) and (2.5.41), these terms cancel each other.

Now writing the Poisson eqn. (1.2.22) as

$$-\varepsilon\nabla_f^2\phi = \rho = \nabla_s \,.\, \mathbf{D}, \tag{4.2.12}$$

and from eqn. (1.2.13)

$$\mathbf{D} = -\varepsilon\nabla\phi, \tag{4.2.13}$$

we have, continuing from eqn. (4.2.11):

$$\begin{aligned}\phi &= \frac{1}{4\pi\varepsilon}\iiint \nabla_s \cdot \frac{\mathbf{D}\,dV}{r} - \frac{1}{4\pi\varepsilon}\iint \frac{\mathbf{D}\cdot d\mathbf{S}}{r} \\ &= \frac{1}{4\pi\varepsilon}\iiint \nabla_s \cdot \frac{\mathbf{D}\,dV}{r} - \frac{1}{4\pi\varepsilon}\iiint \nabla_s \cdot (r^{-1}\mathbf{D})\,dV \qquad \text{by (A13)}, \\ &= \frac{1}{4\pi\varepsilon}\iiint \nabla_s \cdot \frac{\mathbf{D}\,dV}{r} - \frac{1}{4\pi\varepsilon}\iiint [r^{-1}\nabla_s \cdot \mathbf{D} + \mathbf{D}\cdot\nabla_s r^{-1}]\,dV \\ &\qquad\qquad \text{by (A11)}, \\ &= -\frac{1}{4\pi\varepsilon}\iiint \frac{\mathbf{D}\cdot\mathbf{r}}{r^3}\,dV,\end{aligned} \tag{4.2.14}$$

where the subscripts $f$ and $s$ attaching to the Del operator indicate whether the differentiations are at the field point or source point. We have already shown in eqn. (2.5.3) that $\nabla_f r^{-1}$ is $-r^{-3}\mathbf{r}$. It is trivial to verify in like manner that $\nabla_s r^{-1}$ is $+r^{-3}\mathbf{r}$ by writing

$$\mathbf{r} = \mathbf{u}_x(x_f - x_s) + \mathbf{u}_y(y_f - y_s) + \mathbf{u}_z(z_f - z_s), \tag{2.5.1}$$

getting the value of the length $r$ from this and then performing the differentiations with respect to the source coordinates $x_s$, $y_s$, and $z_s$ in computing $\nabla_s r^{-1}$.

By the analogy of eqn. (4.2.10) to eqn. (1.2.22), therefore, we have the solution of eqn. (4.2.10):

$$\Omega = \frac{1}{4\pi}\iiint \frac{\mathbf{M}\cdot\mathbf{r}}{r^3}\,dV, \tag{4.2.15}$$

extrapolating from eqn. (4.2.14). If we now return to the definition of magnetization (4.2.1), we have:

$$\mathbf{M} = \chi\mathbf{H} = \chi[\mathbf{H}_c + \mathbf{H}_m] = \chi[\mathbf{H}_c - \nabla\Omega], \tag{4.2.16}$$

so that, substituting for $\mathbf{H}_c$ from eqn. (4.2.4) and for $\Omega$ from eqns. (4.2.6) and (4.2.15):

$$\mathbf{M} = \frac{\chi}{4\pi}\iiint \frac{\mathbf{J}\times\mathbf{r}}{r^3}\,dV - \frac{\chi}{4\pi}\nabla_f \iiint \frac{\mathbf{M}\cdot\mathbf{r}}{r^3}\,dV, \tag{4.2.17}$$

which may be solved for the only unknown $\mathbf{M}$ in it. In evaluating the term $\nabla_f\Omega$, it is convenient to distinguish between the two cases—first, when the field point is in the permeable part where the term $\mathbf{M}$ of the operand is (so that $r \to 0$) and the second, when it is not. When it is not, the field point is distinct from the source point so that $r \neq 0$, and the evaluation is simple. Observe that $\mathbf{M}$ on the left is sourced by the **M** on the right. Therefore, the right-hand side term **M** belongs to a source point, so that the operator $\nabla_f$ will see it as a constant. We have:

$$\begin{aligned}
\nabla_f \frac{\mathbf{M}\,.\,\mathbf{r}}{r^3} &= \nabla(M_x x r^{-3} + M_y y r^{-3} + M_z z r^{-3}] \qquad \text{by (A2)} \\
&= M_x[(r^{-3}\nabla x + x\nabla r^{-3})] + M_y[(r^{-3}\nabla y + y\nabla r^{-3})] \\
&\quad + M_z[(r^{-3}\nabla z + z\nabla r^{-3})] \\
&= M_x[r^{-3}\mathbf{u}_x - 3xr^{-5}\mathbf{r}] + M_y[r^{-3}\mathbf{u}_y - 3yr^{-5}\mathbf{r}] \\
&\quad + M_z[r^{-3}\mathbf{u}_z - 3zr^{-5}\mathbf{r}] \\
&= r^{-3}[M_x\mathbf{u}_x + M_y\mathbf{u}_y + M_z\mathbf{u}_z] - 3[xM_x + yM_y + zM_z]r^{-5}\mathbf{r} \\
&= r^{-3}\mathbf{M} - 3[\mathbf{r}\,.\,\mathbf{M}]r^{-5}\mathbf{r},
\end{aligned} \tag{4.2.18}$$

where we have evaluated $\nabla x$, $\nabla y$, and $\nabla z$, respectively, as $\mathbf{u}_x$, $\mathbf{u}_y$, and $\mathbf{u}_z$ using (A1), and expanded:

$$\begin{aligned}
\nabla r^{-3} &= \frac{\partial r^{-3}}{\partial r}\frac{\partial r}{\partial x}\mathbf{u}_x + \frac{\partial r^{-3}}{\partial r}\frac{\partial r}{\partial y}\mathbf{u}_y + \frac{\partial r^{-3}}{\partial r}\frac{\partial r}{\partial z}\mathbf{u}_z \\
&= -3r^{-4}(r^{-1}x)\mathbf{u}_x - 3r^{-4}(r^{-1}y)\mathbf{u}_y - 3r^{-4}(r^{-1}z)\mathbf{u}_z \\
&= -3r^{-5}\mathbf{r}.
\end{aligned} \tag{4.2.19}$$

When $r \to 0$, that is, when the field point contains magnetization and is the same as the source point, the evaluation is more complex, and the reader may consult Turner (1973) for it.

Similarly, in a two-dimensional system, the generalization of the solution (2.1.9) to the Poisson equation becomes, in analogy to eqn. (4.2.14):

$$\begin{aligned}
\phi &= \frac{1}{2\pi\varepsilon}\iint \rho ln_e r\, dR + \frac{1}{2\pi\varepsilon}\int \sigma ln_e r\, dS \\
&= \frac{1}{2\pi\varepsilon}\iint \nabla_s\,.\,\mathbf{D} ln_e r\, dR - \frac{1}{2\pi\varepsilon}\int ln_e r(\mathbf{D}\,.\,\mathbf{dS}) \\
&= \frac{1}{2\pi\varepsilon}\iint \nabla_s\,.\,\mathbf{D} ln_e r\, dR - \frac{1}{2\pi\varepsilon}\iint \nabla_s(ln_e r\mathbf{D})\, dR \\
&= \frac{1}{2\pi\varepsilon}\iint \nabla_s\,.\,\mathbf{D} ln_e r\, dR - \frac{1}{2\pi\varepsilon}\iint [ln_e r\nabla_s\,.\,\mathbf{D} + \mathbf{D}\,.\,\nabla_s ln_e r]\, dR \\
&= -\frac{1}{2\pi\varepsilon}\iint \frac{\mathbf{D}\,.\,\mathbf{r}}{r^2}\, dR,
\end{aligned} \tag{4.2.20}$$

where $\nabla_s ln_e r$ has been worked out using eqn. (2.5.7). Therefore, by analogy,

$$\Omega = \frac{1}{2\pi}\iint \frac{\mathbf{M}\,.\,\mathbf{r}}{r^2}\,dR, \tag{4.2.21}$$

and the governing equation for $\mathbf{M}$ that is to be solved becomes

$$\mathbf{M} = \frac{\chi}{2\pi}\iint \frac{\mathbf{J}\times\mathbf{r}}{r^2}\,dR - \frac{\chi}{2\pi}\nabla_f\iint \frac{\mathbf{M}\,.\,\mathbf{r}}{r^2}\,dR. \tag{4.2.22}$$

Again, when the field point $(x, y) \equiv (0, 0)$ contains no magnetization, simplifying $(1/2\pi)\nabla_f(\mathbf{M}\,.\,\mathbf{r}/r^2)$ as in eqn. (4.2.18) and using the fact that $r^2$ is $x^2 + y^2$:

$$\begin{aligned}\frac{1}{2\pi}\nabla_f\frac{\mathbf{M}\,.\,\mathbf{r}}{r^2} &= \frac{1}{2\pi}\{r^{-2}\mathbf{M} - 2r^{-4}[\mathbf{M}\,.\,\mathbf{r}]\mathbf{r}\}\\ &= \frac{1}{2\pi}\{r^{-4}(x^2 + y^2)[\mathbf{u}_x M_x + \mathbf{u}_y M_y]\\ &\quad - 2r^{-4}(xM_x + yM_y)[x\mathbf{u}_x + y\mathbf{u}_y]\}\\ &= -\frac{1}{2\pi r^4}\{u_x[(x^2 - y^2)M_x + 2xyM_y]\\ &\quad + u_y[2xyM_x + (y^2 - x^2)M_y]\}.\end{aligned} \tag{4.2.23}$$

In solving eqn. (4.2.17) numerically in three dimensions or eqn. (4.2.22) in two dimensions, the solution region is the location of permeable materials, and therefore we cannot avoid considering the situation where $r \to 0$. In a very efficient numerical scheme (Newman, Trowbridge, and Turner 1972), the domain is divided into small elements in each of which $\mathbf{M}$ may be assumed constant so as to allow the integrals to be evaluated easily. And generally for a field element $i$ and source element $k$ containing $\mathbf{M}$, for both cases $r = 0$ $(i = k)$ and $r \neq 0$ $(i \neq k)$, there results, say in two dimensions, upon expanding $\nabla_f(\mathbf{M}\,.\,\mathbf{r}/r^2)$:

$$\nabla_f\frac{\mathbf{M}\,.\,\mathbf{r}}{r^2} = -\sum_k\{\mathbf{u}_x[C^{ik}_{xx}M_{kx} + C^{ik}_{xy}M_{ky}] + \mathbf{u}_y[C^{ik}_{yx}M_{kx} + C^{ik}_{yy}M_{ky}]\}, \tag{4.2.24}$$

where for the case $i \neq k$, we have from eqn. (4.2.23):

$$C^{ik}_{xx} = -C^{ik}_{yy} = \frac{1}{2\pi}\iint_k r^{-4}(x^2 - y^2)\,dx\,dy, \tag{4.2.25}$$

and

$$C^{ik}_{xy} = C^{ik}_{yx} = \frac{1}{2\pi}\iint_k 2r^{-4}xy\,dx\,dy. \tag{4.2.26}$$

In the above, the origin of the coordinate system is at the field point $i$. Since $r$ is always nonzero, approximating $r$, $x$, and $y$ to be constant as the integrand roves over the element $k$, these coefficients may easily be evaluated. For the more difficult situation where the self-term $i = k$ is being computed, Turner (1973), by converting the surface integral to a line integral over the boundary of the element $k$, gives for the $n$-noded polygonal element of Figure 4.2.3A, useful for fitting all manner of device shapes,

$$C_{xx}^{ii} = \frac{1}{2\pi}\sum_{k=1}^{n}\left\{\sin^2\phi_k\left[\tan^{-1}\frac{y_k - y_i}{x_k - x_i} - \tan^{-1}\frac{y_{k+1} - y_i}{x_{k+1} - x_i}\right]\right.$$
$$\left. + \sin\phi_k\cos\phi_k[ln_e r_k - ln_e r_{k+1}]\right\} \quad (4.2.27)$$
$$= \frac{1}{2\pi}\sum_{k=1}^{n}\left\{-\beta_k\sin^2\phi_k - \sin\phi_k\cos\phi_k ln_e\frac{r_{k+1}}{r_k}\right\},$$

where the summation is over the vertices $k$ of the polygon, the point $(x_i, y_i)$ is in the middle of the field element $i$, and

$$\phi_k = \tan^{-1}\frac{y_{k+1} - y_k}{x_{k+1} - x_k}, \quad (4.2.28)$$

$$C_{yy}^{ii} = -1 - C_{xx}^{ii}, \quad (4.2.29)$$

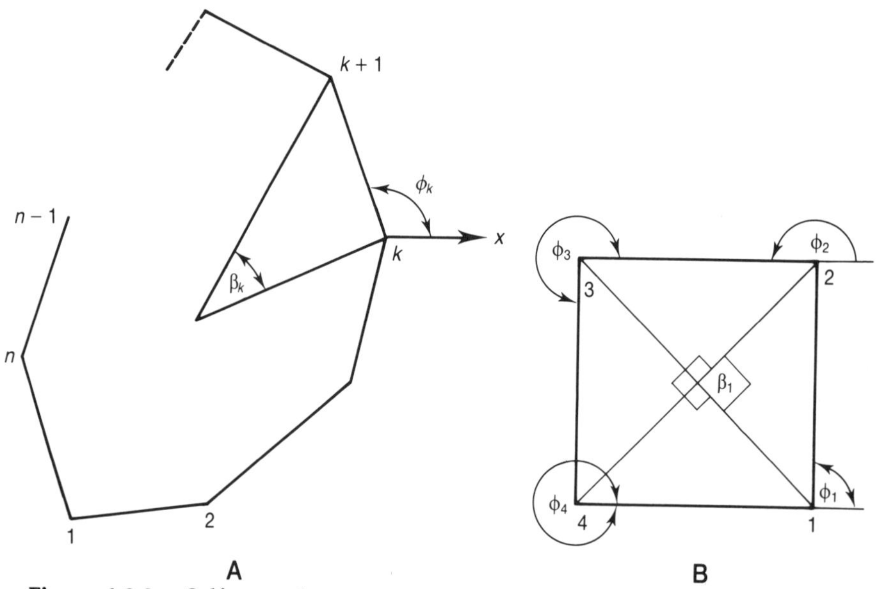


**Figure 4.2.3.** Self-caused term in magnetization computation. **A.** A polygonal element. **B.** A square element.

$$C_{xy}^{ii} = C_{yx}^{ii}$$

$$= \frac{1}{2\pi} \sum_{k=1}^{n} \left\{ -\sin\phi_k \cos\phi_k \left[ \tan^{-1}\frac{y_k - y_i}{x_k - x_i} - \tan^{-1}\frac{y_{k+1} - y_i}{x_{k+1} - x_i} \right] - \cos^2\phi_k [ln_e r_k - ln_e r_{k+1}] \right\} \quad (4.2.30)$$

$$= \frac{1}{2\pi} \sum_{k=1}^{n} - \left\{ \beta_k \sin\phi_k \cos\phi_k + \cos^2\phi_k ln_e \frac{r_{k+1}}{r_k} \right\}.$$

For example, for the square element of Figure 4.2.3B, $n$ is 4, the four $\beta$ values are $\pi/2$ each, and the values $\phi_1$, $\phi_2$, $\phi_3$, and $\phi_4$ are, respectively, $\pi/2$, $\pi$, $3\pi/2$, and $2\pi$. Hence we obtain, $C_{xx}^{ii} = -0.5$ from eqn. (4.2.27), $C_{yy}^{ii} = -0.5$ from eqn. (4.2.29), and $C_{xy}^{ii} = C_{yx}^{ii} = 0$ from eqn. (4.2.30).

Now, so long as we divide the magnetizable parts into polygons, we know how to compute the coefficients C. Putting eqn. (4.2.24) with these computable coefficients into eqn. (4.2.22),

$$\mathbf{u}_x M_{ix} + \mathbf{u}_y M_{iy} = \chi_i [\mathbf{u}_x H_{cix} + \mathbf{u}_y H_{ciy}] + \chi_i \sum_k \{\mathbf{u}_x [C_{xx}^{ik} M_{kx} + C_{xy}^{ik} M_{ky}] + \mathbf{u}_y [C_{yx}^{ik} M_{kx} + C_{yy}^{ik} M_{ky}]\}. \quad (4.2.31)$$

Separating the components in the $x$ and $y$ directions, rearranging, dividing by $\chi_i$, and using $H_i$ in place of $M_i/\chi_i$ and $H_k$ for $M_k/\chi_k$, we have the equations

$$\sum_{k=1}^{n} \{(\chi_k C_{xx}^{ik} - \delta_{ik}) H_{kx} + \chi_k C_{xy}^{ik} H_{ky}\} = -H_{cix}, \quad (4.2.32)$$

$$\sum_{k=1}^{n} \{\chi_k C_{yx}^{ik} H_{kx} + (\chi_k C_{yy}^{ik} - \delta_{ik}) H_{ky}\} = -H_{cix}. \quad (4.2.33)$$

These equations may be put into matrix form and solved easily. Figure 4.2.4A gives the solution of the magnetic flux density within the shield around a square conductor by this method. Figure 4.2.4B shows the corresponding solution of the magnetization, and Figure 4.2.4C shows the magnetic field strength. Notice in Figure 4.2.4D (giving the strength $\mathbf{H}_m$ caused by the magnetization) that, since inside the iron a lower field strength is sufficient to drive the flux, the magnetization is such as to reduce the total field. Symmetry has been used to economize by solving only in the upper half of the shield and conductor, the plots in the lower half being generated through reflection from the solution in the upper half. In performing the integrals of eqns. (4.2.32) and (4.2.33), we must, as in the parallel plate capacitor of section 2.4, allow for the contributions

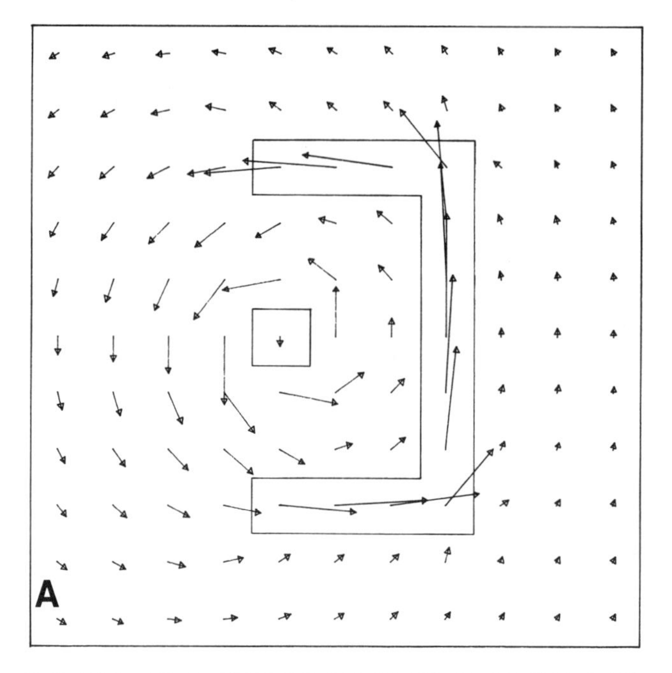


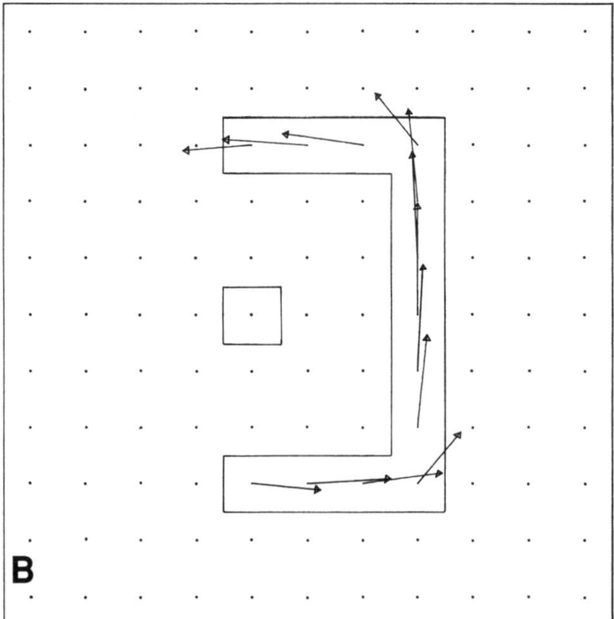


**Figure 4.2.4.** The magnetic field about a shield. **A.** Flux density. **B.** Magnetization. **C.** The field strength. **D.** The field strength on account of magnetization.

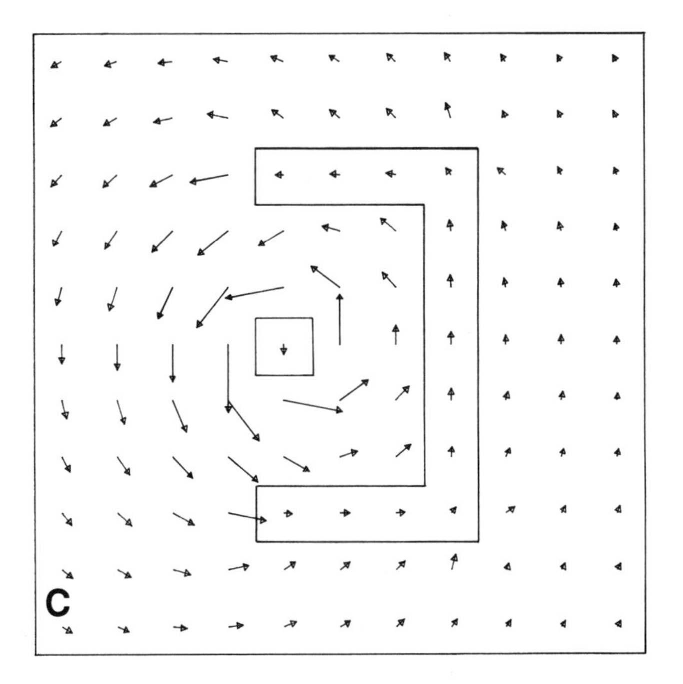


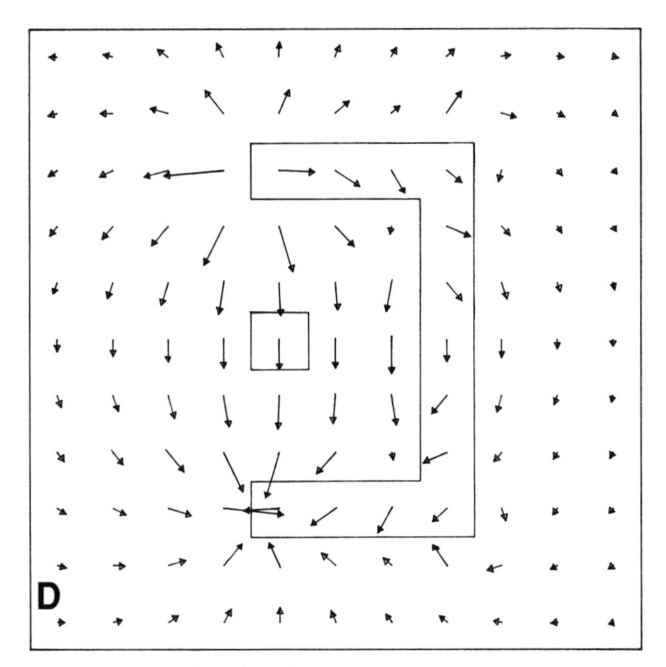


**Figure 4.2.4.** (*continued*)

from the lower half. Symmetry dictates that an element with field strength values $H_{kx}$ and $H_{ky}$ in the $x$ and $y$ directions will have a reflection in the lower half of the shield with the field strengths $-H_{kx}$ and $H_{ky}$. Once the field strengths are solved for, the flux densities for the plot may be obtained by multiplying the answers by the permeability.

Attention is drawn to the fact that the resultant matrix equation is full. However, a major advantage accrues from the fact that, as opposed to finite difference or the finite element method to be discussed in chapter 7, unknowns do not exist everywhere in the solution region, but only where there are magnetic structures. Thus, the size of the matrix is much smaller than with differential methods.

### 4.2.2 The Magnetic Scalar Potential by the Boundary Integral Method

We have just seen how to solve for the magnetization vector by an integral method. One of the disadvantages of the scheme is that two unknowns are involved at each interior location of permeable structures. We may alternatively solve for the magnetic scalar potential $\Omega$ governed by eqn. (4.2.8) by an integral method based on eqn. (2.5.19), which, when extended to eqn. (4.2.8) using the mapping $\varepsilon \rightarrow \mu$, $\phi \rightarrow \Omega$, $\rho \rightarrow \nabla.\ \mu\mathbf{H}_c$ to go from eqn. (1.2.22) to eqn. (4.2.8), gives us:

$$\Omega = \frac{1}{4\pi\mu}\iiint r^{-1}\nabla \cdot \mu\mathbf{H}_c\, dV + \frac{1}{4\pi}\iint [r^{-1}\mathbf{u}_n \cdot \nabla\Omega - \Omega\mathbf{u}_n \cdot \nabla r^{-1}]\, dS \tag{4.2.34}$$

$$= \frac{1}{4\pi}\iint (r^{-1}g - \Omega\mathbf{u}_n \cdot \nabla r^{-1}]\, dS,$$

where $g$ is the normal gradient of the magnetic potential $\Omega$ and we have used the fact that $\mathbf{H}_c$, being a free space magnetic field, has no divergence. As in chapter 2, therefore, we may divide the surface $S$ into small elements, assume $\Omega$ and $g$ to be unknown constants on each element, expand eqn. (4.2.34) in terms of these elements, and then solve for these constants. In the presence of permeable structures, we write the same equation for both regions and use the boundary conditions:

$$\Omega_1 = \Omega_2. \tag{4.2.35}$$

Also, by virtue of eqn. (1.2.17) specifying the normal continuity of magnetic flux here:

$$\mu_1\left[H_{cn1} - \frac{\partial \Omega_1}{\partial n}\right] = \mu_2\left[H_{cn2} - \frac{\partial \Omega_2}{\partial n}\right]. \tag{4.2.36}$$

But in the homogeneous medium of permeability $\mu_1$, without the discontinuity caused by $\mu_2$, we have

$$\mu_1 H_{cn1} = \mu_1 H_{cn2}. \tag{4.2.37}$$

This, therefore, gives us

$$\mu_1 g_1 = \mu_2 g_2 - (\mu_2 - \mu_1)H_{cn}, \tag{4.2.38}$$

where $H_{cn}$ represents the common normal component of $\mathbf{H}_c$ on either side of the discontinuity, and we have written $\partial\Omega/\partial n$ as $g$.

The boundary integral scheme may also be employed to solve for the vector potential governed by the Poission equation in two dimensions. However, in this case, the treatment is no different from the scheme for the electric potential already described in section 2.5.1. All the derivations of that section pertaining to two-dimensional analysis will hold with the mapping, by analogy, $\phi \rightarrow A$, $\rho \rightarrow J$, and $\varepsilon \rightarrow \mu^{-1}$.

# 4.3 DIFFERENTIAL METHODS

## 4.3.1 The Finite Difference Method

We have already seen in section 3.1 the application of the finite difference method to the solution of the Poisson equation for the electric potential. Analogous to eqn. (1.2.22) in electrostatics, in magnetics, the $z$ component of the magnetic vector potential in two-dimensional axisymmetric problems is governed by

$$-\mu^{-1}\nabla^2 A = J. \tag{1.2.32}$$

The analogy then is between $\phi$ and $A$, $\varepsilon$ and $\mu^{-1}$, and $\rho$ and $J$. We may therefore construct an $h \times k$ rectangular finite difference mesh, and the analogous network model corresponding to it. The impedance of a branch along the $x$ direction then becomes, from eqn. (3.1.15),

$$Z = \mu h^2, \tag{4.3.1}$$

and from this the impedance along a material interface with values

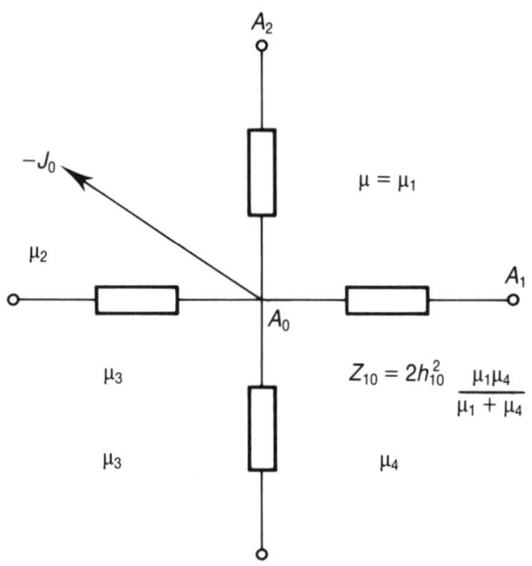


**Figure 4.3.1.** A finite difference node.

$\mu_1$ and $\mu_2$ across, analogous to eqn. (3.1.6), is

$$Z = \frac{2\mu_1\mu_2h^2}{\mu_1 + \mu_2}. \tag{4.3.2}$$

We therefore have, analogous to eqn. (3.1.17), the governing finite difference approximation at a node 0 surrounded by nodes 1, 2, 3, and 4, respectively, to its east, north, west, and south, with materials $\mu_1$, $\mu_2$, $\mu_3$, and $\mu_4$ in the first, second, third, and fourth quadrants about the node 0 of Figure 4.3.1:

$$\frac{A_1 - A_0}{2h_{10}^2\mu_1\mu_4/(\mu_1 + \mu_4)} + \frac{A_2 - A_0}{2h_{20}^2\mu_1\mu_2/(\mu_1 + \mu_2)} + \frac{A_3 - A_0}{2h_{30}^2\mu_2\mu_3/(\mu_2 + \mu_3)}$$
$$+ \frac{A_4 - A_0}{2h_{40}^2\mu_3\mu_4/(\mu_3 + \mu_4)} = -\frac{J_{01}}{4} - \frac{J_{02}}{4} - \frac{J_{03}}{4} - \frac{J_{04}}{4}. \tag{4.3.3}$$

Thus, a given magnetic field problem, provided it is amenable to having its domain discretized into a finite difference mesh, may have a network constructed and solved.

## 4.3.2 The Transmission Line Problem Revisited

As a means of demonstrating the solution of a magnetic field problem, let us take up the problem of a single rectangular transmission line considered in section 2.2. This being a magnetic field problem, if we are given the current in the conductor, let us try to compute the magnetic field inside and outside the conductor.

To define the problem mathematically, copper and the surrounding air having the same permeability as vacuum, we have $\mu = \mu_0$ everywhere, so that this is a problem in a homogeneous medium. The permeabilities being the same, eqn. (4.3.3) simplifies to

$$\frac{A_1 + A_3 - 2A_0}{h^2} + \frac{A_2 + A_4 - 2A_0}{k^2} = -\mu_0 J_0, \tag{4.3.4}$$

analogous to eqn. (3.1.9). $J_0$ is known—assuming a uniform distribution of current in the conductor, $J_0$ is the total current divided by the cross sectional area of the conductor inside the conductor and 0 outside.

This problem can be solved only in a closed region with $A$ or its normal gradient defined along the boundary as shown in Appendix B. From the considerations of section 1.2.4, a faraway boundary will have zero vector potential; but, since a faraway boundary will result in a large solution region with several nodes and an accompanying unbearable and expensive computational burden, we shall as an approximation move this faraway boundary closer, as shown in Figure 4.3.2A. This in effect makes the vector potential fall to zero faster, so that the resulting solutions of the magnetic field will be higher than the actual values. Moreover, by symmetry we may

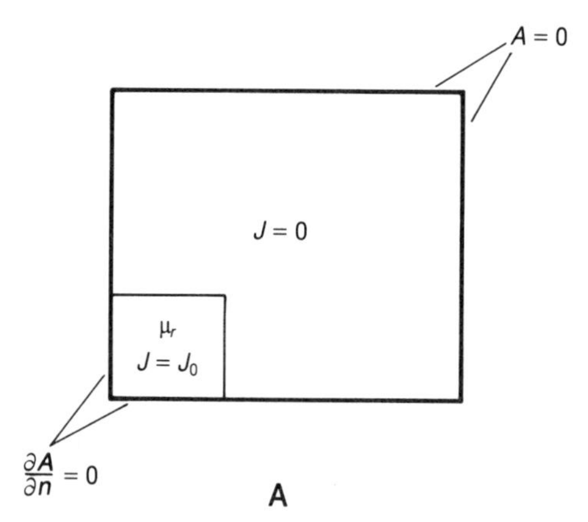


**Figure 4.3.2.** A rectangular conductor. **A.** The boundary value problem. **B.** The solution for relative permeability of 1. **C.** The solution for relative permeability of 10.

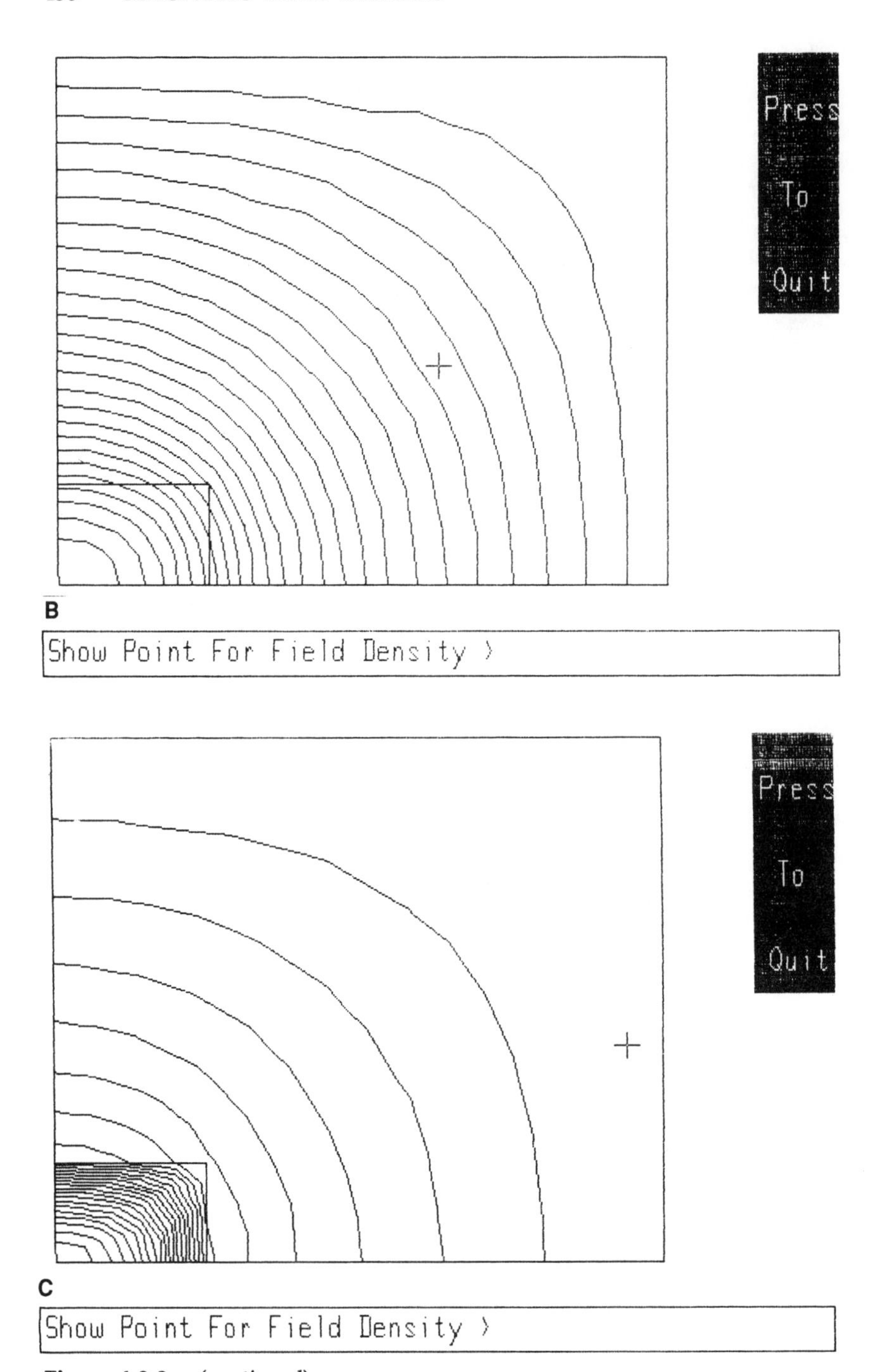


**Figure 4.3.2.** (*continued*)

solve in a quarter of the solution region with Neumann conditions (zero normal gradients) along the lines of symmetry. Note that eqn. (4.3.3), where at a node the term $J_0$ is regarded as contributed by the four quadrants surrounding it, requires taking $J_0$ to be half of the interior current density along the conductor surface and a quarter at the conductor corner; for along the surface only two of the quadrants contribute current, and at a corner only one does. The field within the configuration has been solved for and plotted as shown in Figure 4.3.2B for a relative permeability of 1, and in Figure 4.3.2C for a relative permeability of 10. Observe the flux choosing to flow in the conductor for the latter case in view of the lower reluctance there.

It is left as an exercise for the reader to think of the simplifications that will arise when the conductor is square—not only will $h$ and $k$ be the same, but we may solve within an eighth of the conductor. Even for rectangular conductors, the ratio of the length to breadth of the conductor permitting, by carefully choosing the number of mesh points in the $x$ and $y$ directions, $h$ and $k$ may be made the same.

### 4.3.3 An Electric Machine Slot in Highly Permeable Steel

As another example of solving magnetic field problems, consider the armature reaction flux caused by a conductor in the electric machine slot shown in Figure 4.3.3. Let the load current in the conductor be so small as not to saturate the steel. Then by the considerations of section 1.2.4, the flux ought to meet the steel normally and the boundary value problem may be posed as shown in Figure 4.3.3A. Since the boundary condition is Neumann all around, and we need $A$ to be fixed at least at one point on the boundary for uniqueness according to Appendix B, we shall take $A$ to be zero at an arbitrary point. We have completely avoided solving for the magnetic field in steel by using the fact that flux cannot have a tangential component just outside the steel. The steps in the solution of this problem then become just as in the preceding example and may be traced without difficulty. The solution is presented in Figure 4.3.3B.

### 4.3.4 An Electric Machine Slot in Finitely Permeable Steel

When the slot of the previous example is of finite permeability, we can no longer assume the flux to meet the steel at right angles, and

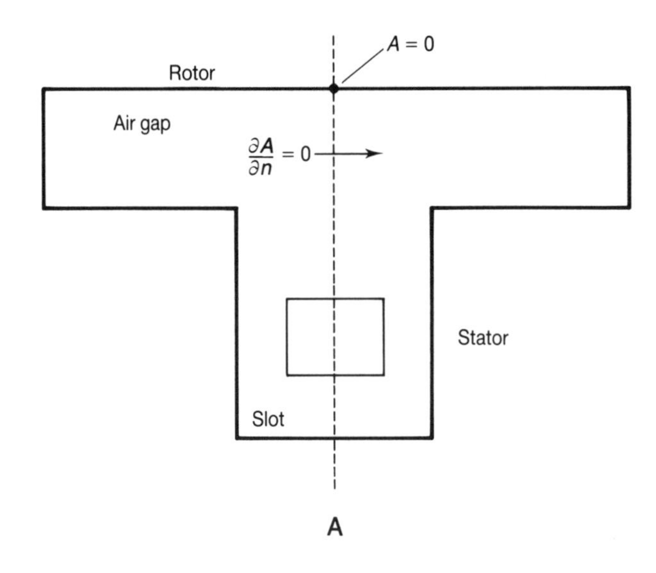


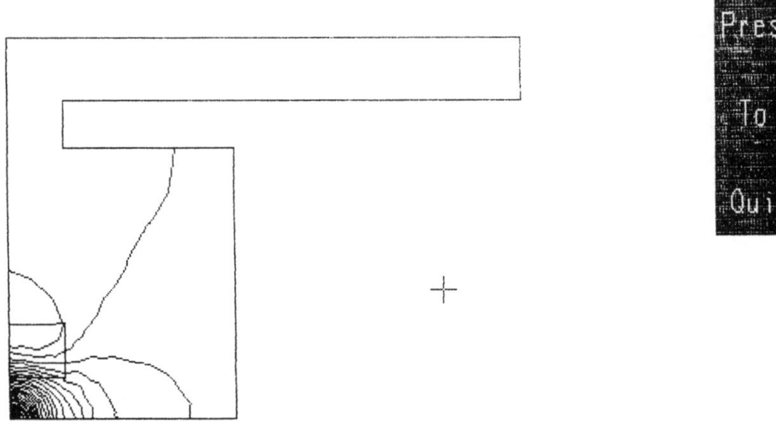

B

Show Point For Field Density >

**Figure 4.3.3.** The armature reaction in a highly permeable slot. **A.** The boundary value problem. **B.** The solution.

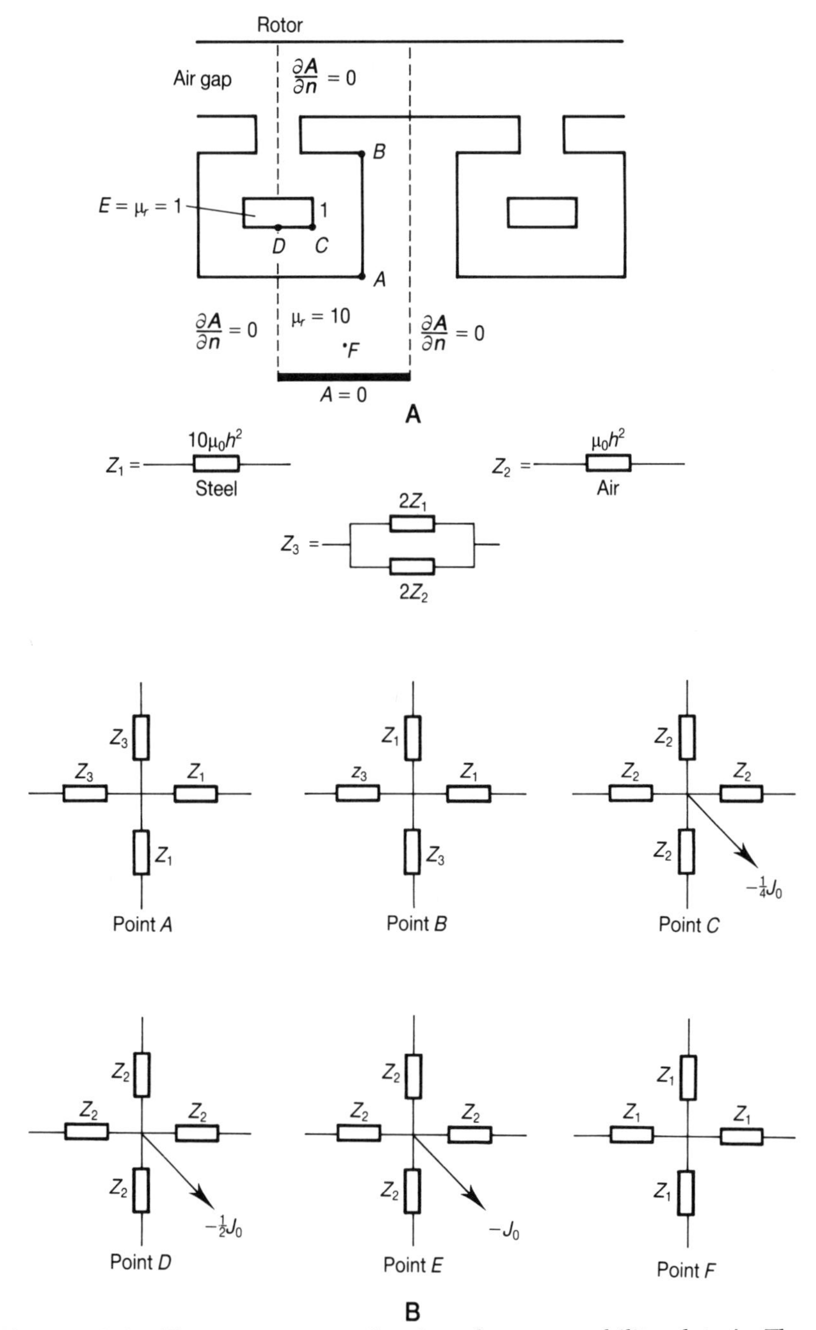


**Figure 4.3.4.** The armature reaction in a low-permeability slot. **A.** The boundary value problem. **B.** Some typical nodal networks. **C.** The solution.

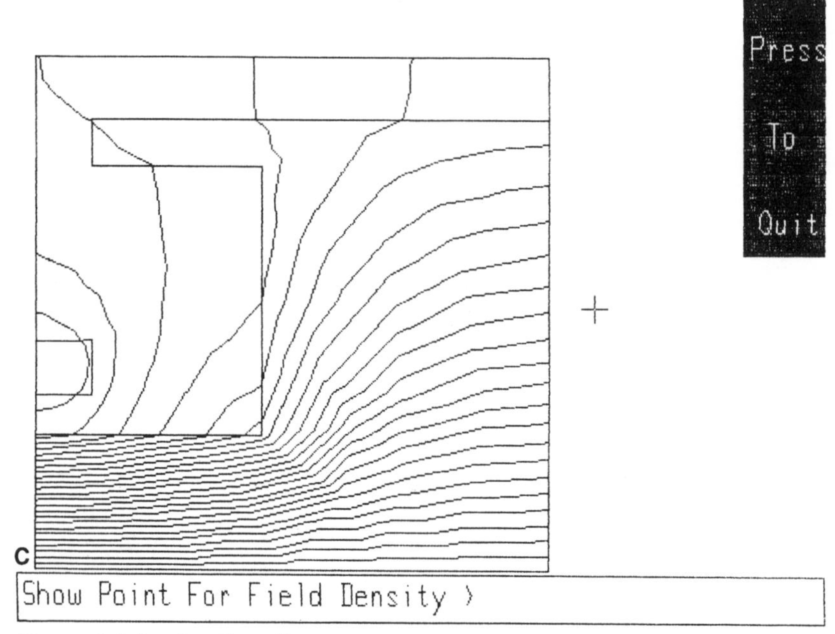


**Figure 4.3.4.** (*continued*)

therefore the solution region must be extended to inside the steel. The boundary value problem is posed as shown in Figure 4.3.4A. We may take the rotor steel, far away from the conductor, to be unsaturated so that it becomes a Neumann boundary. The line of symmetry between two adjacent slots also is a Neumann boundary. We may take a line far below the slot as a flux line where $A$ is constant. This region is divided into a finite difference mesh corresponding to which we write out the analogous network model equations. The currents may be treated as coming from the four quadrants as we did in section 4.3.3; but this time we must also expand the impedances that are shown for some typical points in Figure 4.3.4B, taking the relative permeability in the steel to be 10. The solution corresponding to this was plotted and is shown in Figure 4.3.4C.

An alternative example is in finding the linkage of the conductor in the slot when excited by the flux from the rotor. In this case, there will be no current in the conductor, but the conductor with its permeability (in this case $\mu_0$) has to be part of the model. The exciting flux is imposed by imposing different values of vector

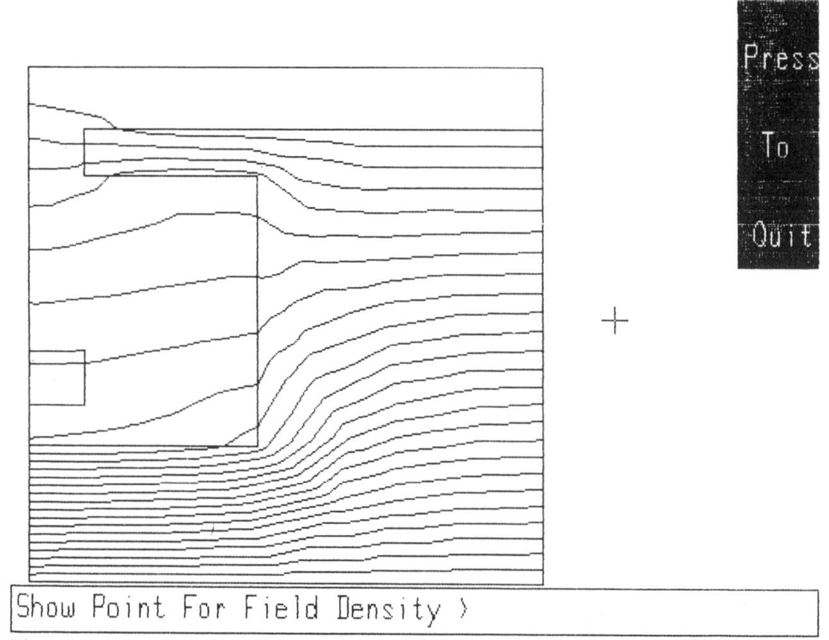


**Figure 4.3.5.** Flux linking a conductor in a slot.

potential $A$ along the rotor surface and the bottom of the slot. The resulting flow of flux is shown in Figure 4.3.5.

# 4.4 INDUCTANCE COMPUTATION

## 4.4.1 Flux Linkage, Inductance, and Vector Potential

We have seen in section 4.2.3 how to compute the mutual inductance between two coils using Neumann's formula, and we encountered some difficulties in computing the self-inductance of a coil. Another limitation of Neumann's method is that it is valid only when the coils are in a uniform medium. In this section we shall encounter a more general method for the computation of inductance, a method that does not suffer from these limitations.

In section 4.1.4 we saw that an alternative definition of inductance is

$$V = L\frac{dI}{dt}. \tag{4.1.25}$$

Combining this with Faraday's law for the voltage induced in a coil of $N$ turns with flux $\psi$ crossing its surface,

$$V = N\frac{d\psi}{dt}, \tag{4.4.1}$$

we have yet another definition of inductance, which we shall use with vector potential formulations:

$$L = \frac{N\psi}{I}. \tag{4.4.2}$$

Thus for a given value of current, if we can determine the flux $\psi$, then we may determine the inductance.

Significantly, with vector potential solutions in two dimensions, the flux is very easy to calculate. To see this, refer to Figure 4.4.1, where two equipotential (or flux) contours at potentials $A_a$ and $A_b$ are drawn and a line $AB$ is drawn in the $x$ direction from one contour to the other. The flux $\psi$ crossing the line $AB$ per unit depth is the flux density times the elemental area integrated from $A$ to $B$. Therefore, using eqn. (1.2.34),

$$\psi = \int_A^B B_y\, dx = \int_A^B -\frac{\partial A}{\partial x}\, dx = -\int_A^B dA = A_a - A_b. \tag{4.4.3}$$

That is, the flux between two points is merely the difference in the vector potential at those two points. We observe from eqn. (1.2.28)

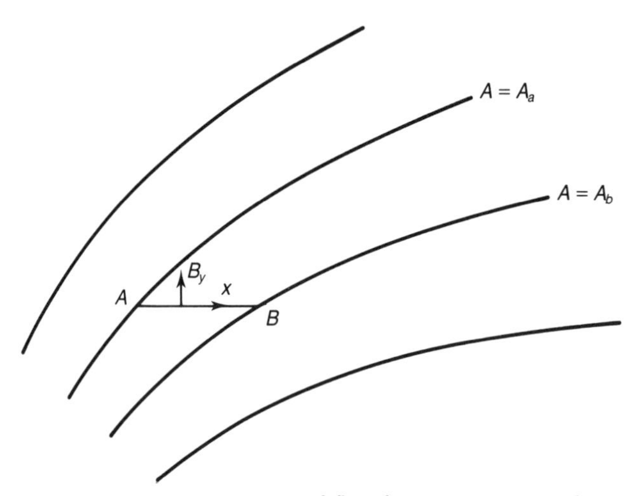


**Figure 4.4.1.** Integration of flux from equipotentials.

that if $\mu$ is constant, the vector potential difference between any two points is directly proportional to the current density $J$ and therefore to the current $I$, making it a linear system. Consequently, the ratio $\psi$ to $I$ in eqn. (4.4.2) is a physical constant independent of the magnetic field, as expected.

Therefore, if we wish to compute the inductance of a coil, we may solve for the vector potential caused by the source current $I$, find the flux $\psi$, and thence the inductance $L$.

A similar result for the flux through a coil applies in three dimensions where the vector potential has three components:

$$\psi = \iint \mathbf{B} \cdot d\mathbf{S} = \iint \nabla \times \mathbf{A} \cdot d\mathbf{S} = \int \mathbf{A} \cdot d\mathbf{C}, \qquad (4.4.4)$$

by the application of Stokes's theorem (A14) to convert the integral over the cross section of the coil to one over the perimeter $C$ of the coil. We therefore perceive eqn. (4.4.3) as a special case of eqn. (4.4.4), since in two dimensions, the integral over the perimeter of the coil is the same as integrating over an infinitely long wire going into the paper at $A$ and returning out at $B$ in Figure 4.4.1.

We may further conclude that when we draw equipotentials between which the potential drop is the same, the same amount of magnetic flux will be flowing between the equipotentials. As a result, when the lines are crowded, the same quantity of flux is flowing through a smaller cross sectional area, so that the flux density has to be high to permit that.

### 4.4.2 The Self- and Leakage Inductances of a Single Phase Transformer

As a demonstration of inductance computations and yet another illustration of the finite difference method, let us consider the single phase transformer shown in Figure 4.4.2, for which we shall calculate the self- and leakage inductances on the primary side. Although the configuration is three-dimensional, along the plane of symmetry on which the figure lies no changes may occur in the normal direction so that along it the problem is two-dimensional (please see the discussion in section 3.3.1, which shows this to be exact in view of the uniform depth of the transformer in the $z$ direction). The exciting primary coil is shown as two rectangles on either side of the left vertical limb. Since the coil goes in on one side of the limb and comes out on the other, the current density in each rectangle will be the total current (current times the number of

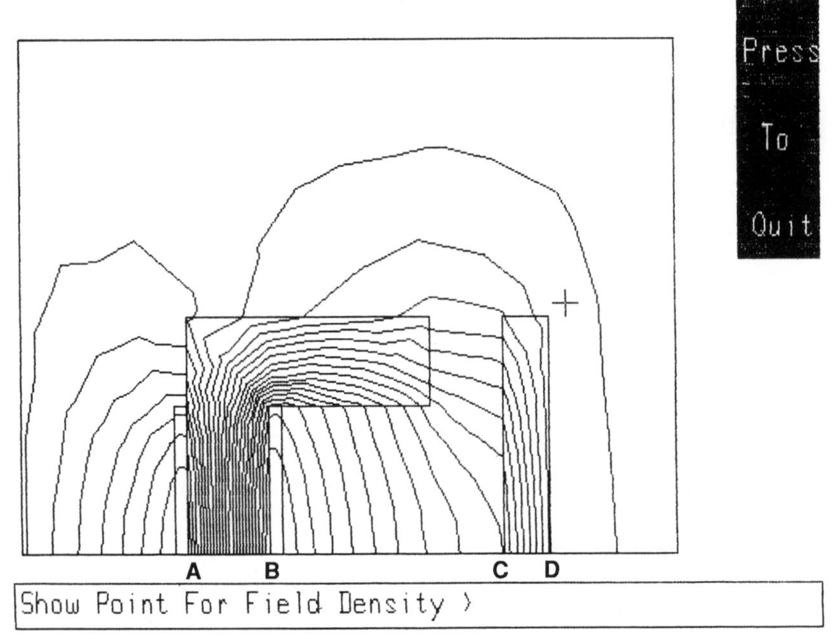


**Figure 4.4.2.** Flux patterns in a single-phase transformer.

turns) divided by the area occupied by the coil. Observe that this requires the signs of $J$ in the two rectangles to be opposite each other. In an ideal transformer there should be no leakage; that is, the flux through the primary coil should flow entirely through the steel to the vertical limb on the other side so as to link the secondary coil not shown in Figure 4.4.2. The amount of leakage determines the leakage inductance of the transformer equivalent circuit, which is required to predict the performance of the machine (McPherson 1981).

The boundary value problem may easily be solved by the finite difference method using the network models already illustrated. The solution corresponding to a relative permeability of 10 in the steel is what is presented in Figure 4.4.2. The difference in vector potential between points $A$ and $B$ in the figure will give the total flux produced by the primary coil, and that may be used in eqn. (4.4.2) to determine the self-inductance. The flux between points $C$ and $D$ is the useful flux that links the secondary coil. The difference between the flux produced $(A_a - A_b)$ and the useful flux $(A_c - A_d)$ is the leakage flux that flows through the air and causes a voltage drop in the transformer. The leakage inductance then may be computed from eqn. (4.4.2).

## 4.5 MATERIAL-BASED NONLINEARITY

### 4.5.1 Magnetic Saturation

A special characteristic of many magnetic materials is nonlinearity. While for low magnetic field strengths the flux density is proportional to the field strength $H$ according to eqn. (1.2.7), higher field strengths produce fewer and fewer increments in flux density. Indeed, flux densities more than 3 T are rarely encountered and a value more than 11 T is very difficult to achieve even with special materials and equipment. This saturation of $B$ with increasing $H$ is shown in Figure 4.5.1, and, correspondingly, the permeability $\mu$ drops with $H$. A more exact representation will involve the hysteresis effect, for which the reader is referred to more advanced literature on the subject (Mayergoyz 1970, 1986; Saito et al. 1983).

In the previous derivations of the magnetic field, we took $\mu$ to be a known constant. Here, however, we have a coupled problem. If we know $\mu$, we know how to compute $B$; if we know $B$ everywhere,

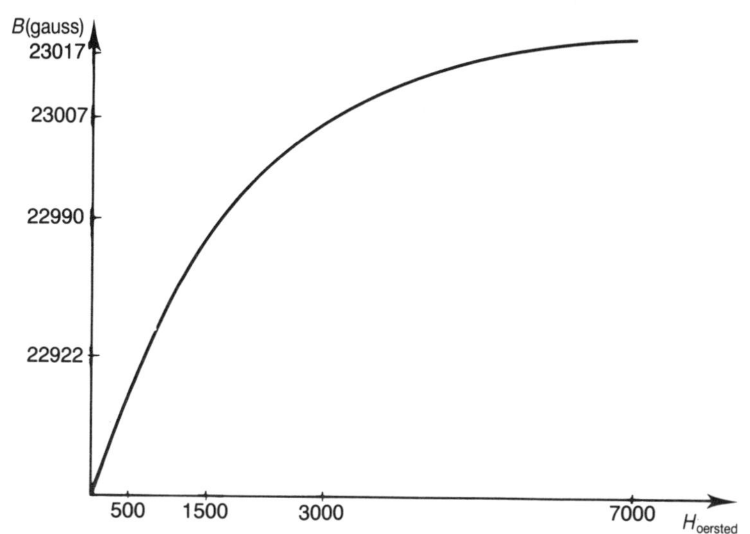


**Figure 4.5.1.** A typical $B$–$H$ curve (AlNiCo steel; 1 At/m = $4\pi \times 10^{-3}$ oersted; 1 tesla = 10,000 gauss).

then we know how to obtain $\mu$ from the material characteristics of the material (such as of Figure 4.5.1), which may be obtained in the laboratory using a sample of the material. In this section we shall see how to obtain the flux density and permeability in a device from the material characteristics and current distributions.

### 4.5.2 Modeling the *B–H* Curve

While the material characteristics such as in Figure 4.5.1 may be obtained in the laboratory, for digital computation we need the characteristics as an equation relating $B$ and $H$, rather than as a graph. As we shall see, the chord method simply requires a value of $\mu$ for any given value of $B$. The Newton method requires the reluctivity

$$\nu = \mu^{-1}, \tag{4.5.1}$$

and its derivative

$$\kappa = \frac{d\nu}{dB^2}, \tag{4.5.2}$$

for any given value of $B$. Moreover, the Newton scheme requires the derivative of $\kappa$ to be a continuous curve for convergence (Chari 1970).

A crude and simple model for the chord method is obtained by interpolation. The chief disadvantage here is that the smoothing effect in drawing the graph that gives a better interpolation between data points is not reflected in the model, which really uses straight-line connections between points; as a result, the first derivative required for the Newton solution of the nonlinear equations is lost. This model consists of a series of, say, $k$ data points from the curve from left to right $(0,0)$, $(B_1, H_1)$, $(B_2, H_2), \ldots, (B_k, H_k)$. Naturally, the more points we have, particularly close to the knee of the curve, the more accurate will our model be. If the point $B$ at which we want the permeability is between data points $i$ and $i+1$, we obtain the corresponding $H$ by interpolation:

$$H = H_i + (H_{i+1} - H_i)\frac{B - B_i}{B_{i+1} - B_i}; \tag{4.5.3}$$

should $B$ be beyond the last data point, we may extrapolate from the last two data points $k$ and $k-1$:

$$H = H_{k-1} + (H_k - H_{k-1})\frac{B - B_{k-1}}{B_k - B_{k-1}}. \tag{4.5.4}$$

Once the value of $H$ corresponding to a given flux density $B$ is computed, $\mu$ is given by the ratio $B:H$.

The smoothing effect may be incorporated by curve fitting. For example, let us assume that:

$$B = x_1 + x_2H + x_3H^2 + x_4H^3. \tag{4.5.5}$$

Of course, by taking higher order terms in $H$, we may improve the model. From the data points, we need to find the unknown coefficients $x_1$, $x_2$, $x_3$, and $x_4$. The error at data point $i$ is

$$e_i = B_i - B = B_i - (x_1 + x_2H_i + x_3H_i^2 + x_4H_i^3). \tag{4.5.6}$$

The sum of the squares of the errors at the node points is then

$$e^2 = \sum_{i=1}^{k} [B_i^2 - 2B_i(x_1 + x_2H_i + x_3H_i^2 + x_4H_i^3) + (x_1 + x_2H_i + x_3H_i^2 + x_4H_i^3)^2]. \tag{4.5.7}$$

Curve fitting then consists of adjusting the coefficients $x_j$, $j = 1, \ldots, 4$ so as to have least mean square error:

$$\frac{\partial e^2}{\partial x_1} = \sum_{i=1}^{k} [-2B_i + 2(x_1 + x_2H_i + x_3H_i^2 + x_4H_i^3)] = 0, \tag{4.5.8a}$$

$$\frac{\partial e^2}{\partial x_2} = \sum_{i=1}^{k} [-2B_iH_i + 2H_i(x_1 + x_2H_i + x_3H_i^2 + x_4H_i^3)] = 0, \tag{4.5.9a}$$

$$\frac{\partial e^2}{\partial x_3} = \sum_{i=1}^{k} [-2B_iH_i^2 + 2H_i^2(x_1 + x_2H_i + x_3H_i^2 + x_4H_i^3)] = 0, \tag{4.5.10a}$$

$$\frac{\partial e^2}{\partial x_4} = \sum_{i=1}^{k} [-2B_iH_i^3 + 2H_i^3 + 2H_i^3(x_1 + x_2H_i + x_3H_i^2 + x_4H_i^3)] = 0. \tag{4.5.11a}$$

Rearranging, we have

$$\sum_{i=1}^{k} [x_1 + x_2H_i + x_3H_i^2 + x_4H_i^3] = \sum_{i=1}^{k} B_i, \tag{4.5.8b}$$

$$\sum_{i=1}^{k} [H_i(x_1 + x_2H_i + x_3H_i^2 + x_4H_i^3)] = \sum_{i=1}^{k} B_iH_i, \tag{4.5.9b}$$

$$\sum_{i=1}^{k} [H_i^2(x_1 + x_2H_i + x_3H_i^2 + x_4H_i^3)] = \sum_{i=1}^{k} B_iH_i^2, \tag{4.5.10b}$$

$$\sum_{i=1}^{k} [H_i^3(x_1 + x_2H_i + x_3H_i^2 + x_4H_i^3)] = \sum_{i=1}^{k} B_iH_i^3. \tag{4.5.11b}$$

Since the terms $H_i$ and $B_i$ are known for all values $i = 1, \ldots, k$ at the data points, eqns. (4.5.8b)–(4.5.11b) are linear in the unknowns $x_j$.

They may be assembled into a matrix equation and solved by Alg. 2.3.1 or Alg. 3.2.1. Observe that the resulting matrix equation is positive definite and symmetric. For example, calling the coefficient matrix $S$, from eqn. (4.5.8), $S_{12}$ is $\sum H_i$; from eqn. (4.5.9), $S_{21}$ is also the same. Positive definiteness arises from our minimizing the square of the error, which is no smaller than zero. As a result, the matrix equation may be efficiently solved by the Cholesky factorization scheme of Alg. 6.2.1 to be described in chapter 6.

It is seen that we may easily generalize eqns. (4.5.8b)–(4.5.11b) to allow for higher-order terms of $H$ in eqn. (4.5.5). One word of caution, though, is in order. Note from eqn. (4.5.8b) that $S_{11}$ is $\sum 1 = k$. On the other hand, from eqn. (4.5.11b) the coefficient $S_{44}$ is $\sum H_i^6$. And since $H_i$, from Figure 4.5.1, takes values from 0 to several thousand ampere turns, the diagonal terms of the matrix will grow from $k$ to very large numbers, as a result of which numerical instability will be encountered. This disability grows if we try to include more terms in eqn. (3.5.5). For instance, if we include the term $x_5H^4$, the fifth diagonal term will be $\sum H_i^8$. To overcome this disadvantage, we may change variables to

$$H^* = \frac{H}{C}, \tag{4.5.12}$$

where $C$ is 1000 or any other suitable number. As a result of this mapping, the stability of the coefficient solution is enhanced. And then to find $B$ in terms of $H$, we have

$$B = x_1 + x_2H^* + x_3H^{*2} + x_4H^{*3} = x_1 + x_2CH + x_3C^2H^2 + x_4C^3H^3. \tag{4.5.13}$$

The reluctivity then follows from

$$\nu = \frac{H}{B} = \frac{H}{(x_1 + x_2CH + x_3C^2H^2 + x_4C^3H^3)}. \tag{4.5.14}$$

If required, the term $\kappa$ is found from:

$$\kappa = \frac{d\nu}{dB^2} = \frac{d\nu/dH}{dB^2/dH}. \tag{4.5.15}$$

A superior procedure that skirts these difficulties is provided by Silvester, Cabayan, and Browne (1973), who directly model $\nu$ as a function of $B^2$ using cubic fits. Under their scheme, the interval over which $\nu$ is to be modeled as a function of $B^2$ is divided into $k$ subintervals, at the ends of each of which $\nu$ and its derivative with respect to $B^2$ are required from the material characteristics of Figure

4.5.1. To obtain $\nu$ and $\kappa$ at any flux density $B$ lying in an interval $i$ with end points $i1$ and $i2$, we use the new variable $x$ ranging from 0 to 1 over the interval,

$$x = \frac{B^2 - B_{i1}^2}{B_{i2}^2 - B_{i1}^2}, \tag{4.5.16}$$

and compute $\nu$ and $\kappa$ from

$$\begin{aligned}\nu(x) = {} & (2x^3 - 3x^2 + 1)\nu_{i1} + (-2x^3 + 3x^2)\nu_{i2} \\ & + (x^3 - 2x^2 + x)\kappa_{i1} + (x^3 - x^2)\kappa_{i2},\end{aligned} \tag{4.5.17}$$

$$\begin{aligned}\kappa(x) = {} & (6x^2 - 6x)\nu_{i1} + (-6x^2 + 6x)\nu_{i2} \\ & + (3x^2 - 4x + 1)\kappa_{i1} + (3x^2 - 2x)\kappa_{i2}.\end{aligned} \tag{4.5.18}$$

Be it noted that at $x = 0$ and $x = 1$, $\nu$ and $\kappa$, respectively, become $\nu_{i1}$ and $\kappa_{i1}$ and $\nu_{i2}$ and $\kappa_{i2}$, thereby providing the continuity of $\nu$ and $\kappa$ over the intervals.

The only difficulty here is in obtaining the values of $\kappa$ at the interval ends from the graphs, using graphical differentiation—a process very much prone to error. Silvester et al. (1973) have overcome this by writing a program to compute the gradients iteratively from input data consisting of the $B$–$H$ curve and values of $H$ at seven equally spaced values of $B$ from the origin. From these values, $\nu$ at the seven points may be computed. The principle behind the computation is that the slope of $\kappa$ with respect to $B^2$ ought to be the same at the interval ends:

$$\frac{d\kappa}{dB^2} = (12x - 6)\nu_{i1} + (-12x + 6)\nu_{i2} + (6x - 4)\kappa_{i1} + (6x - 2)\kappa_{i2} \tag{4.5.19}$$

from eqn. (4.5.18). Therefore, for two adjacent intervals from node $i - 1$ to node $i$ and from node $i$ to node $i + 1$, we have, using eqn. (4.5.19) to evaluate the gradient at the right end at $x = 1$:

$$\left[\frac{d\kappa}{dB^2}\right]_i = 6\nu_{i-1} - 6\nu_i + 2\kappa_{i-1} + 4\kappa_i. \tag{4.5.20}$$

Similarly, evaluating the quantity at node $i$ at $x = 0$ in the second interval:

$$\left[\frac{d\kappa}{dB^2}\right]_i = -6\nu_i + 6\nu_{i+1} - 4\kappa_i - 2\kappa_{i+1}; \tag{4.5.21}$$

equating the gradients of $\kappa$ as required and dividing by 2:

$$\kappa_{i-1} + 4\kappa_i + \kappa_{i+1} = 3\nu_{i+1} - 3\nu_{i-1}. \tag{4.5.22}$$

Thus, for any node with an interval to its left and right, we may use eqn. (4.5.22) to obtain a corresponding equation. However, the leftmost and rightmost nodes require special treatment, since they respectively do not have intervals to their left and right. For the first (or leftmost) node at $B = 0$, we may use the fact that $\nu$ and, therefore, its derivatives, are practically constant at low flux density; that is, $\kappa$ is zero and so does not need to be computed. For the rightmost point, we may not validly make any such assumption; thus, we estimate the slope $\kappa$ there so that in eqn. (4.5.22), when $i$ refers to the point before last, $\kappa_{i+1}$ is known from our estimate. Therefore, eqn. (4.5.22) may be built into a symmetric matrix and solved for $\kappa$ at all the interior points.

However, our computations are based on an estimate of $\kappa$ at the rightmost point and may have some inherent error. It is for this reason that we use iterations. From the computed values, the $B$–$H$ curve is plotted and compared with the original by first computing $\nu$ using eqn. (4.5.17). If a match is not obtained, we scientifically fiddle with the guess of $\kappa$ at the rightmost point and keep repeating this process until a match of the curves is obtained.

A simpler, but less accurate model of $\nu$ versus $B^2$ characteristics has been proposed by Brauer (1975), who uses

$$\nu = k_1 \exp(k_2 B^2) + k_3, \tag{4.5.23}$$

which fits the curve for $\nu$ as a function of $B^2$ fairly well, provided three points from the $B$–$H$ curve, from the linear region and below and above the knee, respectively, are used. Moreover, this expression allows us to use $k_1 k_2 \exp(k_2 B^2)$ as the first derivative $d\nu/dB^2$, so that the finite element Newton iterations may be easily implemented. Although this model of $\nu$ is not as accurate as the third-order spline technique described above, it is much easier to employ, has fine convergence properties, and nearly always has good, unique answers. The only exception to this is when high current sources are involved; with these, the low linear values of $\nu$ at the beginning of the Newton iterations, result in high $B$'s, which make $\nu$ very high for the subsequent iteration, in view of the exponential nature of eqn. (4.5.23). This may be overcome by combining the chord method of dealing with nonlinearity, if and when this happens, at the expense of prolonging the iterations.

A Newton algorithm for finding the coefficients $k_1$, $k_2$, and $k_3$ has been provided by Hoole and Hoole (1986b). Supposing the three points on the $B$–$H$ curve for a material, $B_i$, $H_i$, $i = 1, 2, 3$, are given, then we may easily compute $\nu_i = H_i/B_i$ and $B_i^2$. Now, since the first point is in the linear region, at $B = 0$ in (4.5.23),

$$\nu_1 = k_1 + k_3, \tag{4.5.24}$$

and for the other two points in the nonlinear region,

$$\nu_2 = k_1 e^{k_2 B_2^2} + k_3, \tag{4.5.25}$$

$$\nu_3 = k_1 e^{k_2 B_3^2} + k_3. \tag{4.5.26}$$

These three equations (4.5.24)–(4.5.26) must be solved for the constants $k_1$, $k_2$, and $k_3$. Substituting for $k_3$ using eqn. (4.5.24) in eqns. (4.5.25) and (4.5.26), we have:

$$\nu_2 - \nu_1 = k_1 e^{k_2 B_2^2} - k_1, \tag{4.5.27}$$

$$\nu_3 - \nu_1 = k_1 e^{k_2 B_3^2} - k_1, \tag{4.5.28}$$

and dividing the one by the other, we have:

$$f(k_2) = \frac{e^{k_2 B_2^2} - 1}{e^{k_2 B_3^2} - 1} - \frac{v_2 - v_1}{v_3 - v_1} = 0. \tag{4.5.29}$$

As described in section 3.6.3, to solve this nonlinear equation for $k_2$ by Newton's scheme, we must find the derivative of $f$:

$$\frac{d}{dk_2} f(k_2) = \frac{(e^{k_2 B_3^2} - 1)B_2^2 e^{k_2 B_2^2} - (e^{k_2 B_2^2} - 1)B_3^2 e^{k_2 B_3^2}}{(e^{k_2 B_3^2} - 1)^2}. \tag{4.5.30}$$

Now if $k_2$ is an approximation satisfying $f(k_2) \neq 0$ as in eqn. (4.5.29), with a small error $\Delta k_2$, when the exact solution $k_2 + \Delta k_2$ is put into eqn. (4.5.29) using a Taylor's series expansion, we get, corresponding to eqn. (3.6.11):

$$f(k_2 + \Delta k_2) = f(k_2) + \Delta k_2 \frac{d}{dk_2} f(k_2) = 0. \tag{4.5.31}$$

Since we have, in eqns. (4.5.29) and (4.5.30), expressions to evaluate all the terms but $\Delta k_2$ of eqn. (9), starting with a nonzero guess of $k_2$, $\Delta k_2$ may be computed and $k_2$ successively improved by

$$k_2 = k_2 - \frac{f(k_2)}{df(k_2)/dk_2}, \tag{4.5.32}$$

until $k_2$ changes negligibly. Once $k_2$ is determined, both $k_1$ and $k_3$ may be trivially computed using eqns. (4.5.24) and (4.5.25) or (4.5.26).

The above algorithm may be written into a program to read the three points of any $B-H$ curve from a computer terminal, compute the three coefficients, and write them in a file; this file, in turn, may be read by a nonlinear finite element analysis program, as a result of which no truncation will be inherent to the transfer from one program to the other. Newton's algorithm being quadratically convergent, procedure (4.5.32) takes no more than six or seven iterations to reach the accuracy limits of the computer. For example, consider the $B-H$ values (in $T$ and $At$) of (0.5, 200), (1.3, 709), and 1.65, 2953) fitted by Brauer (1975). Here, for a $B-H$ curve passing through the origin, the first coordinate in the linear region gives us $v_1 = dH/dB = 200/0.5 = 400$, at zero flux, so that the closer the first pair of points is taken to the origin, the more accurate our reluctivity model will be. The answers for $k_1$, $k_2$, and $k_3$ worked out by trial and error by Brauer are 3.8, 2.17, and 396.2, respectively, while those by the Newton method described above, truncated to six digits, are 3.85281, 2.16375, and 396.147. The latter were obtained in five iterations, starting at $k_2 = 1.0$, and stopping when $\Delta k_2/k_2 < 1.e-6$. It is clearly seen that the differences, particularly in $k_2$ at high saturation levels, will make a significant difference to the answers by the finite element program.

Finally, a note of caution is added against the temptation to simplify matters by throwing away the denominator of eqn. (4.5.29) by multiplying throughout by the lowest common denominator; for then, the resulting equation will have multiple solutions for $k_2$ including $k_2 = 0.0$—and the consequent Newton method was always found to converge towards the value of 0.0 for $k_2$.

### 4.5.3 Modeling Permanent Magnets

The $B-H$ characteristics of a permanent magnet are as shown in Figure 4.5.2A. Permanent magnets are increasingly used in devices because of the advances being made in materials technology (Campbell 1986). We address here the problem of determining the magnetic field in a system containing such a magnet. The field strength $H_c$ is developed across the length $L$ of the magnet. This is the equivalent of a magnetomotive force $H_cL$ in the direction of a magnet. Therefore, as done by Fouad, Nehl, and Demerdash (1981), we may replace the magnet by a magnetizable part having the $B-H$ curve of Figure 4.5.2B (which is the curve of Figure 4.5.2A offset to the right by $H_c$) and driven by a current coil of ampere turns ($NI$) equal to $H_cL$.

From this point onwards, the analysis is similar to a regular magnetic field problem.

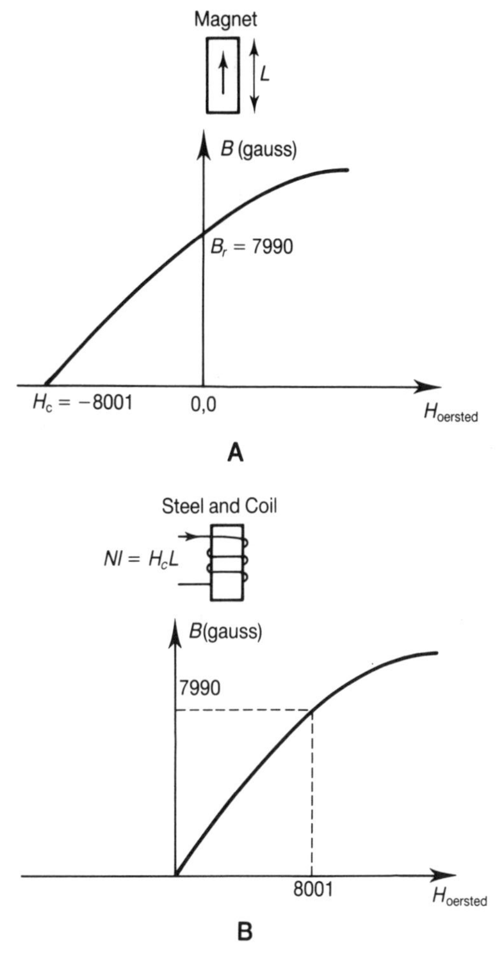


**Figure 4.5.2.** Analysis of a permanent magnet. **A.** Permanent magnet. **B.** Equivalent steel and coil. 1 At/m = $4\pi \times 10^{-3}$ oersted; 1 tesla = 10,000 gauss.

## 4.6 THE ELECTRIC MACHINE SLOT IN NONLINEAR IRON: SOLUTION BY THE CHORD METHOD

Let us take up the solution of the magnetic field in an electric machine slot, as in sections 4.3.3 and 4.3.4. This time, however, let the iron in which the slot lies be nonlinear. We shall further assume that we have already modeled the iron by curve fitting and that the field strength is given by

$$H = 1451B - 178B^3 - 270B^5 + 280B^7. \tag{4.6.1}$$

If we are to solve this by the chord method of section 3.6.2, the algorithm corresponding to eqn. (3.6.1) becomes

$$-\nu(A^{i-1})\nabla^2 A^i = J, \tag{4.6.2}$$

since the current density $J$ remains constant. At the start, corresponding to iteration $i = 0$, we may take $A = 0$ everywhere so that $B$ is zero and eqn. (4.6.1) gives $H = 0$, from which we cannot compute $\nu$. At low values of $B$, therefore, we use the fact that $\nu = dH/dB$, which is 1451 for the model of eqn. (4.6.1). After computing $A^i$, we compute $B^i$ from

$$\mathbf{B} = \mathbf{u}_x \frac{\partial A}{\partial y} - \mathbf{u}_y \frac{\partial A}{\partial x}. \tag{1.2.34}$$

However, in the network models we are using for node $A$ of Figure 4.6.1, to compute the impedances, the value of $\mu$ is required on either side of a finite difference edge. Therefore, to compute the impedance $Z$, for example, we need the flux density $B$ along the edge $AD$, but slightly to its right, so that from $B$ we may find $\nu$ and

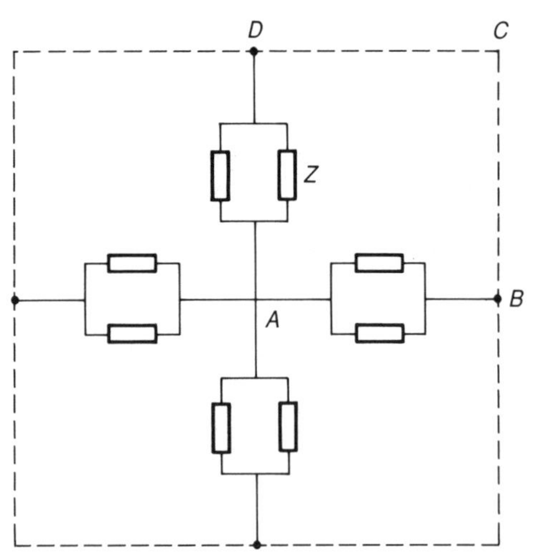


**Figure 4.6.1.** Computation of flux density for nonlinear networks.

thence Z. We have:

$$\frac{\partial A}{\partial x} = \frac{1}{2}\left[\frac{A_b - A_a}{h} + \frac{A_c - A_d}{h}\right], \tag{4.6.3}$$

where we have averaged the derivatives at $A$ and $D$. We also have

$$\frac{\partial A}{\partial y} = \frac{A_d - A_a}{k}. \tag{4.6.4}$$

Observe that the $y$ derivative is the same to the right and left of $AD$ as expected, because this is $B_x$ as seen from eqn. (1.2.34)—if it is a material interface, eqn. (1.2.17) requires the normal flux density to be continuous as reflected in eqn. (4.6.4). Similarly, a little above the horizontal edge $AB$,

$$\frac{\partial A}{\partial x} = \frac{A_b - A_a}{h}, \tag{4.6.5}$$

and

$$\frac{\partial A}{\partial y} = \frac{1}{2}\left[\frac{A_d - A_a}{k} + \frac{A_c - A_b}{k}\right]. \tag{4.6.6}$$

Similar expressions trivially follow for derivatives to the left of vertical branches and below horizontal branches. Once the derivatives are computed, we have for the magnitude of flux density

$$B = \sqrt{[B_x^2 + B_y^2]}. \tag{4.6.7}$$

Once this is put into our reluctivity model, we obtain the required value of $\nu$. We have already seen, however, that unless a deceleration is used, the chord method does not necessarily converge. So before we compute $\nu$, we retard the change in $B$ as we did in eqn. (3.6.7):

$$B^i = B^{i-1} + 0.1(B^i - B^{i-1}), \tag{4.6.8}$$

and then compute the value of $\nu(A^i)$. Now we may go back and compute the vector potential values $A^{i+1}$ of the next iteration. This procedure may be repeated until $A$ does not change more than by what is acceptable from iteration to iteration. A point of caution in making up trial problems is in order. When we arbitrarily set up a problem for testing the convergence of these schemes, we may well take current density values that would result in heavy saturation. As a result, we may be operating numerically in situations that never really obtain in practice, and convergence may not immediately result. In fact, high currents may carry the flux density outside the limits for which the material characteristics were modeled, and this is commonly the cause of nonconvergence. The opposite is also possible. We may set current values so low as not to saturate the iron and, so convergence may result in just a few iterations. In practical problems, the current levels result in fluxes close to the knee of the $B$–$H$ curve so as to make best use of the iron; that is, the quantity of iron is not so large that it is not properly utilized nor is it so small that it cannot properly carry the required fluxes and so saturates. In such problems, the number of iterations for convergence is usually upwards of 50 and can grow to as many as 10,000 for large, complex problems.

The slot problem of section 4.3.4 may be solved by this scheme. By gradually increasing the current, it will be discovered that the flux increasingly tends to flow out of the iron into the air because the iron saturates and offers greater reluctance to the flux.

# 5 TIME VARIATION: ELECTROMAGNETIC FIELDS

## 5.1 THE COUPLING OF ELECTRIC AND MAGNETIC FIELDS

In magnetostatics, an electric field causes a current and the current causes a magnetic field. However, the resulting magnetic field in no way has a feedback effect on the first cause—the electric field. Once time variation is introduced, as in transient and even steady AC systems, the resulting magnetic field does affect the initial electric field via

$$\nabla \times \mathbf{E} = -\frac{\partial \mathbf{B}}{\partial t}. \tag{1.2.6}$$

Similarly, time variation also makes the electric field affect the magnetic field with the extra displacement current term:

$$\nabla \times \mathbf{H} = \mathbf{J} + \frac{\partial \mathbf{D}}{\partial t}. \tag{1.2.4}$$

And this brings us to the subject of electromagnetics.

## 5.2 LOW-FREQUENCY EFFECTS: THE DIFFUSION OR EDDY CURRENT PROBLEM

### 5.2.1 The Diffusion Equation and Phasor Representation

Eddy currents (Chari 1974; Konrad 1985a), in low-frequency systems where we neglect the displacement current, are currents

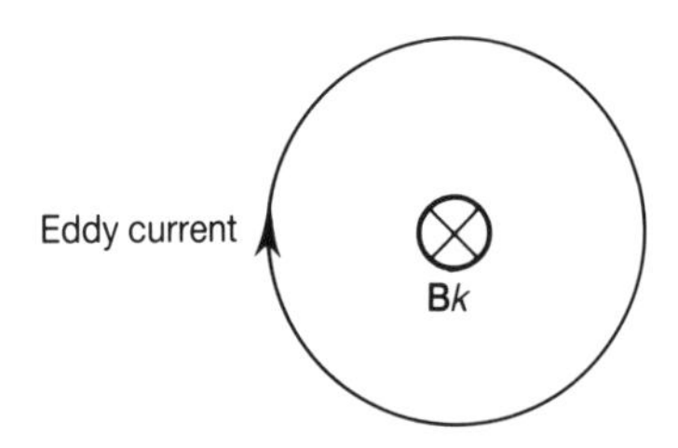


**Figure 5.2.1.** Interaction of electricity and magnetism.

that circulate around the flowing magnetic flux, as a result of the induction defined in eqn. (1.2.6). For example, as flux flows into the paper in Figure 5.2.1 and changes in time, for any contour surrounding the flux

$$V = -\iint \frac{\partial}{\partial t}[\mathbf{B} \cdot \mathbf{dS}] \tag{5.2.1}$$

according to Faraday's law, in which eqn. (1.2.6) has its origin. This shows that a potential difference develops in loops around the flux and so long as the medium has a nonzero conductivity, additional currents will be set up, modifying the current and flux patterns. The circulating nature of these currents confers on them the term *eddy currents.* It will be gathered, roughly, that as the loop is increased in size, the flux crossing the surface will increase as $r^2$, whereas the length of the loop over which the voltage $V$ is dropped goes up only as $r$; consequently, with the electric field increasing as we go out from the middle, the eddy current effect will be stronger in the outer periphery of a conductor carrying flux. We shall show that these eddy currents push the current towards the surface of conductors, making them behave as though they were hollow, and unable to carry any currents in the interior. Hence the alternative term, the *skin effect.*

The skin effect is exploited in power transmission lines. Copper at the transmission frequency of 60 Hz usually has a depth of penetration of about 6 mm, so that the skin effect becomes pronounced when the device dimensions are greater than half an inch, as in power lines. Therefore, in recognition of the fact that no current flows in the middle of a large conductor, hollow conductors are put to use with great economy; only the outer half an inch or so of the conductor being occupied by metal. Alternatively, the conductor is woven from strands of thin wire insulated from each other—each strand being thin, no skin effect is felt in it.

Let us now try to determine the rules that govern eddy currents.

The nondivergent magnetic flux density **B** may be modelled by a vector potential $A$ as in eqn. (1.2.24). Putting this into eqn. (1.2.6), we have:

$$\nabla \times \mathbf{E} = -\frac{\partial \mathbf{B}}{\partial t} = -\frac{\partial \nabla}{\partial t} \times \mathbf{A} = \nabla \times -\frac{\partial \mathbf{A}}{\partial t}. \tag{5.2.2}$$

The two vectors **E** and $-\partial \mathbf{A}/\partial t$ have the same curl, which is possible only if:

$$\mathbf{E} = -\frac{\partial \mathbf{A}}{\partial t} - \nabla \phi, \tag{5.2.3}$$

since, when we take a curl on both sides, $\nabla \times \nabla \phi$ will disappear according to (A4). It is readily realized that the term $-\nabla \phi$ is the imposed electric field as obtains in static systems with no time variation, and $-\partial \mathbf{A}/\partial t$ is the induced electric field. Putting this into eqn. (1.2.4) using $\mathbf{J} = \sigma \mathbf{E}$ and $\mathbf{H} = \mu^{-1}\mathbf{B}$, we have

$$\nabla \times \mu^{-1} \nabla \times \mathbf{A} = -\sigma \nabla \phi - \sigma \frac{\partial \mathbf{A}}{\partial t} \tag{5.2.4}$$

Setting

$$\mathbf{J}_0 = -\sigma \nabla \phi, \tag{5.2.5}$$

the imposed current density, and using (A9) with the gauge eqn. (1.2.26),

$$-\mu^{-1}\nabla^2 \mathbf{A} = \mathbf{J}_0 - \sigma \frac{\partial \mathbf{A}}{\partial t}. \tag{5.2.6}$$

In two dimensions, we have seen in section 1.2 that the vector potential and current density have only one component $\mathbf{u}_z A$ and $\mathbf{u}_z J_0$, so that:

$$-\mu^{-1}\nabla^2 A = J_0 - \sigma \frac{\partial A}{\partial t}. \tag{5.2.7}$$

The solution of this equation for $A$ is rather complicated and is attempted by writing, at time $t^i$,

$$\frac{\partial A}{\partial t} = \frac{A^i - A^{i-1}}{\Delta t^i}, \tag{5.2.8}$$

giving us

$$-\mu^{-1}\nabla^2 A^i = \frac{J_0^i - \sigma[A^i - A^{i-1}]}{\Delta t^i}, \tag{5.2.9}$$

so that at a finite difference node 0 surrounded by nodes 1, 2, 3, and 4 in a uniform $h \times h$ mesh, we have:

$$-\mu^{-1}h^{-2}[A_1^i + A_2^i + A_3^i + A_4^i - 4A_0^i] = J_{00}^i - \frac{\sigma[A_0^i - A_0^{i-1}]}{\Delta t^i}, \quad (5.2.10)$$

$J_0$ being a given function of time such as $\sin \omega t$, $J_0^i$ may be evaluated at any time $t^i$. That is, given $A$ everywhere at a particular time, $\Delta t$ seconds later $A$ may be determined. The difficulties in this scheme are finding a starting solution for the time stepping algorithm, and the build-up of the difference error in eqn. (5.2.10) as time progresses. The texts by Stoll (1974) and Reddy (1984) may be consulted for the alternative Crank–Nicolson scheme, which reduces this difference error.

In a number of linear, steady-state eddy current problems driven by AC supplies, by switching to complex phasors we may achieve a significant simplification. Observe that under nonlinear conditions, although the exciting function $J_0$ is sinusoidal, the resulting vector potential $A$ is not and we therefore cannot use complex analysis. Hoole and Carpenter (1985) give a scheme for time stepping under saturation, and Hoole and Shin (1987) give an algorithm for avoiding the time stepping error. Mayergoyz, Abdel-Kader, and Emad (1984) give analytical schemes for dealing with these problems.

An AC quantity $A$ with, say, cosine variations,

$$A = A^m \cos(\omega t + \theta), \quad (5.2.11)$$

may be represented by the phasor (or complexor)

$$\hat{A} = A^m e^{j\theta}, \quad (5.2.12)$$

the value of $A$ in real time being understood to be given by the real part of

$$\hat{A}e^{j\omega t} = A^m e^{j(\omega t+\theta)} = A^m[\cos(\omega t + \theta) + j \sin(\omega t + \theta)] \quad (5.2.13)$$

by Euler's rule. Observe that a sinusoidal variation may always be represented in the same way by adjusting the value of the phase shift $\theta$. The derivative of $A$ in real time is

$$\frac{\partial A}{\partial t} = -\omega A^m \sin(\omega t + \theta), \quad (5.2.14)$$

which is the real part of the phasor

$$j\omega\hat{A}e^{j\omega t} = j\omega A^m e^{j(\omega t+\theta)} = A^m[-\omega \sin(\omega t + \theta) + j\omega \cos(\omega t + \theta)], \quad (5.2.15)$$

using $j^2 = -1$. Therefore in phasor representation, differentiation is the equivalent of premultiplication by $j\omega$. Since time reference is established by the value $\theta$, let us take the imposed current density as our reference phasor with $\theta = 0$. Using phasor representation, eqn. (5.2.7) is represented by the real part of the complex equation:

$$-\mu^{-1}\nabla^2\hat{A}e^{j\omega t} = \hat{J}_0 e^{j\omega t} - j\omega\sigma\hat{A}e^{j\omega t}, \tag{5.2.16}$$

which upon cancellation of the exponential terms, gives

$$-\mu^{-1}\nabla^2\hat{A} = \hat{J}_0 - j\omega\sigma\hat{A}, \tag{5.2.17}$$

where the terms $\hat{A}$ are complex numbers for which we need to solve. Once we obtain the phasors $\hat{A}$, we know the magnetic field in real time through eqn. (1.2.24) and the electric field through eqn. (5.2.3).

It is instructive to consider the analytical solution of this equation in one dimension. Let an electromagnetic field with **H** in the $y$ direction and **E** in the $z$ direction be traveling in the $x$ direction as shown in Figure 5.2.2, in an infinite medium without impressed currents $J_0$. In view of eqn. (5.2.3), the vector potential **A** will have a component only in the direction of **E**, $\mathbf{u}_z A$. The vastness of the medium guarantees that $\partial/\partial y$ and $\partial/\partial z$ are zero, changes occurring only in the direction of propagation $x$. Eqn. (5.2.7) for the phasors then reduces to

$$\frac{d^2}{dx^2}\hat{A} = j\omega\mu\sigma\hat{A} = jd^{-2}\hat{A} = \gamma^2\hat{A}, \tag{5.2.18}$$

where the depth of penetration $d$ is defined by

$$d^2 = \frac{1}{\omega\mu\sigma}, \tag{5.2.19}$$

and the propagation constant $\gamma$ is defined by

$$\gamma = \alpha + j\beta = d^{-1}\sqrt{j} = d^{-1}\sqrt{e^{j\pi/2}} = d^{-1}e^{j\pi/4} = \frac{1}{d\sqrt{2}} + j\frac{1}{d\sqrt{2}}. \tag{5.2.20}$$

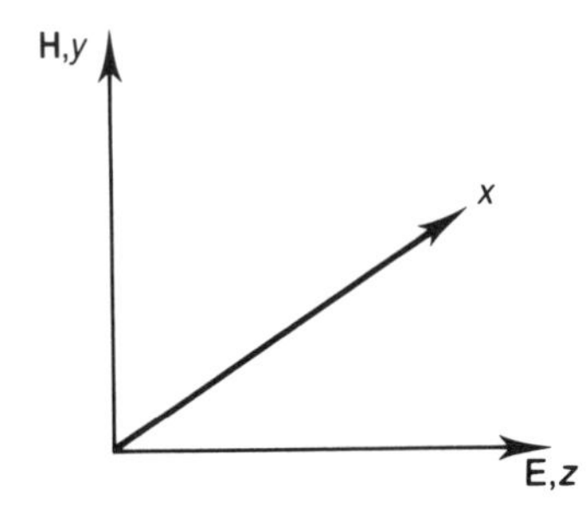


**Figure 5.2.2.** A traveling transverse electromagnetic wave.

The solution of eqn. (5.2.18) is clearly

$$\hat{A} = \hat{A}^{+}e^{\gamma x} + \hat{A}^{-}e^{-\gamma x}, \tag{5.2.21}$$

where $\hat{A}^{+}$ and $\hat{A}^{-}$ are the two constants of integration arising in the integration of a second-order differential equation. And since $\alpha$ in eqn. (5.2.20) is positive and the magnetic field's growing as it propagates is a physical impossibility, $\hat{A}^{+}$ must of necessity vanish. This gives us

$$\hat{A} = \hat{A}^{-}e^{-\alpha x}e^{-j\beta x}, \tag{5.2.22}$$

which shows that the magnitude of the wave $\hat{A}^{-}e^{-\alpha x}$ dies out exponentially as it propagates. The decay increases with $\alpha$, and, from eqn. (5.2.20), increases with decreasing $d$. Hence the term *depth of penetration* for $d$. It is clear from eqn. (5.2.20) that waves cannot propagate in highly permeable structures such as iron, in highly conductive media such as copper and the sea, and at high frequency.

In a uniform $h \times h$ finite difference mesh, with obvious extensions to other meshes, we have at a node $i$ surrounded by nodes 1, 2, 3, and 4

$$-\mu^{-1}h^{-2}[\hat{A}_1 + \hat{A}_2 + \hat{A}_3 + \hat{A}_4 - 4\hat{A}_0] = \hat{J}_{00} - j\omega\sigma\hat{A}_0. \tag{5.2.23}$$

The diagonal term corresponding to $\hat{A}_0$ is $4\mu^{-1}h^{-2} + j\omega\sigma$, or

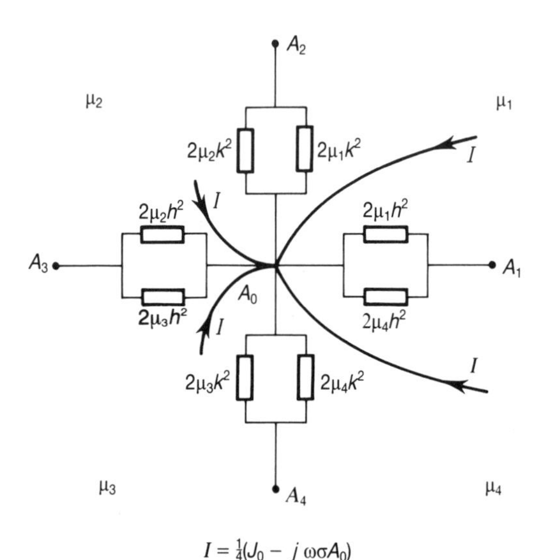


**Figure 5.2.3.** A network model for eddy currents.

$\mu^{-1}[4h^{-2} + jd^{-2}]$, which goes to show that the mesh size $h$ must bear some proportion to the depth of penetration $d$, if the skin effect is to be captured by the numerical solution. Eqn. (5.2.21) indicates an exponential change with $x/d\sqrt{2}$ in $A$, so that we usually use a mesh size $h \approx \frac{1}{3}d$ to get good solutions. The extensions to inhomogeneous problems with rectangular finite difference meshes may be easily worked out as in eqn. (4.3.3), using the network model of Figure 5.2.3. The eddy current boundary value problem in a quarter of a rectangular conductor is shown in Figure 5.2.4A. Parts B–E of Figure 5.2.4 show equiplots of the real component of $\hat{A}$ (the quantity corresponding to real time) with increasing frequency. It is seen that as $d$ decreases, the "skin" along the surface within which most of the magnetic flux flows is so much smaller than the mesh size that the variations are not captured by our computations.

## 5.2.2 Integral Solution of Eddy Current Distributions

The solution of the vector potential distribution has been accomplished in the preceding section using a differential formulation, along with an artificially placed "faraway" boundary. The resulting matrix equation was sparse and easy to solve.

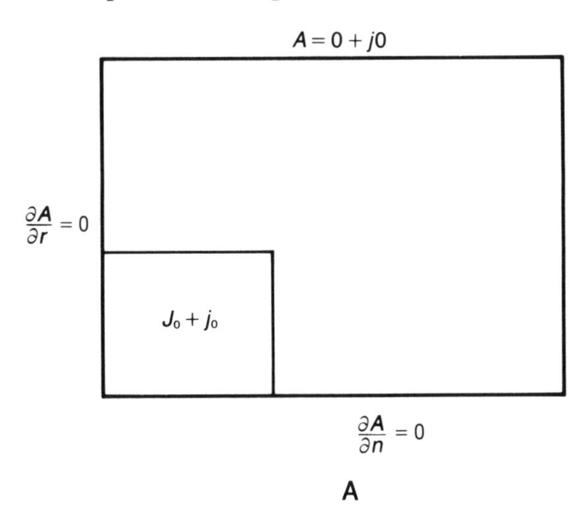


**Figure 5.2.4.** The rectangular conductor—finite difference. **A**. The boundary value problem. **B–E**. The eddy current effect with decreasing depth of penetration: (**B**) at 25 Hz; (**C**) at 50 Hz; (**D**) at 60 Hz; (**E**) solution at 300 Hz as depth of penetration falls below mesh size.

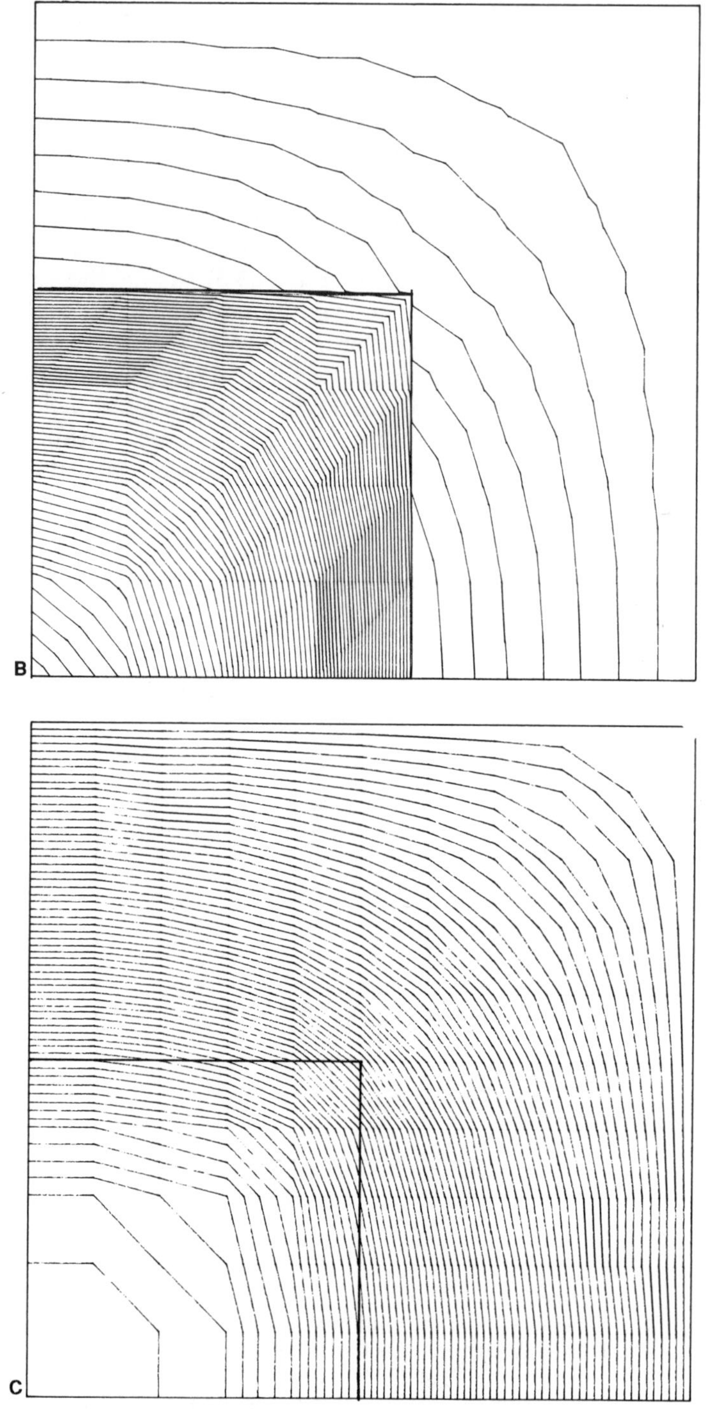


**Figure 5.2.4.** (*continued*)

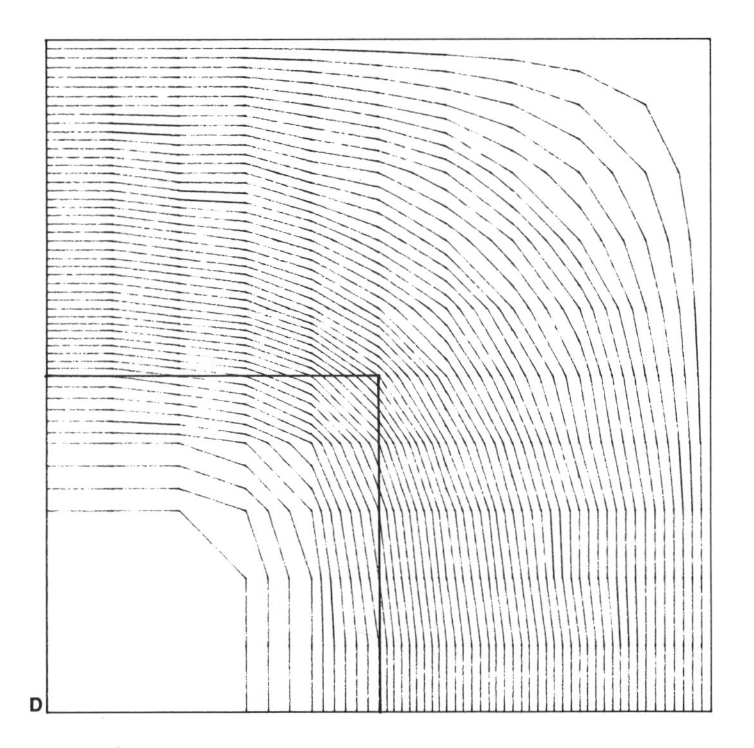


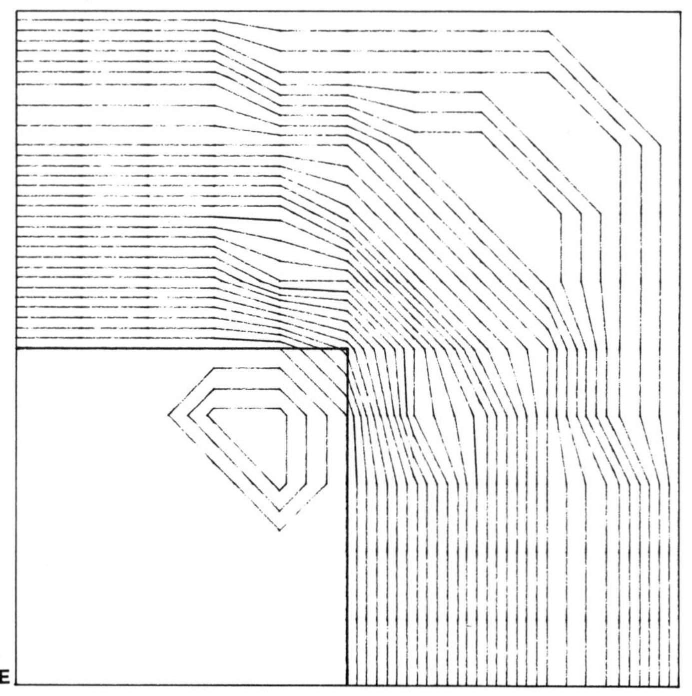


**Figure 5.2.4.** (*continued*)

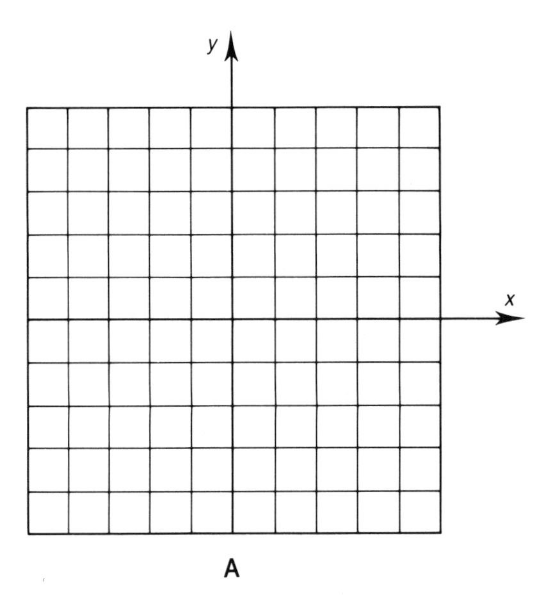


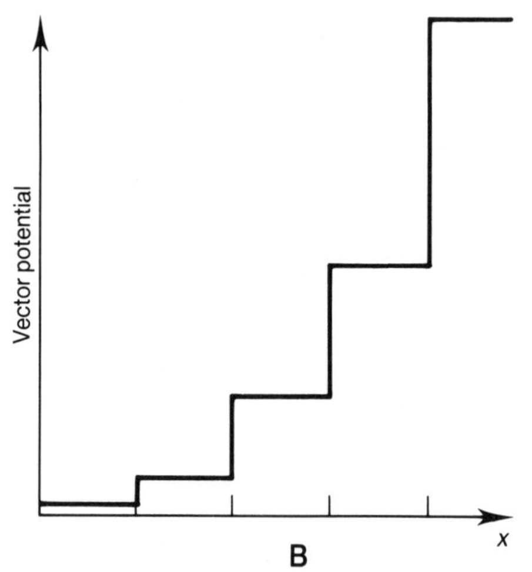


**Figure 5.2.5.** The rectangular conductor–integral method. **A**. Discretization. **B**. The solution along symmetry line.

The same problem may alternatively be solved using an integral formulation that allows for the open nature of the problem, without distorting the solution; the cost, however, is paid in the form of a fully populated matrix. The formulation is based on the fact that the solution to the equation:

$$-\mu^{-1}\nabla^2 A = J, \tag{1.2.32}$$

say in two dimensions, is:

$$A = -\frac{\mu}{2\pi}\iint J ln_e r\, dR. \tag{4.1.4}$$

Multiplying eqn. (5.2.3) by $\sigma$,

$$\hat{J} = \hat{J}_0 - j\omega\sigma\hat{A}, \tag{5.2.24}$$

and substituting eqn. (4.1.4) for $\hat{A}$, we have the requisite integral relationship:

$$\hat{J} = \hat{J}_0 + j\frac{\omega\mu\sigma}{2\pi}\iint \hat{J} ln_e r\, dR. \tag{5.2.25}$$

After dividing the conductor into small pieces over each of which $\hat{J}$ is assumed to be a constant, we may build up the matrix equation for a field element $i$:

$$\hat{J}_i = \sum \hat{J}_j c_{ij} = \hat{J}_{0i}, \tag{5.2.26}$$

where

$$c_{ij} = \frac{j\omega\mu\sigma}{2\pi}\iint_j ln_e r_{ij}\, dR \tag{5.2.27}$$

where the integration is over the source elements $j$, including element $i$. As an approximation, we may take $r_{ij}$ to be the distance between two separate elements $i$ and $j$; and when $i = j$, for square elements of size $h \times h$, we may take $r_{ii}$ to be the geometric mean distance $0.44705h$, as suggested by Silvester (1968). The solution of the same rectangular conductor as in Figure 5.2.4 by this integral scheme is shown plotted in Figure 5.2.5B, along an axis of symmetry. Observe the finer division in Figure 5.2.5A close to the conductor edges to capture the rapid changes in current density. As with the capacitor problems of section 2.4, symmetry has been exploited.

## 5.2.3 The Impedance Boundary Condition

As shown in Appendix B, the unique solution of the scalar boundary value problem requires either the potential or its normal gradient to be specified on the boundary. The impedance boundary condition is one that allows us to eliminate bodies of strong eddy current effect from the solution region, when we are not interested in the solution inside such bodies. This results in a significant improvement in computational efficiency, because, when strong eddy currents are present, the variation in the fields is exponential and several grid points will be required to model the changes. By eliminating such regions, the matrix size is considerably reduced and computational time brought down.

We have seen in section 5.2.1 that eddy currents result in the skin effect and that most of the electromagnetic action is within a few millimeters of the surface when we have high permeability, conductivity, or frequency. Thus, in most devices of dimensions larger than the skin depth $d$, defined in eqn. (5.2.19), the penetration of a surface by an electromagnetic field is one-dimensional. Hoole and Carpenter (1985) have shown this to be true down to a distance of two depths of penetration of corners. That is, in many practical applications, we may use the one-dimensional model of eqns. (5.2.18) to (5.2.22) with profit. Referring to our study there of the penetration of an electromagnetic field into a material semi-infinite in the $y$ and $z$ directions, the vector potential was shown to be $\mathbf{u}_z\hat{A}^- e^{-\gamma x}$ in eqn. (5.2.22). The magnetic field is therefore given by

$$\hat{\mathbf{H}} = \mu^{-1}\hat{\mathbf{B}} = \mu^{-1}\nabla \times \mathbf{u}_z\hat{A}^- e^{-\gamma x} = \mathbf{u}_y \gamma \mu^{-1}\hat{A}^- e^{-\gamma x}, \quad (5.2.28)$$

and the electric field, in the absence of impressed currents, by

$$\hat{\mathbf{E}} = -j\omega\hat{\mathbf{A}} = -\boldsymbol{u}_z j\omega\hat{A}^- e^{-\gamma x}. \quad (5.2.29)$$

Note that at the surface of the material, $x = 0$ and the $y$ and $z$ directions are tangential. Therefore, we may say that the tangential components of **E** and **H** are in the ratio

$$Z_s = \frac{\hat{E}_t}{\hat{H}_t} = \frac{|\hat{\mathbf{E}}|}{|\hat{\mathbf{H}}|} = \frac{j\omega\mu}{\gamma} = \frac{\gamma}{\sigma} = (1 - j)\sqrt{\frac{\omega\mu}{2\sigma}}, \quad (5.2.30)$$

where we have employed eqn. (5.2.18). This quantity, having nothing to do with the magnitude of the impinging field, is referred to as the *surface impedance.*

In solving for the vector potential, the surface impedance gives us an elegant boundary condition. Since, in two dimensions, $\hat{E}_t$ and $\hat{H}_t$ are respectively $j\omega\hat{A}$ and $\mu^{-1}\partial\hat{A}/\partial n$ (from eqns. (5.2.29) and (5.2.30)) exterior to the material and since the tangential components are continuous across the material, the impedance boundary condition effectively states (Hoole, Weeber, and Hoole 1988)

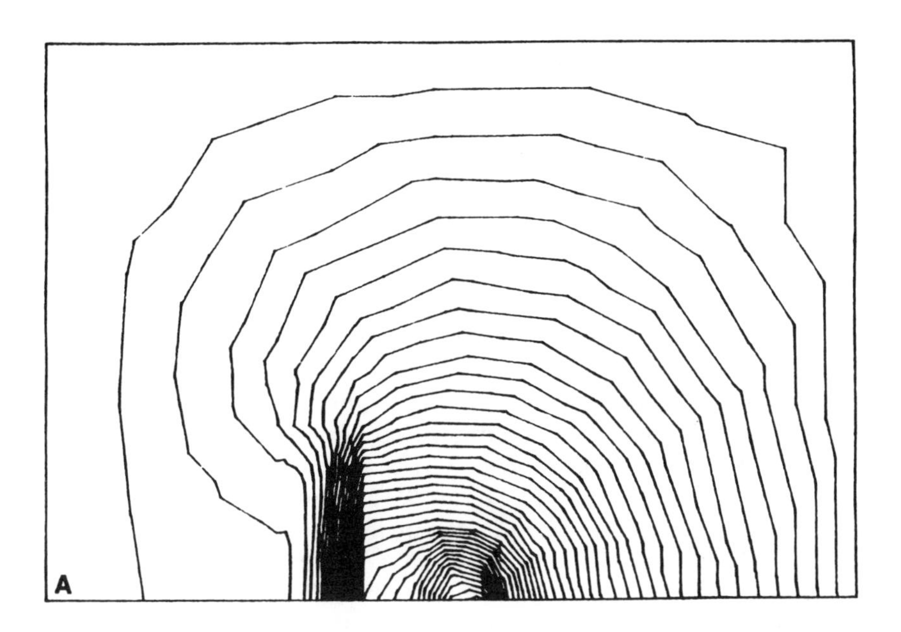


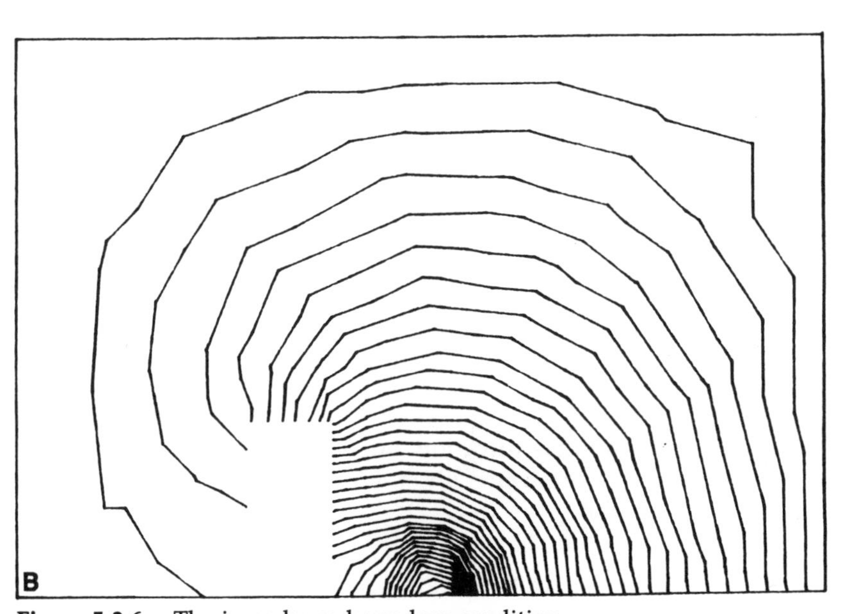


**Figure 5.2.6.** The impedance boundary condition.

that

$$j\omega\hat{A} = \mu^{-1}Z_s\frac{\partial\hat{A}}{\partial n}. \qquad (5.2.31)$$

Consider a finite difference node $(x, y)$ as in Figure 3.1.2, to the right of which we have the material to be eliminated from the solution region. We may use the boundary condition in eqn. (5.2.31) to rewrite the potential at node $(x + h, y)$ that we are eliminating from the solution region, in terms of the exterior node $(x - h, y)$. The surface of the material therefore becomes a boundary.

Likewise, when dealing with the boundary integral scheme, we may use the impedance boundary condition to substitute for $g$ (which is $\partial\hat{A}/\partial n$) in terms of $\hat{A}$ so that, on a surface element, only $\hat{A}$ is unknown in eqn. (2.5.2). Figure 5.2.6 shows the solution of the vector potential about a conductor in the presence of a permeable shield. Figure 5.2.6A shows the usual solution, while Figure 5.2.6B gives the exterior solution using the impedance boundary condition. Other integral formulations using the impedance boundary condition have also been successfully employed, where the magnetic field is directly solved for (Mayergoyz 1981; Fawzi, Ahmed, and Burke 1985; Ali, Ahmed, and Burke 1987).

## 5.2.4 The Boundary Element Solution of Eddy Current Problems

We have seen in section 2.5 the use of the boundary element method in the solution of static field problems. There we used Green's second identity, eqn. (2.5.13), which in two dimensions reduces to

$$\iint [f\nabla^2 g - g\nabla^2 f]\, dR = \int [f\nabla g - g\nabla f]\,.\,\mathbf{dS} = \int \left[f\frac{\partial g}{\partial n} - g\frac{\partial f}{\partial n}\right] dS \qquad (5.2.32)$$

where $n$ is the normal direction on $S$. To restate the boundary element method more generally than in section 2.5, if we are solving the equation

$$-\mu^{-1}\nabla^2 A = J \qquad (1.2.32)$$

we first seek a free-space Green's function $G$, such that

$$\nabla^2 G = \delta(r) \qquad (5.2.33)$$

where the function $\delta$ is one when $r$ is zero and is zero otherwise.

For instance, we know that in two dimensions, $G$ is $-(1/2\pi)ln_e r$, as shown in eqn. (2.5.11). Substituting $A$ for $f$ and $G$ for $g$ in eqn. (5.2.32) and noting that the integral of $f\delta(r)$ over $dR$ is $f$ as shown previously,

$$A = \iint \mu GJ\, dR + \int \left[ A\frac{\partial G}{\partial n} - G\frac{\partial A}{\partial n} \right] dS. \qquad (5.2.34)$$

Thus under the boundary element scheme, we discretize the surface $S$ into elements and solve for $A$ or $\partial A/\partial n$ on the boundary, what is not being solved for being known through Dirichlet or Neumann boundary conditions. $R$, the interior of $S$, has to be subdivided into a second mesh, two-dimensional this time, to define $J$ so that the double integral above may be performed numerically.

In solving the eddy current problem by the boundary element method, following Schneider and Salon (1980), we seek a solution to the equation

$$-\mu^{-1}\nabla^2 A = J - j\omega\sigma A. \qquad (5.2.17)$$

This time, however, we seek a Green's function $G$ satisfying

$$\nabla^2 G = \delta(r) + j\omega\mu\sigma G. \qquad (5.2.35)$$

The solution to this equation is

$$G = \frac{1}{2\pi} K_0(\alpha r), \qquad (5.2.36)$$

$$\alpha^2 = j\omega\mu\sigma, \qquad (5.2.37)$$

where $K_0$ is the modified Bessel function of the second kind of order zero (Abramovitz and Stegun 1964). Substituting $A$ for $f$ and $G$ for $g$ in eqn. (5.2.32),

$$\iint \{A[\alpha^2 G + \delta(r)] - G(\alpha^2 A - \mu J)\}\, dR = \int \left[ A\frac{\partial G}{\partial n} - G\frac{\partial A}{\partial n} \right] dS \qquad (5.2.38)$$

or

$$A = -\iint \mu GJ\, dR + \int \left[ A\frac{\partial G}{\partial n} - G\frac{\partial A}{\partial n} \right] dS, \qquad (5.2.39)$$

which allows us once again to solve for $A$ or its normal gradient on the boundary $S$ by discretizing it into elements.

Some useful properties of these Bessel functions are the following:

$$\frac{\partial K_0}{\partial r}(\alpha r) = \alpha K_1(\alpha r) \qquad (5.2.40)$$

where $K_1$ is the Bessel function of the second kind of order one;

$$\frac{\partial K_1}{\partial r}(\alpha r) = \alpha K_0(\alpha r); \qquad (5.2.41)$$

$$\lim_{r\to 0} K_0(\alpha r) = -ln_e r. \qquad (5.2.42)$$

The latter property is particularly useful in evaluating the self-coefficients (diagonal terms) of boundary elements. This would now follow the same procedures developed in sections 2.5.1 and 2.5.3 for static problems where the integrals of the natural logarithm were detailed. The off-diagonal terms may be evaluated by Gaussian quadrature, using the schemes of section 2.5.1. In these evaluations, most mathematical libraries on computers will allow $K_0$ and $K_1$ to be computed. In a personal computing milieu without such libraries (a very unlikely scenario for now, given the storage requirements in eddy current problems with complex arithmetic, unless large workstations are used), one must write one's own functions to compute the Bessel functions. For this purpose, eqns. 9.8.1, 9.8.2, 9.8.5, 9.8.7, and 9.8.8 from Abramovitz and Stegun (1964) are invaluable.[1] These equations give the functions as series expansions in terms of the argument for different ranges of the argument.

Observe that the current distribution $J$ is in the interior $R$ and requires a second two-dimensional mesh. The experience of this writer is that this term causes large error to accumulate in the solution. Salon and Schneider (1982) skirt this difficulty by transforming $J$, whenever it obeys

$$\nabla^2 J = 0, \qquad (5.2.43)$$

to a boundary term using eqn. (5.2.32). This change, always possible for the usual case of constant impressed $J$, eliminates this source of error. In the following, some typographical errors in the equations of Salon and Schneider in their classic paper (1982) are corrected. To eliminate having to define $J$ inside the solution region using a two-dimensional mesh, we define a function $G_1$ that is the solution of

$$\nabla^2 G_1 = \frac{\delta(r)}{\alpha^2} + G. \qquad (5.2.44)$$

[1] Personal communication from Professor S. J. Salon, Rensselaer Polytechnic Institute, Troy, New York.

It may be shown from eqns. (5.2.39)–(5.2.41) that

$$G_1 = \frac{K_0(\alpha r)}{2\pi\alpha^2} = \frac{G}{\alpha^2}. \tag{5.2.45}$$

Now, writing $J$ for $f$ and $G_1$ for $g$ in eqn. (5.2.32),

$$\iint \left\{J\left[G + \frac{\delta(r)}{\alpha^2}\right] - G_1 \times 0\right\} dR = \int \left[J\frac{\partial G_1}{\partial n} - G_1\frac{\partial J}{\partial n}\right] dS, \tag{5.2.46}$$

or, rearranging,

$$\iint JG\, dR = -\frac{J}{\alpha^2} + \frac{1}{\alpha^2}\int \left[J\frac{\partial G}{\partial n} - G\frac{\partial J}{\partial n}\right] dS. \tag{5.2.47}$$

Using this to eliminate the interior current in eqn. (5.2.38), we have the final equation for solution:

$$A = \frac{\mu J}{\alpha^2} - \frac{\mu}{\alpha^2}\int \left[J\frac{\partial G}{\partial n} - G\frac{\partial J}{\partial n}\right] dS + \int \left[A\frac{\partial G}{\partial n} - G\frac{\partial A}{\partial n}\right] dS. \tag{5.2.48}$$

Thus the same boundary element mesh is used for the current distribution $J$, which now has to be defined only on the boundary. For the common situation of constant currents, the term $\partial J/\partial n$ may be eliminated.

## 5.3 HIGH-FREQUENCY EFFECTS: THE HELMHOLTZ EQUATION

What distinguishes waves at high frequency from those at low frequency is that the displacement current term $\partial \mathbf{D}/\partial t$ becomes significant and can no longer be neglected. Consequently, we have, from eqns. (1.2.2), (1.2.4), (1.2.6), (1.2.7), and (1.2.11):

$$\nabla \times \mathbf{H} = \sigma\mathbf{E} + \varepsilon\frac{\partial \mathbf{E}}{\partial t}, \tag{5.3.1}$$

$$\nabla \times \mathbf{E} = -\mu\frac{\partial \hat{\mathbf{H}}}{\partial t}. \tag{5.3.2}$$

Taking the curl of the first equation,

$$\nabla \times \nabla \times \mathbf{H} = \left(\sigma + \varepsilon\frac{\partial}{\partial t}\right)\nabla \times \mathbf{E} = -\left(\sigma + \varepsilon\frac{\partial}{\partial t}\right)\mu\frac{\partial \mathbf{H}}{\partial t}, \tag{5.3.3}$$

whence, using eqns. (A9) and (1.2.10):

$$\nabla^2\mathbf{H} = \sigma\mu\frac{\partial \mathbf{H}}{\partial t} + \varepsilon\mu\frac{\partial^2 \mathbf{H}}{\partial t^2}. \tag{5.3.4}$$

We have used nought for $\nabla \cdot \mathbf{H}$ from eqn. (1.2.10) since, in a homogeneous region, **H** and **B** always differ only by the same scale factor. Similarly, by taking the curl of eqn. (5.3.2) and using $\rho/\varepsilon$ in $\nabla \cdot \mathbf{E}$ from eqn. (1.2.13), we have:

$$\varepsilon^{-1}\nabla\rho - \nabla^2\mathbf{E} = -\mu\frac{\partial}{\partial t}\left[\sigma\mathbf{E} + \varepsilon\frac{\partial\mathbf{E}}{\partial t}\right], \tag{5.3.5}$$

so that we have

$$\nabla^2\mathbf{E} = \sigma\mu\frac{\partial\mathbf{E}}{\partial t} + \varepsilon\mu\frac{\partial^2\mathbf{E}}{\partial t^2} + \varepsilon^{-1}\nabla\rho. \tag{5.3.6}$$

Eqns. (5.3.4) and (5.3.6) are called the *inhomogeneous wave equations.* When waves propagate through free space, $\sigma$ and $\rho$ are zero and we have the vector Helmholtz wave equations:

$$\nabla^2\mathbf{H} = \varepsilon\mu\frac{\partial^2\mathbf{H}}{\partial t^2}, \tag{5.3.7a}$$

$$\nabla^2\mathbf{E} = \varepsilon\mu\frac{\partial^2\mathbf{E}}{\partial t^2}. \tag{5.3.7b}$$

These equations may be put in the general scalar form of Helmholtz's equation:

$$\nabla^2\phi = c^{-2}\frac{\partial^2\phi}{\partial t^2}, \tag{5.3.8}$$

where $\phi$ stands for the components of **E** and **H**, and

$$c^2 = \frac{1}{\mu\varepsilon} \tag{5.3.9}$$

is a positive constant. We shall see later that $c$ is the velocity of a wave with no component of **E** or **H** in the direction of travel. In general, as with the eddy current problem of eqn. (5.2.17), an equation of the type (5.3.8) for the scalar $\phi$, travelling in the direction, say, $z$, has solution in the form:

$$\phi = f(x, y)g(z)h(t), \tag{5.3.10}$$

which, when put into the governing equation, yields:

$$(\nabla^2_{xy}f)gh + fg''h = c^{-2}fgh'', \tag{5.3.11}$$

where $''$ denotes double differentiation and $\nabla^2_{xy}$ stands for the $\nabla^2$ operator on the $xy$ plane. Dividing throughout by $fgh$, we have a function of $x$, $y$, and $z$ on the left and a function of $t$ only on the right, which is possible only if they are each a constant $k$:

$$f^{-1}\nabla^2 f + g^{-1}g'' = c^{-2}h^{-1}h'' = k. \tag{5.3.12}$$

The equation

$$h''(t) = kc^2 h \tag{5.3.13}$$

has solution

$$h = \begin{cases} At + B & k = 0 \\ A\cos(\omega t + B) & -\omega^2 = kc^2 < 0, \\ A\exp(\omega t) + B\exp(-\omega t) & \omega^2 = kc^2 > 0 \end{cases} \tag{5.3.14}$$

where $A$ and $B$ are constants of integration. Only the second possibility, $k < 0$, is possible for a propagating wave that does not die as time progresses at a spatial position, and we have

$$k = -\frac{\omega^2}{c^2}, \tag{5.3.15}$$

$$f^{-1}\nabla_{xy}^2 f + g^{-1}g'' = -\frac{\omega^2}{c^2}. \tag{5.3.16}$$

Again, since a function of $x$ and $y$ cannot be equal to a function of $z$ only, we must have, by the same argument,

$$g^{-1}g'' = \gamma^2, \tag{5.3.17}$$

$$f^{-1}\nabla_{xy}^2 f = -\gamma^2 - \frac{\omega^2}{c^2}, \tag{5.3.18}$$

$$g = Ae^{-\gamma z} + Be^{\gamma z}. \tag{5.3.19}$$

In general, therefore, a propagating wave, employing the complex phasor notation of the previous section, must be expressible as

$$\hat{\phi} = \hat{f}_1(x, y)e^{j\omega t - \gamma z} + \hat{f}_2(x, y)e^{j\omega t + \gamma z}, \tag{5.3.20}$$

where we have summed the two possibilities of $\omega$ and $\gamma$ being of the same and opposite signs.

Should the propagation constant $\gamma$ have a real part, it will attenuate the wave. When, ideally, there is no attenuation,

$$\gamma = j\beta. \tag{5.3.21}$$

It is quickly recognized that the first term of eqn. (5.3.20) has the same value at the points $(x, y, z, t)$ and $(x, y, z + \Delta z, t + \gamma\Delta z/\omega)$— since

$$\omega t - \beta z = \omega\left(t + \frac{\beta\Delta z}{\omega}\right) - \beta(z + \Delta z). \tag{5.3.22}$$

It therefore represents a wave traveling in the positive $z$ direction with velocity $\omega/\beta$, which is seen to be $c$. The second term will be realized to be a wave traveling in the negative $z$ direction and may therefore be set to zero for open field problems without reflected waves and the now superfluous subscript 1 dropped from $\hat{f}$:

$$\hat{\phi} = f(x, y)e^{j\omega t - \gamma z}. \tag{5.3.23}$$

For a propagating wave, therefore, we must have

$$\frac{\partial}{\partial t} \equiv j\omega \tag{5.3.24}$$

and

$$\frac{\partial}{\partial z} \equiv -\gamma. \tag{5.3.25}$$

And, by convention, a traveling wave such as $\phi$ is written without the term $e^{j\omega t - \gamma z}$, its presence being implicitly understood. That is, we work with $\phi$, a function of $x$ and $y$ only. $\phi$ represents $f$ of the preceding derivations and is governed by eqn. (5.3.18); when we want the actual term $\phi$, we multiply it by the factor $e^{j\omega t - \gamma z}$ as in eqn. (5.3.23) and take the real part of the result. For the scaled $\phi$, therefore, rewriting the governing equation (5.3.18):

$$\nabla^2\phi + \left(\frac{\gamma^2 + \omega^2}{c^2}\right)\phi = 0, \tag{5.3.26}$$

or more conveniently

$$\nabla^2\phi + k^2\phi = 0, \tag{5.3.27}$$

where the new definition of $k$ is

$$k^2 = \gamma^2 + \frac{\omega^2}{c^2}. \tag{5.3.28}$$

Now using eqn. (5.3.24), if we rewrite eqns. (5.3.1) and (5.3.2),

$$\nabla \times \mathbf{E} = -j\omega\mu\mathbf{H} \tag{5.3.29}$$

and

$$\nabla \times \mathbf{H} = j\omega\varepsilon\mathbf{E}. \tag{5.3.30}$$

Expanding into their three components using eqns. (5.3.25) and

(A3), we obtain, respectively,

$$\frac{\partial E_z}{\partial y} + \gamma E_y = -j\omega\mu H_x \tag{5.3.31a}$$

$$-\gamma E_x - \frac{\partial E_z}{\partial x} = -j\omega\mu H_y \tag{5.3.31b}$$

$$\frac{\partial E_y}{\partial x} - \frac{\partial E_x}{\partial y} = -j\omega\mu H_z \tag{5.3.31c}$$

and

$$\frac{\partial H_z}{\partial y} + \gamma H_y = j\omega\varepsilon E_x \tag{5.3.32a}$$

$$-\gamma H_x - \frac{\partial H_z}{\partial x} = j\omega\epsilon E_y \tag{5.3.32b}$$

$$\frac{\partial H_y}{\partial x} - \frac{\partial H_x}{\partial y} = j\omega\varepsilon E_z. \tag{5.3.32c}$$

From eqns. (5.3.31a) and (5.3.32b), upon elimination of $E_y$, we have:

$$H_x = \frac{1}{\gamma^2 + \omega^2/c^2}\left[j\omega\varepsilon\frac{\partial E_z}{\partial y} - \gamma\frac{\partial H_z}{\partial x}\right]. \tag{5.3.33a}$$

Similarly

$$H_y = -\frac{1}{\gamma^2 + \omega^2/c^2}\left[j\omega\varepsilon\frac{\partial E_z}{\partial x} + \gamma\frac{\partial H_z}{\partial y}\right], \tag{5.3.33b}$$

$$E_x = -\frac{1}{\gamma^2 + \omega^2/c^2}\left[\gamma\frac{\partial E_z}{\partial x} + j\omega\mu\frac{\partial H_z}{\partial y}\right], \tag{5.3.34a}$$

$$E_y = \frac{1}{\gamma^2 + \omega^2/c^2}\left[-\gamma\frac{\partial E_z}{\partial y} + j\omega\mu\frac{\partial H_z}{\partial x}\right]. \tag{5.3.34b}$$

Thus, if we can find the electric and magnetic fields $E_z$ and $H_z$ in the direction of propagation, the other four components are quickly extracted using eqns. (5.3.33) and (5.3.34). Three kinds of waves are conveniently distinguished, all waves usually being expressible as a combination of waves belonging to one of these categories. First, we have the transverse electromagnetic waves (TEM) with both the electric and magnetic fields on the transverse plane. That is, both $E_z$ and $H_z$ are zero; from eqns. (5.3.33) and (5.3.34), a nontrivial solution is possible only if

$$\gamma^2 + \frac{\omega^2}{c^2} = 0, \tag{5.3.35}$$

making $\gamma$ a purely imaginary number $j\beta$ so that no attenuation will be possible. It will be seen from eqns. (5.3.33) and (5.3.34) that, with both $E_z$ and $H_z$ zero, **E** and **H** are normal to each other and in the ratio $\sqrt{(\mu/\varepsilon)}$, the impedance of the medium. This is the kind of wave experienced on transmission lines. A more detailed discussion is provided by Schelkunoff (1934).

Second, we have the transverse magnetic waves (TM) with the magnetic field entirely on the transverse plane ($H_z$ zero), so that all the components may be derived from $E_z$, thus lending it the alternative name *E wave*. And third, transverse electric waves (TE or M) have $E_z$ zero.

The cutoff frequency for a guide occurs at the transition of $\gamma^2$ from positive to negative; for, when positive, $\gamma$ is a real number and the factor $e^{-\gamma z}$ serves to attenuate and kill the wave. On the other hand, when negative, $\gamma$ is the purely imaginary number $j\beta$ and the factor $e^{-\gamma z}$ represents only a change in phase, with the magnitude being unaffected. From eqn. (5.3.28), therefore, the cutoff frequency of a guide is given by:

$$\omega_c = ck. \tag{5.3.36}$$

## 5.4 THE COMPUTER-ASSISTED ANALYSIS OF GUIDED WAVES

### 5.4.1 Boundary Conditions on TE and TM Waves

The analysis of waves in guides of any shape may easily be accomplished numerically, by solving for the nontransverse component $\phi = E_z$ in TM waves and $\phi = H_z$ in TE waves. To do so, we need boundary conditions, and these are provided by the fact (as discussed in section 1.2.4) that the electric field tangential to the perfectly conducting guide walls is zero. Therefore, in a two-dimensional guide, $E_z$ pointing into the $xy$ plane containing the guide cross section is tangential to the wall and therefore zero. That is, TM waves have $\phi = 0$ along the walls.

To obtain the boundary conditions on TE waves, consider the $x$ and $y$ axes placed such that the $y$ axis is tangential to the wall and the $x$ axis normal to it—a valid exercise, since the directions of the axes do not change our equations derived above, so long as the $z$ axis along which the wave moves is not changed in direction. Now, since $E_z$ is always zero for a TE wave, $\partial E_z/\partial y$, the rate of change of

$E_z$ along the walls, is also zero. Moreover, $E_y$, being tangential to the walls, is also zero. Therefore, for a TE wave, from eqn. (5.3.34b), $\partial H_z/\partial x$ is zero; that is, the normal gradient of the unknown $H_z$ is zero. The appropriate condition on $\phi$ is therefore Neumann $\partial\phi/\partial n = 0$.

## 5.4.2 The Eigen Value Problem: Iterative Solution

The numerical solution of waveguide problems, we have seen, requires the solution of

$$\nabla^2\phi + \frac{\gamma^2 + \omega^2}{c^2}\phi = 0. \tag{5.4.1}$$

As with the finite difference solution for static fields, the term $\nabla^2\phi$ may be discretized over a finite difference mesh; at a node 0 surrounded by nodes 1, 2, 3, and 4, we have, from eqn. (3.1.9),

$$\frac{1}{h^2}[\phi_1 + \phi_2 + \phi_3 + \phi_4 - 4\phi_0] + \left(\gamma^2 + \frac{\omega^2}{c^2}\right)\phi_0 = 0, \tag{5.4.2}$$

or, rearranging,

$$4\phi_0 - \phi_1 - \phi_2 - \phi_3 - \phi_4 = \lambda\phi_0, \tag{5.4.3}$$

where

$$\lambda = h^2k^2 = h^2\left(\gamma^2 + \frac{\omega^2}{c^2}\right) \tag{5.4.4}$$

and $h$ is the square finite difference mesh size. If eqn. (5.4.3) is built into a matrix, it will yield

$$\mathbf{A}\boldsymbol{\phi} = \lambda\boldsymbol{\phi}, \tag{5.4.5}$$

and $\lambda$ is recognized as an eigenvalue of the matrix $\mathbf{A}$. If $\mathbf{I}$ is a unit matrix of the same size as $\mathbf{A}$, this may be interpreted as stating that $\lambda$, a function of the frequency $\omega$, is such a value that it makes the matrix $[\mathbf{A} - \lambda\mathbf{I}]$ in

$$[\mathbf{A} - \lambda\mathbf{I}]\boldsymbol{\phi} = 0 \tag{5.4.6}$$

singular. For otherwise, with $\mathbf{A}$ nonsingular, $[\mathbf{A} - \lambda\mathbf{I}]^{-1}$ will exist and $\phi$ will be zero and no wave may propagate. That is, only for certain frequencies can $\phi$ be nonzero and waves propagate. The eigenvalues $\lambda$ are then computed either from the determinantal equation

$$|\mathbf{A} - \lambda\mathbf{I}| = 0 \tag{5.4.7}$$

or by using one of the more efficient schemes presented in chapter 6.

In a waveguide, we are usually interested in the solution corresponding to the lowest frequency that passes, and this corresponds to the dominant mode in a guide. This dominant mode may be extracted using an iterative scheme that is a generalization of Liebmann's method presented in section 3.2. Convergence usually is to the dominant of the several possible solutions. The difference here is that in the right-hand side of eqn. (5.4.3), $\lambda$ is not known. This value is updated every few (five is suggested by Silvester, 1968) iterations. If we tried to compute $\lambda$ at each node separately, we would arrive at different answers. The correct computation of $\lambda$ is accomplished by multiplying eqn. (5.4.3) by $\phi$ and integrating over the whole solution region:

$$\gamma^2 + \frac{\omega^2}{c^2} = h^{-2}\lambda = -\frac{\iint \phi \nabla^2 \phi \, dR}{\iint \phi^2 \, dR}$$

$$= \frac{\sum_0 \phi_0[4\phi_0 - \phi_1 - \phi_2 - \phi_3 - \phi_4]}{h^2 \sum_0 \phi_0^2}. \tag{5.4.8}$$

In this integration, we have taken the values of $\phi$ and $\nabla^2\phi$ to extend to an $h \times h$ box surrounding the node 0 over which the summation takes place. A factor of 0.5 should therefore be included for nodes at boundary edges and the factor 0.25 used at nodes at boundary corners. The term $h^{-2}\lambda$ so calculated is called the *Rayleigh coefficient*, and, by the arguments of section 3.1.2, is of a higher order of accuracy than the potentials $\phi$.

Some modifications are required to Alg. 3.2.2 in order to solve the wave equation. First, we must start with an initial guess of $\lambda$, in addition to a guess of $\phi$. Second, an initial guess of $\phi = 0$ is not acceptable, since it is a solution that satisfies the wave equation and, as a result, convergence will be flagged after the first iteration. Third, $\phi$ at a node should be computed from

$$\phi_0 = \frac{\phi_1 + \phi_2 + \phi_3 + \phi_4}{4 - \lambda}, \tag{5.4.9}$$

and then acceleration applied as before. Fourth, after a fixed number of iterations through all the nodes of the mesh, say five, $\lambda$ should be updated using eqn. (5.4.8). Fifth, since, if $\phi$ is a solution of eqn. (5.4.5), for any constant $k$, $k\phi$ is also a solution, to ensure speedy convergence to just one of these several solutions corresponding to a value of $\lambda$, we must scale $\phi$ at the end of every iteration. A convenient means of scaling is to divide throughout the vector $\boldsymbol{\phi}$ by the largest component of $\boldsymbol{\phi}$. Sixth, convergence may be flagged by checking for changes in $\lambda$ instead of $\phi$—this will indeed be a weaker test since $\lambda$ is ahead of $\phi$ in accuracy and should be considered only when we are not interested in the values of $\phi$. Finally, a precautionary measure. Since the equation has multiple solutions corresponding to different values of $\lambda$ and we are seeking the dominant mode corresponding to the lowest value of $\lambda$, it is good to assure ourselves that convergence was indeed to the principal mode. This may be done by repeating the solution with an alternative starting vector. Convergence to one of the secondary modes may occur if, by chance, the starting solution happened to be very close to one of the secondary modes.

### 5.4.3 The Transverse Magnetic Wave: Between Two Parallel Conductors

The previously described numerical procedure is suitable for guides of any shape. However, for purposes of comparison and illustration, let us try to analyze by hand a transverse magnetic wave guided by two infinitely long, parallel plane conductors extending in the positive and negative $y$ and $z$ directions and separated by a distance $a$ in the $x$ direction. This problem has a well-defined closed-form solution, which we may use for comparison. The unknown $\phi = E_z$ can only be a function of $x$, and eqn. (5.3.27) becomes

$$\frac{d^2}{dx^2}\phi = -k^2\phi, \tag{5.4.10}$$

the well-known equation governing simple harmonic motion, with solution

$$\phi = A \sin kx + B \cos kx. \tag{5.4.11}$$

The boundary condition at $x = 0$ is that the tangential electric field $\phi$ must disappear. This makes the constant $B$ vanish. Similarly, the condition at the second plate $x = a$ gives $A = 0$ corresponding to a trivial solution, or alternatively,

$$\sin ka = 0, \tag{5.4.12}$$

making

$$k = \frac{n\pi}{a} \tag{5.4.13}$$

for integral values of $n$. The solution of the electric field then is

$$\phi = A \sin \frac{n\pi x}{a}, \tag{5.4.14}$$

allowing only certain frequencies to pass. The cutoff frequency from eqn. (5.3.36) is

$$\omega_c = \frac{nc\pi}{a}. \tag{5.4.15}$$

For the dominant mode, $n = 1$, and

$$\omega_c = \frac{\pi c}{a} = \frac{3.142c}{a}. \tag{5.4.16}$$

Let us begin with a finite difference mesh with three points, one in the middle, and the other two on the conductor so that they have $\phi = 0$, as shown in Figure 5.4.1A. In two dimensions, from eqn. (3.1.7), the finite difference approximation of eqn. (5.4.9) at a node 0 with nodes 1 and 2, respectively, to the right and left at a distance $h$ along the $x$ axis is

$$h^{-2}[\phi_1 + \phi_2 - 2\phi_0] + k^2\phi_0 = 0 \tag{5.4.17}$$

or

$$-\phi_1 + 2\phi_0 - \phi_2 = h^2k^2\phi_0. \tag{5.4.18}$$

For the $1 \times 1$ "matrix equation" corresponding to the mesh of Figure 5.4.1A,

$$2\phi_1 = h^2k^2\phi_1, \tag{5.4.19}$$

giving us an eigenvalue of 2. It is convenient at this point to express $h$ in terms of the device dimension $a$ and the number $j$ of

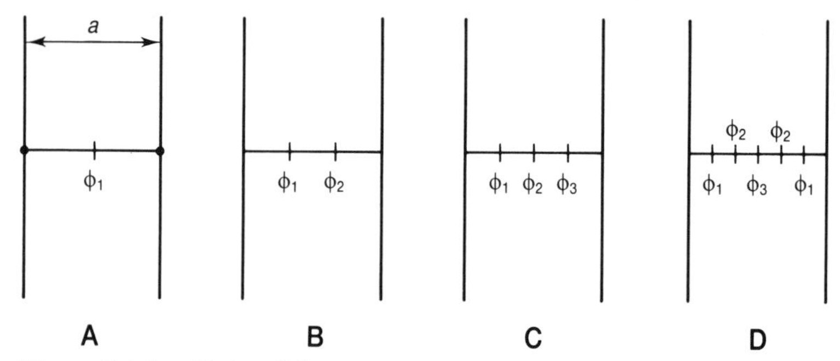


**Figure 5.4.1.** Finite difference meshes for parallel plane guide.

mesh intervals into which it is divided:

$$h = \frac{a}{j}. \tag{5.4.20}$$

This gives us

$$\lambda = h^2k^2 = \frac{a^2k^2}{j^2}, \tag{5.4.21}$$

so that the cutoff frequency from eqn. (5.3.36) is

$$\omega_c = kc = c \cdot \sqrt{\frac{\lambda j^2}{a^2}} = \frac{c}{a} j\sqrt{\lambda}. \tag{5.4.22}$$

Since we have $j = 2$ and $\lambda = 2$, $\omega_c$ is $(c/a) \times 2.828$, much off the actual value of $(c/a) \times 3.142$. This is not surprising, given the crudeness of the mesh. Adding a point to the mesh, as in Figure 5.4.1B, we have

$$\begin{bmatrix} 2 & -1 \\ -1 & 2 \end{bmatrix} \begin{bmatrix} \phi_1 \\ \phi_2 \end{bmatrix} = h^2k^2 \begin{bmatrix} \phi_1 \\ \phi_2 \end{bmatrix}. \tag{5.4.23}$$

By eqn. (5.4.7), the eigenvalues are obtained by solving:

$$\begin{vmatrix} 2-\lambda & -1 \\ -1 & 2-\lambda \end{vmatrix} = 0, \tag{5.4.24}$$

or

$$(2-\lambda)^2 - 1 = 0, \tag{5.4.25}$$

with solutions $\lambda = 1$ or 3. According to eqn. (5.4.22), using $j = 3$, the lowest cutoff frequency corresponding to the lowest value of $\lambda$ is given by $\omega_c = (c/a)3\sqrt{1} = (c/a) \times 3$, an improvement on the previous solution. This also gives us the second mode as having cutoff frequency $(c/a) \times 3\sqrt{3} = (c/a) \times 5.196$, whereas the exact cutoff frequency from eqn. (5.4.17) is $(c/a)n\pi = (c/a) \times 6.283$ for $n = 2$. Adding yet another point, as in Figure 5.4.1C, we have

$$\begin{bmatrix} 2 & -1 & 0 \\ -1 & 2 & -1 \\ 0 & -1 & 2 \end{bmatrix} \begin{bmatrix} \phi_1 \\ \phi_2 \\ \phi_3 \end{bmatrix} = h^2k^2 \begin{bmatrix} \phi_1 \\ \phi_2 \\ \phi_3 \end{bmatrix}. \tag{5.4.26}$$

Solving the determinantal equation

$$\begin{vmatrix} 2-\lambda & -1 & 0 \\ -1 & 2-\lambda & -1 \\ 0 & -1 & 2-\lambda \end{vmatrix} = 0, \tag{5.4.27}$$

we get

$$(2 - \lambda)^3 - 2(2 - \lambda) = 0, \tag{5.4.28}$$

solving which we obtain $\lambda = 0.586$ or 2 or 3.141. The lowest cutoff frequency then is, with $j = 4$, $(c/a) \times 4\sqrt{0.586} = (c/a) \times 3.062$. The second mode has $\omega_c = (c/a) \times 4\sqrt{2} = (c/a) \times 5.657$ and is recognized to have improved. This analysis also gives us the third mode at $\omega_c = (c/a) \times 4\sqrt{3.141} = (c/a) \times 7.089$, whereas the actual answer for the third mode is $(c/a) \times 3\pi = (c/a) \times 9.425$.

Clearly, as we add points to the finite difference mesh we begin to capture more and more modes, and the lower modes have higher accuracy in comparison with the higher modes. We are, moreover, reaching the limits of our abilities in hand computation. We must of necessity transfer our efforts to the computer and use the iterative scheme described above. However, even on the computer, we must seek means to lighten the load on the machine. This is provided by the symmetry of the problem about $x = a/2$ for odd modes. If we introduce a new coordinate system $p$ about the middle, $p = x - a/2$ and the analytical solution $\phi$ of eqn. (5.4.14) is $A \sin[n\pi(p + a/2)/a] = A \sin[n\pi p/a + n\pi/2]$. It is seen that when $n$ is odd, the angle $n\pi/2$ is along the $y$ axis, so that $\phi(p)$ and $\phi(-p)$ are the same. And if it is in the dominant mode with $n = 1$ that we are interested, we may use the mesh of Figure 5.4.1D, and use the potential at node 2 at the node to the right of node 3. We have the matrix equation

$$\begin{bmatrix} 2 & -1 & 0 \\ -1 & 2 & -1 \\ 0 & -2 & 2 \end{bmatrix} \begin{bmatrix} \phi_1 \\ \phi_2 \\ \phi_3 \end{bmatrix} = h^2k^2 \begin{bmatrix} \phi_1 \\ \phi_2 \\ \phi_3 \end{bmatrix}, \tag{5.4.29}$$

giving us the equation

$$(2 - \lambda)^3 - 3(2 - \lambda) = 0, \tag{5.4.30}$$

solving which we obtain $\lambda = 0.267$, 2, or 3.733. $j$ is 6 and we have from eqn. (5.4.15) corresponding values of $\omega_c = (c/a) \times 3.100$, $(c/a) \times 8.486$, and $(c/a) \times 11.593$. We have obviously approximated the first three odd modes. A lesson in this is that symmetry conditions should be used to reduce the work to be done, but with understanding and care lest we lose some modes we might wish to capture.

### 5.4.4 The Transverse Magnetic Wave: In a Rectangular Waveguide

Let us now switch to the more useful two-dimensional system and look at a rectangular guide of size $a \times b$. This is a shape of guide in very common use and for which an analytical solution is available. We need to solve for $\phi = E_z$ governed by eqn. (5.3.27), subject to the boundary conditions $\phi = 0$ at the guide walls $x = 0$, $x = a$, $y = 0$, and $y = b$.

Trying the solution

$$\phi = f(x)g(y) \tag{5.4.31}$$

in eqn. (5.3.27) and dividing by $fg$, we have:

$$f^{-1}f'' + g^{-1}g'' + k^2 = 0. \tag{5.4.32}$$

A function only of $f$ cannot be equal to a function only of $y$, unless both are constants. This gives us:

$$f^{-1}f'' = -k_x^2, \tag{5.4.33}$$

$$g^{-1}g'' = -k_y^2, \tag{5.4.34}$$

$$k^2 = k_x^2 + k_y^2. \tag{5.4.35}$$

The possibilities of $f^{-1}f''$ and $g^{-1}g''$ being zero or positive may be eliminated by showing that for such conditions, the solution $fg$ will be zero, this is left as an exercise to the reader. Therefore

$$f = A_x \cos k_x x + B_x \sin k_x x, \tag{5.4.36}$$

and

$$g = A_y \cos k_y y + B_y \sin k_y y, \tag{5.4.37}$$

giving us

$$\phi = [A_x \cos k_x x + B_x \sin k_x x][A_y \cos k_y y + B_y \sin k_y y]. \tag{5.4.38}$$

Using the boundary condition $\phi = 0$ at $x = 0$,

$$0 = A_x[A_y \cos k_y y + B_y \sin k_y y], \tag{5.4.39}$$

a possibility only if $A_x$ is zero or $A_y$ and $B_y$ are simultaneously zero. The latter, while acceptable, is discounted because it corresponds to a zero solution. Similarly $\phi = 0$ at $y = 0$ gives us $A_y = 0$, leaving us with

$$\phi = A \sin k_x x \times \sin k_y y, \tag{5.4.40}$$

where we have set $A = B_x B_y$. Using $\phi = 0$ at $x = a$ gives us $A = 0$ or

$$\sin k_x a = 0, \tag{5.4.41}$$

meaning

$$k_x = \frac{m\pi}{a} \tag{5.4.42}$$

for all integers $n$. Similarly, $\phi = 0$ at $y = b$, yields

$$k_y = \frac{n\pi}{b}, \tag{5.4.43}$$

so that

$$\phi = E_z = A \sin\frac{m\pi x}{a} \sin\frac{n\pi y}{b}, \tag{5.4.44}$$

$$k^2 = \gamma^2 + \frac{\omega^2}{c^2} = \frac{m^2\pi^2}{a^2} + \frac{n^2\pi^2}{b^2}, \tag{5.4.45}$$

as the only one of the several solution types that passes. The lowest frequency must correspond to $m = n = 1$, since if either $m$ or $n$ is zero, the wave vanishes. The cutoff frequency then is

$$\omega_c = c\pi\sqrt{\left[\frac{m^2}{a^2} + \frac{n^2}{b^2}\right]}, \tag{5.4.46}$$

and the lowest cutoff frequency is

$$\omega_c = c\pi\sqrt{\left[\frac{1}{a^2} + \frac{1}{b^2}\right]}. \tag{5.4.47}$$

Notice is called to zero being unacceptable for $m$ or $n$ since, if this is so, by eqn. (5.4.44), $\phi$ will vanish. Let us try to analyze a square guide using a finite difference mesh. The lowest analytical cutoff frequency with $m = n = 1$ is $(c/a)\pi\sqrt{2} = (c/a) \times 4.44$. With three nodes per edge, as shown in Figure 5.4.2A, the only eigenvalue is 4 and $\omega c = (c/a)j\sqrt{\lambda} = (c/a) \times 4$. For the $3 \times 3$ mesh of Figure

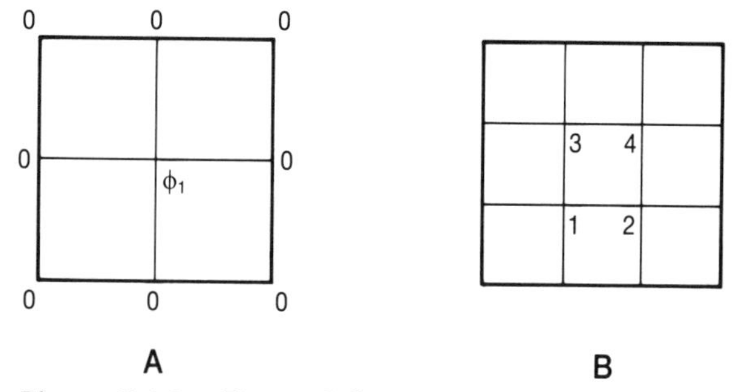


**Figure 5.4.2.** Finite difference meshes for rectangular guide.

5.4.2B, the equation is

$$\begin{bmatrix} 4 & -1 & -1 & 0 \\ -1 & 4 & 0 & -1 \\ -1 & 0 & 4 & -1 \\ 0 & -1 & -1 & 4 \end{bmatrix} \begin{bmatrix} \phi_1 \\ \phi_2 \\ \phi_3 \\ \phi_4 \end{bmatrix} = h^2k^2 \begin{bmatrix} \phi_1 \\ \phi_2 \\ \phi_3 \\ \phi_4 \end{bmatrix}, \quad (5.4.48)$$

for which the eigenvalues are 2, 4, 4, and 6. The corresponding lowest cutoff frequency is $(c/a)3\sqrt{2} = (c/a) \times 4.242$.

It is stressed that although we have worked problems with known analytical answers, the numerical method presented here is equally applicable to other shapes, whereas the analytical scheme of separating the variables does not work for most shapes.

## 5.4.5 The Transverse Magnetic Wave: In a Ridged Waveguide

Having demonstrated the method, it is left as an exercise to the reader to analyze the TM modes in the ridged waveguide of Figure 5.4.3 and obtain the cutoff frequency and flux plots of Figure 5.4.4. The device is of practical importance because by varying the gap length $g$, the cutoff frequency is lowered on account of the capacitive effect at the middle.

## 5.4.6 The Transverse Electric Wave

The derivation of an analytical expression for the TE wave in a rectangular waveguide follows very much along the lines of that for the TM wave, and it is instructive to trace it. We have already shown that

$$\phi = [A_x \cos k_x x + B_x \sin k_x x][A_y \cos k_y y + B_y \sin k_y y]. \quad (5.4.38)$$

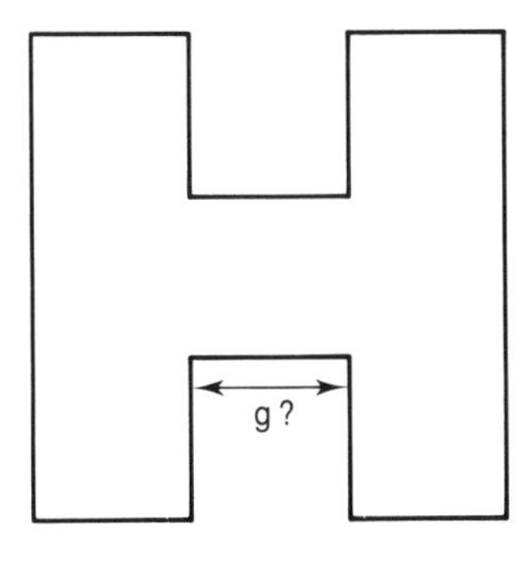


**Figure 5.4.3.** The ridged waveguide.

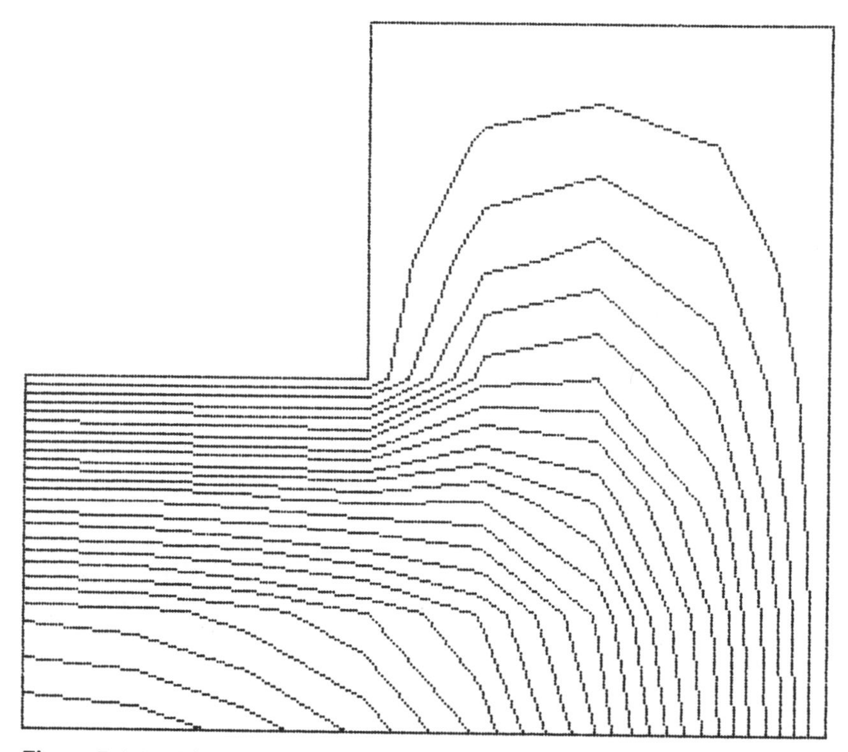

**Figure 5.4.4.** The TM wave in the ridged waveguide.

To solve for $\phi$, we need to find the constants $A_x$, $B_x$, $A_y$, and $B_y$ using the boundary condition that the normal gradient of $\phi$,

$$\frac{\partial \phi}{\partial x} = k_x[-A_x \cos k_x x + B_x \cos k_x x][A_y \cos k_y y + B_y \sin k_y y] \quad (5.4.49)$$

along the walls at $x = 0$ and $x = b$ and

$$\frac{\partial \phi}{\partial y} = k_y[A_x \cos k_x x + B_x \sin k_x x][-A_y \sin k_y y + B_y \cos k_y y] \quad (5.4.50)$$

along the walls at $y = 0$ and $y = b$, shall be zero. Imposing these conditions and eliminating the possibilities leading to trivial solutions as before, we arrive at:

$$\phi = H_z = A \cos \frac{m\pi x}{a} \sin \frac{n\pi y}{b} \quad (5.4.51)$$

The various modes for the different combinations of $m$ and $n$ are referred to as $TE_{mn}$ *waves*. If the wave is not to be evanescent, $m$ may be zero, but $n$ must be at least 1. Therefore, $TE_{01}$ will be the dominant wave with the lowest cutoff frequency. The reader is encouraged to write a program to analyze TE waves and check the program on the rectangular waveguide. Once tested, the program may be used to analyze the guide of Figure 5.4.3 with the TM solution shown in Figure 5.4.4, exploiting symmetry. For the dimensions $a = 1.26 \times 10^{-2}$ m, $b = 3.1 \times 10^{-1}$ m, $H = 0.4a$, and $w = 0.4b$, and a mesh size $h_x = 0.1a$ and $h_y = 0.1b$, a cutoff frequency of $7.04 \times 10^6$ Hz will be obtained. When dealing with a rectangular mesh as here, the terms $h^2$ and $k^2$ should be kept inside the matrix equation, since they do not form a common denominator in the finite difference approximation:

$$\nabla^2\phi = h_x^{-2}[\phi_e + \phi_w - 2\phi_0] + h_y^{-2}[\phi_n + \phi_s - 2\phi_0]. \quad (5.4.52)$$

As a result, the only difference is that the eigenvalue $\lambda$ of the matrix equation will be just $k^2$ and not $h_x^2k^2$ as when the common denominator $h_x^2$, which occurs on the left when $h_x = h_y$, is shifted to the right-hand side.

## 5.5 ANTENNA RADIATION: THE POCKLINGTON EQUATION

In analyzing radiation from antennae, the open nature of the field and the current sources being confined to a few locations in space make integral techniques the natural tool for analysis. We have already used, in section 5.2.2, a method for solving for the current distribution in a conductor under eddy current influence. Here again, if we can find the current distribution down radiating current filaments, we may thence find the pattern of radiation, which is of practical importance. At the time we used the fact that the governing equation:

$$-\mu^{-1}\nabla^2 A = J \quad (1.2.32)$$

in three dimensions has solution

$$\mathbf{A} = \frac{\mu}{4\pi}\iiint r^{-1}\mathbf{J}\, dR. \quad (4.1.3)$$

Here, if we can find an integral solution to the radiating vector potential we can proceed by the same scheme. Under high frequency, we have, after putting eqn. (1.2.24) into eqn. (1.2.4):

$$\nabla \times \mu^{-1}\nabla \times \mathbf{A} = \mathbf{J} + \frac{\partial \mathbf{D}}{\partial t}. \tag{5.5.1}$$

Using eqns. (A9), (1.2.11), and (5.2.3):

$$\mu^{-1}[\nabla\nabla . \mathbf{A} - \nabla^2\mathbf{A}] = \mathbf{J} + \frac{\varepsilon\partial}{\partial t}\left[-\nabla\phi - \frac{\partial}{\partial t}\mathbf{A}\right]. \tag{5.5.2}$$

Since we may select the divergence of **A** arbitrarily, without affecting the physically significant $\nabla \times \mathbf{A}$ (Panofsky and Phillips 1969), we choose the divergence condition known as the Lorentz gauge,

$$\nabla . \mathbf{A} = -\varepsilon\mu\frac{\partial\phi}{\partial t}, \tag{5.5.3}$$

which upon substitution in eqn. (5.5.2) yields the wave equation in **A**:

$$\nabla^2\mathbf{A} = -\mu\mathbf{J} + c^{-2}\frac{\partial^2\mathbf{A}}{\partial t^2}. \tag{5.5.4}$$

We have already proved in section 5.3, however, that waves can propagate only when the time variation is sinusoidal. In fact, this means only that the wave must have a cyclic pattern so that it may be broken up into several sinusoidal components by Fourier analysis. After all, if only a sinusoid may be transmitted, then it would mean that waves cannot be used to send information. Therefore, converting to phasors:

$$\nabla^2\hat{\mathbf{A}} = -\mu\hat{\mathbf{J}} - \omega^2c^{-2}\hat{\mathbf{A}}. \tag{5.5.5}$$

This equation is very similar to eqn. (1.2.32) except for being in complex numbers and the presence of the additional term $-\omega^2c^{-2}\mathbf{A}$. We may therefore expect the solution to the wave equation to reduce to that of the Poisson equation (4.1.3), when $\omega = 0$. Furthermore, we have also shown in section 5.3 that when a wave travels in the $z$ direction without attenuation, the field values vary as $e^{-j\omega z/c}$. And since in spherical coordinates $r$, $\theta$, $\phi$, the wave travels in the direction of increasing $r$, one may expect that

$$\hat{\mathbf{A}} = \frac{\mu}{4\pi}\iiint\frac{\hat{\mathbf{J}}e^{-j\omega r/c}}{r}\,dV \tag{5.5.6}$$

is a good candidate for the solution. To verify our guess, first note that in spherical coordinates, with line elements, $dr$, $r\,d\theta$, and

$r \sin\theta\, d\phi$, the gradient operator is

$$\nabla = \mathbf{u}_r \frac{\partial}{\partial r} + \mathbf{u}_\theta \frac{1}{r}\frac{\partial}{\partial \theta} + \mathbf{u}_\phi \frac{1}{r \sin\theta}\frac{\partial}{\partial \phi}, \tag{5.5.7}$$

and the divergence operation on a vector $\mathbf{A}$ is defined (Panofsky and Phillips 1969) by

$$\nabla \,.\, \mathbf{A} = \frac{1}{r^2} r^2 \frac{\partial A_r}{\partial r} + \frac{1}{r \sin\theta}\frac{\partial}{\partial \theta} \sin\theta A_\theta + \frac{1}{r \sin\theta}\frac{\partial A_\phi}{\partial \phi}. \tag{5.5.8}$$

Observe further that the operator $\nabla^2$ in eqn. (5.5.6) operates only at the field point and sees the source $\mathbf{J}$ as a constant, so that from eqn. (A10),

$$\begin{aligned}
\nabla \hat{J}_x \frac{e^{-j\omega r/c}}{r} &= \hat{J}_x \nabla \frac{e^{-j\omega r/c}}{r} = \hat{J}_x\{r^{-1}\nabla e^{-j\omega r/c} + e^{-j\omega r/c}\nabla r^{-1}\} \\
&= \hat{J}_x\left\{r^{-1}\mathbf{u}_r\left[-\frac{j\omega}{c}e^{-j\omega r/c}\right] + e^{-j\omega r/c}\nabla r^{-1}\right\}, \qquad \text{from eqn. (5.5.7)}, \\
&= -\mathbf{u}_r \hat{J}_x e^{-j\omega r/c}\frac{j\omega}{rc} + \hat{J}_x e^{-j\omega r/c}\nabla r^{-1},
\end{aligned} \tag{5.5.9}$$

whence, using eqn. (A11):

$$\begin{aligned}
\nabla \,.\, \nabla \hat{J}_x \frac{e^{-j\omega r/c}}{r} &= -\nabla \,.\left\{-\mathbf{u}_r \hat{J}_x e^{-j\omega r/c}\frac{j\omega}{rc} + \hat{J}_x e^{-j\omega r/c}\nabla r^{-1}\right\} \\
&= \frac{1}{r^2}\frac{\partial}{\partial r}\left\{-r^2 \hat{J}_x e^{-j\omega r/c}\frac{j\omega}{rc}\right\} + \hat{J}_x\{e^{-j\omega r/c}\nabla^2 r^{-1} + \nabla r^{-1} \,.\, \nabla e^{-j\omega r/c}\} \\
&= -\frac{j\omega \hat{J}_x}{cr^2}\left\{e^{-j\omega r/c} + r\frac{-j\omega}{c}e^{-j\omega r/c}\right\} \\
&\quad + \hat{J}_x\left\{e^{-j\omega r/c}\nabla^2 r^{-1} + (-r^{-2})\frac{-j\omega}{c}e^{-j\omega r/c}\right\} \\
&= -\hat{J}_x \frac{\omega^2}{c^2} r^{-1} e^{-j\omega r/c} + \hat{J}_x e^{-j\omega r/c}\nabla^2 r^{-1}.
\end{aligned} \tag{5.5.10}$$

Integrating both sides over the solution region, we have

$$\frac{4\pi}{\mu}\nabla^2 \hat{A}_x = -\frac{4\pi}{\mu}\frac{\omega^2}{c^2}\hat{A}_x + \iiint \hat{J}_x e^{-j\omega r/c}\nabla^2 r^{-1}\, dV. \tag{5.5.11}$$

But we have in eqn. (2.5.6) that $\nabla^2 r^{-1}$ is zero everywhere except at

$r = 0$, where $e^{-j\omega r/c}$ is 1. Therefore, by eqn. (2.5.10),

$$\iiint \hat{J}_x e^{-j\omega r/c} \nabla^2 r^{-1}\, dV = -4\pi \hat{J}_x. \tag{5.5.12}$$

Substituting in eqn. (5.5.11), we have

$$\nabla^2 \hat{A}_x = -\mu \hat{J}_x - \frac{\omega^2}{c^2} A_x. \tag{5.5.13}$$

Similar expressions for $\hat{A}_y$ and $\hat{A}_z$ may be derived, and summing the three components for $\hat{\mathbf{A}}$, eqn. (5.5.5) will be obtained. Thus, we confirm that eqn. (5.5.6) is indeed the solution to eqn. (5.5.5).

Now, in an AC system, eqns. (5.2.3) and (5.5.3), respectively, reduce to

$$\hat{\mathbf{E}} = -\nabla \hat{\phi} - j\omega\, \hat{\mathbf{A}}, \tag{5.5.14}$$

$$\nabla \,.\, \hat{\mathbf{A}} = -j\varepsilon\mu\omega\hat{\phi}. \tag{5.5.15}$$

Taking the gradient of eqn. (5.5.15) and substituting in eqn. (5.5.14) for $\nabla\phi$, we have:

$$\hat{\mathbf{E}} = \frac{1}{j\varepsilon\mu\omega} \nabla^2 \hat{\mathbf{A}} - j\omega\hat{\mathbf{A}}. \tag{5.5.16}$$

Now substituting for $\mathbf{A}$ using eqn. (5.5.6), we obtain Pocklington's equation:

$$j\omega\varepsilon\hat{\mathbf{E}} = \left(\nabla_f^2 + \frac{\omega^2}{c^2}\right) \iiint_s \frac{\hat{\mathbf{J}} e^{-j\omega r/c}}{4\pi r}\, dR, \tag{5.5.17}$$

where the suffixes $f$ and $s$ indicate whether the integrals and differentiations are over field or source points. Now dividing the antennae into small bits over each of which $\hat{\mathbf{J}}$ is assumed a constant, a matrix equation for the unknown $\hat{\mathbf{J}}$ values in the filaments may be constructed in terms of the driving electric field $\hat{\mathbf{E}}$ and solved.

Pocklington's equation assumes a very simple form for a thin wire antenna lying in, say, the $z$ direction. Then $\hat{\mathbf{J}}$ and, therefore, $\hat{\mathbf{A}}$ have components along the $z$ direction, and eqn. (5.5.6) reduces to

$$j\omega\varepsilon\hat{E}_z = \left(\frac{\partial^2}{\partial z^2} + \frac{\omega^2}{c^2}\right) \iiint_s \frac{\hat{I} e^{-j\omega r/c}}{4\pi r}\, dz. \tag{5.5.18}$$

For constant line currents $\hat{I}_j$ in each element of length $L_j$ of the wire, this may be expressed as

$$\hat{E}_{zi} = \sum A_{ij} \hat{I}_i, \tag{5.5.19}$$

and solved for the unknowns $I_j$ as explained by Silvester and Chan (1972) and Hassan and Silvester (1977).

# 6 MATRIX SOLUTION

## 6.1 THE IMPORTANCE OF EFFICIENCY

In the computer exercises we have come across in the preceding sections, one notices that the matrix equation for $n$ unknowns involves dimensioning an $n \times n$ real matrix with $n^2$ storage locations. Obviously, with this kind of storage even the most powerful machine (let alone small machines) would be on its knees for a simple problem with 1000 unknowns. This situation is unacceptable and we should take into account the architectural details of the computer, the symmetry of the matrix, and whether the matrix is sparse, in the sense that most of the numbers of such a matrix are zero.

Such considerations result in an efficient use of the computing power at our disposal and will allow us to solve problems faster and, indeed, allow us to attempt larger and larger problems.

In the course of the finite difference analysis (or the finite element analysis to be taken up in the next chapter) of electromagnetic fields, among a myriad other sciences, sparse, positive definite, diagonally dominant, and (often) symmetric matrixes must be solved. The matrix solution often requires the most amount of computing in field analysis, and any attempt at making our programs efficient should first be directed to this arena. Two kinds of matrix problem exist in solving the equation

$$\mathbf{S}\mathbf{x} = \mathbf{b} \tag{6.1.1}$$

in field analysis. First, we have the problem of solving for a unique $\mathbf{x}$, with a nonsingular matrix $\mathbf{S}$, which is what we shall be most concerned with here. We have already seen examples of this in the electric and magnetic field problems of chapters 2, 3, 4, and 5. Second, we have the eigenvalue problem of chapter 5, where, for a zero **b** and singular **S**, we need to find **x**. Here, in fact, the matrix **S** is a function of some unknown parameter such as frequency, and we attempt to identify those values of this parameter that will make **S** singular—for only then can a nonzero solution **x** exist.

Several simple schemes exist for gaining efficiency in matrix computation. Most computers, when we dimension an $n \times n$ matrix, store the matrix column by column. Thus, although we may designate, say, **S** as an $n \times n$ matrix, the machine stores it as a vector; i.e., column 1 of the matrix **S** occupies the first $n$ locations of a vector entity **SV** in the machine, column 2 occupies the next $n$ numbers, and so on. In mathematical terms, $S[i,j]$ is stored in a vector **SV** within the computer, at location $(i - 1)^*n + j$. Therefore, whenever a matrix is dimensioned and we call one of its elements $S[i,j]$, the computer identifies **S** as a matrix through the dimension statement and picks out the coefficient of the vector in which it is stored, by evaluating its location $(i - 1)^*n + j$. We may exploit this fact to save some cpu time on the machine by immediately dimensioning all matrices as vectors, instead of letting the machine do it for us; thus, we often spare the machine the burden of computing the locations of terms. In other words, an $n \times n$ matrix **S** is dimensioned directly as a vector **SV** of length $n^2$ and whenever we wish to refer to $S[i,j]$, we devise special algorithms to locate an element without performing the operation $(i - 1)^*n + j$. When this cannot be avoided, we refer to **SV**$[(i - 1)^*n + j]$ after computing $(i - 1)^*n + j$. For example, as shown in Alg. 6.1.1, matrix addition

$$S[i,j] = A[i,j] + B[i,j] \qquad i = 1,\ldots,n; j = 1,\ldots,n \quad (6.1.2)$$

becomes straightforward without ever having to perform a multi-

**Algorithm 6.1.1. Efficient Matrix Addition**

**Procedure MatAdd(S, A, B,** $n, m$**)**
{Function: Adds the $n \times m$ column stored matrices **A** and **B** and returns the sum in **S**
Output: **S**
Inputs: **A, B,** $n, m$}
**Begin**
  **For** $i \leftarrow 1$ **To** $n^*m$ **Do**
    $S[i] \leftarrow A[i] + B[i]$
**End**

**Algorithm 6.1.2. Efficient Matrix Multiplication**

```
Procedure MatMultiply(S, A, B, l, m, n)
{Function: Multiplies the l × m matrix A and the m × n matrix
B and returns the result in the l × n matrix S. A, B, and S are
column stored
Output: S
Inputs: A, B, l, m, n}
Begin
  ij ← 0
  kjStart ← 1 − m
  For j ← 1 To n Do
    kjStart ← kjStart + m
    For i ← 1 To l Do
      ij ← ij + 1
      kj ← kjStart
      ik ← i
      x ← 0.0
      For k ← 1 To m Do
        x ← x + A[ik]*B[kj]
        ik ← ik + 1
        kj ← kj + 1
      S[ij] ← x
End
```

plication for locating a coefficient. Alg. 6.1.2 gives a possible scheme for doing a matrix multiplication

$$\mathbf{S}[i,j] = \sum_{k=1}^{m} \mathbf{A}[i,k]*\mathbf{B}[k,j] \qquad i = 1,\ldots,l;\, j = 1,\ldots,n, \quad (6.1.3)$$

exploiting the way the computer stores numbers. In this algorithm, $ij$ represents the location of $S[i,j]$ in the one-dimensional array **S**; similarly, $ik$ and $kj$ correspond to eqn. (6.1.3). We have used the fact that when we move along a column as in $B[k,j]$, $kj$ will increment by 1, and when along a row as in $A[i,k]$, the indices $ik$ will increment by the number of rows of **A**, which is $l$. A dummy store $x$ is used so as not to cause any page faults (described in the next paragraph) in repeatedly calling the matrix **S**, until the summation (6.1.3) is complete.

Another means of saving time exploits the algorithms we are using. Most machines, such as the DEC-VAX, are allowed only a limited memory in the processor. Thus, while a program may deal with several kilobytes, only a small portion of it remains in core—and the moment we refer through the program to a number that is in secondary storage, what is in core is sent out to secondary storage, and what must be operated on by the computer is called into core. This swapping is called a page fault, because it involves the slow process of reading from and writing on secondary storage, but no cpu time is expended. In matrix operations, for example in a machine that allows 100 words in internal memory, the moment we refer to $S[1,1]$ of a $100 \times 100$ matrix, the computer will call in the first column of the matrix, since 100 real numbers and no more may be kept in core. After operating on $S[1,1]$, if we now refer to $S[2,1]$, since $S[2,1]$ is already in core, the machine will quickly operate on it. However, should we operate on $S[1,2]$ after operating on $S[1,1]$, since $S[2,1]$ is at location 101, which is outside the machine, the machine will have to send out all or a part of the first column and bring in the second column. This is slow and will increase both the page faults on the machine and total running time.

To avoid the disadvantages of this system of paging in a computer, if we look at our SOR algorithm 3.2.1, we will notice that we always refer to the matrix row by row. Thus, running a large matrix solution program with regular matrix or column-by-column storage will yield a large number of page faults. This problem is easily overcome by resorting to row-by-row storage of the matrix. That is, we would dimension the matrix as a vector **SV** of length $n^2$ and store item $\mathbf{S}[i,j]$ in $\mathbf{SV}[(i-1)^*n + j]$. This writer recalls with some amusement (and, needless to say, embarrassment), his days as a graduate student when he tried solving a $600 \times 600$ matrix by the SOR scheme with full storage on a multiuser machine and had to crash the program after two days! Finally, the answer was obtained in two hours with row storage. It is important to stress that this may not be the best scheme when we are using another algorithm in which we do not refer to numbers consecutively along a row.

The procedures described above for enhancing efficiency relate to getting the best efficiency from a given method. However, advances in efficiency may also be made in changing over completely to a different method of equation solution. For finding the solution for nonsingular matrices, two approaches are commonly taken; these are through the direct and iterative schemes. In section 2.3, we were introduced to the direct Gaussian elimination scheme. In section 3.2, we gained some experience with the SOR scheme, which is an iterative scheme. It is seen that in the direct scheme we have a solution at the end of the computational process. The error in the solution will depend on the accuracy of the computer used and is caused by round-off error arising in the arithmetic operations. Thus, as the size of the problem grows, so will the error. In the iterative schemes, on the other hand, we have an approximation to the solution, at any point in the solution process, which is successively improved with the iterations. Thus it is we, the users, who determine when to stop computation. The matrix coefficients are always preserved, so that the error in the solution will not increase with the size of the problem. Iterative schemes have a special advantage when we are able to make a good guess at the solution, as in solving a nonlinear problem or in solving a refined mesh for which a solution corresponding to its cruder predecessor exists; for the approximate solution may be used as a starting solution and the convergence time may be cut down. Moreover, it will be explained in the sections to come that iterative schemes are more amenable to having the sparsity of the matrix exploited. It may be stated in general that as the problem size grows, iterative schemes become more and more preferable. When the transition in economy occurs is a very subjective issue (Hoole, Anandaraj, and Bhatt 1987).

Currently, the two most commonly used methods of matrix solution in field computation are the frontal solution algorithm (Irons 1970), resorting to a direct decomposition of the matrix, and the semi-iterative preconditioned conjugate gradient algorithm (Jennings 1977; Kershaw 1978; Hoole et al. 1986), relying upon finding, for an $n \times n$ matrix, the solution as a sum of $n$ or fewer vectors, each of which is found during an iteration.

Sparse $n \times n$ matrices using full storage require memory growing as $n^2$, although the nonzero coefficients usually grow only as $n$. For example, in a finite difference mesh, the equation at each node is expressed in terms of the four surrounding nodes, so that only five coefficients appear in a row, giving us $5n$ entries in the matrix. Where the problem is surrounded by a Dirichlet boundary, the matrix equation is symmetric in addition to being sparse, so that only the lower or upper half of the matrix needs to be stored, and numbers from the half that is not stored may be obtained from that which is stored. If we decide to store the lower half on a particular row, the stored coefficients will correspond to columns numbered lower than the row. So out of the four nodes surrounding a node, probabilistically taking two to have lower numbers, we need to store only three coefficients, counting the diagonal terms, giving us $3n$ entries. As will be seen in the next chapter, the finite element matrices are always symmetric. In an ideal, first-order, optimally triangular finite element mesh, all triangles will have similar error, and therefore triangles will be either equilateral or close to it. Therefore, each node will be approximately connected to

**A**

| | | | | | | | |
|---|---|---|---|---|---|---|---|
| 1 | | | | | | | |
| 2 | 3 | | | | | | |
| 0 | 4 | 5 | | | | | |
| 0 | 0 | 0 | 6 | | | | |
| 0 | 7 | 8 | 0 | 9 | | | |
| 10 | 0 | 0 | 11 | 0 | 12 | | |
| 0 | 0 | 13 | 0 | 14 | 0 | 15 | |
| 16 | 17 | 0 | 18 | 0 | 19 | 0 | 20 |

**B**

| | | | | | | | |
|---|---|---|---|---|---|---|---|
| 1 | | | | | | | |
| 2 | 3 | | | | | | |
| 0 | 4 | 5 | | | | | |
| 0 | 0 | 0 | 6 | | | | |
| 0 | 7 | 8 | 9 | 10 | | | |
| 11 | 12 | 13 | 14 | 15 | 16 | | |
| 0 | 0 | 17 | 18 | 19 | 20 | 21 | |
| 22 | 23 | 24 | 25 | 26 | 27 | 28 | 29 |

**C**

**Col** = [1, 1,2, 2,3, 4, 2,3,5, 1,4,6,
3,5,7, 1,2,4,6,8]

**Dia** = [1,3,5,6,9,12,15,20]

**SV** = [1.0,
0.5,1.0,
0.2,1.0,
1.0,
-0.1,0.2,1.0,
-0.15,0.4,1.0,
-0.32,0.15,1.0,
0.32,-0.24,0.16,-0.16,1.0]

**D**

| | | | | | | | |
|---|---|---|---|---|---|---|---|
| 1 | | | | | | | |
| 2 | 3 | | | | | | |
| 4 | 5 | 6 | | | | | |
| 0 | 7 | 8 | 9 | | | | |
| 0 | 0 | 10 | 11 | 12 | | | |
| 0 | 0 | 0 | 0 | 13 | 14 | | |
| 0 | 0 | 0 | 0 | 15 | 16 | 17 | |
| 0 | 0 | 0 | 0 | 0 | 18 | 19 | 20 |

**E**

**FC** = [1, 1, 2, 4, 2, 1, 3, 1]

**Dia** = [1, 3, 5, 6, 9,12,15,20]

**SV** = [1.0,
0.5,1.0,
0.2,1.0,
1.0,
-0.1,0.2,**0.0,**1.0,
-0.15,**0.0,0.0,**0.4,**0.0,**1.0,
-0.32,**0.0,**0.15,**0.0,**1.0,
0.32,-0.24,**0.0,**0.16,**0.0,**-0.16,**0.0,**1.0]

**Figure 6.1.1.** Storage in sparse matrices. **A.** Sparse symmetric matrix and storage scheme. **B.** Dense nature of inverse fill-in within band. **C.** Data structures for sparse symmetric storage corresponding to **A**. **D.** Renumbered matrix of **A**. **E.** Data structures for symmetric profile storage corresponding to **A**.

six others; i.e., an equation corresponding to a particular node will have seven coefficients relating the potential at that node to those at the other six nodes. Thus, arguing as before, if the symmetry and sparsity of the system are exploited, storing the diagonal coefficient and three off-diagonal ones in the lower half, only $4n$ coefficients need be stored. More generally, the storage requirement for the coefficients is $cn$ where $c$ is the approximate connectivity of a node, or, with symmetric matrices, half of the average connectivity of a node. However, storing only the nonzero coefficients also means storing the rows and columns of the coefficients. Special schemes exist (Silvester, Cabayan, and Browne 1973) for storing all the information of the matrix with $nc$ real numbers and $(n + 1)c$ integers. This is shown in Figure 6.1.1, where Figure 6.1.1D gives the data structure corresponding to the matrix of Figure 6.1.1A. The 20 nonzero coefficients of the $8 \times 8$ matrix are stored in a vector **SV** of real numbers, and **Col**, an array of 20 integers, gives the corresponding column numbers of the coefficients. **Diag**, an array of eight integers, gives the positions in **SV** of the diagonal terms. The precise use of these will be taken up in detail in section 6.4 below.

Therefore, in solving sparse matrices a fair deal of time must first be expended in generating the data structure as in Figure 6.1.1C. We go through the matrix equation, or alternatively the physical system that produces the equation (such as a finite element grid), and, using the connectivities, compute the arrays **Col** and **Diag** or similar arrays coming from some variant of the storage scheme. This process can take as much as 30% of total solution time, depending on the efficiencies of the algorithms used for sparsity computation and matrix solution.

Although computing the data structures for sparse storage takes up time, we undertake it for two reasons. First, by reducing storage requirements to the order of the matrix size from the previous square, the solution of large problems is made practicable. Second, by not storing the nonzero coefficients, several multiplications by zero (which on a computer take the same time as multiplications with nonzero numbers) are avoided, thus more than offsetting the time lost in computing the sparsity.

## 6.2 THE CHOLESKY FACTORIZATION SCHEME

The Cholesky factorization technique is a direct scheme for solving the matrix equation (6.1.1) and is particularly suited to symmetric matrices **S**. This relies on the factorization

$$\mathbf{S} = \mathbf{L}\mathbf{L}^t, \tag{6.2.1}$$

where **L** is a lower triangular matrix. The solution method relies on splitting eqn. (6.1.1) into two:

$$\mathbf{L}\mathbf{z} = \mathbf{b} \tag{6.2.2}$$

and

$$\mathbf{L}^t\mathbf{x} = \mathbf{z}, \tag{6.2.3}$$

which is the same as

$$\mathbf{L}\mathbf{L}^t\mathbf{x} = \mathbf{b}. \tag{6.2.4}$$

Thus if we can find **L**, from eqn (6.2.2), by forward elimination **z** may be computed and then from eqn. (6.2.3) in which **z** is now known, **x** may be computed by back substitution. For example, if we had

$$\begin{bmatrix} 4 & 2 \\ 2 & 5 \end{bmatrix} \begin{bmatrix} x_1 \\ x_2 \end{bmatrix} = \begin{bmatrix} 1 \\ 1 \end{bmatrix}, \tag{6.2.5}$$

we may use the factorization

$$\begin{bmatrix} 4 & 2 \\ 2 & 5 \end{bmatrix} = \begin{bmatrix} 2 & 0 \\ 1 & 2 \end{bmatrix} \begin{bmatrix} 2 & 1 \\ 0 & 2 \end{bmatrix}, \tag{6.2.6}$$

so that corresponding to eqn. (6.2.2),

$$\begin{bmatrix} 2 & 0 \\ 1 & 2 \end{bmatrix} \begin{bmatrix} z_1 \\ z_2 \end{bmatrix} = \begin{bmatrix} 1 \\ 1 \end{bmatrix}. \tag{6.2.7}$$

Using the first row we have $2z_1 = 1$ or $z_1 = \frac{1}{2}$, and from the second row we have $z_1 + 2z_2 = 1$, which gives us $z_2 = \frac{1}{4}$. Now corresponding to eqn. (6.2.3):

$$\begin{bmatrix} 2 & 1 \\ 0 & 2 \end{bmatrix} \begin{bmatrix} x_1 \\ x_2 \end{bmatrix} = \begin{bmatrix} \frac{1}{2} \\ \frac{1}{4} \end{bmatrix}. \tag{6.2.8}$$

From the last row we now have $2x_2 = \frac{1}{4}$ or $x_2 = \frac{1}{8}$, and from the first row we have $2x_1 + x_2 = \frac{1}{2}$, using the now known value of $x_1 = \frac{3}{16}$.

It is seen that once the Cholesky factorization of the matrix is made, the computation of **x** is trivial. To determine the rule for computing **L**, consider:

$$\begin{bmatrix} L_{11} & 0 & 0 & \cdots \\ L_{21} & L_{22} & 0 & \cdots \\ L_{31} & L_{32} & L_{33} & \cdots \\ | & | & | & \end{bmatrix} \begin{bmatrix} L_{11} & L_{21} & L_{31} & \cdots \\ 0 & L_{22} & L_{32} & \cdots \\ 0 & 0 & L_{33} & \cdots \\ | & | & | & \end{bmatrix} = \begin{bmatrix} S_{11} & S_{21} & S_{31} & \cdots \\ S_{21} & S_{22} & S_{23} & \cdots \\ S_{31} & S_{32} & S_{33} & \cdots \\ | & | & | & \end{bmatrix}. \tag{6.2.9}$$

To obtain an element $\mathbf{S}[i, j]$ of **S**, we multiply the $i$th row $\mathbf{L}^i$ of **L** with the $j$th column of $\mathbf{L}^t$, which is $\mathbf{L}^j$, the $j$th row of **L**, both of $n$ terms. Now since **S** is symmetric, the equation for $S_{12}$ will be the same as that for $S_{21}$, etc. So confining ourselves to the lower half of **S** where $i \geq j$:

$$\mathbf{L}^i = [L_{i1} \quad L_{i2} \quad L_{i3} \ldots \quad L_{ij} \ldots \quad L_{ii} \quad 0 \quad 0 \ldots], \tag{6.2.10}$$

$$\mathbf{L}^j = [L_{j1} \quad L_{j2} \quad L_{j3} \ldots \quad L_{jj} \quad 0 \quad 0 \quad 0 \ldots], \tag{6.2.11}$$

where the diagonal or last nonzero term of $\mathbf{L}^j$ appears not later than its counterpart in $\mathbf{L}^i$ and both vectors go to $n$ terms. Performing the multiplication, we need to consider two cases. First $i = j$:

$$S_{ii} = \mathbf{L}^i\mathbf{L}^i = L_{i1}^2 + L_{i2}^2 + L_{i3}^2 + \ldots + L_{ii}^2 = \sum_{k=1}^{i} L_{ik}^2, \tag{6.2.12}$$

and second $i > j$:

$$S_{ij} = \mathbf{L}^i\mathbf{L}^j = L_{i1}L_{j1} + L_{i2}L_{j2} + L_{i3}L_{j3} + \ldots + L_{ij}L_{jj} = \sum_{k=1}^{j} L_{ik}L_{jk}. \tag{6.2.13}$$

Rearranging, we have

$$L_{ij} = \begin{cases} \sqrt{\left[S_{ij} - \sum_{k=1}^{i-1} L_{ik}^2\right]} & i = j > 1 \\ \dfrac{S_{ij} - \sum_{k=1}^{j-1} L_{ik}L_{jk}}{L_{jj}} & i > j \end{cases} \tag{6.2.14}$$

Observe also that, by multiplying $\mathbf{L}^i$ and $\mathbf{L}^{t1}$ in eqn. (6.2.9),

$$S_{11} = L_{11}^2. \tag{6.2.15}$$

Knowing $L_{11}$, then, we have from eqn. (6.2.14):

$$L_{21} = \frac{S_{21}}{L_{11}}, \tag{6.2.16}$$

$$L_{22} = \sqrt{[S_{22} - L_{21}^2]}, \tag{6.2.17}$$

$$L_{31} = \frac{S_{31}}{L_{11}}, \tag{6.2.18}$$

$$L_{32} = \frac{S_{32} - L_{31}L_{21}}{L_{22}}, \tag{6.2.19}$$

$$L_{33} = \sqrt{[S_{33} - L_{31}^2 - L_{32}^2]}, \tag{6.2.20}$$

and so on. It should be noted that once $L_{ij}$ is computed, $S_{ij}$ is never required. This allows us to write $L_{ij}$ in the same storage location as assigned to $S_{ij}$ at great savings in memory. The chief disadvantage in the method is that as the matrix size becomes large, the memory requirements become excessive. Moreover, on account of large round-off errors in the solution for large matrices, so as to quantify and check the round-off error, the need to store **S** becomes greater with increasing matrix size, so that the storage savings realized by overwriting **L** on **S** vanish.

## 6.3 THE CONJUGATE GRADIENTS ALGORITHM

The conjugate gradients scheme (Kershaw 1978; Meijerink and van der Vost 1977) is used for the solution of sparse positive definite symmetric matrix equations of the type seen in eqn. (6.1.1). In this method, the solution is found as a series of vectors $\mathbf{p}_i$:

$$\mathbf{x} = \alpha_1\mathbf{p}_1 + \alpha_2\mathbf{p}_2 + \ldots + \alpha_m\mathbf{p}_m, \tag{6.3.1}$$

where $m$ is no larger than the matrix size $n$ and depends on the starting direction $\mathbf{p}_1$ and the scatter between the eigenvalues of **S**, or the spectral radius of **S**. Hence, although this is an iterative scheme, it is guaranteed to converge in $n$ or fewer iterations; the term *semi-iterative* is therefore commonly employed.

The underlying principle of the conjugate gradients scheme is the existence of $n$ vectors $\mathbf{p}_i$ orthogonal to the matrix of coefficients:

$$\mathbf{p}_i^t\mathbf{S}\mathbf{p}_j \begin{cases} = 0 & i \neq j \\ \neq 0 & i = j. \end{cases} \tag{6.3.2}$$

Thus, by substituting eqn. (6.3.1) in eqn. (6.1.1) and premultiplying by $\mathbf{p}_i^t$:

$$\mathbf{p}_i^t\mathbf{S}[\alpha_1\mathbf{p}_1 + \alpha_2\mathbf{p}_2 + \ldots + \alpha_m\mathbf{p}_m] = \mathbf{p}_i^t\mathbf{b}, \tag{6.3.3}$$

which gives us

$$\alpha_i = \frac{\mathbf{p}_i^t\mathbf{b}}{\mathbf{p}_i^t\mathbf{S}\mathbf{p}_i}. \tag{6.3.4}$$

Thus, the problem of finding the unknown $\mathbf{x}$ reduces to one of finding the orthogonal vectors $\mathbf{p}_i$, having which, using eqn. (6.3.4), we may compute the terms $\alpha_i$ and then $\mathbf{x}$ from eqn. (6.3.1). Since the set of vectors $\mathbf{p}$ is not unique, we are free to assume any value for $\mathbf{p}_1$. In the algorithm for finding the other $\mathbf{p}$'s, there is no need to store all of them, so that the demands on computer memory are not heavy.

$$\mathbf{p}_1 = \mathbf{b} - \mathbf{S}x_1 \tag{6.3.5}$$

is a good value to take for the starting vector $\mathbf{p}_i$, because what we now need to add to the starting value $x_1$, namely a multiple $\alpha$ of $\mathbf{p}_1$, must have components parallel to the residual. In general, at a step $i$ in the iterative process,

$$\mathbf{x}_{i+1} = \mathbf{x}_i + \alpha_i \mathbf{p}_i. \tag{6.3.6}$$

We need at this point to find $\alpha_i$. If we regard the operation at this iterative step as solving the matrix equation

$$\mathbf{S}\mathbf{e}_i = \mathbf{b} - \mathbf{S}\mathbf{x}_i = \mathbf{r}_i \tag{6.3.7}$$

for the vector residual

$$\mathbf{e}_i = \mathbf{x} - \mathbf{x}_i = \alpha_i \mathbf{p}_i + \alpha_{i+1}\mathbf{p}_{i+1} + \ldots \tag{6.3.8}$$

in the current solution $\mathbf{x}_i$, then according to eqn. (6.3.4):

$$\alpha_i = \frac{\mathbf{p}_i^t \mathbf{r}_i}{\mathbf{p}_i^t \mathbf{S} \mathbf{p}_i}. \tag{6.3.9}$$

After computing $\alpha$, we may update the solution to $\mathbf{x}_{i+1}$, using eqn. (6.3.8), corresponding to which we will have a new residual $\mathbf{r}_{i+1}$. Now we need to find $\mathbf{p}_{i+1}$, to make another improvement in the solution, and, as before, we may ideally take $\mathbf{p}_{i+1}$ as the new residual $\mathbf{r}_{i+1}$. However, having accounted for all components in the solution parallel to $\mathbf{p}_i$, we do not wish $\mathbf{p}_{i+1}$ to have any components parallel to $\mathbf{p}_i$, with which no more gains in accuracy may be made in the solution. Therefore, we need to take off from the residual an appropriate multiple $\beta$:

$$\mathbf{p}_{i+1} = \mathbf{r}_{i+1} - \beta_i \mathbf{p}_i. \tag{6.3.10}$$

The value of $\beta_i$ is found by requiring the orthogonality relationship eqn. (6.3.2):

$$\mathbf{p}_i^t \mathbf{S} \mathbf{p}_{i+1} = \mathbf{p}_i^t \mathbf{S} \mathbf{r}_{i+1} - \beta_i \mathbf{p}_i^t \mathbf{S} \mathbf{p}_i = 0, \tag{6.3.11}$$

**Algorithm 6.3.1. The Conjugate Gradients Algorithm**

Guess $\mathbf{x}$
$\mathbf{r} \leftarrow \mathbf{b} - \mathbf{S}\mathbf{x}$
$\mathbf{p} \leftarrow r$
$\varepsilon \leftarrow$ Permitted Error
$e \leftarrow$ Norm($\mathbf{r}$)
**While** $e > \varepsilon$ **Do**
$\alpha \leftarrow \mathbf{p}^t\mathbf{p}/\mathbf{p}^t\mathbf{S}\mathbf{p}$
$\mathbf{x} \leftarrow \mathbf{x} + \alpha\mathbf{p}$
$\mathbf{r} \leftarrow \mathbf{b} - \mathbf{S}\mathbf{x}$
$e \leftarrow$ Norm($\mathbf{r}$)
$\beta \leftarrow -\mathbf{r}^t\mathbf{S}\mathbf{p}/\mathbf{p}^t\mathbf{S}\mathbf{p}$
$\mathbf{p} \leftarrow \mathbf{r} - \beta\mathbf{p}$

whence

$$\beta_i = \frac{\mathbf{p}_i^t\mathbf{S}\mathbf{r}_{i+1}}{\mathbf{p}_i^t\mathbf{S}\mathbf{p}_i}. \tag{6.3.12}$$

Now we may compute $\mathbf{p}_{i+1}$ and we are ready for the next iterative improvement in $\mathbf{x}$. This procedure is algorithmically stated in Alg. 6.3.1.

In its initial form, as proposed by Hestenes and Stiefel (1952), the conjugate gradients method was not competitive with other schemes of solution. Thus, it did not come into its own until the introduction of the preconditioning schemes of recent years, relying upon solving an equivalent matrix equation with improved convergence properties. It may be shown that the number of conjugate gradient steps required for convergence is the same as the number of independent eigenvalues in the matrix of coefficients $\mathbf{S}$. Preconditioning, therefore, is an attempt to solve an equivalent matrix equation with clustered eigenvalues. The ideal matrix equation has the identity matrix as the coefficient matrix, since it has all $n$ eigenvalues the same, given by the solution of $[\lambda - 1]^n = 0$. The method becomes competitive with direct methods in its preconditioned form. In general, the modified equation takes the form

$$\mathbf{B}\mathbf{S}\mathbf{B}^t[B^{-t}\mathbf{x}] = \mathbf{B}\mathbf{b}, \tag{6.3.13}$$

where the preconditioning matrix $\mathbf{B}$ is any matrix that clusters the eigenvalues of the new coefficient matrix $\mathbf{B}\mathbf{S}\mathbf{B}^t$, or makes it as close to the unit matrix as possible. Attention is drawn to the use of the term $\mathbf{B}^t\mathbf{B}^{-t}$ and the change of unknowns from $\mathbf{x}$ to $\mathbf{B}^{-t}\mathbf{x}$ so as to keep the coefficient matrix symmetric in keeping with the requirements of the conjugate gradients algorithm. Thus, in the incomplete Cholesky preconditioning, **B** is the inverse of an approximate Cholesky factor **L** defined by eqn. (6.2.1). If **B** were made exactly equal to $\mathbf{L}^{-1}$, then the coefficient matrix is $\mathbf{L}^{-1}\mathbf{L}\mathbf{L}^t\mathbf{L}^{-t}$ or the identity matrix, but the cost of computation will be high; and besides, there will be no need to use the conjugate gradients algorithm, since we may easily compute the solution from **L**. The approximate Cholesky factor places no large memory load and is easy to compute. It is arrived at by ignoring terms at locations where **S** has zeros. Since the exact factor will have nonzero terms only to the right of the leading nonzero of the profile of **S**, any approximation will be within the profile. In Evans's preconditioning, a similar splitting of the matrix **S** is employed, after scaling eqn. (6.1.1) to have unit terms on the diagonal, using the factorization

$$\mathbf{S} = \mathbf{DAD}, \tag{6.3.14}$$

where the diagonal matrix **D** is

$$D_{ij} = \begin{cases} \sqrt{S_{ij}} & i = j \\ 0 & i \neq j \end{cases}, \tag{6.3.15}$$

and

$$A_{ij} = \frac{S_{ij}}{(D_{ii}D_{jj})}. \tag{6.3.16}$$

For example, the splitting of the following matrix is

$$\begin{bmatrix} 4 & 1 \\ 1 & 9 \end{bmatrix} = \begin{bmatrix} \sqrt{4} & 0 \\ 0 & \sqrt{9} \end{bmatrix} \begin{bmatrix} \dfrac{4}{\sqrt{4}\sqrt{4}} & \dfrac{1}{\sqrt{4}\sqrt{9}} \\ \dfrac{1}{\sqrt{9}\sqrt{4}} & \dfrac{9}{\sqrt{9}\sqrt{9}} \end{bmatrix} \begin{bmatrix} \sqrt{4} & 0 \\ 0 & \sqrt{9} \end{bmatrix}. \tag{6.3.17}$$

Note that only the diagonal terms of **D** need to be stored. And since we know that the diagonal terms of **A** are 1, we may store them here and overwrite the rest of **A** upon **S**. Whenever necessary, **S** may be recovered from **A**, so that no extra memory is taken up by **A** or **D**. Eqn. (6.1.1) therefore reduces by this scaling to

$$\mathbf{A}(\mathbf{Dx}) = (\mathbf{D}^{-1}\mathbf{b}) \tag{6.3.18}$$

or

$$\mathbf{Az} = \mathbf{c}, \tag{6.3.19}$$

where, using the fact that **D** is diagonal, the new variable **z** is defined by

$$z_i = D_{ii}x_i = x_i\sqrt{S_{ii}}, \tag{6.3.20}$$

and the new right-hand side by

$$c_i = \frac{b_i}{D_{ii}} = \frac{c_i}{\sqrt{S_{ii}}}. \tag{6.3.21}$$

**c**, too, may be overwritten on **b**. Once the solution **z** is found, **x** may be found through the application of eqn. (6.3.20).

Interestingly, with this scaling, even if we do not use any preconditioning and employ the simple conjugate gradients iterations, convergence will be found to have been improved. Apparently, when the diagonal terms are all of the same size, the eigenvalues must be getting clustered. In Evans's or SOR preconditioning for matrices with 1 everywhere along the diagonal, **L** is assumed to be

$$L_{ij} = \begin{cases} =S_{ij} & \text{if } i > k \\ =S_{ii} = 1 & \text{if } i = j \\ =0 & \text{if } i < j \end{cases} \tag{6.3.22}$$

and as a result, no extra storage is required for the matrix **L**. It is elementary to verify that this approximately fits eqn. (6.2.1) if the diagonal dominance of **S** is accounted for; since off diagonal terms are smaller than unity, their squares may be assumed negligible. For example, for the diagonally dominant symmetric matrix **A** scaled to have 1's along the diagonal:

$$\begin{bmatrix} 1.0 & 0.1 & 0.2 \\ 0.1 & 1.0 & 0.3 \\ 0.2 & 0.3 & 1.0 \end{bmatrix},$$

taking the lower triangular part as its Cholesky factor **L**, we have:

$$\begin{bmatrix} 1.0 & 0.0 & 0.0 \\ 0.1 & 1.0 & 0.0 \\ 0.2 & 0.3 & 1.0 \end{bmatrix}\begin{bmatrix} 1.0 & 0.1 & 0.2 \\ 0.0 & 1.0 & 0.3 \\ 0.0 & 0.0 & 1.0 \end{bmatrix} = \begin{bmatrix} 1.0 & 0.1 & 0.2 \\ 0.1 & 1.01 & 0.32 \\ 0.2 & 0.32 & 1.13 \end{bmatrix} \tag{6.3.23}$$

or approximately

$$\mathbf{LL}^t \approx \mathbf{A}, \tag{6.3.24}$$

as expected.

Thus, it is seen that both methods of preconditioning assume some approximate Cholesky splitting of **S**; while the incomplete Cholesky scheme has the advantage of computing **L** more accurately, Evans's method uses a splitting whose coefficients are already found in **S** and so does not require any extra storage or computation. Significantly, as we shall see, both methods assume the same sparsity pattern for the Cholesky splits as the original matrix **S**, and so ignore all elements of the decomposition factor which fall within the profile.

With a preconditioning matrix **B**, the conjugate gradients method has been shown to reduce to Alg. 6.3.2 (Kershaw 1978). Observe that when **B** is lower triangular as with Evans's or Cholesky preconditioning, the computation of $\mathbf{B}^{-1}\mathbf{r}$ and $\mathbf{B}^{-t}\mathbf{r}$ is, respectively, by forward elimination and back substitution, so that the inverses never need to be computed explicitly. That is, we in turn solve the equations $\mathbf{Bx} = \mathbf{r}$ (which gives us $\mathbf{B}^{-1}\mathbf{r}$) and $\mathbf{B}^t\mathbf{y} = \mathbf{x}$ (which gives us $\mathbf{B}^{-t}\mathbf{B}^{-1}\mathbf{r}$). In the algorithm the vector **p** is made to contain $\mathbf{B}^{-t}\mathbf{B}^{-1}\mathbf{p}$ where **p** is the vector of the theory above. The algorithm is therefore considerably simplified by our never having to compute the coefficient matrix $\mathbf{BSB}^t$.

**Algorithm 6.3.2. The Preconditioned Conjugate Gradients Algorithm**

$\mathbf{p} \leftarrow 0$
$\gamma \leftarrow 1$
$\mathbf{x} \leftarrow \mathbf{B}^{-1}\mathbf{b}$ {Guess **x**}
$\mathbf{x} \leftarrow \mathbf{B}^{-t}\mathbf{x}$
$\varepsilon \leftarrow$ Permitted Error
**Repeat**
  $\mathbf{r} \leftarrow \mathbf{b} - \mathbf{Sx}$
  $\mathbf{r} \leftarrow \mathbf{B}^{-1}\mathbf{r}$
  $e \leftarrow \text{Norm}(\mathbf{r})$
  $\theta \leftarrow \mathbf{r}^t\mathbf{r}$
  $\beta \leftarrow -\theta/\gamma$
  $\gamma \leftarrow \theta$
  $\mathbf{r} \leftarrow \mathbf{B}^{-t}\mathbf{r}$
  $\mathbf{p} \leftarrow \mathbf{r} - \beta\mathbf{p}$
  $\delta \leftarrow \mathbf{p}^t\mathbf{Sp}$
  $\alpha \leftarrow \gamma/\delta$
  $\mathbf{x} \leftarrow \mathbf{x} + \alpha\mathbf{p}$
**Until** $e < \varepsilon$

## 6.4 SYMMETRIC AND SPARSE STORAGE SCHEMES

The exploitation of the symmetry of a matrix is very straightforward. Since, for a symmetric matrix **S**, $S[i,j] = S[j,i]$, we need to store only the diagonal and the upper or lower half. The storage of the diagonal and the lower half is as shown in Figure 6.4.1. It is seen that $S[1,1]$ is stored as $SV[1]$ of a vector **SV**, $S[2,1]$ and $S[1,2]$ as $SV[2]$, $S[2,2]$ as $SV[3]$, and so on. Now, on row 1, we have one number; on row 2, two numbers; on row 3, three numbers; and on row $i-1$, $i-1$ numbers. $S[i,j]$ and $S[j,i]$ for $i \geq j$ will then be at a location given by all the numbers up to row $i-1$ plus $j$ numbers. Since $1+2+\ldots+i-1$ is $(i-1)*i/2$, we have $S[i,j]$ stored at $SV[(i-1)*i/2+j]$.

To describe this algorithmically, we may write a procedure or subroutine to determine the location Loc of $S[i,j]$ as described in Alg. 6.4.1 in FORTRAN. Thus, in the program, we would store the $n \times n$ matrix **S** as a vector **SV** of length $(n+1)*n/2$ and save approximately half the space originally used. Thereafter, wherever we find $S[i,j]$ in the program, we would replace this by $SV[\text{Loc}(i,j)]$; here, Loc$(i,j)$ will call the function Loc, which in turn will compute and return in Loc the integer location $ij$ of $S[i,j]$ in **SV**.

A sparse matrix is one in which most of the numbers are zero, and it is the exploitation of this that allows us to solve large problems by the finite element method. Although by using the property of symmetry we may cut down storage by approximately half, storage requirements are still of the order of $n^2$. For a 1000* 1000 matrix, with symmetry, we may bring storage from $10^6$ numbers to (1000 + 1)* 1000/2 = 500,500, still a considerable number, despite the savings. However, the use of sparsity brings

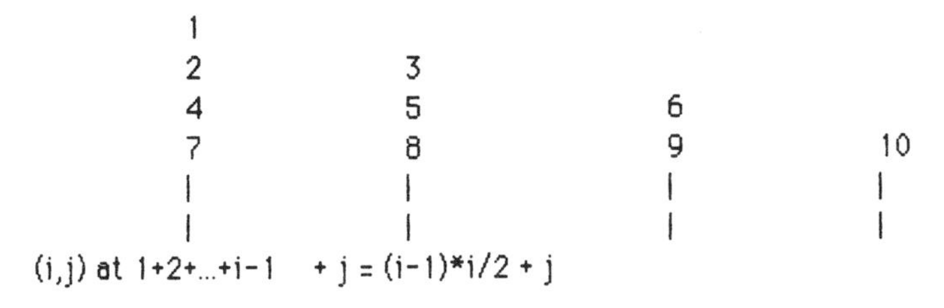


**Figure 6.4.1.** Symmetric matrix storage.

**Algorithm 6.4.1. Locating Coefficients in Symmetric Storage**

```
Integer Function Loc(i, j)
Integer i, j, ij
If(i . LT . j) Then   !Number in Upper Half, Stored at (j, i)
  ij = (j - 1)*j/2 + i
Else  !Number in Lower Half, Stored at (i, j)
  ij = (i - 1)*i/2 + j
EndIf
Loc = ij
Return
End
```

storage down to the order of $n$—for a first-order finite element mesh, for example, an equation for the potential at a node will be in terms of that potential itself and the potentials at those nodes that are linked to that node through triangle edges. For a regular mesh, triangles will tend to be equilateral, so that each node will be surrounded by approximately six others. Probabilistically, of these six, three will have smaller numbers than that node, and therefore the corresponding coefficients in the lower half that need to be stored are three in number. In total then, including the diagonal, we would have four real numbers per row to store, giving us a total of $4n$ real numbers to store. But, because the nonzero coefficients appear in arbitrary locations, we also need to store the row and column corresponding to each element of the matrix so that we might know where it belongs in the matrix. That is, total storage is $4n$ real numbers, $4n$ integers for row locations, and $4n$ integers for the column locations—a storage requirement growing as $n$. For the example of 1000 nodes, we then need to store 4000 real numbers and 8000 integers.

How are data accessed when we have this storage? Let the nonzero coefficients of a matrix **S** be stored as a vector **SV**, such that item **S**$[i, j]$ of **S** is **SV**$[ij]$, $i$ = **Row**$[ij]$, and $j$ = **Col**$[ij]$. As an illustration, consider the $10 \times 10$ symmetric matrix **S** of Figure 6.4.2A, coming from a finite element discretization, with nodes 1, 9, and 10 being on Dirichlet boundaries and the others in the interior of the region of solution. This matrix is stored as the real vector **SV** and integer vectors **Row** and **Col** and the number of nonzero elements NElems, as shown in Figure 6.4.2B. The $10 \times 10$ matrix has been stored with 28 real numbers and 56 integers. In terms of savings this example does not show the large gain usually attendant upon this storage method, because it is for large $n$ that the

| | | | | | | | | | | |
|---|---|---|---|---|---|---|---|---|---|---|
| | 4 | -1 | 0 | 0 | -1 | -1 | 0 | 0 | 0 | 0 |
| | -1 | 4 | -1 | 0 | -1 | 0 | 0 | -1 | 0 | 0 |
| | 0 | -1 | 4 | 0 | 0 | -1 | -1 | 0 | 0 | -1 |
| | 0 | 0 | 0 | 4 | -1 | -1 | -1 | 0 | -1 | 0 |
| | -1 | -1 | 0 | -1 | 4 | -1 | 0 | 0 | 0 | 0 |
| | -1 | 0 | -1 | -1 | 0 | 4 | 0 | -1 | 0 | 0 |
| | 0 | 0 | -1 | -1 | 0 | 0 | 4 | 0 | -1 | -1 |
| | 0 | -1 | 0 | 0 | 0 | -1 | 0 | 4 | -1 | -1 |
| | 0 | 0 | 0 | -1 | 0 | 0 | -1 | -1 | 4 | 0 |
| **A** | 0 | 0 | -1 | 0 | 0 | 0 | -1 | -1 | 0 | 4 |

| | SV = | Col = | Row = | NElems = |
|---|---|---|---|---|
| | [4, | [1, | [1, | 28 |
| | -1,4 | 1,2, | 2,2, | |
| | -1,4 | 2,3 | 3,3 | |
| | 4 | 4, | 4, | |
| | -1,-1,-1,4, | 1,2,4,5, | 5,5,5,5, | |
| | -1,-1,-1,4, | 1,3,4,6, | 6,6,6,6, | |
| | -1,-1,4, | 3,4,7, | 7,7,7, | |
| | -1,-1,4, | 2,6,8 | 8,8,8, | |
| | -1,-1,-1,4, | 4,7,,8,9 | 9,9,9,9, | |
| **B** | -1,-1,-1,4] | 3,7,8,10] | 10,10,10,10] | |

**C** **Diag** = [1,3,5,6,10,14,17,20,24,28]

**Figure 6.4.2.** A sparse symmetric matrix and storage. **A.** The matrix. **B.** Possible storage scheme. **C.** Replacement of vector **Row** by vector **Diag**.

difference is apparent—since in one scheme the requirements increase as $n$ and in the other as $n^2$.

A more efficient storage scheme naturally results from our noticing that the vector **Row** does not change for each row. That is, it is sufficient to denote where the changes take place, since for a given row the value of **Row** is that row number. We therefore change our data structure and replace the vector **Row** by the vector of integers **Diag** of the same length as the matrix size $n$. Item $i$ of **Diag** tells us where in **SV** the last nonzero item of row $i$ of **S** (the diagonal term, commonly) is stored. For the example selected above, **Diag** is of length 10 and is given in Figure 6.4.2C. This states that the last item of row 1 is **SV**[1], of row 2 is **SV**[3], of row 3 is **SV**[5], and so on. Another example of this storage is given in Figure 6.1.1. We have, by these means, for the example of the first-order regular finite element mesh, cut down storage requirements from $4n$ real numbers + $8n$ integers, to $4n$ real numbers for **SV** + $4n$ integers for **Col** + $n$ integers for **Diag**. And how do we identify **S**$[i, j]$ from this data structure? We use the fact that all numbers of row $i$ are stored between locations **Diag**$[i - 1] + 1$ and **Diag**$[i]$ in **SV**; for example, in the matrix above, all items of row 5 are between the seventh (**Diag**[5 − 1]) and 10th (**Diag**[5]) locations of **SV**. The algorithm for this operation of identifying the location is given in Alg. 6.4.2, where a procedure Locate identifies the location $ij$ in the vector **SV** at which the nonzero coefficient of the matrix **S**$[i, j]$ is stored. A Boolean variable Found is used to flag true when the location row $i$, column $j$ exists in the unstored matrix **S**.

With this procedure called Locate, one may replace those parts of a program with full storage where the statement **S**$[i, j]$ appears with a call to Locate followed by using for **S**$[i, j]$ the number **SV**$[ij]$ when Found is true and 0.0 when Found is false. However, this is wasteful of computer power since in a large matrix most coefficients are zero and Locate will be called unnecessarily several times, returning Found = False. We would then not be maximizing the second advantage of sparse storage, besides memory savings—that is, time saving through the avoidance of multiplications with zeros. By this scheme, although we may avoid a multiplication, much time will be expended in searching for zeros.

However, under certain circumstances, it is possible to enhance efficiency by running sequentially through every element of **SV** and thereby skipping all the zeros. Conventionally, on the other hand, the sequence of operations will be through the rows, as represented by the first Do loop of Alg. 3.2.1, for example. Naturally, this entails some modification of the algorithm. In the typical Liebmann (or Gauss–Seidel) Alg. 3.2.1 for iterative matrix equation solution, we improve the unknown **x**$[i]$ of every row $i$ as we run down the rows of the matrix. In computing **x**$[i]$, we step through the column numbers $k$ and perform operations using **S**$[i, j]$. But because we store only the lower triangular part of a symmetric matrix, the coefficients **S**$[i, j]$ will be stored sequentially up to column $j = i$; thereafter, since the coefficients belong to the upper triangular part, their reflections along the diagonal in the lower triangle will need to be used. When we move sequentially along a row in the upper triangle, however, we will be moving down a column in the lower triangle. That is, these are stored in sequential rows and not columns, so that they are not stored sequentially in the array **SV**. This will cause some searching as well as page faults. It is therefore more efficient to convert to the original and slower Gaussian

**Algorithm 6.4.2. An Efficient Sparse Symmetric Storage**

```
    Subroutine Locate(ij, Found, Diag, Col, NElems, i, j)
        Integer Row(NPoints), Col(NElems), ij, NElems, i, j, p, q
        Logical Found
    Cmmnt: Is (i, j) in lower or upper half of S?
        If (i . Ge . j) Then
    Cmmnt: We are in the lower half
          p = i
          q = j
        Else
    Cmmnt: We are in the unstored upper half. Pick equal
    number at (j, i)
          p = j
          q = i
        EndIf
        Cmmnt: Pick location of Row p, Column q
        If (p.Eq.1) Then
          ij = 1
          Found =. True.
          Go To 2
        Else
          ij = Diag(p - 1) + 1
    1     If(Col(ij).Eq.q) Then
            Found =. True.
            Else
            If((Col(ij).Gt.q).Or.(ij.Eq.Diag(p)) Then
    Cmmnt: We have passed S(p, q) without finding it
    Cmmnt: so that S(p, q) is zero.
              Found =. False.
            Else
              ij = ij + 1
              Go To 1
            EndIf
          EndIf
        EndIf
    2   Return
        End
```

iteration where we improve the unknown vector **x** only at the end of an iteration. This is presented in Alg. 6.4.3. Observe that if the matrix of coefficients is preconditioned as in eqn. (6.3.4), convergence will be faster and there will be no need to divide the residual in the array **Temp** by the diagonal coefficients, which are 1.

**Algorithm 6.4.3. Gauss Iterations for Sparse Matrices**

```
Procedure Liebmann(x, A, B, n, Precis, Col, Diag, NElems)
{Function: solves the matrix equation Ax = B
Output: x = The solution, an n vector
Inputs: A = The n × n matrix of coefficients stored as a vector
without zeros; B = The right-hand side, an n vector; n = The
size of the equation; Precis = The percentage precision to
which convergence is required; Col, Diag, NElems = sparse
matrix data structures}
Begin
  Repeat
    For k ← 1 To nDo
      Temp[i] ← B[i]
    Row ← 2
    For ij ← 2 To NElems Do
        Column ← Col[ij]
        If Column = Row
        Then
          Row ← Row + 1
        Else
          Temp[Row] ← Temp[Row] − A[ij]*x[Column]
          Temp[Column] ← Temp[Column] − A[ij]*x[Row]
    MaxX ← 0
    MaxChange ← 0
    For k ← 1 To n Do
      Old ← x[i]
      x[i] ← Temp[i]/A[Diag[i]] {Not necessary if diagonals of
      A are 1 as in eqn. (6.3.3)}
      Change ← x[i] − Old
      x[i] ← Old + 1.25*Change {Acceleration}
      AbsChange ← Abs(Change)
      AbsX ← Abs(x[i])
      If AbsChange > MaxChange
        Then MaxChange ← AbsChange
      If AbsX > MaxX
        Then MaxX ← AbsX
    Error ← 100*MaxChange/MaxX
  Until (Error < Precis)
End
```

**Algorithm 6.4.4. Forward Elimination for Sparse Matrices**

```
Procedure Forward(L, x, b, Diag, Col, NElems, MatSize)
{Function: solves the equation Lx = b by forward elimination.
L is a lower triangular sparse matrix stored as a 1-D array}
Begin
  x[1] ← b[1]/L[1]
  ij ← 2
  For Row ← 2 To MatSize Do
    Sto ← b[Row]
    Column ← Col[ij]
    While Column < Row Do
      Sto ← Sto − L[ij]*x[Column]
      ij ← ij + 1
      Column ← Col[ij]
    x[Row] ← Sto/L[ij]
End
```

**Algorithm 6.4.5. Back Substitution for Sparse Matrices**

```
Procedure Backward(L, x, b, Diag, Col, NElems, MatSize)
{Function: solves the equation L^t x = b by back substitution. L
is a lower triangular sparse matrix stored as a 1-D array}
Begin
  For i ← 1 To MatSize Do
    x[i] ← b[i]
  ij ← NElems + 1
  Row ← MatSize + 1
  For i ← 2 To MatSize Do
    Row ← Row − 1
{or For Row ← MatSize DownTo 2 Do in place of the last three
lines}
    ij ← ij − 1
    x[Row] ← x[Row]/L[ij]
    Stop ← Diag[Row − 1] + 1
    While ij > Stop Do
      ij ← ij − 1
      Column ← Col[ij]
      x[Column] ← x[Column] − L[ij]*x[Row]
  x[1] ← x[1]/L[1]
End
```

**Algorithm 6.4.6. Performing the Multiplication $P^tSP$ for Sparse Matrices**

```
Function TransPeeSPee(S, p, Diag, Col, NElems, MatSize): Real
{Function: Computes p^t Sp where S is a sparse matrix and p is a
column vector}
Begin
  Answer ← 0.0
  RowSum ← 0.0
  Row ← 1
  For ij ← 1 To NElems Do
    Column ← Col[ij]
    RowSum ← RowSum + S[ij]*p[Column]
  If Column = Row
      Then
        Answer ← Answer + p[Row]*(2*RowSum −
          S[ij]*p[Row])
        Row ← Row + 1
        RowSum ← 0.0
  TransPeeSPee ← Answer
End
```

**Algorithm 6.4.7. Procedures for ICCG Algorithm**

A. Sparse Matrix Multiplication

```
Procedure SparseMatMultiply(S, X, b, Diag, Col, NElems,
MatSize)
{Function: Evaluates b = Sx, where S is a sparse matrix and b
is a vector}
Begin
  For Row ← 1 To MatSize Do
    b[Row] ← 0.0
    Row ← 1
    For ij ← 1 To NElems Do
      Column ← Col[ij]
      b[Row] ← b[Row] + S[ij]*x[Column]
      If Row = Column
        Then Row ← Row + 1
        Else b[Column] ← b[Column] + S[ij]*x[Row]
End
```

B. The Infinity Norm of a Vector

```
Function Norm(r, n)
{Function: Finds the largest component of the vector r of length
n}
Begin
  Answer ← 0.0
  For i ← 1 To n Do
    Temp ← Abs(r[i])
    If Temp > Answer
      Then Answer ← Temp
  Norm ← Answer
End
```

**Algorithm 6.4.8. The ICCG Algorithm with Sparse Matrices**

```
p ← 0
γ ← 1
Forward(S, x, b, Diag, Col, NElems, MatSize)
Backward(S, x, b, Diag, Col, NElems, MatSize)
ε ← Permitted Error
Repeat
  SparseMatMultiply(S, x, r, Diag, Col, NElems, MatSize)
  For i ← 1 To MatSize Do
    r[i] ← b[i] − r[i]
  Forward(S, r, r, Diag, Col, NElems, MatSize)
  e ← Norm(r, MatSize)
  θ ← 0.0
  For i ← 1 To MatSize Do
    θ ← θ + r[i]*r[i]
  β ← −θ/γ
  γ ← θ
  Backward(S, r, r, Diag, Col, NElems, MatSize)
  For i ← 1 To MatSize Do
    p[i] ← r[i] − β*p[i]
  δ ← TransPeeSPee(S, p, Diag, Col, NElems, MatSize)
  α ← γ/δ
  For i ← 1 To MatSize Do
    x[i] ← x[i] − α*p[i]
Until e < ε
```

Similarly, the procedures of forward elimination, back substitution, and computation of the product $\delta = \mathbf{p}^t\mathbf{S}\mathbf{p}$ in the preconditioned conjugate gradients algorithm (Alg. 6.3.2) may be based on sequentially taking up the stored nonzero coefficients of the matrix. The procedures are presented in Algs. 6.4.4, 6.4.5, and 6.4.6, respectively. Observe that in solving by back substitution, after determining a term **x**[*i*], we are free to back substitute in advance using all coefficients of the matrix at column *i* above row *i*. Note also that in Algs. 6.4.4 and 6.4.5, **x** may be overwritten on **b** so that in the calling statement the same array may stand for **x** and **b**. In Alg. 6.4.6 we use the fact that a coefficient **S**[*ij*] standing for **S**[Row, Column] appears twice if it is off the diagonal and only once if Row = Column—thus, we subtract the diagonal contribution from the term RowSum since it would have been added once too often.

In Alg. 6.4.8 (which uses Alg. 6.4.7), the same principles are used in rewriting the Incomplete Cholesky Conjugate Gradients Algorithm of Alg. 6.3.2 for sparse matrices. Because Evans's preconditioning is used, it is assumed that the diagonal coefficients of the matrix **S** are 1 and that the lower triangular part of **S** (i.e., the stored part of **S**) is the incomplete Cholesky factor **B**.

## 6.5 PROFILE STORAGE

When direct methods of solution such as Cholesky factorization are employed to solve a sparse symmetric matrix of the type of Figure 6.1.1A, in the process of decomposition, subtracting a row from subsequent ones, the sparse structure of the matrix is disturbed. For example, in Figure 6.1.1A, to make all terms of the first column beyond row 1 zero, we would subtract an appropriate multiple of row 1 from all subsequent ones—it is seen that doing this for row 6 will result in some factor of the coefficient (1, 2) appearing in location (6, 2), originally zero. This process whereby a zero coefficient becomes nonzero is called a *fill-in* and occurs only at zero locations to the right of the first nonzero term on a row. Therefore in direct schemes, storage is allocated for the whole profile as in Figure 6.1.1B, thus requiring 29 locations in place of the 20, as shown in the example. For larger problems, this difference is much greater.

For the matrix of Figure 6.1.1A, profile storage would replace the vector **Col**, giving the column numbers of the nonzeros, by the vector **FC**, which gives the first column on a row occupied by a nonzero, as shown in Figure 6.1.1E. Observe that, as opposed to the sparse storage of Figure 6.1.1C, the vector **SV** containing the coefficients of the matrix **S** now has several zeros packed into it, shown in bold lettering in Figure 6.1.1E. In using Cholesky factorization for solution, the Cholesky factor **L** of the matrix **S** is usually overwritten on **S** to save storage. These additional zeros then are the places where fill-in may be accommodated.

**Algorithm 6.5.1. Cholesky Factorization with Profile Storage**

```
Procedure CholeskyFactor(L, S, FC, Diag, MatSize)
{Function: Computes the lower Cholesky factor L of the matrix
S of size MatSize by MatSize. Profile storage is used for S, with
FC giving the first column of each row and Diag the location of
the diagonal entry of each row. In calling, L and S may be made
the same}
Begin
  L[1] ← Sqrt(S[1])
  Row ← 2
  ij ← 2
  Repeat
    For column ← FC[Row] To Row Do
      Sum ← S[ij]
      If Column > 1
        Then
          If FC[Row] > FC[Column]
            Then kStart ← FC[Row]
            Else kStart ← FC[Column]
          rk ← Diag[Row − 1] + 1 + kStart − FC[Row]
          ck ← Diag[Column − 1] + 1 + kStart −
            FC[Column]
          For k ← kStart To Column − 1 Do
            Sum ← Sum − L[rk]*L[ck]
            rk ← rk + 1
            ck ← ck + 1
      If Row = Column
        Then
          L[ij] ← Sqrt(Sum)
          Row ← Row + 1
        Else
          L[ij] ← Sum/L[Diag[Column]]
      ij ← ij + 1
  Until Row > MatSize
End
```

Alg. 6.5.1 shows how the Cholesky factor **L** is computed using the data structures **FC** and **Diag**. Note how the term at row $i$, column $j$ is identified. Row $i$ begins at location **Diag**$[i-1]+1$, occupied by column **FC**$[i]$. Since all columns to the right of this are occupied up to the diagonal, column $j$ is $j-$ **FC**$[i]$ numbers away. So, the term $i, j$ of **S** is at location **Diag**$[i-1]+1+j-$ **FC**$[i]$ in the vector **SV**. Algs. 6.5.2 and 6.5.3, respectively, give forward elimination and back substitution schemes with profile storage.

Observe from the foregoing, therefore, that the more zeros there are within the profile, the more costly matrix solution will be, both in terms of storage requirements and computing time. For this reason, in direct schemes, a renumbering of the variables is done so as to minimize the bandwidth of the matrix and reduce the number of zeros in the band so that storage space for the inverse may be saved. The renumbered matrix of Figure 6.1.1A is shown in Figure 6.1.1D, where the variables $[1,2,3,4,5,6,7,8]$ have now been renumbered as $[4,5,6,1,7,2,8,3]$. It is seen that now there is no zero within the profile, so that the inverse will take only 20 locations, like the original, as opposed to 29 before renumbering. While for this small matrix we have achieved the ideal of no extra memory for the inverse, it has been shown by Silvester et al. (1973) that for large first-order finite element schemes, the inverse takes about four times as much space as the original matrix and, partly as

**Algorithm 6.5.2. Forward Elimination for Profile Stored Matrices**

```
Procedure Forward(L, x, b, FC, Diag, MatSize)
{Function: Solves the equation Lx = b by forward elimination.
L is a lower triangular matrix stored as a 1-D array using profile
data structures FC and Diag}
Begin
  x[1] ← b[1]/L[1]
  ij ← 2
  For Row ← 2 To MatSize Do
    Sto ← b[Row]
    Column ← FC[Row]
    While Column < Row Do
      Sto ← Sto − L[ij]*x[Column]
      ij ← ij + 1
      Column ← Column + 1
    x[Row] ← Sto/L[ij]
End
```

**Algorithm 6.5.3. Back Substitution for Profile Stored Matrices**

```
Procedure Backward(L, x, b, FC, Diag, MatSize)
{Function: Solves the equation L^t x = b by back substitution. L
is a lower triangular profile matrix stored as a 1-D array, using
FC and Diag}
Begin
  For i ← 1 To MatSize Do
    x[i] ← b[i]
  ij ← Diag[MatSize] + 1
  Row ← MatSize + 1
  For i ← 2 To MatSize Do
    Row ← Row-1
{or For Row ← MatSize Down To 2 Do in place of the last three
lines}
    ij ← ij − 1
    x[Row] ← x[Row]/L[j]
    Stop ← Diag[Row − 1] + 1
    Column ← Row
    While ij > Stop Do
      ij ← ij − 1
      Column ← Column − 1
      x[Column] ← x[Column] − L[ij]*x[Row]
  x[1] ← x[1]/L[1]
End
```

a result of the relevance of this fact to a microcomputing milieu, direct schemes are tending to be used less, although they are still very popular.

## 6.6 RENUMBERING OF VARIABLES: THE CUTHILL–McKEE ALGORITHM

Renumbering of the variables is normally done with direct schemes so as to reduce the band of the matrix and thereby save on storage. That is, we minimize

$$K = \sum_i D_i \qquad \text{where } D_i = \max[i - j] \text{ for every } j \text{ connected to } i. \tag{6.6.1}$$

In the foregoing, $D_i$ is clearly the largest distance between the diagonal term of row $i$ and any other term $j$ on that row. Of the many renumbering schemes available, the most commonly used is the Cuthill–McKee algorithm (Cuthill and McKee 1969). This is

| Node | No. of Connections | Connections to | New Number | New Connections to |
|---|---|---|---|---|
| 1 | 3 | 2,6,8 | 4 | 5,2,3 |
| 2 | 4 | 1,3,5,8 | 5 | 4,6,7,3 |
| 3 | 3 | 2,5,7 | 6 | 5,7,8 |
| 4 | 2 | 6,8 | 1 | 2,3 |
| 5 | 3 | 2,3,7 | 7 | 5,6,8 |
| 6 | 3 | 1,4,8 | 2 | 4,1,3 |
| 7 | 2 | 3,5 | 8 | 6,7 |
| 8 | 4 | 1,2,4,6 | 3 | 4,5,1,2 |

**Figure 6.6.1.** Cuthill–McKee renumbering algorithm implementation connecting matrix of Figure 6.1.1A to that of Figure 6.1.1D.

demonstrated in Figure 6.6.1 for the matrix of Figure 6.1.1A and yields the matrix of Figure 6.1.1D. The scheme is highly systematic and lends itself well to computer implementation.

We start the renumbering scheme by searching for the connectivities between variables and making the second and third columns of Figure 6.6.1. The second column says, for example, that variable 1 is connected to three others, and these three are given as variables 2, 6, and 8 in the third column.

The rule for renumbering is that we start with a least connected variable, 4 or 7 in this case, and call it 1. Had we chosen to call variable 4 our new variable 1 as in the example, we would enter it under column 4; and variable 6, the least connected of the connections of 4 (i.e. connections 6 and 8 of column 3) becomes 2; and variable 8, the other variable connected to 4, becomes 3. Now we do the same thing to the new node 2 (old 6). Of its connections 1, 4, and 8, 4 and 8 are already renumbered, and therefore 1 takes the new number 4. Proceeding thus, all the variables are renumbered as in column 4. It has been shown (George 1973) that reversing the algorithm results in more efficient renumbering. It is also known that as a result of the arbitrariness in numbering equally connected variables, different strategies result in different storage requirements. But we will not delve into this aspect, which is not of direct relevance here.

After renumbering, we have generated all the connectivities in Figure 6.6.1 and these may be profitably used to compute the sparsity data structure of Figure 6.1.1C, which otherwise would take a large portion of the total computing time. That is, we have found that the time for sparsity computation is only marginally smaller than the time for renumbering plus the time to compute sparsity using the tables generated in Figure 6.6.1.

To compute the sparsity of the renumbered matrix using Figure 6.6.1, we generate column 5 of the tables, which is really overwritten on column 3. For example, looking at the new variable 4 of the first row, which was the old variable 1 and connected to the old variables 2, 6, and 8 of column 3, we would say that it is now connected to 5, 2, and 3 of the last column, which is overwritten on the third; this is obtained by saying that 2 is now 5, 6 is now 2, and so on. Having generated this column, we then reorder it to go in increasing integers 2, 3, 5 (which really is the array **Col** of Figure 6.4.1C) once we drop 5 belonging to the upper triangle of the matrix, which is not stored in view of symmetry. Thus, including the diagonal term, 2, 3, 4 will belong to that part of **Col** corresponding to row 4 in the renumbered matrix.

## 6.7 RENUMBERING AND PRECONDITIONING

In this section, we show that although there is nothing to be gained by renumbering in terms of memory, in the preconditioned conjugate gradients algorithm, there is much to be gained in terms of time, when preconditioning methods are employed using approximate inverses having the same profile as the matrix solved. We will also show that the data on the connectivity of the matrix to be computed for renumbering may be used with profit to allocate storage for the sparse matrix in the manner of Figure 6.1.1C, so that the time expended in renumbering may be compensated for by savings in time for allocation of storage.

We have already seen that in the two commonly used methods of preconditioning, Evans's and Incomplete Cholesky, the approximate inverse is assumed to have the same sparsity structure as the original; that is, we ignore those terms of the inverse that take the positions of the zero coefficients within the profile.

Therefore, the more zero terms there are within the profile of the matrix, the more terms of the inverse we will be ignoring. As seen in the example of Figure 6.1.1, the inverse of the matrix of Fig. 6.1.1A with 20 coefficients requires 29 locations, and therefore the approximate inverse will be neglecting nine terms. However, had we renumbered the matrix of Figure 6.1.1A to get that of Figure 6.1.1D, before solving, the inverse would take only 20 locations. Thus, using the same sparsity structure as the original neglects no number. Although, in this example, renumbering results in a storage requirement of only 20, commonly this will be slightly above the number of nonzero coefficients, since for large problems renumbering rarely results in no zeros within the profile as here. This error coming from neglecting terms builds up as we compute the incomplete Cholesky factor, since we start from the top of the matrix (Silvester and Ferrari 1983) and keep working down. Thus, as we go down we will be using less and less accurate values of the decomposed factor to compute those at current locations.

What happens, therefore, in renumbering, is that the approximate inverse is improved so that the matrix equation we solve has a better approximation to the unit matrix. That is, the radius of the matrix has been reduced by clustering the eigenvalues, and therefore fewer search directions may be expected in the conjugate gradients process.

Test runs of the renumbered and unrenumbered conjugate gradients algorithm were made using Evans's method of preconditioning. When the same problem is taken and run with increasing refinement, the total time for solution against matrix size on a log–log graph was obtained as in Figure 6.7.1A. This clearly confirms the theory of the foregoing. The total time for solution is by itself not of significance, since it is clearly a function of the

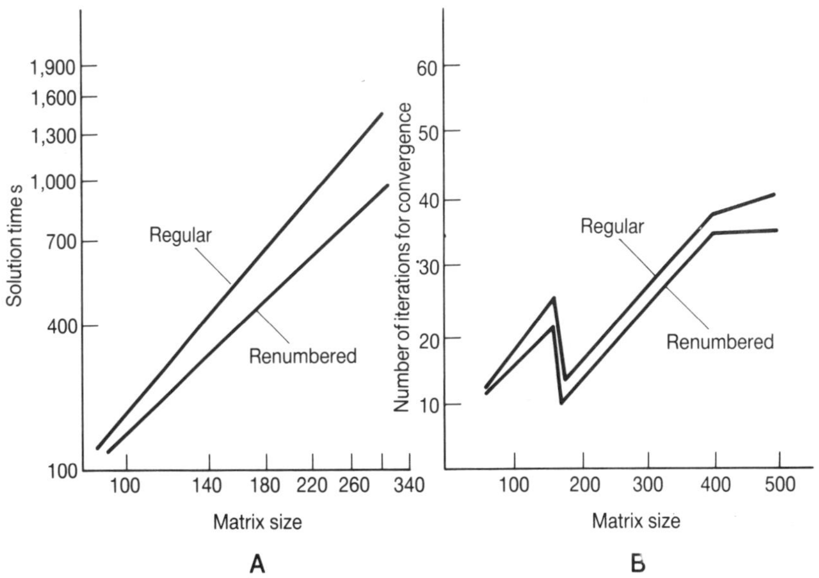


**Figure 6.7.1.** Efficiency through renumbering. **A.** Solution time against matrix size. **B.** Conjugate gradients iterations against matrix size.

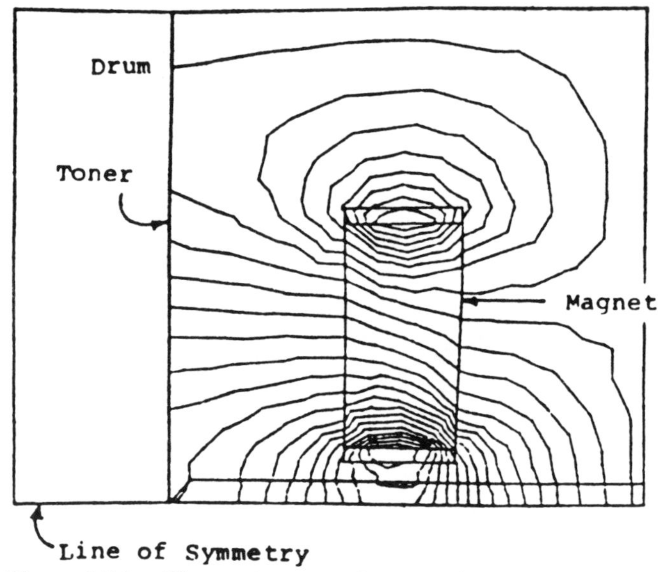

**Figure 6.7.2.** The nonimpact printer mechanism.

computer and operating system used (HP Pascal in this case) and the efficiency of the algorithms. What is important is that the renumbered curve is below the unrenumbered curve. Figure 6.7.1B for the number of conjugate gradient iterations with different matrix sizes, taken from variegated physical problems, shows that while the number of iterations for convergence is not strictly a function of matrix size (but rather of the starting solution and the problem and how the fields change within the region of solution), the renumbered scheme clearly produces fewer iterations for convergence.

The nonimpact printer mechanism of Figure 6.7.2 was solved to demonstrate an example from the finite element method to be described in the next chapter, to produce the flux plots of Figure 6.7.2. Figures 6.7.3A and 6.7.3B show the mesh before and after renumbering. It is seen that the renumbering has started in a region of low connectivity in the top right corner, where node 10 has been made 1. This problem resulted in a 165 × 165 matrix which took 25 iterations to converge before renumbering, for a total time of 508 s, whereas after renumbering it took 20 iterations to converge and 432 s.

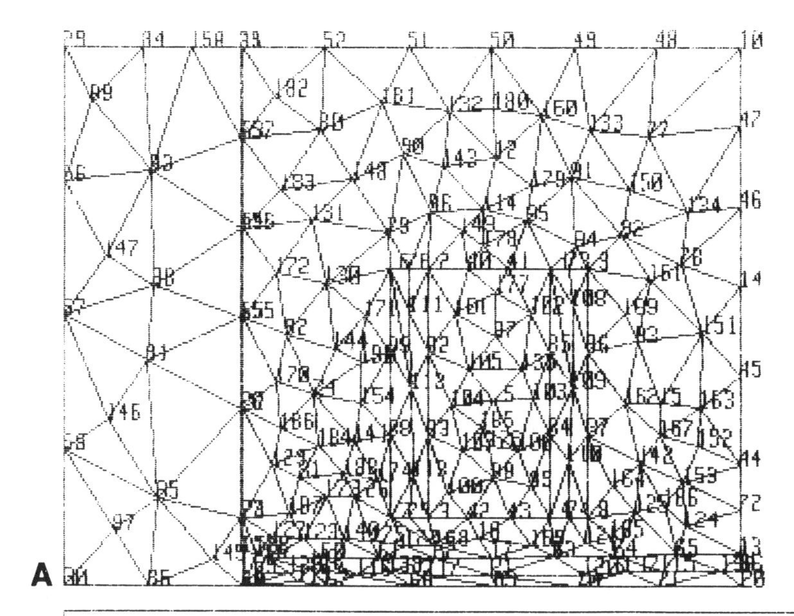


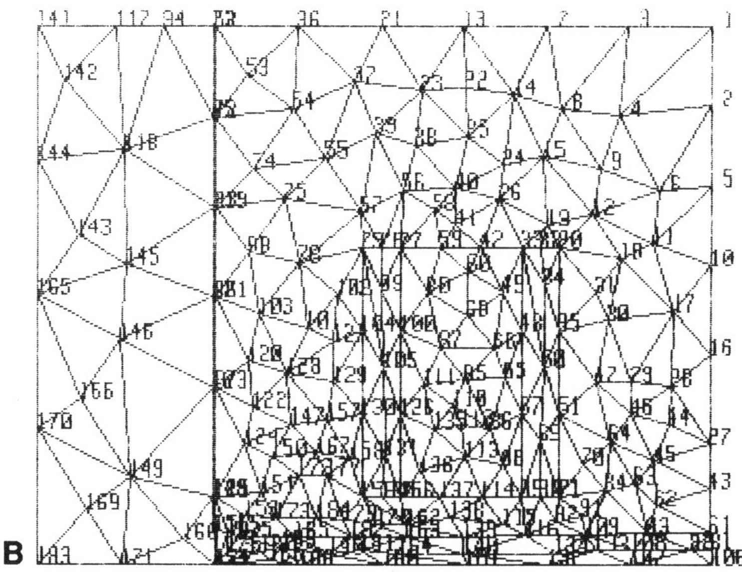


**Figure 6.7.3.** Renumbering (Cuthill–McKee algorithm). **A.** Original node numbers for mesh of printer, before renumbering. **B.** The renumbered mesh.

## 6.8 COMPUTATION OF EIGENVALUES

### 6.8.1 Jacobi Diagonalization Scheme

In section 5.4 we worked out by hand the eigenvalues of matrices for finding the cutoff frequencies of waveguides. We also used an iterative scheme to find the dominant mode of waveguide. But should we be interested in all the propagating modes, we cannot use the iterative scheme, nor is it possible for us to evaluate the determinant $|\mathbf{A} - \lambda\mathbf{I}|$ and then obtain all the eigenvalues by solving the ensuing polynomial equation, since the cost is prohibitive.

The Jacobi diagonalization scheme is a simple, albeit a slightly inefficient method for getting all the eigenvalues of a symmetric matrix. It is based on the fact that similarity transformations of a matrix, defined by

$$\mathbf{A}^* = \mathbf{B}^{-1}\mathbf{A}\mathbf{B}, \qquad (6.8.1)$$

leave the eigenvalues unchanged. If $\lambda$ is an eigenvalue with corresponding eigenvector $\mathbf{q}$, then

$$\mathbf{A}\mathbf{q} = \lambda\mathbf{q}. \qquad (6.8.2)$$

Using eqn. (6.8.1) to substitute for $\mathbf{A}$,

$$\mathbf{B}\mathbf{A}^*\mathbf{B}^{-1}\mathbf{q} = \lambda\mathbf{q}, \qquad (6.8.3)$$

or

$$\mathbf{A}^*(\mathbf{B}^{-1}\mathbf{q}) = \lambda(\mathbf{B}^{-1}\mathbf{q}). \qquad (6.8.4)$$

That is, $\mathbf{A}^*$ has the same eigenvalue $\lambda$, but a different eigenvector $\mathbf{B}^{-1}q$. Jacobi's scheme transforms the matrix $\mathbf{A}$ until it becomes diagonal, and it is trivial to confirm that the eigenvectors of a diagonal matrix are the diagonal terms. Moreover, if we choose our transformation matrix $\mathbf{B}$ such that it is an orthogonal matrix

$$\mathbf{B}^t\mathbf{B} = I, \qquad (6.8.5)$$

then since $\mathbf{B}^t$ is $\mathbf{B}^{-1}$, the transformed matrix $\mathbf{B}^t\mathbf{A}\mathbf{B}$ is symmetric so long as $\mathbf{A}$ too is symmetric. And this property may be used to save storage when dealing with symmetric systems of equations.

Jacobi's scheme sequentially and systematically eliminates the largest off-diagonal terms of a matrix $\mathbf{A}$ by transformation, until the matrix becomes diagonal. Let $a_{pq}$ be the largest component of a matrix **A**. Following the description of Jennings (1976), we then take

$$
\begin{array}{cl}
 & \begin{array}{cccccccccc} \phantom{0} & \phantom{0} & \phantom{\cdots} & \text{Col } p & \phantom{0} & \phantom{0} & \phantom{\cdots} & \text{Col } q & \phantom{\cdots} & \phantom{0} \end{array} \\
\mathbf{B} = & \left[\begin{array}{cccccccccc}
1 & 0 & \cdots & 0 & 0 & 0 & \cdots & 0 & \cdots & 0 \\
0 & 1 & \cdots & 0 & 0 & 0 & \cdots & 0 & \cdots & 0 \\
\cdots & \cdots & \cdots & \cdots & \cdots & \cdots & \cdots & \cdots & \cdots & \cdots \\
0 & 0 & \cdots & \cos\alpha & 0 & 0 & \cdots & -\sin\alpha & \cdots & 0 \\
0 & 0 & \cdots & 0 & 1 & 0 & \cdots & 0 & \cdots & 0 \\
0 & 0 & \cdots & 0 & 0 & 1 & \cdots & 0 & \cdots & 0 \\
\cdots & \cdots & \cdots & \cdots & \cdots & \cdots & \cdots & \cdots & \cdots & \cdots \\
0 & 0 & \cdots & \sin\alpha & 0 & 0 & \cdots & \cos\alpha & \cdots & 0 \\
\cdots & \cdots & \cdots & \cdots & \cdots & \cdots & \cdots & \cdots & \cdots & \cdots \\
0 & 0 & \cdots & \cdots & \cdots & \cdots & \cdots & \cdots & \cdots & 1
\end{array}\right]
\begin{array}{l}
\\ \\ \\ \text{Row } p \\ \\ \\ \\ \text{Row } q \\ \\ \\
\end{array}
\end{array}
\tag{6.8.6}
$$

It is easy to verify that this is orthogonal because it satisfies eqn. (6.8.5). We must now choose $\cos\alpha$ and $\sin\alpha$ such that the element at location $(p, q)$ in the transformed matrix is zero. By performing the multiplications $\mathbf{B}^t\mathbf{A}\mathbf{B}$, we obtain, using the fact that **B** has only diagonal terms plus two terms at locations $(p, q)$ and $(q, p)$, in general:

$$
A^*_{ij} = \sum_k \sum_m B^t_{ik} A_{km} B_{mj} = \sum_k \sum_m B_{ki} A_{km} B_{mj}. \tag{6.8.7}
$$

Only rows $p$ and $q$ and columns $p$ and $q$ are modified by this transformation, and therefore we need only consider the special cases when $i$ or $j$ is $p$ or $q$. Therefore, when neither $i$ nor $j$ is equal to $p$ or $q$, only $B_{ii}$ and $B_{jj}$ will exist in the summation, and both are 1:

$$
\begin{aligned}
A^*_{ij} &= \sum_k \sum_m B_{ki} A_{km} B_{mj} \\
&= \sum_m B_{ii} A_{im} B_{mj} = B_{ii} A_{ij} B_{jj} = A_{ij}.
\end{aligned}
\tag{6.8.8}
$$

When $i = j = p$, we have $B_{ii} \neq \cos\alpha$.

$$
\begin{aligned}
A^*_{pp} &= \sum_k \sum_m B_{kp} A_{km} B_{mp} \\
&= \sum_m \{[B_{pp} A_{pm} + B_{qp} A_{qm}] B_{mp}\} \\
&= [B_{pp} A_{pp} + B_{qp} A_{qp}] B_{pp} + [B_{pp} A_{pq} + B_{qp} A_{qq}] B_{qp} \\
&= \cos^2 \alpha A_{pp} + \sin^2 \alpha A_{qq} + 2 \sin\alpha \cos\alpha A_{pq},
\end{aligned}
\tag{6.8.9}
$$

$$
A^*_{qq} = \cos^2 \alpha A_{qq} + \sin^2 \alpha A_{pp} - 2 \sin\alpha \cos\alpha A_{pq}. \tag{6.8.10}
$$

Similarly, if $i = p$ and $j \neq p$ or $q$, $B_{kp}$ is nonzero only for $k = p$ and

$k = q$, and $B_{mj}$ is nonzero only for $m = j$, for which case it is 1:

$$\begin{aligned} A^*_{pj} &= \sum_k \sum_m B_{kp} A_{km} B_{mj} \\ &= \sum_m [B_{pp} A_{pm} + B_{qp} A_{qm}] B_{mj} \\ &= [B_{pp} A_{pj} + B_{qp} A_{qj}] B_{jj} \\ &= \cos \alpha A_{pj} + \sin \alpha A_{qj}. \end{aligned} \tag{6.8.11}$$

Doing the same thing:

$$A^*_{qj} = -\sin \alpha A_{qj} + \cos \alpha A_{qj}. \tag{6.8.12}$$

If we know $\cos \alpha$ and $\sin \alpha$, then all the coefficients of the modified matrix may be computed. Let us now consider the case $i = p$ and $j = q$ for $A_{pq}$, from which, by setting it to zero, we shall determine $\cos \alpha$ and $\sin \alpha$. $B_{kp}$ is nonzero only when $k = q$, and $k = p$ when it takes the values $\sin \alpha$ and $\cos \alpha$. $B_{mq}$ is nonzero only when $m = p$, and $m = q$ when it is, respectively, $-\sin \alpha$ and $\cos \alpha$. We therefore have:

$$\begin{aligned} A^*_{pq} &= \sum_k \sum_m B_{kp} A_{km} B_{mq} \\ &= \sum_m [B_{qp} A_{qm} + B_{pp} A_{pm}] B_{mq} \\ &= [B_{qp} A_{qp} + B_{pp} A_{pp}] B_{pq} + [B_{qp} A_{qq} + B_{pp} A_{pq}] B_{qq} \\ &= [-A_{pp} + A_{qq}] \sin \alpha \cos \alpha + A_{pq} [\cos^2 \alpha - \sin^2 \alpha] = 0. \end{aligned} \tag{6.8.13}$$

Squaring this result and substituting $1 - \sin^2 \alpha$ for $\cos^2 \alpha$,

$$[-A_{pp} + A_{qq}]^2 \sin^2 \alpha (1 - \sin^2 \alpha) = A^2_{pq} [1 - 2 \sin^2 \alpha]^2, \tag{6.8.14}$$

whence we get

$$\begin{aligned} &[(A_{pp} - A_{qq})^2 + 4A^2_{pq}] \sin^4 \alpha \\ &\qquad - [(A_{pp} - A_{qq})^2 + 4A^2_{pq}] \sin^2 \alpha + A^2_{pq} = 0, \end{aligned} \tag{6.8.15}$$

and solving this,

$$\sin^2 \alpha = \frac{1}{2} - \frac{A_{pp} - A_{qq}}{2r}, \tag{6.8.16}$$

where

$$r^2 = (A_{pp} - A_{qq})^2 + 4A_{pq}^2, \tag{6.8.17}$$

$$\cos^2 \alpha = \frac{1}{2} + \frac{A_{pp} - A_{qq}}{2r}, \tag{6.8.18}$$

and

$$\sin \alpha \cos \alpha = \frac{A_{pq}}{r}. \tag{6.8.19}$$

Observe that the expressions for $\sin^2 \alpha$ and $\cos^2 \alpha$ may be interchanged. We may conveniently let $\alpha$ and $\sin \alpha$ be positive and allow $\cos \alpha$ to have the same sign as $A_{pq}$. After computing $\cos \alpha$ from eqn. (6.8.18) when $A_{pp} > A_{qq}$, or $\sin \alpha$ from eqn. (6.8.16) when $A_{pp} < A_{qq}$, the uncomputed term is obtained from eqn. (6.8.19), thereby avoiding floating-point error and a square-root computation. The modified matrix may then be computed from eqns. (6.8.8)–(6.8.12).

As the matrix is progressively transformed, some of the locations at which the coefficients were made to vanish may once again have components reappear as other locations are made zero. These new fill-ins, however, will be smaller than the originals, and, finally, the matrix will become diagonal. The algorithm, therefore, involves repeating a cycle of computations until the largest off-diagonal coefficient is acceptably small. The cycle itself will consist of searching for the largest off-diagonal term and identifying the row and column positions $p$ and $q$ and then computing $\cos \alpha$ and $\sin \alpha$ and transforming the matrix according to the above rules. $\mathbf{A}^*$ is overwritten on $\mathbf{A}$. The diagonal terms are then the eigenvalues. If we are interested in the eigenvectors, then knowing the eigenvalues, they may easily be obtained by Liebmann's iteration.

## 6.8.2 The Power Method

The power method provides a quick means for obtaining the eigenvalue with largest magnitude and its corresponding eigenvector. It relies on our choosing an arbitrary vector (perhaps using random number generators), repeatedly premultiplying it by the matrix whose largest eigenvalue we want, and scaling the result so as to have the largest component equal to 1. The principle is based on the fact that an arbitrary vector $\mathbf{x}$ may be expressed as a linear combination of the $n$ eigenvectors $\mathbf{x}_1, \mathbf{x}_2, \ldots, \mathbf{x}_n$ of the symmetric matrix with eigenvalues $\lambda_1 > \lambda_2 > \lambda_3 \ldots > \lambda_n$. This is equally true of unsymmetric matrices, provided that that matrix with the eigenvectors $\mathbf{x}$ as columns is nonsingular. Thus,

$$\mathbf{x} = c_1\mathbf{x}_1 + c_2\mathbf{x}_2 + \ldots + \lambda_n\mathbf{x}_n. \tag{6.8.20}$$

Performing the required operations, we have

$$\mathbf{A}\mathbf{x} = c_1\lambda_1\mathbf{x}_1 + c_2\lambda_2\mathbf{x}_2 + \ldots + c_n\lambda_n\mathbf{x}_n \tag{6.8.21}$$

$$\mathbf{A}^2\mathbf{x} = c_1\lambda_1^2\mathbf{x}_1 + c_2\lambda_2^2\mathbf{x}_2 + \ldots + c_n\lambda_n^2\mathbf{x}_n \tag{6.8.22}$$

$$\cdots\cdots\cdots\cdots\cdots\cdots\cdots\cdots\cdots\cdots$$

$$\mathbf{A}^i\mathbf{x} = c_1\lambda_1^i\mathbf{x}_1 + c_2\lambda_2^i\mathbf{x}_2 + \ldots + c_n\lambda_n^i\mathbf{x}_n \approx c_1\lambda_1^i\mathbf{x}_1 \tag{6.8.23}$$

for large $i$. Thus after several multiplications, only $\mathbf{x}_1$ will remain. Clearly, the success of the method depends on the arbitrary vector containing the eigenvector $\mathbf{x}_1$ corresponding to the largest eigenvalue. The purpose of scaling is to throw away the term $c_1\lambda_1^i$ after each multiplication to avoid numerical buildup. The iterations are stopped when the scaled vector that results stops changing significantly in value. To illustrate, consider the matrix of eqn. (5.4.48) for a rectangular waveguide carrying a TM wave. We have already computed the eigenvalues as 2, 4, 4, and 6 from the determinant. The application of the power method should give us the eigenvalue 6. Let us choose [1 0 0 1] as our arbitrary vector (a better choice being given by filling up the decimal places with more numbers). Performing the repeated premultiplications and scaling:

$$\begin{bmatrix} 4 & -1 & -1 & 0 \\ -1 & 4 & 0 & -1 \\ -1 & 0 & 4 & -1 \\ 0 & -1 & -1 & 4 \end{bmatrix}\begin{bmatrix} 1 \\ 0 \\ 0 \\ 1 \end{bmatrix} = \begin{bmatrix} 4 \\ -2 \\ -2 \\ 4 \end{bmatrix} = 4 \times \begin{bmatrix} 1 \\ -0.5 \\ -0.5 \\ 1 \end{bmatrix}, \tag{6.8.24}$$

$$\begin{bmatrix} 4 & -1 & -1 & 0 \\ -1 & 4 & 0 & -1 \\ -1 & 0 & 4 & -1 \\ 0 & -1 & -1 & 4 \end{bmatrix}\begin{bmatrix} 1 \\ -0.5 \\ -0.5 \\ 1 \end{bmatrix} = \begin{bmatrix} 5 \\ -4 \\ -4 \\ 5 \end{bmatrix} = 5 \times \begin{bmatrix} 1 \\ -0.8 \\ -0.8 \\ 1 \end{bmatrix}, \tag{6.8.25}$$

$$\begin{bmatrix} 4 & -1 & -1 & 0 \\ -1 & 4 & 0 & -1 \\ -1 & 0 & 4 & -1 \\ 0 & -1 & -1 & 4 \end{bmatrix}\begin{bmatrix} 1 \\ -0.8 \\ -0.8 \\ 1 \end{bmatrix} = \begin{bmatrix} 5.6 \\ -5.2 \\ -5.2 \\ 5.6 \end{bmatrix} = 5.6 \times \begin{bmatrix} 1 \\ -0.9285 \\ -0.9285 \\ 1 \end{bmatrix}. \tag{6.8.26}$$

On the next multiplication we will get the vector $[1, -1, -1, 1]$ and thereafter the result will always be $6[1, -1, -1, 1]$ so that we have 6 as the eigenvalue and also have the corresponding eigenvector. *Purification* is the term used for now throwing away the component

$\mathbf{x}_1$ from the arbitrary vector and then applying the scheme to compute $\lambda_2$ (Jennings 1976).

### 6.8.3 The Shifted Power Method

The power method gives us the eigenvalue of the largest modulus. But in waveguide and several other eigenvalue problems in science, it is the smallest eigenvalue that is physically the most significant. The shifted power method is based on the fact that if the matrix **A** has an eigenvector **x** and corresponding eigenvalue $\lambda$ (i.e., $\mathbf{Ax} = \lambda\mathbf{x}$), then the modified matrix $\mathbf{A} - \mu\mathbf{I}$ has the same eigenvector **x**, but the eigenvalue would be shifted by $\mu$ to $\lambda - \mu$:

$$[\mathbf{A} - \mu\mathbf{I}]\mathbf{x} = \mathbf{Ax} - \mu\mathbf{Ix} = (\lambda - \mu)\mathbf{x}. \tag{6.8.27}$$

If we know the largest eigenvalue $\lambda_1$, then the matrix $\mathbf{A} - 0.9\lambda_1\mathbf{I}$ will have eigenvalues $0.1\lambda_1$, $\lambda_2 - 0.9\lambda_1$, $\lambda_3 - 0.9\lambda_1, \ldots, \lambda_n - 0.9\lambda_1$. Although these eigenvalues are in descending order of magnitude, the last one has the highest magnitude, provided these eigenvalues are not greater than $0.9\lambda_1$. Therefore, the power method should give use the value $\lambda_n - 0.9\lambda_1$. Returning to the same example matrix of section 6.8.2, since we know that $\lambda_1$ is 6, let us shift the eigenvalues of the matrix by $0.9 \times 6 = 5.4$. (If it does not work, we may try shifting by a larger number such as $0.98 \times 6$ or even 6.) Changing the sign of the matrix and thereby the signs of the resulting eigenvalues, we have:

$$\begin{bmatrix} 1.4 & 1 & 1 & 0 \\ 1 & 1.4 & 0 & 1 \\ 1 & 0 & 1.4 & 1 \\ 0 & 1 & 1 & 1.4 \end{bmatrix} \begin{bmatrix} 1 \\ 0 \\ 0 \\ 1 \end{bmatrix} = \begin{bmatrix} 1.4 \\ 1 \\ 1 \\ 1.4 \end{bmatrix} = 1.4 \times \begin{bmatrix} 1 \\ 0.7142 \\ 0.7142 \\ 1 \end{bmatrix}, \tag{6.8.28}$$

$$\begin{bmatrix} 1.4 & 1 & 1 & 0 \\ 1 & 1.4 & 0 & 1 \\ 1 & 0 & 1.4 & 1 \\ 0 & 1 & 1 & 1.4 \end{bmatrix} \begin{bmatrix} 1 \\ 0.7142 \\ 0.7142 \\ 1 \end{bmatrix} = \begin{bmatrix} 2.828 \\ 2.999 \\ 2.999 \\ 2.828 \end{bmatrix} = 2.999 \times \begin{bmatrix} 0.943 \\ 1.0 \\ 1.0 \\ 0.943 \end{bmatrix}. \tag{6.8.29}$$

Proceeding thus, we shall soon arrive at the eigenvalue 3.4. Since we shifted all values by 5.4, we identify the lowest eigenvalue of the original matrix as $5.4 - 3.4 = 2$.

### 6.8.4 The Linearized Eigenvalue Problem

As we shall see when dealing with the finite element analysis of guided waves in chapter 10, the eigenvalue problem is often posed by the matrix equation

$$\mathbf{Ax} = \lambda \mathbf{Tx}, \tag{6.8.30}$$

where $\mathbf{T}$ is a nonsingular symmetric matrix. The cutoff frequencies $\omega$ are proportional to $\lambda^{-1/2}$, so that we need to identify the lowest eigenvalues. This problem may be tackled by the power method, and to this end, we split $\mathbf{T}$ into its Cholesky factors:

$$\mathbf{T} = \mathbf{LL}^t. \tag{6.8.31}$$

This is used to recast the problem in linearized symmetric form, giving us:

$$\mathbf{L}^{-1}\mathbf{AL}^{-t}(\mathbf{L}^t\mathbf{x}) = \lambda(\mathbf{L}^t\mathbf{x}) \tag{6.8.32}$$

with the new eigenvectors $\mathbf{L}^t\mathbf{x}$, but still the same eigenvalues $\lambda$.

To apply the power method, we need repeatedly to premultiply an arbitrary vector $\mathbf{v}$ by $\mathbf{L}^{-1}\mathbf{AL}^{-t}$ and scale the result, until the scaled result no longer changes. Fortunately, this does not require our computing the inverse matrices $\mathbf{L}^{-1}$ and $\mathbf{L}^{-t}$. We know from section 6.2 how to split $\mathbf{T}$ into its Cholesky factors $\mathbf{L}$ and $\mathbf{L}^t$ and write $\mathbf{L}$ in the locations originally reserved for $\mathbf{T}$. The premultiplication requires three steps. First, we need to evaluate the result $\mathbf{v}_1$ of the multiplication

$$\mathbf{v}_1 = \mathbf{L}^{-t}\mathbf{v}, \tag{6.8.33}$$

which may be accomplished by solving the equation

$$\mathbf{L}^t\mathbf{v}_1 = \mathbf{v} \tag{6.8.34}$$

by back substitution. Second, we need to perform the multiplication

$$\mathbf{v}_2 = \mathbf{Av}_1, \tag{6.8.35}$$

which is straightforward since we are in possession of $\mathbf{A}$. And finally, we need to compute

$$\mathbf{v}_3 = \mathbf{L}^{-1}\mathbf{v}_2, \tag{6.8.36}$$

which is done by solving the equation

$$\mathbf{Lv}_3 = \mathbf{v}_2 \tag{6.8.37}$$

by forward elimination. Once this is done, $\mathbf{v}_3$ is scaled and put in the place of $\mathbf{v}$ for the next iteration. After identifying the largest eigenvalue, we may use the shifted power method to get the smallest.

When dealing with TM waves, since $\phi = E_z$ is zero along the boundary, the matrix $\mathbf{A}$ is nonsingular, so that we may alternatively split

$$\mathbf{A} = \mathbf{L}\mathbf{L}^t \tag{6.8.38}$$

and in place of eqn. (6.8.32) solve

$$\mathbf{L}^{-1}\mathbf{T}\mathbf{L}^{-t}(\mathbf{L}^t\mathbf{x}) = \lambda^{-1}(\mathbf{L}^t\mathbf{x}) \tag{6.8.39}$$

for the eigenvalue $\lambda^{-1}$. And since the smallest eigenvalue $\lambda$ of eqn. (6.8.32) corresponds to the largest eigenvalue $\lambda^{-1}$ of eqn. (6.8.39), the first application of the power method will give us the desired dominant mode.

# 7 THE FINITE ELEMENT METHOD

## 7.1 THE FINITE ELEMENT METHOD

The finite element method allows us to solve large-scale, complex electromagnetic field problems, posed through simple and general data structures. For this reason, it is one of today's best-accepted methods. The method, as a result, is highly amenable to automation. Interestingly, despite its generality, the concepts that underlie the finite element method are mathematically elegant in their simplicity. In view of the method's importance, we shall devote some extra space to the topic.

## 7.2 THE HISTORY OF THE FINITE ELEMENT METHOD

Finite elements made their earliest appearance in 1941. Hrenikoff (1941) and later McHenry (1943) introduced the concept of replacing a continuum by a latticelike assembly of bars, in analogy with a structure of steel struts. In consequence, although the method is generally applicable to the solution of any differential equation in a continuum, the terminology of the science of finite elements is laced with the terminology of structural analysis.

Proper finite elements (proper in the sense in which they are perceived today, however) made the scene only 15 years later in 1956, when they were introduced by Turner et al (1956). In this form, a variable inside an element is expressed in terms of interpolated nodal variables. These ideas were picked up by Melosh

(1961) and Zienkiewicz and Cheung (1965) for their work in stress analysis. Subsequently, the finite element method was championed by Zienkiewicz, who, it may be said, was the one person who both advocated the method and caused it to come to the attention of the scientific community at large, to the extent that the 1970s witnessed an explosion in the scientific literature dealing with the subject.

In electrical engineering, the first work was by Winslow (1967). He was very advanced for his time, when most scientists were using the finite difference method. In designing magnetic lenses, Winslow dealt with subdivision of the solution region into elements (or subdomains), trial functions continuous within and across domains, the Ritz formulation, treatment of nonlinearity, and all those concepts that we now associate with finite elements (Mikhlin and Smolitsky 1967; Zienkiewicz 1977; Heubner and Thornton 1982). However, coming from a finite difference background, he chose to call it "finite differences for triangles," and, as a costly result, not many today recognize him as the father of finite elements in electrical engineering (Demerdash and Nehl 1976). Just as Zienkiewicz popularized finite elements in civil engineering, it was Silvester who, together with his colleagues, developed the method to new heights in electrical engineering and brought it to the notice of the electrical engineering community. He and his coworkers widely applied the method to waveguides (Silvester 1969; Cendes and Silvester 1970), electrical machines (Silvester and Chari 1970; Chari and Silvester 1971), magnetoelluric modeling (Silvester and Haslam 1972), antennae (Silvester and Chan 1972), and axisymmetric problems (Silvester and Konrad 1973; Weiss, Konrad, and Silvester 1982). He introduced high-order polynomial triangular elements (Silvester 1969), the idea of ballooning for open boundary problems (Silvester, Lowther, Carpenter, and Wyatt 1977), and the concept of universal matrices (Silvester 1978, 1982a, 1982b) on which most of the current finite elements research effort in electrical engineering is based. Thanks largely to his endeavors, by the middle of the 1970s the method was acknowledged as one of the best for the solution of large field problems in electromagnetics. Curiously, development in the area of electron devices has been divorced from the happenings in magnetics (Adachi, Yoshii, and Sudo 1979; Barnes and Lomax 1974, 1977; Fichtner, Rose, and Bank 1983; Mayergoyz 1986a). Some of the other areas of application of finite elements are electroheat problems (Armor and Chari 1976; Lavers 1986; Namjoshi and Biringer 1985), hyperthermia (Sathiaseelan, Iskander, Howard, and Bleehen 1986), magnetic domain pattern studies (Seshan and Cendes 1985), and permanent magnet systems (Kaminga 1975; Nakata and Takahasi 1983).

The finite element method has now been developed into a very sophisticated form with fast, powerful, and easy–to-use general purpose software for the solution of field problems. Indeed, expert computer aided design (CAD) systems are now being constructed to all but eliminate the user in interactive computer-aided design (Hoole 1984; Lowther, Saldhana, and Choy 1985). The principal advantages of the method are easy formulation with a simple data structure for complex geometry, the ability to increase accuracy in regions of special interest using fine meshes or high-order approximations, a sparse, symmetric, and positive definite matrix for solution, and inherent natural boundary conditions. One area in which the finite element method is yet to establish clearly its superiority over the competing boundary element method is three-dimensional magnetostatics, and this topic is the focus of much current research. The reader is referred to the works by Armor and Chari (1976); Zienkiewicz, Lyness, and Owen (1977); Ferrari and Maile (1978); Simkin and Trowbridge (1979); Trowbridge (1981); Coulomb (1981); Chari et al (1981); Cendes, Weiss, and Hoole (1981); Demerdash et al (1981); Kotiuga and Silvester (1982); Mohammed et al (1982); Mishra et al (1983); Nakata, Takahashi, and Kawase (1985); Chari et al (1985); Hoole and Cendes (1985); Mayergoyz (1986); and Simkin (1986).

More recently, emphasis seems to have shifted to combining integral and differential formulations so as to get the best of both worlds (Silvester and Hsieh 1971; McDonald and Wexler 1972; Salon 1985), inverse problems (Salon and Istfan 1986), and expert systems (Lowther, Saldhana, and Choy 1985).

## 7.3 THE FINITE ELEMENT METHOD IN ONE DIMENSION

### 7.3.1 A Simple Demonstrative Problem from Electrostatics

The finite element method for solving differential equations and what it means are best demonstrated by a simple example from electrostatics in one dimension. Consider the simple problem configuration of Figure 7.3.1, where we have two long parallel plates 10 m apart, at voltages 0 and 100 V with a charge of constant density equal to the permittivity in between. This problem in a more generalized form with a jump in permittivity is of considerable interest to the oil industry, where the long plates will

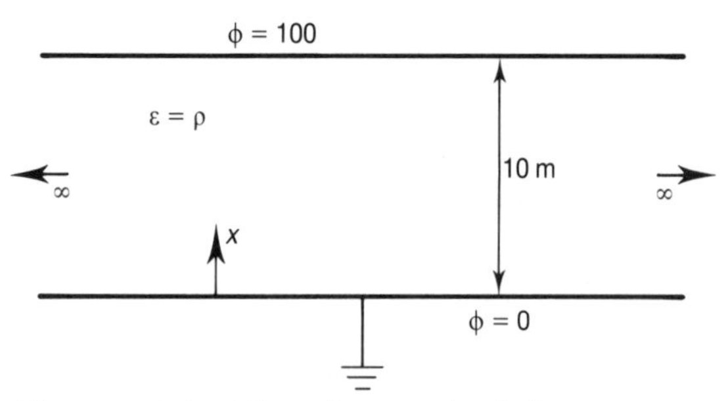


**Figure 7.3.1.** The charge cloud between two plates.

really be the walls of a pipe, the lower part of the capacitor will consist of a liquid, and the upper part, of charged vapor. We wish to determine the potential in a region far removed from plate edges. Here, by virtue of the large size of the plates, any changes in potential can take place only in the $x$ direction, going from plate to plate. Since we have seen that the electric potential obeys the Poissons eqn. (1.2.22), for this problem in one dimension with $\partial/\partial y \equiv \partial/\partial z \equiv 0$, the governing equation becomes

$$-\varepsilon \frac{d^2}{dx^2}\phi = \rho \tag{7.3.1a}$$

with boundary conditions

$$x = 0 \rightarrow \phi = 0; \qquad x = 10 \rightarrow \phi = 100, \tag{7.3.2}$$

obtained from the plate potentials. Of course, this is a trivial problem with a closed-form solution and needs no recourse to approximation schemes. The purpose of selecting this example, however, is to demonstrate the finite element method in one dimension, which, once we have grasped the essential ingredients of the method, may be generalized to complex equations that have no closed-form solution, such as when we have arbitrary material jumps and charge distributions as in the oil industry problem. The closed-form solution also allows us to compare the approximate solution with whatever we may obtain by numerical techniques. To get the exact closed-form solution, first note that for our example $\rho = \varepsilon$, so that eqn. (7.3.1a) reduces to

$$-\frac{d^2}{dx^2}\phi = 1. \tag{7.3.1b}$$

Integrating eqn. (7.3.1b) twice, we get

$$\phi = -\tfrac{1}{2}x^2 + ax + b, \tag{7.3.3}$$

where $a$ and $b$ are constants of integration. Putting in the boundary conditions of eqn. (7.3.2), we get $b = 0$ from the first and $a = 15$ from the second, so that

$$\phi = -\tfrac{1}{2}x^2 + 15x \tag{7.3.4}$$

is the exact solution as shown in Figure 7.3.2.

To introduce the finite element method, we shall pursue the more easily understood variational method (Gould 1957; Gelfand and Fomin 1963). In a variational approach, we first identify some functional (i.e., a function of the unknown function $\phi$), which is at its minimum at the point of solution. It is easily shown that

$$\mathscr{L}[\phi] = \tfrac{1}{2}\int \left[\varepsilon \frac{d}{dx}\phi\right]^2 dx - \int \phi\rho\, dx \tag{7.3.5a}$$

is that functional that at its minimum satisfies eqn. (7.3.1a), provided $\phi$ or its first derivative is fixed at either end of the solution region. This is required to obtain a unique solution, with the two constants of integration coming from a second-order differential equation pegged down. For our particular example of eqn. (7.3.1b), this functional reduces to

$$\mathscr{L}[\phi] = \tfrac{1}{2}\int \left[\frac{d}{dx}\phi\right]^2 dx - \int \phi\, dx. \tag{7.3.5b}$$

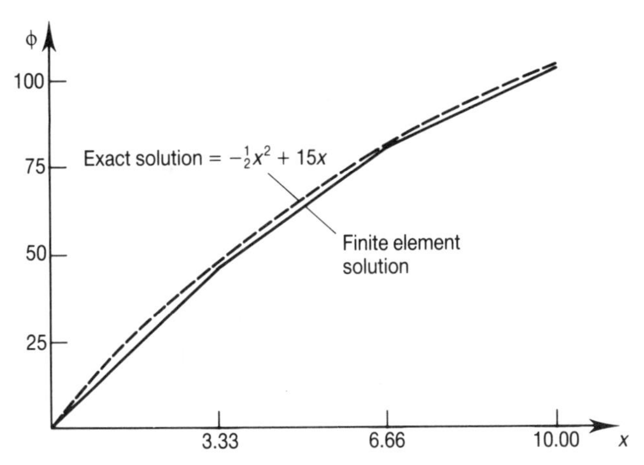


**Figure 7.3.2.** The exact and finite element solutions.

To see the validity of eqn. (7.3.5a), let $\phi$ take a small excursion to $\phi + \delta\phi$, about the exact solution. If $\mathscr{L}$ is truly a functional satisfying the differential equation (7.3.1a) at its minimum, then $\delta\mathscr{L}$ ought to be zero in the limit as we shall show:

$$\delta\mathscr{L} = \tfrac{1}{2}\int \varepsilon\left[\frac{d}{dx}(\phi + \delta\phi)\right]^2 dx - \int(\phi + \delta\phi)\rho\, dx$$

$$- \tfrac{1}{2}\int \varepsilon\left[\frac{d}{dx}\phi\right]^2 dx + \int \delta\rho\, dx$$

$$\approx \int\left[\varepsilon\frac{d}{dx}\phi\frac{d}{dx}\delta\phi\right] dx - \int \delta\phi\rho\, dx, \qquad \text{(neglecting } (\delta\phi)^2\text{)}$$

$$= \int \varepsilon\frac{d}{dx}\left[\frac{d\phi}{dx}\delta\phi\right] dx - \int \varepsilon\delta\phi\frac{d^2}{dx^2}\phi\, dx - \int \delta\phi\rho\, dx,$$

(by the chain rule)

$$= \left[\varepsilon\frac{d\phi}{dx}\delta\phi\right]_2 - \left[\varepsilon\frac{d\phi}{dx}\delta\phi\right]_1 - \int \delta\phi\left[\varepsilon\frac{d^2}{dx^2}\phi + \rho\right] dx, \qquad (7.3.6)$$

where the subscripts $_2$ and $_1$, respectively, refer to the value of the quantity within the square brackets evaluated at the end and beginning of the interval of integration. These will naturally be zero for our problem since $\phi$ is fixed at the limits $x = 0$ and $x = 10$ so that $\delta\phi = 0$ at those points. Indeed, these terms would be zero even if $d\phi/dx$ had been alternatively zero at the limits of integration. Thus, it is seen that if the differential equation (7.3.1a) is satisfied, then $\delta\mathscr{L}$ tends to zero as $\delta\phi \to 0$ so that $\mathscr{L}$ must be at an extremum. We may look at it in another way that is relevant to our numerical scheme. Let us assume a trial function $\phi$, which satisfies boundary conditions that make the boundary terms of eqn. (7.3.6) go to zero. If we put any such $\phi$ into $\mathscr{L}$ and extremize $\mathscr{L}$ (that is, set $\delta\mathscr{L}/\delta\phi$ to zero) with respect to the variational parameters, then all the terms of eqn. (7.3.6) except for the last will be zero. In this term, since $\delta\phi$ is arbitrary, the differential eqn. (7.3.1a) will be optimally satisfied in the solution region.

Let us apply this theory by constructing the simple trial function shown in Figure 7.3.3. Here, we have divided the solution region into three parts, our finite elements, with four interpolation nodes at $x = 0, 3\frac{1}{3}, 6\frac{2}{3}$, and 10. The values of the potential $\phi$ at these nodes are $\phi_1$, $\phi_2$, $\phi_3$, and $\phi_4$. The first and last of these are known and given by the boundary conditions. We need to determine $\phi_2$ and $\phi_3$, and these are the variational parameters that may vary to

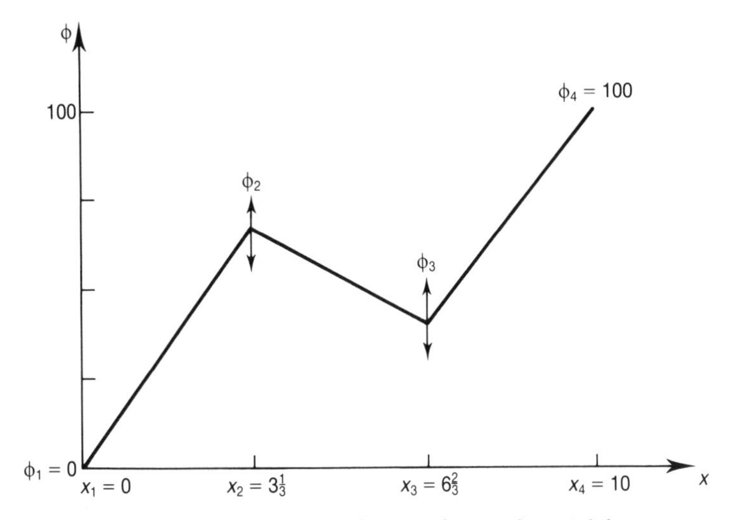


**Figure 7.3.3.** A simple three element first order trial function.

satisfy the condition that the functional $\mathscr{L}$ should be at a minimum. Now our functional involves an integration over the whole solution space going from $x = 0$ to $x = 10$. This integration may be replaced by summing the integrals over the three finite elements. In general, over an element from $x = x_i$ to $x_j$, we have assumed a linear variation of $\phi$:

$$\phi = \phi_i + (\phi_j - \phi_i)\frac{x - x_i}{x_j - x_i}, \tag{7.3.7}$$

so that

$$\frac{d}{dx}\phi = \frac{\phi_j - \phi_i}{x_j - x_i}. \tag{7.3.8}$$

Putting these into eqn. (7.3.5b), the contribution to the functional made by an element $i$ beginning at node $x = x_i$ is

$$\begin{aligned}\mathscr{L}_i &= \tfrac{1}{2}\int_{x_i}^{x_j}\left[\frac{\phi_j - \phi_i}{x_j - x_i}\right]^2 dx - \int_{x_i}^{x_j}\left[\phi_i + (\phi_j - \phi_i)\frac{x - x_i}{x_j - x_i}\right] dx \\ &= \frac{1}{2}\frac{(\phi_j - \phi_i)^2}{x_j - x_i} - \tfrac{1}{2}(\phi_i + \phi_j)(x_j - x_i).\end{aligned} \tag{7.3.9}$$

Therefore, summing for the three elements,

$$\begin{aligned}\mathscr{L} &= \mathscr{L}_1 + \mathscr{L}_2 + \mathscr{L}_3\\ &= \frac{1}{2}\frac{(\phi_2 - \phi_1)^2}{x_2 - x_1} - \tfrac{1}{2}(\phi_1 + \phi_2)(x_2 - x_1)\\ &+ \frac{1}{2}\frac{(\phi_3 - \phi_2)^2}{x_2 - x_1} - \tfrac{1}{2}(\phi_2 + \phi_3)(x_3 - x_2)\\ &+ \frac{1}{2}\frac{(\phi_4 - \phi_3)^2}{x_4 - x_3} - \tfrac{1}{2}(\phi_4 + \phi_3)(x_4 - x_3).\end{aligned} \tag{7.3.10}$$

In this expression, $\phi_1$ and $\phi_4$ being known, only $\phi_2$ and $\phi_3$ are variational parameters. We have seen in eqn. (7.3.6) that it is when $\delta\mathscr{L}$ is zero and the boundary conditions are satisfied that the differential equation is satisfied. To make $\delta\mathscr{L}$ zero, we must extremize eqn. (7.3.10) with respect to the free variables:

$$\frac{\partial\mathscr{L}}{\partial\phi_2} = \frac{\phi_2 - \phi_1}{x_2 - x_1} - \tfrac{1}{2}(x_2 - x_1) + \frac{\phi_2 - \phi_3}{x_3 - x_2} - \tfrac{1}{2}(x_3 - x_2) = 0, \tag{7.3.11a}$$

$$\frac{\partial\mathscr{L}}{\partial\phi_3} = \frac{\phi_3 - \phi_2}{x_3 - x_2} - \tfrac{1}{2}(x_3 - x_2) + \frac{\phi_3 - \phi_4}{x_4 - x_3} - \tfrac{1}{2}(x_4 - x_3) = 0. \tag{7.3.12a}$$

Putting in the values for the coordinates and for $\phi_1$ and $\phi_4$, we get:

$$\frac{3\phi_2}{5} - \frac{3\phi_3}{10} = \frac{10}{3}, \tag{7.3.11b}$$

$$\frac{-3\phi_2}{10} + \frac{3\phi_3}{5} = \frac{100}{3}. \tag{7.3.12b}$$

Solving, we get:

$$\phi_2 = \frac{400}{9} = 44.444, \tag{7.3.11c}$$

and

$$\phi_3 = \frac{700}{9} = 77.778. \tag{7.3.12c}$$

To compare with the analytical solution eqn. (7.3.4), $\phi_2 = -\tfrac{1}{2}*100/9 + 15*10/3$, which is 44.444, and $\phi_3 = -\tfrac{1}{2}*400/9 + 15*20/3 = 77.778$! Although we have seemingly got the exact solution, this is not so. What we have got is the best possible solution for the trial function of Figure 7.3.3. The heights of the graph at the interpolation nodes have been variationally set and the solution is a straight-line variation from one interpolation node to the next, as seen in Figure 7.3.2. That the values at the interpolation nodes coincide with the exact values, is a mere fortuitous accident. The finite element solution, however, is different from the exact solution.

Clearly in the above process, we could equally well have placed more interpolation nodes and this would have resulted in a more accurate solution, at the greater computational effort of solving for more variational parameters. The reader is encouraged to try this. Equally, instead of trying a straight line variation, we could have assumed a solution of the form:

$$\phi = a + bx + cx^2 \tag{7.3.13a}$$

over the whole interval [0, 10], where the constants $a$, $b$, and $c$ are the three degrees of freedom our trial function has. For this trial function, applying the boundary conditions of eqn. (7.3.2), the first boundary condition gives $a = 0$ and the second one gives $b = 10 - 10c$. This reduces our trial function to

$$\phi = 10(1 - c)x + cx^2, \tag{7.3.13b}$$

with only one independent variational parameter, or one degree of freedom. Our functional eqn. (7.3.5b) therefore becomes

$$\begin{aligned}\mathscr{L} &= \frac{1}{2}\int_0^{10} [10(1 - c) + 2cx]^2\, dx - \int_0^{10} [10(1 - c)x + cx^2]\, dx \\ &= \frac{1}{2}\left[100(1 - c)^2 x + 20(1 - c)cx^2 + \frac{4c^2x^3}{3}\right]_0^{10} \\ &\qquad - \left[\frac{10(1 - c)x^2}{2} + \frac{cx^3}{3}\right]_0^{0} \\ &= 500(1 - 2c + c^2) + 1000(c - c^2) + \frac{2000c^2}{3} - 500(1 - c) - \frac{1000c}{3} \\ &= \frac{500c^2}{3} + \frac{500c}{3}.\end{aligned} \tag{7.3.14}$$

Extremizing this with respect to the only free variable $c$, we get

$$\frac{\partial \mathscr{L}}{\partial c} = \frac{1000c}{3} + \frac{500}{3} = 0, \tag{7.3.15a}$$

so that

$$c = -\tfrac{1}{2} \tag{7.3.15b}$$

and

$$\phi = 15x - \tfrac{1}{2}cx^2, \tag{7.3.13c}$$

a solution that exactly matches the analytical solution of eqn. (7.3.4) everywhere in the solution region! Interestingly, in the trial function of eqn. (7.3.13b) we have only one degree of freedom in $c$ and yet we have obtained a better solution than from the trial function of Figure 7.3.3, where we have two degrees of freedom in $\phi_2$ and $\phi_3$ and therefore more work. And what does this teach us? It is that a trial function must be judiciously chosen and that the finite element method gives us the best possible shape for the trial function we choose.

### 7.3.2 Symmetry and Natural Boundary Conditions

In the handworked example of the parallel plate capacitor above, we had Dirichlet boundary conditions with the potential $\phi$ fixed at both ends $x = 0$ and $x = 10$ of the domain or interval of solution. These conditions we imposed through the trial functions by saying $\phi_1 = 0$ and $\phi_4 = 100$. Such boundary conditions, which are forced to be satisfied exactly through the trial functions, are said to be *strongly imposed.* We saw in eqn. (7.3.6) that it was necessary to enforce these boundary conditions exactly so that the residual $[\varepsilon(d^2\phi/dx^2) + \rho]$ of the governing Poisson equation would vanish. That is, $\delta\phi_2$ and $\delta\phi_1$ being zero when the functional $\mathscr{L}$ of eqn. (7.3.5a) is extremized making $\delta\mathscr{L}$ zero, the residual of the differential equation (7.3.1a) we are solving disappears everywhere in the interval, so that eqn. (7.3.1a) is satisfied.

Likewise, when we are solving a differential equation with a Neumann boundary condition with $d\phi/dx$ vanishing at one end of the solution interval, we may force the boundary condition through the trial function. For example, let us suppose that we are solving eqn. (7.3.1b) subject to the new boundary conditions, in place of the old conditions given in eqn. (7.3.2):

$$x = 0 \rightarrow \phi = 0; \qquad x = 10 \rightarrow \frac{d}{dx}\phi = 0. \tag{7.3.16a}$$

To solve this problem by the variational principle of eqn. (7.3.6), we may postulate the trial function of Figure 7.3.4 in place of Figure 7.3.3. In the new trial function, we have forced the Neumann condition $d\phi/dx$ at $x_4$ by making $\phi_3 = \phi_4$ so that the graph is compulsorily made to be flat. This, according to eqn. (7.3.6), will again make the residual vanish because now the term $[\varepsilon(d\phi/dx)\delta\phi]_2$ vanishes because $d\phi_2/dx$ is zero, and $[\varepsilon(d\phi/dx)\delta\phi)]_1$ vanishes because $\delta\phi_1$ is zero. Unfortunately, however, this strong

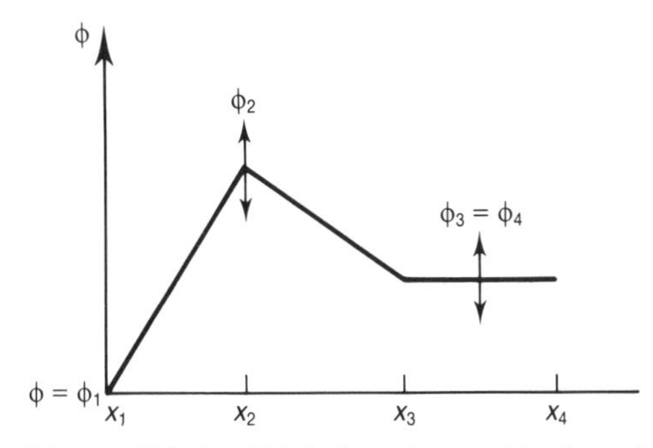


**Figure 7.3.4.** Trial function with forced Neumann conditions.

imposition of the Neumann condition not only makes $d\phi/dx$ zero at $x_4$, but it also makes it zero throughout the last element from $x = x_3$ to $x = x_4$. As a result, the satisfaction of eqn. (7.3.1b) in that interval becomes poor. True, as we refine the mesh into finer and finer elements, the last element becomes negligibly small, so that ultimately the finite element solution will converge towards the exact solution with mesh refinement.

However, the Neumann condition may be weakly imposed through the extremization of the functional (7.3.5a). That is, if we take as our trial function the graph of Figure 7.3.5A, where the $\phi_3$ and $\phi_4$ are completely free and the Neumann boundary condition is totally ignored, and put the trial functional into the functional of eqn. (7.3.5a) and extremize with respect to the free parameters $\phi_2$, $\phi_3$, and $\phi_4$ of the trial function, we would have a simultaneous satisfaction of the differential equation in the interval of solution and the Neumann condition at $x_4$. In other words, in eqn. (7.3.6), when $\delta\mathscr{L}$ vanishes because of extremization and $\delta\phi_1$ because of $\phi_1$ being fixed, the gradient $d\phi_2/dx$ and the residual $[\varepsilon(d^2\phi/dx^2) + \rho]$ will both be zero. Regrettably, this is not obvious from eqn. (7.3.6). In fact all that we can say from eqn. (7.3.6) when we extremize $\mathscr{L}$ of eqn. (7.3.5a), ignoring the Neumann condition, is that

$$\left[\varepsilon\frac{d\phi}{dx}\delta\phi\right]_2 - \int \delta\phi\left[\varepsilon\frac{d^2}{dx^2}\phi + \rho\right]dx = 0. \tag{7.3.17}$$

How then are these Neumann conditions natural to the variational principle? The natural conditions result from the strong analogy that exists between natural Neumann conditions and

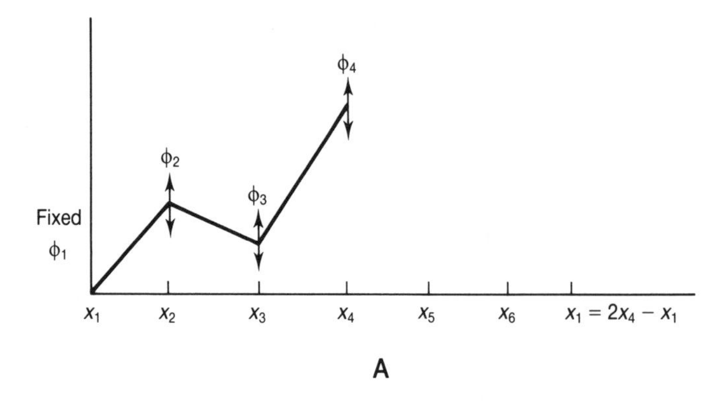


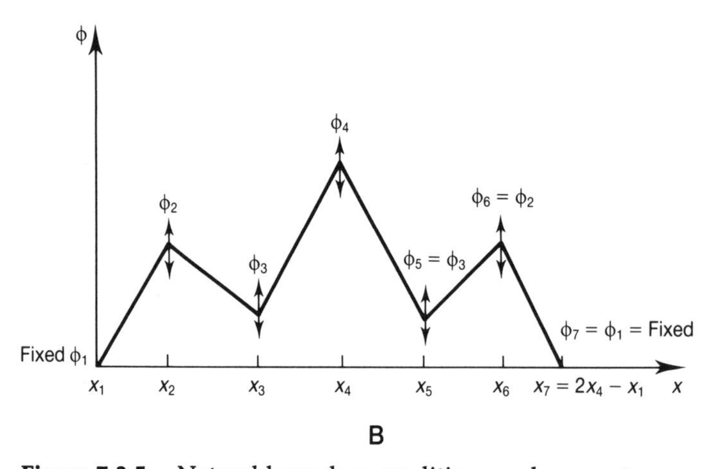


**Figure 7.3.5.** Natural boundary conditions and symmetry.

symmetry. To see this, consider a problem defined by eqn. (7.3.1a) over the interval $[x_1, x_4]$ with a defined charge distribution $\rho(x)$ in that interval with boundary conditions:

$$x = x_1 \rightarrow \phi = \phi_1; \qquad x = x_4 \rightarrow \frac{d}{dx}\phi = 0. \tag{7.3.16b}$$

The corresponding trial functions for this, with subdivision of the interval into three equal parts, are shown in Figure 7.3.5A. Now take the extended symmetric problem over the interval $[x_1, 2x_4 - x_1]$—that is, the interval consisting of the original domain of the problem plus its symmetric reflection about $x_4$. Let the governing equation be eqn. (7.3.1a) over the extended interval with the charge density being defined by

$$\begin{aligned}\rho(x) &= \rho(x) \qquad && x_1 \le x \le x_4 \\ &= \rho(2x_4 - x) \qquad && x_4 < x,\end{aligned} \tag{7.3.18}$$

where $\rho(x)$ is the charge distribution of the original problem defined by eqns. (7.3.1a). Let us also impose symmetry through the trial functions for the now extended problem as shown in Figure 7.3.5B, with symmetric subdivision of elements and $\phi_3 = \phi_5$, $\phi_2 = \phi_6$, and $\phi_1 = \phi_7$. Now since $\phi_1$ is fixed, the boundary conditions for the extended problem are

$$x = x_1 \to \phi = \phi_1; \qquad x = x_7 = 2x_4 - x_1 \to \phi = \phi_1. \tag{7.3.19}$$

From eqn. (7.3.6), since $\delta\phi$ is zero at the two ends of the interval, we may say that putting the trial function of Figure 7.3.5B into the functional of eqn. (7.3.5a) and extremizing the functional will make the residual $\varepsilon d^2\phi/dx^2 + \rho$ of eqn. (7.3.1a) vanish everywhere over $[x_1, x_7]$. But by the symmetry of our trial functions and the charge distribution (7.3.18) about $x_4$, the functional of the extended problem with the trial function of Figure 7.3.5B is exactly double the functional of the original problem defined by eqns. (7.3.1a) and (7.3.16b) with the trial function of Figure 7.3.5A. Therefore, in the interval $[x_1, x_4]$, the results from the extended problem will exactly match the results from putting the trial function of Figure 7.3.5A into the functional (7.3.5a). But the results of the extended problem will have $\phi_3 = \phi_5$, which is to say that $[d\phi/dx]_4 = (\phi_5 - \phi_3)/(x_5 - x_4)$ will be zero. Therefore, we may conclude that ignoring the Neumann boundary condition and using the functional of eqn. (7.3.5a) over the solution interval will naturally result in the satisfaction of the differential equation and the Neumann condition simultaneously.

Allowing the Neumann condition to turn out naturally is referred to as a weak imposition, because, as can be seen from Figure 7.3.5B, the gradient at $x_4$ is not exactly flat. Clearly, though, the curve will become more flat as we resort to a finer subdivision of the interval into elements. But the weak formulation, of which we shall hear more in chapter 8, satisfies the differential equation and the boundary condition in an optimal sense—so that although neither is exactly satisfied, both are satisfied as best as can be within the limits of the trial function being used. Strongly enforcing the boundary conditions skews the solution towards the boundary condition so that the differential equation is poorly satisfied (particularly in the elements near the forced Neumann condition). The reader is encouraged to solve eqn. (7.3.1b) subject to the boundary condition (7.3.16a) analytically and then by the finite element method, first using the trial function of Figure 7.3.4 and then that of Figure 7.3.5A and then comparing the solutions to see whether the answers bear out the observations of this section.

## 7.4 A GENERAL DEFINITION OF THE FINITE ELEMENT METHOD

Having worked an example by the finite element method, we are now able to identify in a meaningful way the essential ingredients of the method. These ingredients may be summarized as follows.

**1. Division of the Solution Region into Elements or Subdomains**

In the example above, we divided the solution region from $x = 0$ to 10 into three elements from 0 to $3\frac{1}{3}$, from $3\frac{1}{3}$ to $6\frac{2}{3}$, and from $6\frac{2}{3}$ to 10. These intervals need not have been equal and could have been of different sizes, as may be seen from the working of the previous problem. Indeed, as in the trial function eqn. (7.3.13a), we could have had one element covering the entire interval of solution. This flexibility is useful in having finer element subdivisions in regions of high change so that models of the form of Figure 7.3.3 would better fit the actual answer using our straight-line approximations.

**2. Postulation of a Trial Function**

We need to postulate a trial function as our solution, over the solution region with free parameters, which are to be determined. Once these free parameters are found, the trial function (which is our solution) is known. In our example, we considered a piecewise continuous linear trial function, as well as a second-order function over the whole interval. In fact, we just as well might have used any other trial function that struck our minds, keeping in mind that the final answer is only as good as the trial function we use and how it may bend using the free parameters to fit the exact answer to the equation being solved. Be it noted that resorting to a finer subdivision of the solution region, such as by placing additional interpolation nodes in Figure 7.3.3, is in effect using a better trial function. We shall repeatedly use this fact in finite elements where the typical analysis involves a cycle. In this cycle a subdivision of the solution region into elements is made, we get the corresponding solution and examine it, and, if the resulting graph of the form of Figure 7.3.2 is not smooth enough, knowing that it is less accurate where it is less smooth, we refine the elements by imposing additional nodes in the jagged regions and solve the problem again. For our example of Figure 7.3.1, from the solution presented in Figure 7.3.2, we may impose additional nodes at $x = 5$, 8, and 9 and try again, to get a better answer.

### 3. Identification of an Optimality Criterion

We must identify an objective criterion that in some manner optimizes the free parameters of our trial function. In our example, we identified the variational criterion in eqn. (7.3.5), for optimization; it will be shown in the next section that this expression is really the energy of an electrostatic system, and in extremizing it, we are actually imposing the condition that the energy of the system shall be a minimum. We could also have imposed a least-square error criterion on the governing equation (eqn. (7.3.1)) being solved. In the next chapter, we shall encounter the generalized Galerkin method, another optimality criterion that is a very powerful tool in finite elements. It should be noted also that our optimality criterion should be consistent with our trial function. Supposing, for example, we had assumed a step function as our trial function, then in every interval $\phi$ would be a constant, so that when we put the trial function into our functional of eqn. (7.3.5), the term $d\phi/dx$ would be zero and we would not be able to solve the problem at all. Thus, for the functional of eqn. (7.3.5), the trial function ought to be at least of order 1 in $x$. Similarly, if we wish to use a least-square criterion on the residual of eqn. (7.3.1), the trial function ought to be at least of order 2 for the error functional to be useful.

### 4. Solution of a Set of Linear Equations

These linear equations for solution, such as eqns. (7.3.11a) and (7.3.12a), arise from applying our optimality criterion to our trial functions. For large problems this involves the application of matrix solution routines. As will be seen from our example above, the nature of the term $(d\phi/dx)^2$ in our optimality criterion of eqn. (7.3.5) ensures that the matrix equation shall be positive definite and symmetric as evidenced in eqns. (7.3.11b) and (7.3.12b). It should also be noted from the derivation of eqn. (7.3.9) that a term $\phi_i$ will appear in the evaluation of the Lagrangian (eqn. (7.3.5)) over the whole interval, only as a coefficient of the potentials $\phi$ from elements on which they appear together as interpolation nodes. Thus, it will be observed from eqn. (7.3.10) that in the Lagrangian $\mathscr{L}$, $\phi_1$ will be a coefficient only of the term $\phi_2$; $\phi_2$ will be a coefficient of $\phi_1$ and $\phi_3$; $\phi_3$ will be a coefficient of $\phi_2$ and $\phi_4$; and $\phi_4$ a coefficient only of $\phi_3$. As a direct consequence, in extremizing the Lagrangian with respect to $\phi_2$, only $\phi_1$ and $\phi_3$ will appear in eqn. (7.3.11a) for $\phi_2$, and in extremizing with respect to $\phi_3$, only $\phi_2$ and $\phi_4$ will appear in eqn. (7.3.12a) for $\phi_3$. For our example, therefore, however many interpolation nodes there may be, the resulting matrix equation will be symmetric and sparse with at most only three coefficients per row; this obviously is of greatly advantageous significance to storage requirements and computational speed, as we have seen in chapter 6. However, these computational advantages will largely depend on the trial function and optimality criterion we choose; for example, a direct Galerkin optimality criterion presented in section 8.2 will not result in symmetric matrices, nor will a trial function like eqn. (7.3.13) with more higher order coefficients in $x$ result in sparsity. Therefore, in making such choices the import of our choice to computational efficiency should be borne in mind.

## 7.5 FIRST-ORDER TRIANGULAR FINITE ELEMENTS

### 7.5.1 Two-Dimensional Systems

In the preceding sections we have been introduced to the concept of finite element analysis through a one-dimensional example. One-dimensional examples occur comparatively rarely in practice. Two-dimensional analysis, on the other hand, is the most commonly performed study in the design environment for several reasons, chief among which are the computational expense and effort required for three-dimensional computations and the absence of freely available (or easily written) general purpose software for three-dimensional problems. Hence, we try to make do with lower-dimensional analysis, which is useful so long as we know what its limitations are.

Although all devices in practice are truly three-dimensional, two-dimensional studies often give us sufficient information for most, but certainly not all, practical purposes. For the parallel plate capacitor example of section 7.3, if the breadth and width of the plates are large compared to the plate separation, then in the region close to the middle of the plates the problem is one-dimensional. If, say, only the width of the plate is large, then in the middle of the width the problem is two-dimensional, with the field varying from plate to plate and along the breadth. In either event, the analysis will not be valid close to the edges of the plates, where fringing will take place. Thus, the results of one- or two-dimensional analysis will be applicable only to certain parts of the capacitor. Another useful example is a transmission line that has uniform cross section and comparatively infinite length, so that a two-dimensional model may be used for the entire length except the terminals.

In rotating electric machine problems, the air gap between the rotor and the stator is so small that for most of the length of the machine, except the end regions, the device is practically two-dimensional in operation; this makes the use of two-dimensional models for analysis purposes reasonably accurate. It is not necessary that a reduction in the dimensionality of the problem be achieved only where comparatively large device dimensions exist. Rotational symmetry makes several problems two-dimensional. An example immediately springing to mind is the insulator string by which transmission lines are hung from towers (Figure 1.3.1). If the problem is viewed in cylindrical coordinates $(r, \theta, z)$, it will be found that the electric voltage will be constant as we move in the direction in which only $\theta$ changes, keeping $r$ and $z$ constant—so that a two-dimensional analysis on the $r$–$z$ plane would suffice and the results would be generally applicable to all parts of the device. Most electron gun systems with lenses to focus the beam also have this rotational symmetry. In fact, in a long single-wire cylindrical transmission line, already a two-dimensional problem because of the large transmission length, the rotational symmetry reduces the problem to a one-dimensional one—obviously, along a cross section normal to the direction of transmission, if we move perpendicularly to any radius, no changes will be seen.

Even for devices which do not have relatively long dimensions or rotational symmetry permitting a reduction in the dimensionality of the field analysis problem, some method may commonly be found to employ two-dimensional analysis and profitably use the results. For example, if we had a three-limbed transformer with a short depth, the problem would be truly three-dimensional; while a two-dimensional solution cannot be associated with a region within the device, the solution at the plane of symmetry cutting the transformer into two halves along the middle of the depth will have the same solution as a fictitious transformer having the same cross section, but infinite depth; therefore, analyzing this fictitious machine by two-dimensional methods will give us the field of the real device at its plane of symmetry. This information at the plane of cross-section will also tell us something of the nearby magnetic fields.

On a final note, it ought to be realized that despite the possible loss in accuracy associated with the use of two-dimensional techniques, the savings in cost and time make them useful and necessary for the time being.

Therefore, to extend the application of the finite element method to two-dimensional problems, from the basic ingredients we have extracted from the one-dimensional example, we must (1) discretize a plane into elements; (2) define (preferably interpolatory) trial functions over the solution region; (3) pick an optimization criterion that will give the free parameters of the trial function that fit best the differential equation being solved; and (4) devise rules for implementing the optimization criterion.

### 7.5.2 First-Order Triangles

To discretize a two-dimensional solution region into elements we first must define a suitably shaped element type, unlike in one dimension where there was only one choice—the line element. Several element shapes are in use, the principal ones being the triangle, the quadrilateral, and curvilinear shapes (Zienkiewicz 1977). The latter are commonly used in structural analysis where bent plates must be modeled, and are of no particular importance in electrical engineering where we always deal with planar shapes. A good use in electrical engineering of these curvilinear elements is in modeling curved boundaries, such as the perimeter of a cylindrical conductor (this is known to increase accuracy). But we shall dispense with this element shape at this point for two reasons. First, curved boundaries are easily modeled by small straight lines which may be the edges of quadrilateral or triangular elements. Second, curvilinear elements are difficult to employ in general purpose software, since the curves of the elements may be of different shapes.

This leaves us a choice between the triangle and the quadrilateral, both of which could be used to fit any shape; therefore in the rest of this book we will employ the triangle as our standard element, as has become the practice in electrical engineering. A triangle is a simple, flexible type of element shape that allows us to model all manner of shapes and has the added advantage of being symmetric around the three vertices, so as to yield some advantage

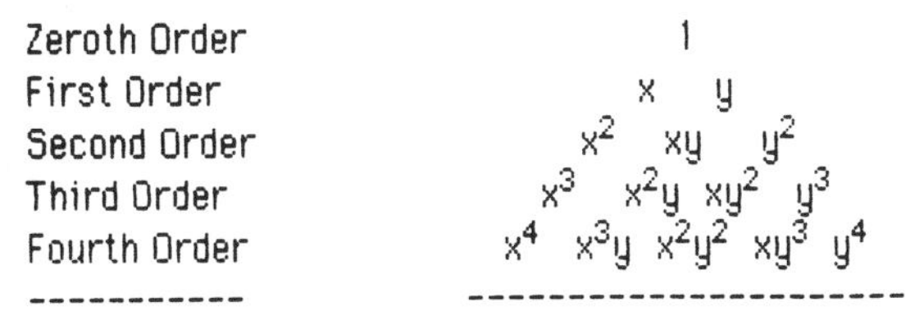

Zeroth Order 1
First Order $x$ $y$
Second Order $x^2$ $xy$ $y^2$
Third Order $x^3$ $x^2y$ $xy^2$ $y^3$
Fourth Order $x^4$ $x^3y$ $x^2y^2$ $xy^3$ $y^4$


**Figure 7.5.1.** The Pascal triangle.

in mathematical manipulation. The quadrilateral, on the other hand, does not possess this symmetry and cannot be used to fit different shapes with the same flexibility as the triangle. Moreover, in higher-order finite element modeling (Silvester 1969b, 1969c, 1978) where we employ highly accurate trial functions for better finite element solutions, the shape of the Pascal triangle in Figure 7.5.1 for high-order polynomials and its correspondence to a

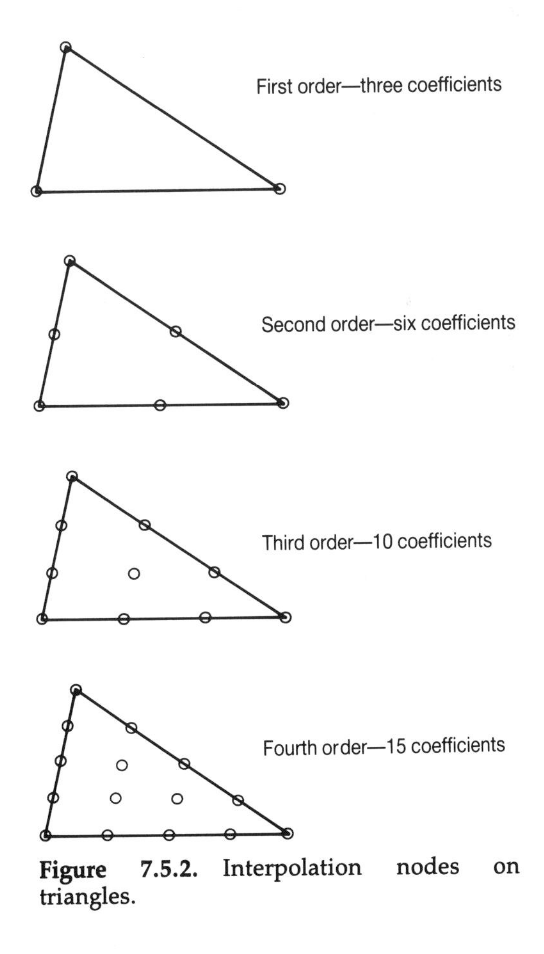


**Figure 7.5.2.** Interpolation nodes on triangles.

triangle shape give us an excellent choice of locations symmetric about the three vertices, for placing interpolation nodes on a triangle. For example, the Pascal triangle tells us that for a first-order trial function we need three terms 1, $x$, and $y$ in our trial function. Therefore, the trial function (for the variable $\phi$, say) may alternatively be written in terms of the vertex values $\phi_1$, $\phi_2$, $\phi_3$ as shown in Figure 7.5.2A. If we wished to use a higher-order approximation using second-order terms, then according to the Pascal triangle, we need to use three more interpolation nodes to correspond to the additional terms $x^2$, $xy$, and $y^2$, and these may be placed at the midsides of the edges, just as indicated by the first three rows of the Pascal triangle of Figure 7.5.1 and shown in Figure 7.5.2. For the quadrilateral, on the other hand, no such correspondence may be obtained. And for a first-order quadrilateral, for example, if we had the interpolation nodes at the four vertices, since only three terms are required for first-order interpolation, one of the nodes would become dependent on the other three and therefore become superfluous. Additional computational profit accrues from the rotational mathematical symmetry of the triangle about the vertices (Silvester 1982b).

### 7.5.3 First-Order Interpolations Using Triangular Coordinates

For two-dimensional analysis we have decided to use the triangle as our basic element shape, and we now must decide on a suitable trial function. Considering the triangle of Figure 7.5.3, the symmetry of the triangle results upon our changing to triangular coordinates defined by:

$$\zeta_i = \frac{h_i}{H_i}, \tag{7.5.1}$$

$$(x, y) \equiv (\zeta_1, \zeta_2, \zeta_3). \tag{7.5.2}$$

The three $\zeta$'s, although dimensionless, are first-order functions of distance since they vary with the first power of altitude $h_i$ in eqn. (7.5.1). Notice that $\zeta_1$ is zero along the edge 1 running along $2 \rightarrow 3$ since $h_1$ is zero there; and $\zeta_1$ is 1 at vertex 1 since $h_1 = H_1$ at that point; similarly, $\zeta_2$ and $\zeta_3$, respectively, along edges 2 and 3 are 0 and at nodes 2 and 3, respectively, are 1. That is, the three vertices have triangular coordinates $(1, 0, 0)$, $(0, 1, 0)$, and $(0, 0, 1)$, so that the best and natural first-order interpolation to use within the triangle

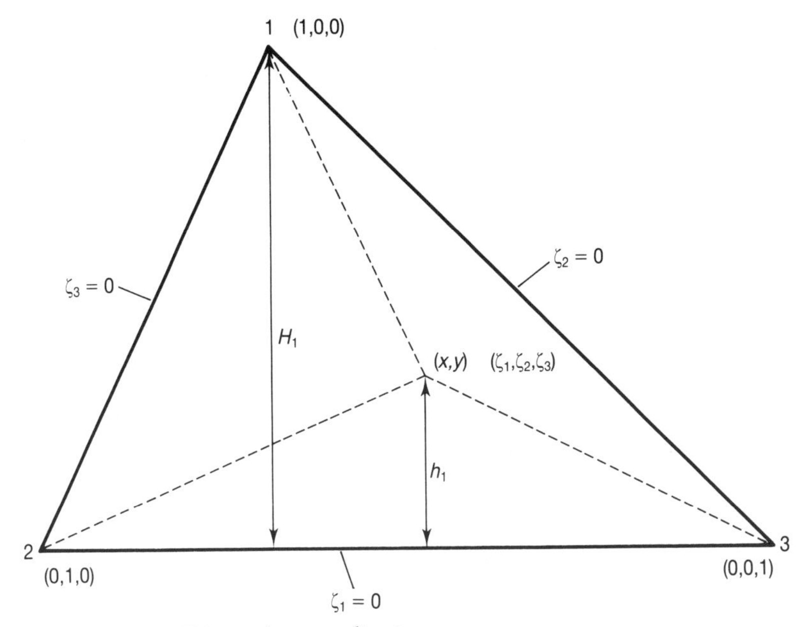


**Figure 7.5.3.** Triangular coordinates.

is

$$\phi(x, y) = \zeta_1\phi_1 + \zeta_2\phi_2 + \zeta_3\phi_3 = \tilde{\alpha}\underset{\sim}{\Phi}, \tag{7.5.3}$$

where the row vector

$$\tilde{\alpha} = [\zeta_1 \quad \zeta_2 \quad \zeta_3], \tag{7.5.4}$$

and the column vector

$$\underset{\sim}{\Phi} = [\phi_1 \quad \phi_2 \quad \phi_3]^t. \tag{7.5.5}$$

It is seen that eqn. (7.5.3) at vertex 1 will give $\phi_1$; vertex 2, $\phi_2$; and vertex 3, $\phi_3$. Be it noted that along, say, edge 3 along $1 \to 2$, where $\zeta_3 = 0$, the interpolation function of eqn. (7.5.3) becomes $\zeta_1\phi_1 + \zeta_2\phi_2$, which is independent of $\phi_3$; as a result, so long as the same first-order trial function is used in the adjacent triangle abutting along edge $1 \to 2$, we would have continuity of the trial function from triangle to triangle. It follows that the derivatives of this type of trial function are also tangentially continuous from triangle to triangle. The normal derivative of $\phi$ at edge 3 in triangle 123, however, will be expressed in terms of $\phi_3$, while the same derivative on the opposite side of that edge will involve the vertex value of $\phi$ opposite the edge in the neighbouring triangle. Therefore, the normal derivatives will not match. This absence of the continuity of the first derivatives (technically referred to as $C^1$ continuity) was first recognized by Irons and Draper (1965). This must be expected at least for our first-order triangles, since we model the function $\phi$ as a first-order one in the coordinates so that the derivative in each triangle will be a constant.

Since the triangular coordinates give us perfect interpolation functions, we should be able to express them in terms of $x$ and $y$ so as to be able to use them. To find the values of the three triangular coordinates in terms of $x$ and $y$, we need three equations. Clearly, referring to eqn. (7.5.2), we cannot obtain three independent triangular coordinates from two independent Cartesian coordinates. The relationship expressing the mutual dependence of the $\zeta$'s may be determined by summing the areas of the three triangles formed by joining the vertices of the triangle to the point $(x, y)$, as shown in Figure 7.5.3:

$$\tfrac{1}{2}h_1L_1 + \tfrac{1}{2}h_2L_2 + \tfrac{1}{2}h_3L_3 = A, \tag{7.5.6}$$

where $A$ is the total area of the triangle. Now rewriting $h_1$, $h_2$, and $h_3$ using eqn. (7.5.1), we have:

$$\tfrac{1}{2}\zeta_1H_1L_1 + \tfrac{1}{2}\zeta_2H_2L_2 + \tfrac{1}{2}\zeta_3H_3L_3 = A. \tag{7.5.7}$$

But, for every $i = 1,2,3$,

$$A = \tfrac{1}{2}H_iL_i, \tag{7.5.8}$$

so that

$$\zeta_1 + \zeta_2 + \zeta_3 = 1. \tag{7.5.9}$$

To obtain two more relationships between the triangular coordinates, in terms of the Cartesian coordinates, since $(x, y)$ is of order one, these coordinates may be fitted exactly, with no inherent approximation, by the interpolation of eqn. 7.5.3:

$$x_1\zeta_1 + x_2\zeta_2 + x_3\zeta_3 = x, \tag{7.5.10}$$

$$y_1\zeta_1 + y_2\zeta_2 + y_3\zeta_3 = y, \tag{7.5.11}$$

A more apparent way of justifying the above two eqns. (7.5.10) and (7.5.11) is to say (taking eqn. (7.5.10), for example) that since $x$ and the three $\zeta$'s are of order 1 in distance, $x$ should be expressible as a linear combination of the three $\zeta$'s: $c_1\zeta_1 + c_2\zeta_2 + c_3\zeta_3$. Now by point matching at the three vertices of triangular coordinates $(1, 0, 0)$, $(0, 1, 0)$, and $(0, 0, 1)$, and having $x$ coordinates $x_1$, $x_2$, and $x_3$, respectively, we will obtain eqn. (7.5.10); likewise eqn. (7.4.11).

Solving the three eqns. (7.5.9)–(7.5.11) we obtain the triangular coordinates in terms of the rectilinear coordinates:

$$\zeta_i = \frac{\begin{vmatrix} 1 & 1 & 1 \\ x & x_{i1} & x_{i2} \\ y & y_{i1} & y_{i2} \end{vmatrix}}{\begin{vmatrix} 1 & 1 & 1 \\ x_1 & x_2 & x_3 \\ y_1 & y_2 & y_3 \end{vmatrix}} \tag{7.5.12}$$

where $i = 1, 2, 3$ and $i$, $i1$, and $i2$ are a cyclic permutation of 1, 2, and 3. Expanding this, we have:

$$\zeta_k = \frac{a_i + b_i x + c_i y}{\Delta} \qquad i = 1,2,3, \tag{7.5.13}$$

where

$$a_i = x_{i1} y_{i2} - x_{i2} y_{i1}, \tag{7.5.14}$$

$$b_i = y_{i1} - y_{i2}, \tag{7.5.15}$$

$$c_i = x_{i2} - x_{i1}, \tag{7.5.16}$$

$$\Delta = \begin{vmatrix} 1 & 1 & 1 \\ x_1 & x_2 & x_3 \\ y_1 & y_2 & y_3 \end{vmatrix}$$

$$= x_2 y_3 - x_3 y_2 + x_3 y_1 - x_1 y_3 + x_1 y_2 - x_2 y_1 = 2\mathrm{Abs}(A), \tag{7.5.17}$$

and $A$ is the area of the triangle by classical definition in coordinate geometry. It is left as an exercise to show that $\Delta$ is $+2A$ when the vertices 1, 2, and 3 are numbered anticlockwise, and it is $-2A$ when the numbering is clockwise.

### 7.5.4 Energy Functional for a Poissonian System

In this introductory chapter on finite elements, we shall show, using energy-based arguments, that the functional corresponding to the commonly occurring Poisson equation

$$-\varepsilon \nabla^2 \phi = \rho \tag{1.2.22}$$

is

$$\mathscr{L}[\phi] = \iint \left\{ \frac{1}{2} \varepsilon [\nabla \phi]^2 - \rho \phi \right\} dR. \tag{7.5.18}$$

That is, when $\mathscr{L}$ is a minimum, eqn. (1.2.22) is satisfied. A more rigorous proof is left to section 8.1. For the moment we shall accept the validity of eqn. (7.5.18) as being a functional satisfying eqn. (1.2.22) at its minimum, on mere physical reasoning. The stored energy density in an electrostatic system, analogous to eqn. (4.1.26) for magnetics, is $\frac{1}{2}\mathbf{D} \cdot \mathbf{E}$, or, using eqns. (1.2.11) and (1.2.21) to substitute for $\mathbf{D}$ and $\mathbf{E}$, it is $\frac{1}{2}[\varepsilon\nabla\phi] \cdot [\nabla\phi]$, or $\frac{1}{2}\varepsilon[\nabla\phi]^2$ in simpler notation; this makes $\frac{1}{2}\varepsilon[\nabla\phi]^2\,dR$ the stored energy in an elemental portion $dR$ of the solution region. Moreover, by the definition of electric potential $\phi$ at a point, it is the work expended in moving a unit charge from infinity to that point. Therefore, the work done in moving the charge $\rho\,dR$ to its location at potential $\phi$ is $\rho\phi\,dR$.

In essence, therefore, $\mathscr{L}$ of eqn. (7.5.18) stands for the total energy of the system. And we know that the energy of any physical system is at its minimum at steady-state operation. We also know that an electrostatic system at steady state satisfies eqn. (1.2.22), so that we may say that $\mathscr{L}$ being at its minimum is tantamount to the satisfaction of the Poisson eqn. (1.2.22). In the absence of a more rigorous justification, particularly regarding the boundary conditions that ought to apply, we shall use the analogy of the one-dimensional system of section 7.3, which we put in a more rigorous setting, to make ad hoc assumptions on the boundary conditions. There we showed that when the functional $\mathscr{L}$ of eqn. (7.3.5a) (which is eqn. (7.5.18) reduced to one-dimension) is at its minimum, at that time eqn. (7.3.1a) or eqn. (1.2.22) in one-dimension is satisfied—provided $\phi$ is explicitly fixed or its normal gradient $d\phi/dx$ is set to zero, at the boundary of the solution region. We also learned that if $d\phi/dx$ is not set to zero through the trial functions, it will naturally turn out to be so. It was further seen that by allowing $d\phi/dx = 0$ to turn out naturally, we were able to get better answers than by imposing the condition through the trial functions.

By extension, therefore we shall make the same assumptions here and deal with the subject more rigorously in section 8.3. That is, we shall assume that in minimizing the functional of eqn. (7.5.18) all Dirichlet boundaries on which $\phi$ is specified should be set through the trial function; i.e., triangle vertices falling on such boundaries should have the value of $\phi$ at such vertices specified. We shall further assume that when Neumann boundaries where the normal gradient $\partial\phi/\partial n$ is zero are encountered, we should just ignore the condition and that it will naturally turn out to be so in the solution.

For the magnetostatic problem, by exact numerical analogy of the governing eqn. (1.2.32)

$$-\nu\nabla^2 A = J, \tag{1.2.32}$$

with eqn. (1.2.22) we may assert that the corresponding energy functional is

$$\mathscr{L}[A] = \iint \{\tfrac{1}{2}\nu[\nabla A]^2 - JA\}\, dR. \tag{7.5.19}$$

## 7.6 FUNCTIONAL MINIMIZATION

We have thus far determined to use triangular element shapes and have decided on first-order interpolation functions defined in eqn. (7.5.3), where we have the nodal values of $\phi$ as the free variational parameters. We have also identified a functional corresponding to eqn. (1.2.22) in eqn. (7.5.18). Now we need to put our trial functions into the functional and determine the conditions that result on the nodal potentials $\phi$ when we impose the restriction that $\mathscr{L}$ shall be at a minimum. We shall assume for the moment that the right-hand-side term $\rho$ is constant in every triangle, given by

$$\rho(x, y) = \rho_0. \tag{7.6.1}$$

This is often valid in magnetostatic problems such as in the analysis of electric machinery, where the current density $J$ is a specified constant in conductors and zero elsewhere; this would also be true trivially in Laplacian problems where $\rho$ is zero everywhere. In electrostatic problems where charges exist on conductors, such as in an overhead transmission line, the charge region is commonly excluded from the solution region by removing the inside of the conductor and replacing it by the surface of the conductor as boundary, at a prescribed potential; this is perfectly valid so long as the charge-carrying part is of infinite conductivity so that it cannot have a potential difference across it. Here, although the charge distribution within the conductor is not of constant density, we circumvent the problem of handling varying $\rho$ by excluding the charges from the solution region, which may not always be possible. When charge clouds exist in imperfect conductors, as in an electron device, the charge density $\rho$ is not necessarily a constant and we cannot validly say that the whole region is at one potential and thereby exclude the region from the solution. We may therefore have to take the average value for $\rho_0$. This inherent approximation becomes more and more valid as we use smaller and smaller triangles.

However in the finite element method there is no need to assume that $\rho$ is constant and we shall extend the method to varying sources $\rho$ in a later section. Here, to keep the mathematics simple and thereby promote greater clarity and understanding, we shall stick to a constant $\rho$.

Putting our trial function of eqn. (7.5.3) into our functional and using the fact that any integral over a domain is the sum of the integrals over the subdomains (or elements) making up that domain, we have, from eqn. (A1) for the operator $\nabla$ defined as $\mathbf{u}_x(\partial/\partial x) + \mathbf{u}_y(\partial/\partial y)$:

$$\begin{aligned}\mathscr{L}[\phi] &= \sum_{\Delta} \iint_{\Delta} \{\tfrac{1}{2}\varepsilon[\nabla\tilde{\boldsymbol{\alpha}}\underset{\sim}{\boldsymbol{\Phi}}]^2 - [\tilde{\alpha}\underset{\sim}{\boldsymbol{\Phi}}]\rho_0\}\, dR \qquad (7.6.2)\\ &= \sum_{\Delta} \iint_{\Delta} \left\{\frac{1}{2}\left[\left(\frac{\partial\tilde{\boldsymbol{\alpha}}\underset{\sim}{\boldsymbol{\Phi}}}{\partial x}\right)^2 + \left(\frac{\partial\tilde{\boldsymbol{\alpha}}\underset{\sim}{\boldsymbol{\Phi}}}{\partial y}\right)^2\right] - [\tilde{\boldsymbol{\alpha}}\underset{\sim}{\boldsymbol{\Phi}}]\rho_0\right\} dR,\end{aligned}$$

where the integral is over a typical element and the summation is over all the elements. Notice that all the terms within the integral are scalars, although they may have been written in vector notation. From eqns. (7.5.4) and (7.5.13),

$$\frac{\partial}{\partial x}\tilde{\boldsymbol{\alpha}} = \frac{[b_1 \quad b_2 \quad b_3]}{\Delta} = \tilde{\mathbf{b}}, \qquad (7.6.3)$$

$$\frac{\partial}{\partial y}\tilde{\boldsymbol{\alpha}} = \frac{[c_1 \quad c_2 \quad c_3]}{\Delta} = \tilde{\mathbf{c}}. \qquad (7.6.4)$$

The above row matrices $\tilde{\mathbf{b}}$ and $\tilde{\mathbf{c}}$ are often respectively denoted $\mathbf{D}_x$ and $\mathbf{D}_y$ and are referred to as *first-order differentiation matrices,* because they effectively reduce differentiation to premultiplicatory matrix operations. Therefore, in view of eqns. (7.6.3) and (7.6.4) being constants just as we expect when differentiating a first-order polynomial, the first term of the integral eqn. (7.6.2) will turn out to be the area of the element. We also need to deal with the second term of the integral $\tilde{\boldsymbol{\alpha}}\, dR$ in eqn. (7.6.2), involving integrals of the type $\zeta_i\, dR$. To determine what these are, consider a coordinate system as shown in Figure 7.6.1, with the $x$ axis directed along the edge $i$, so that $y$ and $h_i$ are the same. $dR$ then becomes $dx\, dy = dx\, dh_i$

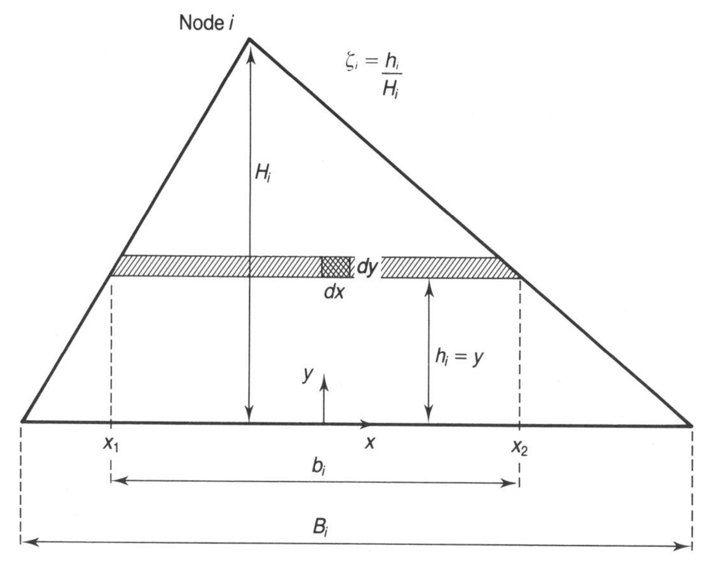


**Figure 7.6.1.** Integrating a triangular coordinate.

and using eqn. (7.5.1) to substitute for $\zeta_i$ in terms of $h_i$,

$$\iint_{\Delta} \zeta_i \, dR = \int_{h_i=0}^{H_i} \int_{x=x_1}^{x_2} \frac{h_i dx \, dh_i}{H_i} = \int_{h_i=0}^{H_i} [x_2 - x_1] \frac{h_i \, dh_i}{H_i} = \int_{h_i=0}^{H_i} \frac{b_i h_i \, dh_i}{H_i}. \tag{7.6.5}$$

But by similarity of triangles we have,

$$\frac{b_i}{B_i} = \frac{H_i - h_i}{H_i}, \tag{7.6.6}$$

so that

$$\iint_{\Delta} \zeta_i \, dR = \int_{h_i=0}^{H_i} h_i(H_i - h_i) \frac{dh_i B_i}{H_i^2} = \tfrac{1}{3}\,\tfrac{1}{2} H_i B_i = \tfrac{1}{3} A. \tag{7.6.7}$$

That is,

$$\iint_{\Delta} \tilde{\boldsymbol{\alpha}} \, dR = \iint_{\Delta} [\zeta_1 \quad \zeta_2 \quad \zeta_3] \, dR = [\tfrac{1}{3}A \quad \tfrac{1}{3}A \quad \tfrac{1}{3}A] = \mathbf{T}^{0,1} A, \tag{7.6.8}$$

where $\mathbf{T}$ is called a *metric tensor*, and the superscripts 0 and 1 arise from the fact that $\mathbf{T}$ is a multiple of the first-order approximation vector $\tilde{\boldsymbol{\alpha}}$ used for $\phi$ and the zeroth-order (constant) approximation used for $\rho$.

Putting the relationships in eqns. (7.6.3), (7.6.4), and (7.6.7) into eqn. (7.6.2):

$$\mathscr{L}[\phi] = \sum_{\Delta} \tfrac{1}{2}\varepsilon A[(\tilde{\mathbf{b}}\boldsymbol{\Phi}) \cdot (\tilde{\mathbf{b}}\boldsymbol{\Phi}) + (\tilde{\mathbf{c}}\boldsymbol{\Phi})(\tilde{\mathbf{c}}\boldsymbol{\Phi})] - \rho_0(\mathbf{T}^{0,1}\boldsymbol{\Phi})A$$

$$= \sum_{\Delta} \tfrac{1}{2}\varepsilon A[(\boldsymbol{\Phi}^t\tilde{\mathbf{b}}^t) \cdot (\tilde{\mathbf{b}}\boldsymbol{\Phi}) + (\boldsymbol{\Phi}^t\tilde{\mathbf{c}}^t)(\tilde{\mathbf{c}}\boldsymbol{\Phi})] - \rho_0(\boldsymbol{\Phi}^t\mathbf{T}^{1,0})A$$

(by writing the scalars $\tilde{\mathbf{b}}\boldsymbol{\phi}$ and $\mathbf{T}^{0,1}\boldsymbol{\phi}$ as their transposes)

$$= \sum_{\Delta} \tfrac{1}{2}\varepsilon A\boldsymbol{\Phi}^t[\tilde{\mathbf{b}}^t\tilde{\mathbf{b}} + \tilde{\mathbf{c}}^t\tilde{\mathbf{c}}]\boldsymbol{\Phi} - A\boldsymbol{\Phi}^t\mathbf{T}^{1,0}\rho_0, \tag{7.6.9a}$$

where the brackets between the scalars are now no longer necessary because the terms as matrix multiplications are perfectly compatible. It should be noted that $\boldsymbol{\phi}$, along with $\mathbf{T}^{1,0}$, is $3 \times 1$ and both $\tilde{\mathbf{b}}$ and $\tilde{\mathbf{c}}$ are $1 \times 3$, so that $\mathscr{L}$ is $1 \times 1$ or a scalar.

When treating devices where the source $\rho$ is also a variable, we may assume it to vary linearly within a triangle:

$$\rho(x, y) = \zeta_1\rho_1 + \zeta_2\rho_2 + \zeta_3\rho_3 = \tilde{\boldsymbol{\alpha}}\boldsymbol{\rho}. \tag{7.6.10}$$

This may in fact be an approximation if $\rho$ varies as a curve instead of as a straight line between any two points. As we shall see later in chapter 8, higher-order approximations are also possible for greater accuracy. For the moment we shall be content with evaluating $\rho_1$, $\rho_2$, and $\rho_3$ at the vertices from the distribution $\rho(x, y)$ and using them here. It should be realized that the $\rho$'s are associated with triangles and not nodes, despite eqn. (7.6.10) expressing the nodal values $\rho_1$, $\rho_2$, and $\rho_3$. For example, consider two neighboring triangles 123 and 142 shown in Figure 7.6.2, the former in a transistor with a charge distribution $\rho(x, y)$ and the latter in charge free air. Therefore, when treating triangle 142, all three values $\rho_1$, $\rho_4$, and $\rho_2$ are zero but when treating triangle 123, they take the respective values $\rho(x_1, y_1)$, $\rho(x_2, y_2)$, and $\rho(x_3, y_3)$—otherwise we may be assuming a charge in triangle 142 as a result of eqn. (7.6.10).

Substituting eqn. (7.6.10) in the functional, the term that will require different treatment is

$$\iint \rho\phi\, dR = \iint (\tilde{\boldsymbol{\alpha}}\boldsymbol{\rho})(\tilde{\boldsymbol{\alpha}}\boldsymbol{\phi})\, dR = \iint (\boldsymbol{\phi}^t\tilde{\boldsymbol{\alpha}}^t)(\tilde{\boldsymbol{\alpha}}\boldsymbol{\rho})\, dR = \boldsymbol{\phi}^t\left\{\iint \tilde{\boldsymbol{\alpha}}^t\tilde{\boldsymbol{\alpha}}\, dR\right\}\boldsymbol{\rho}$$

$$= A\boldsymbol{\phi}^t\mathbf{T}^{1,1}\boldsymbol{\rho}. \tag{7.6.11}$$

where the metric tensor (or Gram or Mass matrix) $\mathbf{T}^{1,1}$ arises from integrating the product of two first-order approximation vectors

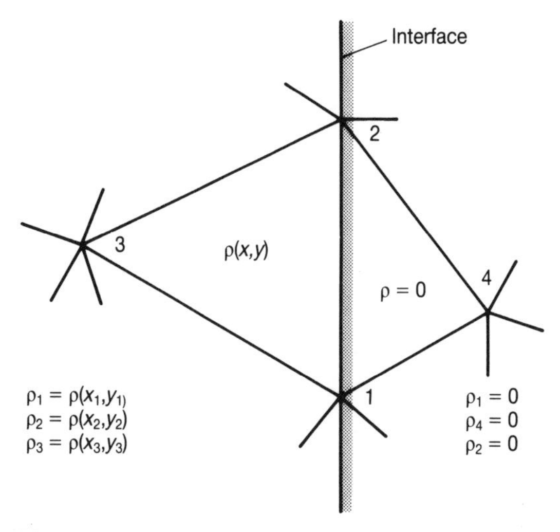


**Figure 7.6.2.** Nodal sources.

using the integral relationship for triangular coordinates proved in Appendix D:

$$\iint_\Delta \zeta_1^i \zeta_2^j \zeta_3^k \, dR = \frac{i!\,j!\,k!\,2!\,A}{(i+j+k+2)!}, \tag{D1}$$

where $A$ is the area of the triangle $\Delta$ over which the integration is being performed. That is,

$$AT^{1,1} = \iint_\Delta \tilde{\boldsymbol{\alpha}}^t \tilde{\boldsymbol{\alpha}} \, dR = \iint \begin{bmatrix} \zeta_1^2 & \zeta_1\zeta_2 & \zeta_1\zeta_3 \\ \zeta_1\zeta_2 & \zeta_2^2 & \zeta_2\zeta_3 \\ \zeta_1\zeta_3 & \zeta_2\zeta_3 & \zeta_3^2 \end{bmatrix} dR$$

$$= \frac{A}{12} \begin{bmatrix} 2 & 1 & 1 \\ 1 & 2 & 1 \\ 1 & 1 & 2 \end{bmatrix}. \tag{7.6.12}$$

The difference this makes to eqn. (7.6.9a) is that it now becomes

$$\mathscr{L}[\phi] = \sum_\Delta \{\tfrac{1}{2}\varepsilon A \underset{\sim}{\mathbf{\Phi}}^t [\tilde{\mathbf{b}}^t \tilde{\mathbf{b}} + \tilde{\mathbf{c}}^t \tilde{\mathbf{c}}] \underset{\sim}{\mathbf{\Phi}} - A \underset{\sim}{\mathbf{\Phi}}^t \mathbf{T}^{1,1} \underset{\sim}{\boldsymbol{\rho}}\}. \tag{7.6.9b}$$

Notice that eqn. (7.6.7) could have been derived using eqn. (D1) and that when $\rho_1$, $\rho_2$, and $\rho_3$ are equal, say, to $\rho_0$, eqn. (7.6.9b) reduces to eqn. (7.6.9a). The term $\varepsilon A[\tilde{\mathbf{b}}^t \tilde{\mathbf{b}} + \tilde{\mathbf{c}}^t \tilde{\mathbf{c}}]$ is referred to as the *Dirichlet* or *stiffness matrix* in the literature and will be seen to be symmetric if we take the transpose of it, which will leave it unchanged. Eqn. (7.6.9) may be generally written as

$$\mathcal{L}[\phi] = \sum_{\Delta} \{\tfrac{1}{2}\boldsymbol{\Phi}^t \mathbf{P} \boldsymbol{\Phi} - \boldsymbol{\Phi}^t \mathbf{q}\}, \tag{7.6.13}$$

where $\mathbf{P}$ is a $3 \times 3$ symmetric matrix and $\mathbf{q}$ is a $3 \times 1$ column vector. Now we must impose conditions of extremum upon this by differentiating with respect to the free variables, which are those components of $\boldsymbol{\Phi}$ that are not along those portions of the boundary where $\phi$ is prescribed.

The formation of $\mathbf{P}$ and $\mathbf{q}$ for a triangle of vertex coordinates $x[1], y[1]$; $x[2], y[2]$; and $x[3], y[3]$ with material constant $\varepsilon$ and $\rho$ of values $\rho[1]$, $\rho[2]$ and $\rho[3]$ at the three vertices may be summarized by Algs. 7.6.1 and 7.6.2, which are based on eqns. (7.5.14)–(7.5.17), (7.6.3), (7.6.4), (7.6.12), and (7.6.9). The main algorithm, Alg. 7.6.2, first calls the procedure Triangle defined in Alg. 7.6.1 to construct the first order differentiation matrices $\tilde{\mathbf{b}}$ and $\tilde{\mathbf{c}}$ and then constructs the local matrices $\mathbf{P}$ and $\mathbf{q}$.

To see how the extremization of the functional works, consider the contribution of the first triangle to $\mathcal{L}$:

$$\mathcal{L}_1 = \tfrac{1}{2}[\phi_1 \quad \phi_2 \quad \phi_3] \begin{bmatrix} P_{11} & P_{12} & P_{13} \\ P_{21} & P_{22} & P_{23} \\ P_{31} & P_{32} & P_{33} \end{bmatrix} \begin{bmatrix} \phi_1 \\ \phi_2 \\ \phi_3 \end{bmatrix} - [\phi_1 \quad \phi_2 \quad \phi_3] \begin{bmatrix} q_1 \\ q_2 \\ q_3 \end{bmatrix} \tag{7.6.14}$$

If we expand this using the symmetric property $\mathbf{P}[i,j] = \mathbf{P}[j,i]$, we will obtain:

$$\begin{aligned} \mathcal{L}_1 = \tfrac{1}{2}\{&\phi_1^2 P_{11} + \phi_1\phi_2 P_{12} + \phi_1\phi_2 P_{21} + \phi_1\phi_3 P_{13} + \phi_1\phi_3 P_{31} + \phi_2^2 P_{22} \\ &+ \phi_2\phi_3 P_{23} + \phi_2\phi_3 P_{32} + \phi_3^2 P_{33}\} - \phi_1 q_1 - \phi_2 q_2 - \phi_3 q_3. \end{aligned} \tag{7.6.15}$$

Now:

$$\begin{aligned} \frac{\partial \mathcal{L}_1}{\partial \phi_1} = \tfrac{1}{2}\{&P_{11}\phi_1 + P_{12}\phi_2 + P_{13}\phi_3 \\ &+ P_{11}\phi_1 + P_{21}\phi_2 + P_{31}\phi_3\} - q_1, \end{aligned} \tag{7.6.16a}$$

$$\begin{aligned} \frac{\partial \mathcal{L}_1}{\partial \phi_2} = \tfrac{1}{2}\{&P_{21}\phi_1 + P_{22}\phi_2 + P_{23}\phi_3 \\ &+ P_{12}\phi_1 + P_{22}\phi_2 + P_{32}\phi_3\} - q_2, \end{aligned} \tag{7.6.16b}$$

$$\begin{aligned} \frac{\partial \mathcal{L}_1}{\partial \phi_3} = \tfrac{1}{2}\{&P_{31}\phi_1 + P_{32}\phi_2 + P_{33}\phi_3 \\ &+ P_{13}\phi_1 + P_{23}\phi_2 + P_{33}\phi_3\} - q_3, \end{aligned} \tag{7.6.16c}$$

**Algorithm 7.6.1. Computing First-Order Differentiation Matrices**

```
Procedure Triangle (b, c, A, x, y)
{Function: To compute the first order differentiation matrices
Outputs: b,c = The 1 × 3 first-order differentiation matrices
defined in eqns. (7.6.3) and (7.6.4); A = Area of triangle
Inputs: x,y = 3 × 1 vectors containing the coordinates of the
three vertices}
Begin
  Delta ← x[2]*y[3] − x[3]*y[2] + x[3]*y[1]
            − x[1]*y[3] + x[1]*y[2] − x[2]*y[1]      !Eqn. (7.5.17)
  A ← Abs(Delta)/2      !Eqn. (7.5.17)
  For i ← 1 To 3 Do
  i1 ← Mod(i, 3) + 1 {Or i Mod 3 + 1}
  i2 ← Mod(i1, 3) + 1
  b[i] ← (y[i1] − y[i2])/Delta                 !Eqns. (7.5.15), (7.6.3)
  c[i] ← (x[i2] − x[i1])/Delta                 !Eqns. (7.5.16), (7.6.4)
End
```

**Algorithm 7.6.2. Forming Local Matrices for First-Order Triangle**

```
Procedure FirstOrderLocalMats (P, q, x, y, Eps, Rho)
{Function: To compute the first-order differentiation matrices
Outputs: P = The 3 × 3 first order local element matrix—Eqn.
(7.6.9); q = The 3 × 1 local right hand side vector—Eqn. (7.6.12)
or (7.6.3)
Inputs: x,y = 3 × 1 vectors containing the coordinates of the
three vertices; Eps = The constant material value ε in the
triangle—Eqn. (1.2.22); Rho = 3 × 1 vector giving source ρ at
the three vertices (or a constant if ρ is a constant)
Required: Procedure Triangle}
Begin
  Triangle (b, c, A, x, y)
  For i ← 1 To 3 Do
    q[i] ← A*(2*Rho[i] + Rho[i1] + Rho[i2])/12
      {Or if Rho is a constant q[i] ← A*Rho/3
      in place of the previous two lines}
    For j ← 1 To 3 Do
      P[i, j] ← Eps*A*(b[i]*b[j] + c[i]*c[j])         {Eqn. (7.6.9)}
End
```

so that we may write in matrix notation for differentiation by a vector $\underset{\sim}{\mathbf{\Phi}}$

$$\frac{\partial \mathscr{L}_1}{\partial \underset{\sim}{\mathbf{\Phi}}} = \left[\frac{\partial \mathscr{L}_1}{\partial \phi_1} \frac{\partial \mathscr{L}_1}{\partial \phi_2} \frac{\partial \mathscr{L}_1}{\partial \phi_3}\right]^t$$
$$= \tfrac{1}{2}\{\mathbf{P} + \mathbf{P}^t\}\underset{\sim}{\mathbf{\Phi}} - \underset{\sim}{\mathbf{q}} = \mathbf{P}\underset{\sim}{\mathbf{\Phi}} - \underset{\sim}{\mathbf{q}}, \tag{7.6.17}$$

because $\mathbf{P}$ is symmetric. This result may be generally extended to differentiation by a vector. For a first-order polynomial $\underset{\sim}{\mathbf{\Phi}}^t\underset{\sim}{\mathbf{q}}$,

$$\frac{\partial}{\partial \underset{\sim}{\mathbf{\Phi}}} \underset{\sim}{\mathbf{\Phi}}^t\underset{\sim}{\mathbf{q}} = \underset{\sim}{\mathbf{q}}, \tag{7.6.18}$$

and for a second-order polynomial $\underset{\sim}{\mathbf{\Phi}}^t\mathbf{P}\underset{\sim}{\mathbf{\Phi}}$ in the coefficients of $\phi$:

$$\frac{\partial}{\partial \underset{\sim}{\mathbf{\Phi}}} \underset{\sim}{\mathbf{\Phi}}^t\mathbf{P}\underset{\sim}{\mathbf{\Phi}} = [\mathbf{P} + \mathbf{P}^t]\underset{\sim}{\mathbf{\Phi}}, \tag{7.6.19}$$

where $\mathbf{P}$ is a matrix and $\underset{\sim}{\mathbf{q}}$ is a column vector.

We cannot apply extremization to the contribution of one element and say that eqn. (7.6.17) is equal to zero. The reason is that the total energy of the system is the sum of the contributions coming from the various triangles, and what is at a minimum is the total energy and not the energy of a particular element. Eqn. (7.6.17) may therefore be summed for the whole region of solution $R$ made up by all the triangles of the finite element mesh. When the global equation is assembled and solved, we have the solution of the field at the vertices, say $n$ in number, of the finite element mesh—$\phi_1$, $\phi_2, \ldots, \phi_n$. However, although the solution of the matrix equation explicitly returns only the vertex potentials, the finite element method, unlike the finite difference method, gives us the solution everywhere in the solution region, for we actually optimize the trial function of eqn. (7.5.3) and not the vertex potentials. Thus, once we obtain the vertex potentials, we know through the application of eqn. (7.5.3) the solution everywhere.

Now, how are we to sum the contributions of the type of eqn. (7.6.17) for the whole mesh? This is seen by examining the contributions to energy from the example of two adjacent triangles, say, made up by the nodes 1, 3, 51, and 3, 1, 4. Let us also say, by way of illustration, that we have 60 nodes of which the first 50 are unknown and the last 10, being on Dirichlet parts of the boundary are known. Now, we may evaluate the (scalar) energy contribution $\mathscr{L}_1$ from the first triangle, using eqn. (7.6.14). Suppose that for the first triangle, when we evaluate using the coordinates of the vertices and the values of $\varepsilon$ and $\rho$ in that triangle, we obtain an energy contribution given by eqn. (7.6.14). Since the first, second, and third vertices of triangle 1 are 1, 3, and 51, making the mapping $1 \to 1$, $2 \to 3$, and $3 \to 51$, we have

$$\begin{aligned}\mathscr{L}_1 = {} & \tfrac{1}{2}\phi_1^2 P_{11} + \phi_1\phi_3 P_{12} + \phi_1\phi_{51} P_{13} + \tfrac{1}{2}\phi_3^2 P_{22} \\ & + \phi_3\phi_{51} P_{23} + \tfrac{1}{2}\phi_{51}^2 P_{33} - \phi_1 q_1 - \phi_3 q_2 - \phi_{51} q_3.\end{aligned} \tag{7.6.20}$$

Similarly, if we evaluate the energy contribution from triangle 2, because the shape and properties of the triangle need not be the same, we may get a different matrix **U** in place of **P** and coefficients **v** in place of **q**:

$$\begin{aligned}\mathscr{L}_2 = {} & \tfrac{1}{2}\phi_3^2 U_{11} + \phi_3\phi_1 U_{12} + \phi_3\phi_4 U_{13} + \tfrac{1}{2}\phi_1^2 U_{22} \\ & + \phi_1\phi_4 U_{23} + \tfrac{1}{2}\phi_4^2 U_{33} - \phi_3 v_1 - \phi_1 v_2 - \phi_4 v_3.\end{aligned} \tag{7.6.21}$$

How the $3 \times 3$ matrix derivatives (as in eqn. (7.6.17)) are made to apply to the whole mesh, is seen from summing $\mathscr{L}_1$ and $\mathscr{L}_2$. It is easily shown that

$$\begin{aligned}\mathscr{L}_1 + \mathscr{L}_2 = {} & [\phi_1 \quad \phi_3 \quad \phi_4 \quad \phi_{51}]\mathbf{W}[\phi_1 \quad \phi_3 \quad \phi_4 \quad \phi_{51}]^t \\ & - [\phi_1 \quad \phi_3 \quad \phi_4 \quad \phi_{51}]\,(q_1 + v_2)(q_2 + v_1)\,(v_3)\,(q_3)]^t,\end{aligned} \tag{7.6.22}$$

where the matrix **W** is given by

$$\mathbf{W} = \begin{bmatrix} (P_{11} + U_{22}) & (P_{12} + U_{12}) & (U_{23}) & (P_{13}) \\ (P_{12} + U_{12}) & (P_{22} + U_{11}) & (U_{13}) & (P_{23}) \\ (U_{23}) & (U_{13}) & (U_{33}) & (0) \\ (P_{13}) & (P_{23}) & (0) & (P_{33}) \end{bmatrix}. \tag{7.6.23}$$

In the global context, the last column and row correspond to $\phi_{51}$, which is known, and therefore small variations cannot be taken with respect to it. For this reason, the energy $\mathscr{L}$ is not extremized with respect to $\phi_{51}$ and the other known potentials $\phi_{52}$, $\phi_{53}, \ldots, \phi_{60}$, so that the global matrix for our example will be just $50 \times 50$. In the final matrix equation, then, what will happen to columns corresponding to the knowns? To see the answer, making the mapping $1 \to 1$, $2 \to 3$, and $3 \to 51$ for the derivative in eqn. (7.6.16a), we have

$$\frac{\partial \mathscr{L}_1}{\partial \phi_1} = P_{11}\phi_1 + P_{12}\phi_3 + P_{13}\phi_{51} - q_1, \tag{7.6.24}$$

so that since $\phi_{51}$ is known, the term $P_{13}\phi_{51}$ is also known and may be lumped together with $q_1$. That is, the coefficients of the last column of eqn. (7.6.22), which multiply with the knowns such as $\phi_{51}$, are multiplied and taken to the right. The resulting functional takes the form, with all elemental contributions added:

$$\mathscr{L}[\Phi] = \tfrac{1}{2}\underset{\sim}{\mathbf{\Phi}}^t \mathbf{P}^g \underset{\sim}{\mathbf{\Phi}} - \underset{\sim}{\mathbf{\Phi}}^t \underset{\sim}{\mathbf{q}}^g, \tag{7.6.25}$$

which by our rules of eqns. (7.6.18) and (7.6.19) for differentiation by a vector (really several differentiations by every coefficient of the vector) gives us, after using the symmetry condition $\mathbf{P}^{gt} = \mathbf{P}^g$:

$$\frac{\partial \mathscr{L}}{\partial \underset{\sim}{\mathbf{\Phi}}} = \mathbf{P}^g \underset{\sim}{\mathbf{\Phi}} - \underset{\sim}{\mathbf{q}}^g = 0, \tag{7.6.26}$$

which matrix equation has to be solved for the vector $\underset{\sim}{\mathbf{\Phi}}$ containing the values of $\phi$ at all the unknown interpolation nodes.

The placing of derivative contributions of the type $\mathbf{P}\underset{\sim}{\boldsymbol{\phi}} - \underset{\sim}{\mathbf{q}}$, as in eqn. (7.6.17), in the global equation for the mesh $\mathbf{P}^g\underset{\sim}{\mathbf{\Phi}} = \underset{\sim}{\mathbf{q}}^g$ may therefore be generalized from the preceding considerations. For a triangle with vertices **v**[1], **v**[2], **v**[3] of a mesh of NUnk unknowns,

### Algorithm 7.6.3. Placing a Local Matrix in a Global Matrix

```
Procedure GlobalPlace (PG, qG, P, q, v, Phi)
{Function: Adds the local matrices to the global matrices
Outputs: PG = the global matrix of size NUnk × NUnk; qG =
the global right-hand side of size NUnk × 1
Inputs: P = 3 × 3 local matrix, q = 3 × 1 local right-hand side
vector; NUnk = number of unknown nodes in mesh numbered
before the known nodes; v = a 3-vector containing the node
numbers of the triangle; Phi = a vector as long as there are
finite element nodes. The first NUnk elements are unknown
and to be determined by the finite element method, and the rest
of the elements are known through the boundary conditions}
Begin
  !For Row ← 1 To 3 Do
    If v[Row] ≤ NUnk
    Then !Else Row Corresponds to a Known; Ignore it
        qG[v[Row]], ← qG[v[Row]] + q[Row]
        For Col ← 1 To 3 Do
        If v[Col] ≤ NUnk
          Then !Add to Global Matrix
            PG[v[Row]v[Col]] ← PG[v[Row],v[Col]] +
              P[Row, Col]
          Else !Column Corresponds to a Known. Shift Right
            qG[v[Row]] ← qG[v[Row]] −
              P[Row, Col]*Phi[v[Col]]
End
```

the generalization may be expressed as the heuristic procedure GlobalPlace defined in Alg. 7.6.3, written without considerations of any efficiency that may be gained by exploiting symmetry, speed of access, etc.

These principles are best understood by turning now to the handworked example of section 7.7.

## 7.7 THE ELECTRIC AND MAGNETIC FIELDS IN A COAXIAL CABLE: A HANDWORKED EXAMPLE

### 7.7.1 The Magnetic Field

To explain through computation the theory expounded so far, we shall take up for hand solution the problem of a two-conductor cable system, shown in Figure 7.7.1. Here we have an infinitely long cable with a rectangular stranded inner conductor carrying the forward current of 24 A, which is returned through the outside sheath. This is, then a two-dimensional problem. We wish to determine the magnetic field within the cable system; an obvious reason for doing so is to compute the mutual inductance between the inner and outer conductors.

Although the system of currents may be alternating, for purposes of computing the self inductance, defined as flux linkage divided by the causing current, we may employ a DC system.

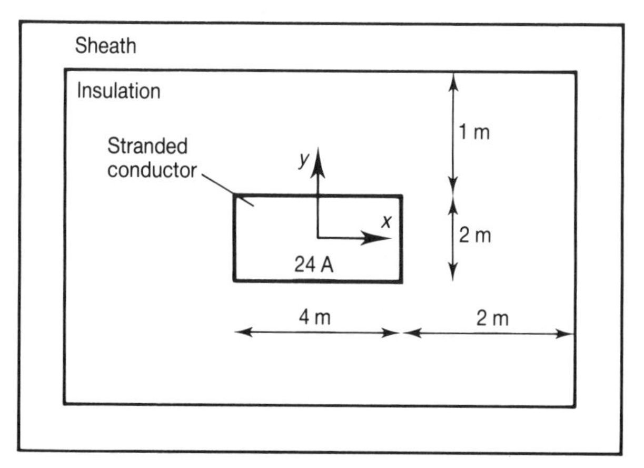


**Figure 7.7.1.** A cable system.

Usually AC flux linkage is different from DC linkage because of eddy current effects, which tend to force the currents to flow along the outer regions of the conductor—the so-called skin effect. It has been shown in chapter 5 that in AC systems, the electromagnetic quantities of flux and current exponentially decay with distance of penetration into the body of a material as we proceed from the surface; the rate of decay is measured by the depth of penetration varying as the negative half power of frequency, permeability, and conductivity. Therefore the skin effect is especially pronounced in conductors and magnetically permeable materials and at high frequencies.

In this problem, as in most practical systems, the inner conductor is stranded (that is, made up of several tightly wound strands of conductor insulated from each other) so as to force the currents to flow through the entire cross section of the conductor and thereby utilize the material better. The cross section of each strand is small compared with the depth of penetration so that no skin effect is possible within a strand. Hence the validity of assuming uniform current density as in DC systems. Now in the AC system which we are modeling, as show in section 1.2.4, no normal magnetic flux density $B_n$ may be present in a good conductor. What we have then is the exact equivalent of a two-dimensional DC system with the rectangular sheath forced to be a flux line. Therefore, in the air bordering the outer sheath, by the normal continuity of flux density (eqn. (1.2.17)), flux will flow tangentially to the surface. As shown in section 1.2.4, this line will be a line of constant vector potential, which we may take as our reference potential $A = 0$. For such a system we have seen that the governing equation is

$$-\nu\nabla^2 A = J. \tag{1.2.32}$$

As shown in Appendix B, since the entire boundary is Dirichlet with $A = 0$, the boundary value problem has a unique solution, which we may attempt by the finite element method. For this purpose we need to discretize the solution region as shown in Figure 7.7.2A, where all the known nodes on the boundary have been numbered last. We have 32 triangular elements and 25 nodes of which 9 are unknown. And solving a $9 \times 9$ matrix equation, let alone forming it, is truly a tedious task. Notice the discretization at the four corners, such as to cut each corner in two. For example, instead of triangles 5, 15, 16 and 5, 16, 17, if we had chosen the alternative discretization into triangles 17, 15, 16 and 17, 5, 15, we would effectively assume a zero potential in triangle 17, 15, 16

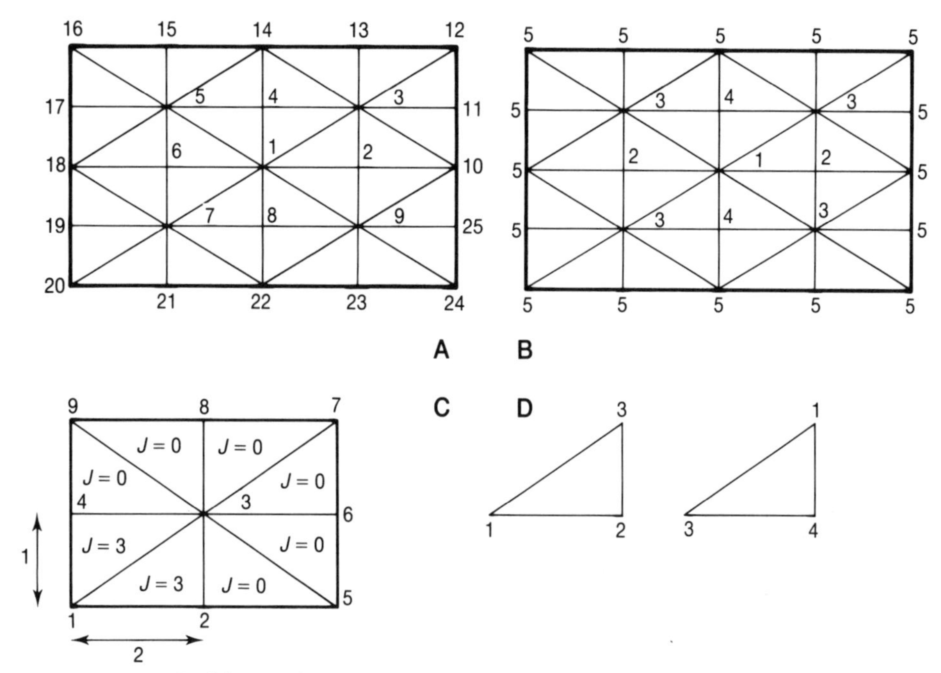


**Figure 7.7.2.** Problem reduction.

because all three of its vertices 17, 15, and 16 are along the Dirichlet boundary and their potentials are therefore known, so that our trial function eqn. (7.5.3) would give us a zero potential inside. To so assume a zero potential through the trial functions is to distort the solution and must therefore be avoided.

Before rushing headlong into computations it is good to consider whether we may in any way reduce the computational load. We know by the corkscrew rule (Skitek and Marshal 1982) that the flux will circle clockwise around the conductor so that equipotential lines (lines along which $A$ has the same value) will form rings about the conductor, which need not necessarily be circular in shape. By symmetry, therefore, we may say that nodes 2 and 6 are at the same potential; similarly, the sets $(4, 8)$, $(3, 5, 7, 9)$, and $(10, 11, \ldots, 25)$ consist of nodes at the same potential. This leads us to the new numbering of Figure 7.7.2B, where we still have 32 elements, but only five node numbers, of which four are unknown. While the reduction of the matrix size from 9 to 4 accompanies a significant reduction in work, we still need to form the global matrix $\mathbf{P}^g$ from 32 elements. A further complication arises in that although, say, nodes 3, 5, 7, and 9 of Figure 7.7.2A all have the number 3 in Figure 7.7.2B because of their having the same potential, in forming the local matrix **P** of eqn. (7.6.17) for each element, the correct coordinates must be employed. That is, node 3 has different coordinates depending on its location, and this complicates the data structure.

The work in forming the $4 \times 4$ matrix equation from 32 elements may be reduced and the attendant complexity of having nodes with many locations may be avoided by noting that in Figure 7.7.2A, the conditions $A_3 = A_5$ and $A_3 = A_g$, respectively, imply that the vertical and horizontal lines of symmetry have the condition $\partial A/\partial n = 0$, making them Neumann boundaries of the smaller problem of Figure 7.7.2C where we have only eight elements and four unknown nodes. We may validly solve a quarter of the problem, since this has either Neumann or Dirichlet conditions along the boundary so that the solution is unique (as shown in Appendix B). The Dirichlet boundary nodes have been given separate node numbers because although they are of the same $A$ values, they are also known, so that giving them the same numbers does not reduce the work but reduces the clarity involved in recognizing the locations of the node numbers.

To perform the finite element analysis we must first add the contributions of the eight elements. Consider the first element 1, 2, 3 with coordinates $(x_1, y_1) = (0, 0)$, $(x_2, y_2) = (2, 0)$, and $(x_3, y_3) = (0, 1)$, as shown in Figure 7.7.2C.

For this triangle, according to eqns. (7.5.14)–(7.5.17), we have

$$\Delta = 2A = x_2 y_3 - x_3 y_2 + x_3 y_1 - x_1 y_3 + x_1 y_2 - x_2 y_1 = 2, \quad (7.7.1)$$

$$b_1 = y_2 - y_3 = -1, \quad (7.7.2)$$

$$b_2 = y_3 - y_1 = 1, \quad (7.7.3)$$

$$b_3 = y_1 - y_2 = 0, \quad (7.7.4)$$

$$c_1 = x_3 - x_2 = 0, \quad (7.7.5)$$

$$c_2 = x_1 - x_3 = -2, \quad (7.7.6)$$

$$c_3 = x_2 - x_1 = 2. \quad (7.7.7)$$

Therefore, according to eqns. (7.6.3) and (7.6.4),

$$\tilde{\mathbf{b}} = [-0.5 \quad 0.5 \quad 0.0], \quad (7.7.8)$$

$$\tilde{\mathbf{c}} = [0.0 \quad -1.0 \quad 1.0], \quad (7.7.9)$$

so that from eqn. (7.6.9), with a relative permeability for copper of

$\varepsilon = 1$, and, for 24 A over 8 m$^2$, a current density $\rho_0 = 3$:

$$\mathcal{L}_1 = \tfrac{1}{2}\varepsilon A \underset{\sim}{\Phi}^t[\tilde{\mathbf{b}}^t\tilde{\mathbf{b}} + \tilde{\mathbf{c}}^t\tilde{\mathbf{c}}]\underset{\sim}{\Phi} - A\underset{\sim}{\Phi}^t T^{1,0}\rho_0$$

$$= \tfrac{1}{2}[\phi_1 \quad \phi_2 \quad \phi_3]\begin{bmatrix} 0.25 & -0.25 & 0.0 \\ -0.25 & 1.25 & -1.00 \\ 0.00 & -1.00 & 1.00 \end{bmatrix}\begin{bmatrix} \phi_1 \\ \phi_2 \\ \phi_3 \end{bmatrix} \tag{7.7.10}$$

$$- [\phi_1 \quad \phi_2 \quad \phi_3]\begin{bmatrix} 1.0 \\ 1.0 \\ 1.0 \end{bmatrix}$$

Now we may take, arbitrarily, 134 as shown in Figure 7.7.3B, as the ordering of our second element and this would be perfectly all right in general-purpose software where for every element we compute the local matrices **P** and **q**. In hand calculations, however, we should make an attempt to choose element shapes and node ordering judiciously, since this can make a huge difference in the time we spend on the calculations. As evident from the above working, our choice of right angled triangles has made the computation of the local matrices easy. Similarly, if we reorder our element 134 and call node 3 our first node, then our element is 341, of the same shape as our first element 1, 2, 3, as seen in Figure

**A**

| $V_1$ | $V_2$ | $V_3$ | $\varepsilon$ | $\rho_0$ |
|---|---|---|---|---|
| 1 | 2 | 3 | 1.0 | 3.0 |
| 3 | 4 | 1 | 1.0 | 3.0 |
| 5 | 2 | 3 | 1.0 | 0.0 |
| 3 | 6 | 5 | 1.0 | 0.0 |
| 3 | 4 | 9 | 1.0 | 0.0 |
| 9 | 8 | 3 | 1.0 | 0.0 |
| 7 | 8 | 3 | 1.0 | 0.0 |
| 3 | 6 | 7 | 1.0 | 0.0 |

**B**

| X | Y |
|---|---|
| 0.0 | 0.0 |
| 2.0 | 0.0 |
| 2.0 | 1.0 |
| 0.0 | 1.0 |
| 4.0 | 0.0 |
| 4.0 | 1.0 |
| 4.0 | 2.0 |
| 2.0 | 2.0 |
| 0.0 | 2.0 |

**C**

No. of Unknowns = 4
No. of Points = 9
No. of Triangles = 8.

**Figure 7.7.3.** Data for cable problem. **A.** Triangle data. **B.** Nodal coordinate data. **C.** General information.

7.7.2D, so that the local matrices of eqn. (7.7.10) would still be valid, except that in place of $\phi_1$, $\phi_2$, and $\phi_3$ we would now respectively have $\phi_3$, $\phi_4$, and $\phi_1$. If we number all elements like this, our node orderings and element data will be as given in Figure 7.7.3A and the nodal data as in Figure 7.7.3B. When solving this problem by a computer program, general information as in Figure 7.7.3C may also be given or, alternatively, may be obtained using "Until End of File" file reading commands. Observe that we have abandoned the anticlockwise renumbering so as to simplify the hand arithmetic.

For these data, we have:

$$\mathcal{L} = \mathcal{L}_1 + \mathcal{L}_2 + \ldots + \mathcal{L}_8$$

$$= \tfrac{1}{2}[\phi_1 \quad \phi_2 \quad \phi_3] \begin{bmatrix} 0.25 & -0.25 & 0.00 \\ -0.25 & 1.25 & -1.00 \\ 0.00 & -1.00 & 1.00 \end{bmatrix} \begin{bmatrix} \phi_1 \\ \phi_2 \\ \phi_3 \end{bmatrix}$$

$$- [\phi_1 \quad \phi_2 \quad \phi_3] \begin{bmatrix} 1.0 \\ 1.0 \\ 1.0 \end{bmatrix}$$

$$+ \tfrac{1}{2}[\phi_3 \quad \phi_4 \quad \phi_1] \begin{bmatrix} 0.25 & -0.25 & 0.00 \\ -0.25 & 1.25 & -1.00 \\ 0.00 & -1.00 & 1.00 \end{bmatrix} \begin{bmatrix} \phi_3 \\ \phi_4 \\ \phi_1 \end{bmatrix}$$

$$- [\phi_3 \quad \phi_4 \quad \phi_1] \begin{bmatrix} 1.0 \\ 1.0 \\ 1.0 \end{bmatrix}$$

$$+ \tfrac{1}{2}[\phi_5 \quad \phi_2 \quad \phi_3] \begin{bmatrix} 0.25 & -0.25 & 0.00 \\ -0.25 & 1.25 & -1.00 \\ 0.00 & -1.00 & 1.00 \end{bmatrix} \begin{bmatrix} \phi_5 \\ \phi_2 \\ \phi_3 \end{bmatrix}$$

$$- [\phi_5 \quad \phi_2 \quad \phi_3] \begin{bmatrix} 0.0 \\ 0.0 \\ 0.0 \end{bmatrix}$$

$$+\tfrac{1}{2}[\phi_3 \quad \phi_6 \quad \phi_5]\begin{bmatrix} 0.25 & -0.25 & 0.00 \\ -0.25 & 1.25 & -1.00 \\ 0.00 & -1.00 & 1.00 \end{bmatrix}\begin{bmatrix} \phi_3 \\ \phi_6 \\ \phi_5 \end{bmatrix}$$

$$-[\phi_3 \quad \phi_6 \quad \phi_5]\begin{bmatrix} 0.0 \\ 0.0 \\ 0.0 \end{bmatrix}$$

$$+\tfrac{1}{2}[\phi_3 \quad \phi_4 \quad \phi_9]\begin{bmatrix} 0.25 & -0.25 & 0.00 \\ -0.25 & 1.25 & -1.00 \\ 0.00 & -1.00 & 0.00 \end{bmatrix}\begin{bmatrix} \phi_3 \\ \phi_4 \\ \phi_9 \end{bmatrix}$$

$$-[\phi_3 \quad \phi_4 \quad \phi_9]\begin{bmatrix} 0.0 \\ 0.0 \\ 0.0 \end{bmatrix}$$

$$+\tfrac{1}{2}[\phi_9 \quad \phi_8 \quad \phi_3]\begin{bmatrix} 0.25 & -0.25 & 0.00 \\ -0.25 & 1.25 & -1.00 \\ 0.00 & -1.00 & 1.00 \end{bmatrix}\begin{bmatrix} \phi_9 \\ \phi_8 \\ \phi_3 \end{bmatrix}$$

$$-[\phi_9 \quad \phi_8 \quad \phi_3]\begin{bmatrix} 0.0 \\ 0.0 \\ 0.0 \end{bmatrix}$$

$$+[\phi_7 \quad \phi_8 \quad \phi_3]\begin{bmatrix} 0.25 & -0.25 & 0.00 \\ -0.25 & 1.25 & -1.00 \\ 0.00 & -1.00 & 1.00 \end{bmatrix}\begin{bmatrix} \phi_7 \\ \phi_8 \\ \phi_3 \end{bmatrix}$$

$$-[\phi_7 \quad \phi_8 \quad \phi_3]\begin{bmatrix} 0.0 \\ 0.0 \\ 0.0 \end{bmatrix}$$

$$+[\phi_3 \quad \phi_6 \quad \phi_7]\begin{bmatrix} 0.25 & -0.25 & 0.00 \\ -0.25 & 1.25 & -1.00 \\ 0.00 & -1.00 & 1.00 \end{bmatrix}\begin{bmatrix} \phi_3 \\ \phi_6 \\ \phi_7 \end{bmatrix}$$

$$-[\phi_3 \quad \phi_6 \quad \phi_7]\begin{bmatrix} 0.0 \\ 0.0 \\ 0.0 \end{bmatrix}. \tag{7.7.11}$$

To see how these local matrices add up to a global polynomial, note that our global finite element (column) vector is:

$$\begin{aligned} \underset{\sim}{\mathbf{\Phi}} &= [\phi_1 \quad \phi_2 \quad \phi_3 \quad \phi_4 \mid \phi_5 \quad \phi_6 \quad \phi_7 \quad \phi_8 \quad \phi_9]^t \\ &= [\underset{\sim}{\mathbf{\Phi}}_{uk}^t \quad \underset{\sim}{\mathbf{\Phi}}_k^t]^t, \end{aligned} \tag{7.7.12}$$

where we have split $\underset{\sim}{\mathbf{\Phi}}$ into an unknown part $\underset{\sim}{\mathbf{\Phi}}_{uk}$ and a known part $\underset{\sim}{\mathbf{\Phi}}_k$. Writing $\mathscr{L}_1$ of eqn. (7.7.10) in terms of the global polynomial, we have

$$\mathscr{L}_1 = \tfrac{1}{2}\underset{\sim}{\mathbf{\Phi}}^t \times$$

$$\begin{bmatrix} 0.25 & -0.25 & 0.0 & 0.0 & 0.0 & 0.0 & 0.0 & 0.0 & 0.0 \\ -0.25 & 1.25 & -1.0 & 0.0 & 0.0 & 0.0 & 0.0 & 0.0 & 0.0 \\ 0.0 & -1.0 & 1.0 & 0.0 & 0.0 & 0.0 & 0.0 & 0.0 & 0.0 \\ 0.0 & 0.0 & 0.0 & 0.0 & 0.0 & 0.0 & 0.0 & 0.0 & 0.0 \\ 0.0 & 0.0 & 0.0 & 0.0 & 0.0 & 0.0 & 0.0 & 0.0 & 0.0 \\ 0.0 & 0.0 & 0.0 & 0.0 & 0.0 & 0.0 & 0.0 & 0.0 & 0.0 \\ 0.0 & 0.0 & 0.0 & 0.0 & 0.0 & 0.0 & 0.0 & 0.0 & 0.0 \\ 0.0 & 0.0 & 0.0 & 0.0 & 0.0 & 0.0 & 0.0 & 0.0 & 0.0 \\ 0.0 & 0.0 & 0.0 & 0.0 & 0.0 & 0.0 & 0.0 & 0.0 & 0.0 \end{bmatrix} \underset{\sim}{\mathbf{\Phi}} - \underset{\sim}{\mathbf{\Phi}}^t \begin{bmatrix} 1.0 \\ 1.0 \\ 1.0 \\ 0.0 \\ 0.0 \\ 0.0 \\ 0.0 \\ 0.0 \\ 0.0 \end{bmatrix} \tag{7.7.13}$$

Now also writing $\mathscr{L}_2$ in similar form, in keeping with the rules of eqns. (7.6.20)–(7.6.23), we have

$$\mathscr{L}_2 = \tfrac{1}{2}\underset{\sim}{\mathbf{\Phi}}^t \times$$

$$\begin{bmatrix} 1.0 & 0.0 & 0.0 & -1.0 & 0.0 & 0.0 & 0.0 & 0.0 & 0.0 \\ 0.0 & 0.0 & 0.0 & 0.0 & 0.0 & 0.0 & 0.0 & 0.0 & 0.0 \\ 0.0 & 0.0 & 0.25 & -0.25 & 0.0 & 0.0 & 0.0 & 0.0 & 0.0 \\ -1.0 & 0.0 & -0.25 & 1.25 & 0.0 & 0.0 & 0.0 & 0.0 & 0.0 \\ 0.0 & 0.0 & 0.0 & 0.0 & 0.0 & 0.0 & 0.0 & 0.0 & 0.0 \\ 0.0 & 0.0 & 0.0 & 0.0 & 0.0 & 0.0 & 0.0 & 0.0 & 0.0 \\ 0.0 & 0.0 & 0.0 & 0.0 & 0.0 & 0.0 & 0.0 & 0.0 & 0.0 \\ 0.0 & 0.0 & 0.0 & 0.0 & 0.0 & 0.0 & 0.0 & 0.0 & 0.0 \\ 0.0 & 0.0 & 0.0 & 0.0 & 0.0 & 0.0 & 0.0 & 0.0 & 0.0 \end{bmatrix} \underset{\sim}{\mathbf{\Phi}} - \underset{\sim}{\mathbf{\Phi}}^t \begin{bmatrix} 1.0 \\ 0.0 \\ 1.0 \\ 1.0 \\ 0.0 \\ 0.0 \\ 0.0 \\ 0.0 \\ 0.0 \end{bmatrix} \tag{7.7.14}$$

Adding the two together:

$$\mathscr{L}_1 + \mathscr{L}_2 = \tfrac{1}{2}\underset{\sim}{\mathbf{\Phi}}^t \left[\begin{array}{ccccccccc} 0.25 & -0.25 & 0.0 & 0.0 & 0.0 & 0.0 & 0.0 & 0.0 & 0.0 \\ +1.0 & +0.0 & +0.0 & -1.0 & & & & & \\ \hline -0.25 & 1.25 & -1.0 & 0.0 & 0.0 & 0.0 & 0.0 & 0.0 & 0.0 \\ \hline 0.0 & -1.0 & 1.0 & 0.0 & 0.0 & 0.0 & 0.0 & 0.0 & 0.0 \\ +0.0 & +0.0 & +0.25 & -0.25 & & & & & \\ \hline -1.0 & 0.0 & -0.25 & 1.25 & 0.0 & 0.0 & 0.0 & 0.0 & 0.0 \\ \hline 0.0 & 0.0 & 0.0 & 0.0 & 0.0 & 0.0 & 0.0 & 0.0 & 0.0 \\ \hline 0.0 & 0.0 & 0.0 & 0.0 & 0.0 & 0.0 & 0.0 & 0.0 & 0.0 \\ \hline 0.0 & 0.0 & 0.0 & 0.0 & 0.0 & 0.0 & 0.0 & 0.0 & 0.0 \\ \hline 0.0 & 0.0 & 0.0 & 0.0 & 0.0 & 0.0 & 0.0 & 0.0 & 0.0 \\ \hline 0.0 & 0.0 & 0.0 & 0.0 & 0.0 & 0.0 & 0.0 & 0.0 & 0.0 \end{array}\right] \underset{\sim}{\mathbf{\Phi}} - \underset{\sim}{\mathbf{\Phi}}^t \left[\begin{array}{c} 1.0 \\ +1.0 \\ \hline 1.0 \\ \hline 1.0 \\ +1.0 \\ \hline 1.0 \\ \hline 0.0 \\ \hline 0.0 \\ \hline 0.0 \\ \hline 0.0 \\ \hline 0.0 \end{array}\right] \tag{7.7.15}$$

Proceeding thus, we may construct the global polynomial, which in block matrix form may be written as:

$$\begin{aligned} \mathscr{L}_1 &= \tfrac{1}{2}[\underset{\sim}{\mathbf{\Phi}}_{uk}^t \underset{\sim}{\mathbf{\Phi}}_k^t] \begin{bmatrix} \mathbf{P}^g & U \\ \mathbf{U}^t & V \end{bmatrix} \begin{bmatrix} \underset{\sim}{\mathbf{\Phi}}_{uk} \\ \underset{\sim}{\mathbf{\Phi}}_k \end{bmatrix} - [\underset{\sim}{\mathbf{\Phi}}_{uk}^t \quad \underset{\sim}{\mathbf{\Phi}}_k^t] \begin{bmatrix} \mathbf{q}^g \\ \mathbf{w}^g \end{bmatrix} \qquad (7.7.16) \\ &= \tfrac{1}{2}\underset{\sim}{\mathbf{\Phi}}_{uk}^t \mathbf{P}^g \underset{\sim}{\mathbf{\Phi}}_{uk} + \underset{\sim}{\mathbf{\Phi}}_{uk}^t \mathbf{U} \underset{\sim}{\mathbf{\Phi}}_k + \tfrac{1}{2}\underset{\sim}{\mathbf{\Phi}}_k^t \mathbf{V} \underset{\sim}{\mathbf{\Phi}}_k - \underset{\sim}{\mathbf{\Phi}}_{uk}^t \mathbf{q}^g - \underset{\sim}{\mathbf{\Phi}}_k^t \mathbf{w}^g, \end{aligned}$$

where $\mathbf{P}^g$ and $\mathbf{U}$ are both symmetric matrices. Now, since the elements of $\underset{\sim}{\mathbf{\Phi}}_k$ are known, the only variational quantities that may adjust themselves to permit $\mathscr{L}$ to take a lower value are the elements of $\underset{\sim}{\mathbf{\Phi}}_{uk}$. Performing differentiations by the rules detailed in eqns. (7.6.18) and (7.6.19), we get:

$$\frac{\partial \mathscr{L}}{\partial \underset{\sim}{\mathbf{\Phi}}_{uk}} = \mathbf{P}^g \underset{\sim}{\mathbf{\Phi}}_{uk} + \mathbf{U} \underset{\sim}{\mathbf{\Phi}}_k - \underset{\sim}{\mathbf{q}}^g = 0. \tag{7.7.17}$$

Thus, it is seen that we have no use at all for the last row block of the polynomial expression of eqn. (7.7.16), so that happily we do not need to form those rows. The term $\mathbf{U}\mathbf{\Phi}_k$ being a column vector, even the matrix $\mathbf{U}$ need not be formed. In our example problem in particular, since all the known potentials $\phi$ are zero, $\underset{\sim}{\mathbf{\Phi}}_k$ vanishes, so that we get the $4 \times 4$ matrix equation

$$\mathbf{P}^g \underset{\sim}{\mathbf{\Phi}}_{uk} = \underset{\sim}{\mathbf{q}}^g. \tag{7.7.18}$$

For purposes of illustration we give the equation that results through functional minimization corresponding to eqn. (7.7.17):

$$
\frac{\partial \mathscr{L}}{\partial \tilde{\boldsymbol{\Phi}}_{uk}} =
\left[\begin{array}{ccccccccc}
\begin{array}{c}0.25^1\\+1.0^2\end{array} & -0.25^1 & \begin{array}{c}0.0^1\\0.0^2\end{array} & -1.0^2 & & & & & \\
\hline
-0.25^1 & \begin{array}{c}1.25^1\\+1.25^3\end{array} & \begin{array}{c}-1.0^1\\-1.0^3\end{array} & & -0.25^3 & & & & \\
\hline
\begin{array}{c}0.0^1\\+0.0^2\end{array} & \begin{array}{c}-1.0^1\\-1.0^3\end{array} & \begin{array}{c}1.0^1\\+0.25^2\\+1.0^3\\+0.25^4\\+0.25^5\\+1.0^6\\+1.0^7\\+0.25^8\end{array} & \begin{array}{c}-0.25^2\\-0.25^2\end{array} & \begin{array}{c}0.0^3\\+0.0^4\end{array} & \begin{array}{c}-0.25^4\\-0.25^8\end{array} & \begin{array}{c}0.0^7\\0.0^8\end{array} & \begin{array}{c}-1.0^6\\-1.0^7\end{array} & \begin{array}{c}0.0^5\\+0.0^6\end{array} \\
\hline
-1.0^2 & & \begin{array}{c}-0.25^2\\-0.25^5\end{array} & \begin{array}{c}1.25^2\\+1.25^5\end{array} & & & & & -1.0^5
\end{array}\right]
\begin{bmatrix} A_1 \\ A_2 \\ A_3 \\ A_4 \\ A_5 \\ A_6 \\ A_7 \\ A_8 \\ A_9 \end{bmatrix}
=
\left[\begin{array}{c}
1.0^1\\+1.0^2\\ \hline
1.0^1\\+0.0^3\\ \hline
1.0^1\\+1.0^2\\+0.0^3\\+0.0^4\\+0.0^5\\+0.0^6\\+0.0^7\\+0.0^8\\ \hline
1.0^2\\+0.0^5
\end{array}\right]
\tag{7.7.19}
$$

The superscripts refer to the element from which each contribution to the global matrix is made. The block matrix on the left is recognized as $\mathbf{P}^g$ and that on the right as $\mathbf{U}$. We need to subtract $\mathbf{U}\boldsymbol{\Phi}_k$ from both sides, but because $\boldsymbol{\Phi}_k$ is zero this is unnecessary. The finite element matrix equation we get is therefore:

$$
\begin{bmatrix}
1.2500 & -0.2500 & 0.0000 & -1.0000 \\
-0.2500 & 2.5000 & -2.0000 & 0.0000 \\
0.0000 & -2.0000 & 5.0000 & -0.5000 \\
-1.0000 & 0.0000 & -0.5000 & 2.5000
\end{bmatrix}
\begin{bmatrix} A_1 \\ A_2 \\ A_3 \\ A_4 \end{bmatrix}
=
\begin{bmatrix} 2.0000 \\ 1.0000 \\ 2.0000 \\ 1.0000 \end{bmatrix},
\tag{7.7.20}
$$

and this we shall solve by Gaussian elimination, discussed earlier in section 2.3. We first divide the first equation by a factor of 1.25 to make the first diagonal term unity, then subtract an appropriate multiple of the resulting first row from the subsequent rows to make the first column below the diagonal term zero to get:

$$\begin{bmatrix} 1.0000 & -0.2000 & 0.0000 & -0.8000 \\ 0.0000 & 2.4500 & -2.0000 & -0.2000 \\ 0.0000 & -2.0000 & 5.0000 & -0.5000 \\ 0.0000 & 0.2000 & -0.5000 & 1.7000 \end{bmatrix} \begin{bmatrix} A_1 \\ A_2 \\ A_3 \\ A_4 \end{bmatrix} = \begin{bmatrix} 1.6000 \\ 1.4000 \\ 2.0000 \\ 2.6000 \end{bmatrix}. \tag{7.7.21}$$

Proceeding thus, as illustrated in the handworked example of section 2.4.1, we get the upper triangle matrix equation

$$\begin{bmatrix} 1.0000 & -0.2000 & 0.0000 & -0.8000 \\ 0.0000 & 1.0000 & -0.81633 & 0.08163 \\ 0.0000 & 0.0000 & 1.0000 & -0.19697 \\ 0.0000 & 0.0000 & 0.0000 & 1.55303 \end{bmatrix} \begin{bmatrix} A_1 \\ A_2 \\ A_3 \\ A_4 \end{bmatrix} = \begin{bmatrix} 1.6000 \\ 0.5714 \\ 0.93333 \\ 3.33333 \end{bmatrix}. \tag{7.7.22}$$

Now from the last equation, we have

$$A_4 = \frac{3.33333}{1.55303} = 2.14634. \tag{7.7.23}$$

From the last equation but one:

$$A_3 = 0.93333 + 0.1969A_4 = 1.35610. \tag{7.7.24}$$

By such back substitution

$$A_2 = 0.571429 + 0.81633A_3 + 0.08163A_4 = 1.85366, \tag{7.7.25}$$

and

$$A_1 = 0.2A_2 + 0.8A_4 + 1.6 = 3.6878. \tag{7.7.26}$$

## 7.7.2 The Electric Field

In the preceding section we have just finished working out by hand the magnetic vector potential in a coaxial cable. Now we wish to work out the electric field in the same cable. In computing the electric field we ought to avoid going into the cable, because we do not know the charge distribution inside—in fact, most of the charge would accumulate on the surface, being attracted by Coulomb force towards its counterparts on the outer sheath of different voltage. But, as explained in section 1.2.4, the inner line, being made of a perfect conductor, has the same voltage all through its cross section,

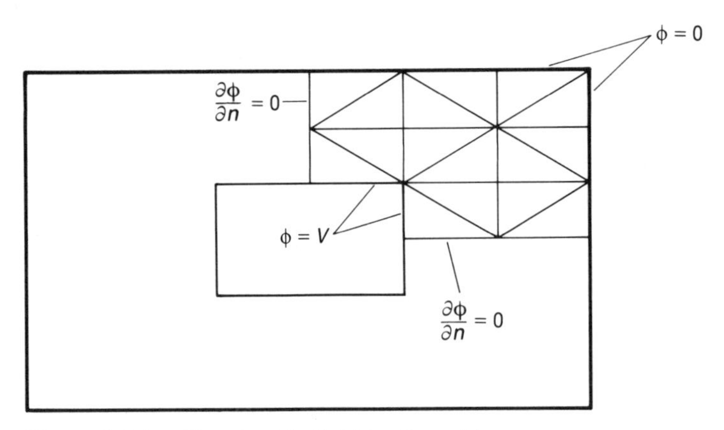


**Figure 7.7.4.** The electric field in the cable.

and this voltage is the known voltage of transmission. We may therefore solve the field problem only in the region between the two conductors, by closing off the inner section of the inner conductor by replacing it with its surface of known voltage as shown in Figure 7.7.4. The region of solution has been discretized into triangular finite elements. Enough material has been provided so far to allow the reader to solve for the voltage distribution between the two conductors, and the working is left as an exercise.

## 7.8 HAND PLOTTING RESULTS

At the end of a finite element computation, we obtain a set of numbers that represent the values of the electric potential or magnetic vector potential at the finite element nodes. For example, in the previous example we determined that the vector potential was 3.6878, 1.85366, 1.35610, and 2.14634, respectively, at nodes 1, 2, 3, and 4. This upon immediate glance has no meaning. The best way of interpreting the numbers is through some graphical means. A quick, clear method of postprocessing the results is by getting equipotential contours—these tell us the paths along which the magnetic flux will flow.

In this section we demonstrate, through the example of the already computed magnetic vector potential in a coaxial cable of section 7.7.1, the method of hand plotting such equipotential contours. In Figure 7.8.1, we see the coaxial cable drawn to scale, with the dotted lines indicating the positioning of the finite element mesh. At each node the computed vector potential is

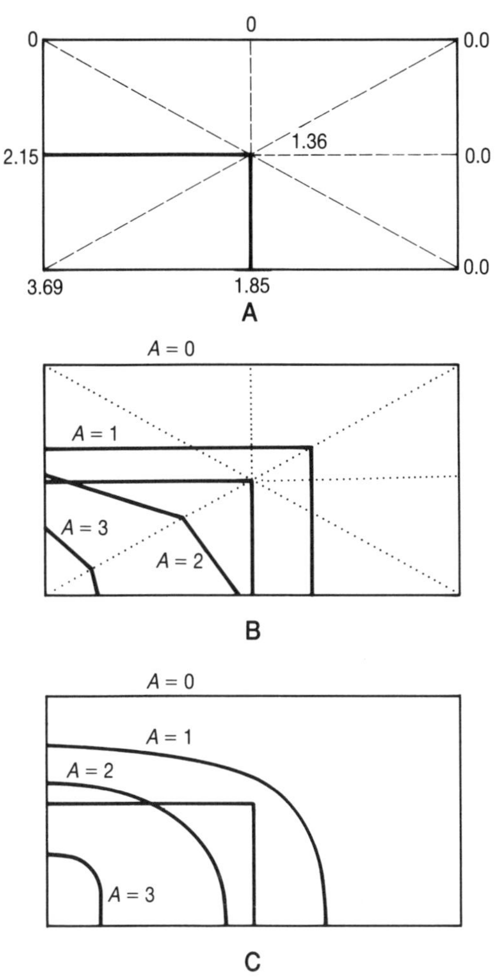


**Figure 7.8.1.** Plotting in the cable. **A.** The mesh. **B.** The exact plot. **C.** The smoothed plot.

marked. We have a vector potential solution that ranges from 0 to 3.69. For simplicity let us decide, therefore, to draw four contours at potentials 0, 1, 2, and 3.

Consider the first element 123, with edges 12, 23, and 31. Taking up the edge 12 first, we have a potential going from 3.69 down to 1.85, from coordinate $(0, 0)$ to $(2, 0)$. According to our first-order approximation eqn. (7.5.3), this potential goes down linearly. Therefore, this edge will contain the contours at potentials 2 and 3. By linear interpolation the contour at a potential 2 will cross this edge at $x$ coordinate $0 + (3.69 - 2.0)*(2.0 - 0.0)/(3.69 - 1.85) = 1.84$ and $y$ coordinate 0.0. In general, for an edge from $(x_1, y_1)$ to $(x_2, y_2)$ with potentials from $\phi_1$ to $\phi_2$, the potential contour $\phi$ will cross at the point:

$$(x, y) \equiv \left(x_1 + \frac{(x_2 - x_1)(\phi - \phi_1)}{(\phi_2 - \phi_1)}, y_1 + \frac{(y_2 - y_1)(\phi - \phi_1)}{(\phi_2 - \phi_1)}\right). \quad (7.8.1)$$

Now this contour at potential 2 may run to either side 23 or 31. We see that only side 31 contains the range 1.36 to 3.69 containing the potential 2. Employing eqn. (7.8.1), we may say that this contour ought to meet edge 31 at point $2 + (0 - 2)(2 - 1.36)/(3.69 - 1.36)$, $1 + (0 - 1)(2 - 1.36)/(3.69 - 1.36) \equiv 1.45, 0.7254$. As it is a first-order trial function, we draw the contour as a straight line from $(1.84, 0)$ to $(1.45, 0.7254)$, as shown in Figure 7.8.1B. Now if we are merely interested in a rough idea of the behavior of the fields, then we need not get these coordinates accurately and may visually estimate the locations of these coordinates and draw the straight lines. We similarly draw the other contour at potential 3 from edge 12 to edge 31. Now we take up edge 23. This contains none of the contours we are plotting. The third edge 31 need not be considered, since any contour crossing it has to cross edge 13 or 23 and all contours crossing them have been drawn. Now consider the next element 341. Taking up the first edge 34, we only have the contour at potential 2 to draw. This crosses the edge at coordinate $2 - (0 - 2)(2 - 1.36)/(2.15 - 1.36)$, $1.0 \equiv 0.37$, $1.0$. We already know from our treatment of triangle 123 that this crosses edge 31 at $(1.45, 0.7254)$, and we draw the appropriate straight line. Proceeding thus, we take up every element, draw the contours which lie in it, and obtain the equipotential plot of Figure 7.8.1B. That the lines of symmetries have turned out to be Neumann boundaries with the contours meeting them normally is seen.

We also know from section 1.2.4 that these lines represent the paths taken by the magnetic flux and that the magnetic flux is greater where equipotential contours of equal steps, as here, are crowded. This fits the fundamental laws from which we may say that flux should flow around a conductor, that it should be the greatest close to the edge of the conductor where more current is linked, and that it should fall as we move away from the conductor. What we are unable to see here is that flux increases as we proceed from the middle of the conductor towards the boundary as more and more current is linked; this is because we are using just one element from the middle of the conductor to the boundary. If we refine the mesh with an additional node at the middle of edge 12 and solve the problem again, we find that the fall in potential from node 1 to that node is lower than the fall from there to node 2; i.e., more flux flows between that node and node 2 than between that node and node 1. This conclusion follows from eqn. (1.2.34) and the considerations of section 4.4.1, where it has been shown that in two dimensions, the flux flow between two points is the difference in the single component vector potential between those two points.

To get a more accurate estimate of flux densities we know that we must go back and improve the trial functions by imposing more nodes by mesh refinement. This would be tedious with a much larger matrix to be assembled and solved. However, a shortcut exists that permits a slight improvement in the solution, without recourse to mesh refinement. This uses the fact that flux contours, according to eqns. (1.2.15) and (1.2.17), ought to be smooth, except at permeability discontinuities, which we do not have in our example. We may also employ the fact that the contours should meet the symmetry boundaries exactly normally. Keeping these thoughts in mind, we may connect the points along the edges where the contours meet, smoothly by hand, to get the improved flux pattern of Figure 7.9.1C.

## 7.9 A SIMPLE FINITE ELEMENT PROGRAM

The simple ideas of first-order finite element analysis expounded above may easily be programmed. The input to the program is usually given in a separate file, which may easily be created using a screen editor. The input to the example problem of the coaxial cable is as shown in Figure 7.7.3. This is designed so that any new problem may be run by the same program without modification—only the input file changing from problem to problem. In the example of Figure 7.7.3, the inputs being read are the number of nodes, the number of known nodes, the number of elements, the series of elements (each defined by three nodes), the material $\varepsilon$ of eqn. (1.2.22), the constant source density $\rho_0$ within the triangle, and a series of coordinates corresponding to the nodes.

This program, after initializing the global matrices **PG** and **QG** and reading the general information about the numbers of triangles, nodes, and unknowns, should read every element, form the local matrices **P** and **q** as in Alg. 7.6.2, and place them in the global matrices **PG** and **QG**. Having formed the global matrix, the matrix

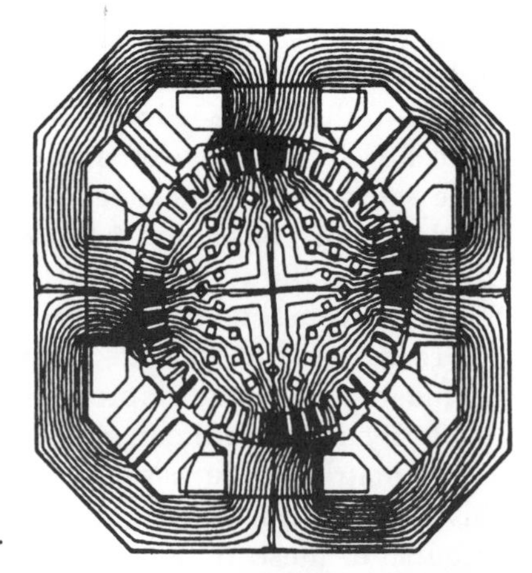

**Figure 7.9.1.** The DC machine on load.

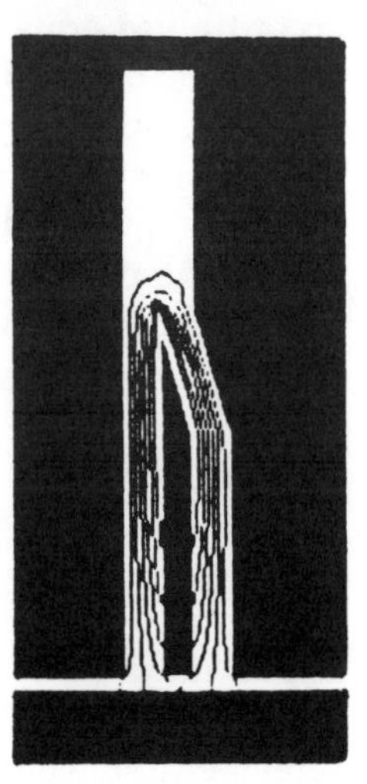

**Figure 7.9.2.** The magnetic recording head.

equation may be solved employing either the Gaussian elimination scheme (Alg. 2.3.1) or the SOR algorithm (Alg. 3.2.1). The nodal values of potential may then be output.

Once we write our own program, we are not restricted by the limitations of hand computations, and much larger problems may be solved; only the input data having to be changed at every run. The reader is encouraged to discretize the coaxial cable problem into a finer mesh and solve it using this program to see the increase in flux density as we move closer to the boundary of the conductor from inside.

Figure 7.9.1 gives the solution of a DC machine on load by the finite element method, and Figure 7.9.2 that of a recording head, both taken from Konrad (1985b).

# 8 FINITE ELEMENTS: DEEPER CONSIDERATIONS

## 8.1 THE VARIATIONAL APPROACH TO FINITE ELEMENTS

We have already used and seen from energy-related arguments in chapter 7 that the satisfaction of the Poisson equation

$$-\varepsilon\nabla^2\phi = \rho, \tag{1.2.22}$$

in a region with Neumann or Dirichlet boundary conditions, is the equivalent of extremizing the functional

$$\mathscr{L} = \iint \{\tfrac{1}{2}\varepsilon[\nabla\phi]^2 - \rho\phi\}\, dR. \tag{7.5.18}$$

In the variational approach to finite elements, we always must identify a functional whose minimization corresponds to the satisfaction of the differential equation being solved. In this section we shall prove that eqn. (1.2.22) is satisfied when the Lagrangian of eqn. (7.5.18) is at a minimum, in a more rigorous mathematical setting. To do so, consider an excursion $\kappa\eta(x, y)$ in $\phi$; i.e., $\phi \to \phi + \kappa\eta$, where $\kappa$ is an arbitrary constant and $\eta$ is a function of $x$ and $y$. Then we have

$$\mathscr{L} + \delta\mathscr{L} = \iint \{\tfrac{1}{2}\varepsilon[\nabla(\phi + \kappa\eta)]^2 - \rho(\phi + \kappa\eta)\}\, dR. \tag{8.1.1}$$

Therefore, subtracting eqn. (7.5.18) from eqn. (8.1.1),

$$\delta\mathscr{L} = \iint \{\varepsilon\kappa[\nabla\phi \,.\, \nabla\eta] + \kappa^2[\nabla\eta]^2 - \rho\kappa\eta\}\, dR. \tag{8.1.2}$$

Now, if we integrate the modified identity (A11),

$$\nabla . \{\eta \nabla \phi\} = \eta \nabla^2 \phi + \nabla \phi . \nabla \eta, \qquad (8.1.3)$$

over the region of solution $R$ and apply Gauss's theorem (A13) to the left-hand side,

$$\int \eta \nabla \phi . \mathbf{dS} = \iint \{\eta \nabla^2 \phi + \nabla \phi . \nabla \eta\}\, dR. \qquad (8.1.4)$$

Putting this into eqn. (8.1.2), we get

$$\delta \mathscr{L} = \int \varepsilon \kappa \eta \frac{\partial \phi}{\partial n}\, dS - \iint \kappa \eta \{\varepsilon \nabla^2 \phi + \rho\}\, dR + \iint \kappa^2 \varepsilon [\nabla \eta]^2\, dR. \qquad (8.1.5)$$

We see that the first integral vanishes when the Dirichlet condition applies on the boundary (where since $\phi$ is fixed, $\eta$ is zero) or the Neumann condition applies there (so that $\partial \phi / \partial n$ is zero). The second integral vanishes when eqn. (1.2.22) is satisfied and the last term is positive and of order $\kappa^2$, which shows that $\mathscr{L}$ is indeed at a minimum and also that when changes in $\phi$ are of order $\kappa$, changes in the functional are even smaller and of order $\kappa^2$. This is of immense practical significance because in many practical systems, $\mathscr{L}$ is energy. In electrostatics, for example, $-\nabla \phi$ is the electric field intensity **E** and $-\varepsilon \nabla \phi$ is the flux density **D**, so that $\frac{1}{2}\varepsilon[\nabla \phi]^2$ of eqn. (7.5.18) is $\frac{1}{2}\mathbf{D} . \mathbf{E}$, which is the stored energy density. Similarly, $\rho \phi$ is the work expended in moving the charge $\rho\, dR$ to its present location. Eqn. (8.1.5) states, therefore, that energy is of one higher order of accuracy than potential; that is, after finding the potential $\phi$, should we use that to compute energy-related quantities such as power losses, inductances, impedances, etc., the resulting answer would be more accurate than our potential solution.

The variational approach, therefore, in fact allows us to obtain energy-related values of one higher order and, by the conversion of the $\nabla^2 \phi$ term to the term $[\nabla \phi]^2$ in eqn. (7.5.18), allows us to use the simple first-order trial functions for $\phi$, which will not be allowed when dealing with $\nabla^2 \phi$, which will reduce the second derivative to zero. Moreover, the $[\nabla \phi]^2$ term yields the symmetry behind the finite element matrices, as an examination of chapter 7 will show.

Be that as it may, the variational approach has one strong disadvantage—the need to identify a functional with a differential equation. This is not always possible, unless one is ready to use such frequently inconvenient methods as a least-square minimization of the residual of the differential equation.

# 8.2 THE GENERALIZED GALERKIN APPROACH TO FINITE ELEMENTS

The Galerkin method of approximately solving differential equations is a powerful and general one, the details of which are much beyond the scope of this book. Indeed, it may be shown that the above variational formulation may also be arrived at by the Galerkin method. In fact, while the variational method cannot be applied unless we are able to identify a functional, no such restriction applies to the Galerkin method.

In solving the operator equation

$$A\phi(x, y) = \rho(x, y), \tag{8.2.1}$$

the actual solution $\phi$ in general has no restrictions on it other than the boundary conditions. But $\phi^*$, the approximate solution we obtain, belongs to a restricted subspace of the universal space—the approximation space. For instance, in eqn. (7.5.3) we have restricted the solution to belong to the subspace of linear polynomials. The Galerkin method provides us with the best of the various solutions within the approximation space. Thus, the approximate solution is limited by our trial function and gives us the best fit of the governing equation.

Supposing we seek an approximation in a space of order $n$, with basis functions $\alpha_1, \ldots, \alpha_n$, and the right-hand side $\rho$, or an approximation of it $\rho^*$, belongs to a space of order $m$, with basis $\beta_1, \ldots, \beta_m$; then

$$\phi^* = \sum_{i=1}^{n} \alpha_i \phi_i, \tag{8.2.2}$$

$$\rho^* = \sum_{i=1}^{m} \beta_i \rho_i. \tag{8.2.3}$$

At this point it is in order to give a rather simplistic definition of basis functions, drawing upon an analogy with basis vectors. A one-dimensional vector space, we know, has the unit vector $\mathbf{u}_x$ in the direction of the dimension as basis, so that any vector is expressed as $a\mathbf{u}_x$. Similarly in two dimensions, we have a basis consisting of $\mathbf{u}_x$, $\mathbf{u}_y$, so that any vector from a two-dimensional subspace is defined by $a\mathbf{u}_x + b\mathbf{u}_y$ for unique values of $a$ and $b$. Analogously, function spaces are generated by basis functions. For example, the subspace of constants is generated by the basis function 1, so that any constant is $c \,.\, 1$; the subspace of first-order polynomials in two-dimensional space is generated by the three basis functions 1, $x$, and $y$, so that first-order polynomials in $x$ and $y$ belong to a subspace of dimension 3. Second-order polynomials in $x$ and $y$ are generated by 1, $x$, $y$, $xy$, $x^2$, and $y^2$, so that they belong to a subspace of dimension 6. Please note that just as basis vectors are not unique, basis functions too are nonunique. For example, $\mathbf{u}_x$ and $\frac{1}{2}\mathbf{u}_y$ also form a basis for two-dimensional vectors, just as the set 1, $x$, and $x + y$ is a basis for first-order polynomials in $x$ and $y$. However, since the dimension of the subspace cannot vary, any two basis sets must of necessity have the same number of elements. Moreover, it is easily shown that the elements of a basis set must be independent for the expansion of an element from the subspace to be unique. That is, an element of a basis set cannot be expressed by a combination of the other elements.

With this definition, we may define the *generalized Galerkin method*, which, once we have decided on a trial function $\phi^*$ lying in a subspace of dimension $n$, allows us to get that of the various functions in that subspace that has least error. It states that, for any basis $\gamma_1, \ldots, \gamma_n$ of the space of $\phi^*$ (not necessarily the same as the basis set $\alpha_1, \alpha_2, \ldots, \alpha_n$), the best approximation to the equation being solved (8.2.1) is when

$$\sum_{i=1}^{n} \phi_i \langle \gamma_k \mid A\alpha_i \rangle = \sum_{i=1}^{m} \rho_i \langle \gamma_k \mid \beta_i \rangle \qquad k = 1, \ldots, n, \tag{8.2.4}$$

where $\langle \mid \rangle$ denotes what is called an inner product, which is commonly defined as:

$$\langle f(x, y) \mid g(x, y) \rangle = \int fg \, dR. \tag{8.2.5}$$

Be it noted that eqn. (8.2.4) is a matrix equation, since the inner products are known once we decide upon the approximation spaces and, therefore, the basis sets to employ. We have $n$ such equations for every $\gamma_k$, for $k = 1, \ldots, n$, so that eqn. (8.2.4) is in reality an $n \times n$ matrix equation for the $n$ unknowns $\phi_k$. Moreover, since the function elements of a basis are independent, the coefficient matrix will be nonsingular and therefore invertible. The matrix form of the equation is

$$\mathbf{P\Phi} = \mathbf{q}, \tag{8.2.6}$$

where the matrix $\mathbf{P}$ and the vector $\mathbf{q}$, respectively of size $n \times n$ and $n \times 1$, and the column vector $\mathbf{\Phi}$ of length $n$ are defined by and computed from

$$\mathbf{P}(i,j) = \int \gamma_i A\alpha_j \, dR \qquad i = 1, \ldots, n; j = 1, \ldots, n, \tag{8.2.7}$$

$$A^r\mathbf{T}(i,j) = \int \gamma_i \beta_j \, dR \qquad i = 1, \ldots, n; j = 1, \ldots, m, \tag{8.2.8}$$

$$\mathbf{q} = A^r\mathbf{Tr}, \tag{8.2.9}$$

$$A^r = \int dR = \text{Area}. \tag{8.2.10}$$

The elements of $\mathbf{\Phi}$ and $\mathbf{r}$ are, respectively, $\phi$ and $\rho$; that is, $\mathbf{\Phi}(i) = \phi_i$ and $\mathbf{r}(i) = \rho_i$. In finite element analysis, the region of solution is often divided into several smaller subregions over each of which a different trial function with interelement relationships is defined. Thus, the integrations are performed over the different elements and then summed to get the total integral over the solution region. This summation is similar to the building of the global matrix from the local matrix demonstrated with the variational method in section 7.6.

The generalized Galerkin method, in West European and North American literature, is also known as the method of weighted residuals or the method of moments, because the functions $\gamma$ serve as weighting functions. While the method may be justified in elaborately rigid mathematical terms, a simplistic explanation of how it works again makes recourse to the analogy with vector spaces; here we regard the Galerkin method as setting every component of the residual to zero, in the approximation space. For the approximations of eqns. (8.2.2) and (8.2.3), the residual from eqn. (8.2.1) is

$$\mathcal{R}(\phi) = \sum_{i=1}^{n} \phi_i A\alpha_i - \sum_{i=1}^{m} \rho_i \beta_i, \tag{8.2.11}$$

which is still being generated by the basis vectors $\alpha_1, \ldots, \alpha_n$. This residual lies in the approximation space having the independent functions $\zeta_1, \ldots, \zeta_n$ as basis. Thus, if we multiply the residual by each $\zeta_i$ and set the integral over the solution region $R$ to zero, we would ensure the orthonormality of the residual to each element of a basis of the approximation space; i.e., if the residual in the approximation space is expanded in terms of the $\zeta$'s, the coefficient corresponding to each $\zeta$ is made zero. To draw on our analogy with vector spaces again, if, corresponding to the residual $\mathcal{R}$, we had the vector $\boldsymbol{\mathcal{R}}$ given by $f_1\mathbf{u}_1 + f_2\mathbf{u}_2 + \ldots + f_i\mathbf{u}_i + \ldots + f_n\mathbf{u}_n$, then to set the component $f_i$ of $\mathcal{R}$ in the direction $\mathbf{u}_i$ to zero, we would, using the dot (or scalar) product, set $\mathbf{u}_i \cdot \mathcal{R} = f_i = 0$. Thus, to set $\mathcal{R}$ to zero in every direction, we would do the same thing with each of the basis elements, $\mathbf{u}_1, \ldots, \mathbf{u}_n$. In dealing with function spaces as opposed to vector spaces, however, we must set $\mathcal{R} \cdot \zeta_i$ to zero for every $i$. However, the resulting terms being functions of space, an overall optimal reduction of the residual is achieved by the integration of the product over the solution region, as in eqn. (8.2.4).

Since the approximation space has different bases, the choice of basis would affect the final approximation of $\phi$, and hence the individual components $\zeta_i$ weight the solution in the different directions. In this sense, the basis set $\zeta$ employed gives the generalized Galerkin method different names. For example, if all three bases $\alpha$, $\beta$, and $\gamma$ are the same, the method is known as the *Bubnov–Galerkin method.* On the other hand, if we take $\gamma_i = \partial\mathcal{R}/\partial\phi_i = A\alpha_i$, $i = 1, \ldots, n$, as the basis $\gamma$, we have the *least-squares method.* In the least-squares method, we say that the nonnegative quantity $\mathcal{R}^2$ should be a minimum. When we impose this condition on eqn. (8.2.6), after squaring and differentiating with respect to each $\phi_i$, we obtain

$$\sum_{i=1}^{n} \phi_i \langle A\alpha_k \mid A\alpha_i \rangle = \sum_{i=1}^{m} \rho_i \langle A\alpha_k \mid \beta_i \rangle \qquad k = 1, \ldots, n. \tag{8.2.12}$$

The same Galerkin method becomes the so-called *collocation method* when we set the residual $\mathcal{R}$ to zero at $n$ preselected points called *collocation points,* and thereby obtain $n$ equations for solution. Here the choice of the location of the points will determine the accuracy of the solution, and it may be shown that collocation falls under the generalized Galerkin method, with $\gamma_i$ taken as the Dirac–Delta function $\delta(x - x_i)$ at the collocation point $x = x_i$.

A further generalization of the Galerkin method may be made for a function governed by two or more equations. For example, if $\phi$ is governed by eqn. (8.2.1) and the equation

$$P\phi(x, y) = \sigma(x, y), \tag{8.2.13}$$

then the corresponding generalized Galerkin approximation for the equation pair, by applying the same principle to the sum of the residuals from the two governing equations when the trial functions are put in, is

$$\sum_{i=1}^{n} \phi_i \langle \gamma_k \mid (A + \lambda P)\alpha_i \rangle = \sum_{i=1}^{m} (\rho_i + \lambda\sigma_i) \langle \gamma_k \mid \beta_i \rangle \qquad k = 1, \ldots, n, \tag{8.2.14}$$

where $\lambda$ is a weighting function, often called a *Lagrangian multiplier.* For large $\lambda$, the solution would tend to satisfy eqn. (8.2.13) more (in the sense of the norm of the residual), and for small $\lambda$, the answer would match eqn. (8.2.1) better.

Another way in which the Galerkin method is described in the literature is, corresponding to eqn. (8.2.4),

$$\int \psi[A\phi - \rho]\, dR = 0, \tag{8.2.15}$$

where $\phi$ is the unknown in the approximation space and $\psi$ too is a complete function in the same approximation space. By *complete* is meant a function that may be described only by all the basis functions of the subspace. Thus, $\psi$ becomes a sum of the $\zeta$ functions and, equating each of the $n$ components of the expression eqn. (8.2.15), we obtain eqn. (8.2.4). Similarly, corresponding to eqn. (8.2.14), we have

$$\int \psi[A\phi - \rho]\, dR + \lambda \int \psi[P\phi - \sigma]\, dR = 0. \tag{8.2.16}$$

# 8.3 NORMAL GRADIENT BOUNDARY CONDITIONS IN FINITE ELEMENTS

## 8.3.1 Forced and Natural Boundary Conditions

In solving a differential equation, we may impose the boundary conditions through trial functions. For example, if we are using a first-order trial function such as eqn. (7.5.3) in a triangle where we have node 1 at a Dirichlet boundary with potential 100, then we merely set $\phi_1$ to 100 in eqn. (7.5.3), and automatically, the Dirichlet boundary condition is satisfied at that node. A Dirichlet boundary condition is referred to as an *essential* boundary condition, because at least one point on the boundary is required to be a Dirichlet boundary as shown in Appendix B.

Similarly, Neumann boundaries also may be imposed through trial functions. For the example of eqn. (7.5.3), if a particular side of the triangle is a Neumann boundary, then we would explicitly impose a relationship between the three vertex potentials so as to satisfy the Neumann condition. This would result in two potentials, instead of the three independently variable potentials $\phi_1$, $\phi_2$, and

$\phi_3$ as before in the triangle to be solved. Correspondingly, the approximation space within the triangle will no longer be spanned by the functions $\zeta_1$, $\zeta_2$, and $\zeta_3$, but rather by two appropriate functions, which explicitly satisfy the Neumann condition. This way of explicitly imposing Neumann conditions through the trial functions is known as a *strong imposition*. Since our solution space is already approximate and of limited dimension, strongly asserted conditions further reduce the freedom of the solution to fit the governing equation and are generally considered undesirable; for we have a situation where a quantity must obey two laws (the governing equation and the boundary conditions), and our approximation does not allow that quantity to obey them both completely, since the actual answer will not necessarily take the form of our trial function. The finite element solution can fit the governing equation only in an optimal sense and exactly only when, by chance, we happen to take a trial function having the same form as the exact solution. Therefore, we will distort the solution (already distorted by the trial functions) further, by forcing the solution to obey one of the laws completely. What is preferred is to impose both laws in the same way so that the approximation may obey them both as much as possible, and, within the limits imposed by the approximation space, arrive at a solution that satisfies both conditions as best as it can, or optimally.

When we do not explicitly impose the Neumann conditions exactly, the procedure is called a *weak formulation*. Eqn. (8.2.11) gives us a natural way of imposing the Neumann condition. For the Poisson equation, the residual of the first equation is $[-\varepsilon\nabla^2\phi - \rho]$ operating within the region of solution $R$, and the residual of the Neumann boundary condition, specifying that the normal gradient of the field is $\sigma$ along a part $\Gamma$ of the total boundary $S$, is $[\partial\phi/\partial n - \sigma]$, so that the Galerkin formulation, corresponding to eqn. (8.2.16), is

$$\iint_R \psi[-\varepsilon\nabla^2\phi - \rho]\, dR + \int_\Gamma \lambda\psi\left[\frac{\partial}{\partial n}\phi - \sigma\right] dS = 0. \qquad (8.3.1)$$

The choice of $\lambda$ would determine toward which of the two equations the solution is to be weighted, and $\lambda = \varepsilon$ would be the most natural choice without biasing the solution towards either equation, since the operator $-\nabla^2$ is already scaled by the factor $\varepsilon$. Using the identity (A11) with $\phi = \psi$ and $\mathbf{A} = \nabla\phi$, integrating over the solution region $R$ and applying Gauss's theorem (A13), we obtain

$$\iint_R [\varepsilon\nabla\psi \,.\, \nabla\phi - \psi\rho]\, dR - \int_S \varepsilon\psi \frac{\partial}{\partial n}\phi\, dS + \int_\Gamma \varepsilon\psi\left[\frac{\partial}{\partial n}\phi - \sigma\right] dS = 0. \tag{8.3.2}$$

Now $\psi$ is a sum of the components of a basis spanning the approximation space, and is such as to strongly satisfy the boundary conditions along the Dirichlet portions of $S$; i.e., along portions of $S$ excluding $\Gamma$. On these Dirichlet portions of $S$, since the value of $\phi$ is imposed through the trial functions, the approximation space, having no freedom there, will have no contribution to make to $\psi$. Therefore, the first surface integral over $S$ will be zero over the Dirichlet boundary $(S - \Gamma)$, the nonzero portion of the integral being only over $\Gamma$. The first surface integral will therefore cancel the first part of the second surface integral over $\Gamma$, giving:

$$\iint_R [\varepsilon\nabla\psi \,.\, \nabla\phi - \psi\rho]\, dR - \int_\Gamma \varepsilon\psi\sigma\, dS = 0. \tag{8.3.3}$$

If we expand $\phi$ as a sum of the $n$ terms $\zeta_i\phi_i$, where the $\phi_i$ are to be determined, and $\rho$ as a sum of the $m$ terms $\beta_i\rho_i$; and equate every component $\gamma_i$, making up $\psi$, to zero, we obtain:

$$\sum_{i=1}^{n} \phi_i\langle\nabla\gamma_k \,|\, \nabla\zeta_i\rangle = \sum_{i=1}^{m} \rho_i\langle\gamma_k \,|\, \beta_i\rangle + \varepsilon\sigma\langle\gamma_k \,|\, 1\rangle \qquad k = 1, \ldots, m. \tag{8.3.4}$$

A very important relationship between the variational derivation of eqn. (7.5.18) and the generalized Galerkin expression of eqn. (8.3.4) may be seen here. If we express $\phi$ as a complete unknown in the approximation space in eqn. (7.5.18), we obtain:

$$\mathscr{L}[\phi] = \tfrac{1}{2}\sum_{i=1}^{n}\sum_{k=1}^{n} \varepsilon\phi_i\langle\nabla\zeta_k \,|\, \nabla\zeta_i\rangle\phi_k - \sum_{i=1}^{m}\sum_{k=1}^{n} \phi_k\langle\beta_i \,|\, \zeta_k\rangle\rho_i, \tag{8.3.5}$$

$$\frac{\partial}{\partial\phi_k}\mathscr{L}[\phi] = \sum_{i=1}^{n} \varepsilon\phi_i\langle\nabla\zeta_k \,|\, \nabla\zeta_i\rangle - \sum_{i=1}^{m} \rho_i\langle\zeta_k \,|\, \beta_i\rangle = 0 \qquad k = 1, \ldots, n, \tag{8.3.6}$$

which is exactly the same as eqn. (8.3.4) but for $\gamma_i = \zeta_i$ and $\sigma$, the specified normal gradient at Neumann boundaries being zero. Therefore, the variational formulation presented above is the same as the Bubnov–Galerkin method for zero Neumann derivative.

This similarity has an important implication for the variational formulation. What this equivalence states is that whenever the normal gradient of the field is zero, we simply do not bother about it when we employ the variational formulation. It will automatically turn out to be so. This is, therefore, called a *natural* boundary condition. In eqn. (8.1.5) we saw that when the Neumann condition is imposed and the energy $\mathscr{L}$ is minimized, the Poisson equation is satisfied. What we have just learned is that we need not impose simple Neumann conditions at all. We merely extremize the energy and we will simultaneously satisfy the simple Neumann condition on non-Dirichlet boundaries and the Poisson equation everywhere within the region of solution. That this condition turns out automatically in one-dimensional problems from symmetry conditions has already been shown in section 7.3 using the energy functional. Here we have used the Galerkin method to prove the same thing in two dimensions as well. An extension of the proof for two dimensions using the energy functional is given by Hoole (1988).

Needless to say, since we have by choosing an approximation space limited the ability of our solution to satisfy these equations exactly, the solution we obtain will be that function of our approximation space that satisfies them best; i.e., while our solution will neither fit the Poisson equation nor the simple Neumann boundary exactly, it will satisfy both in an optimal sense. This optimality will be in terms of the energy $\mathscr{L}$, which we are minimizing; and that is why, as was discussed in section 8.1, the energy computed from the solution for $\phi$ will be of an order of accuracy higher than $\phi$.

We have already seen from the previous chapter how all the terms but $\int \varepsilon\psi\sigma\, dS$ in eqn. (8.3.3) are handled. To determine the handling of this term, consider a triangle 123 with edge 23 forming a part of the boundary on which $\sigma$ is specified. Since $\zeta_1$ is zero along this edge, the unknown $\phi$ is $\zeta_2\phi_2 + \zeta_3\phi_3$ for first-order interpolation, so that $\gamma_k$ will assume the values $\zeta_2$ and $\zeta_3$ in eqn. (8.3.4). The relevant terms of eqn. (8.3.4), then, corresponding to the three weighting functions $\zeta_1$, $\zeta_2$, and $\zeta_3$ are, respectively, zero,

$$\int \varepsilon\sigma\zeta_2\, dS = \varepsilon\sigma \int_0^1 \zeta_2 L_{23}\, d\zeta_2 = \tfrac{1}{2}\varepsilon\sigma L_{23}, \tag{8.3.7}$$

and

$$\int \varepsilon\sigma\zeta_3\, dS = \varepsilon\sigma \int_0^1 \zeta_3 L_{23}\, d\zeta_3 = \tfrac{1}{2}\varepsilon\sigma L_{23}, \tag{8.38}$$

where $L_{23}$ is the length of the edge 23. This means that zero, $\frac{1}{2}\varepsilon\sigma L_{23}$, and $\frac{1}{2}\varepsilon\sigma L_{23}$, respectively, must be added to the right-hand-side terms of $q$ of eqn. (7.6.7). This additional term is

$$\mathbf{v} = \tfrac{1}{2}\varepsilon\sigma L_{23}[0 \quad 1 \quad 1]^t. \tag{8.3.9}$$

In variational terms this corresponds to minimizing the functional

$$L(\phi) = \iint \{\tfrac{1}{2}\varepsilon[\nabla\phi]^2 - \phi\rho\}\, dR + \int \varepsilon\sigma\phi\, dS. \tag{8.3.10}$$

## 8.3.2 The Electric Fuse

In field problems, this specified normal gradient boundary condition is useful if the input flux into the solution region is specified. In electric field problems, this may correspond to a boundary charge specification. In magnetic field problems, after getting the solution of the magnetization interior to permeable materials, we may use the normal gradient of the magnetic field on the surface of such materials to obtain the field in the exterior free space.

For example, consider the electric fuse of Figure 8.3.1. This is a typical current flow problem as defined in section 3.3.2. When current exceeds a predesigned amount, the fuse melts and opens at the constriction on account of the high current density and attendant

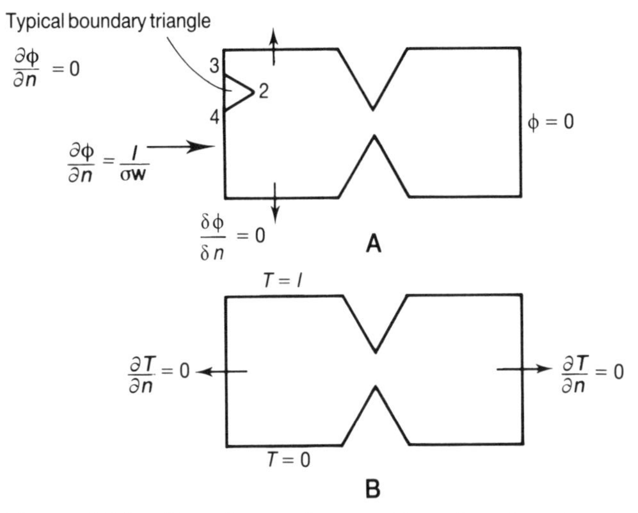


**Figure 8.3.1.** Boundary value problem for the electric fuse. **A.** The scalar potential formulation. **B.** The vector potential formulation.

heating there. Supposing we are given, instead of the input and output voltages, the total flow of current $I$. Then we have at the input, normal component of current density $J$ as:

$$\sigma \nabla \phi \, . \, \mathbf{u}_n = \frac{I}{w}, \tag{8.3.11}$$

where $\sigma$ is the conductivity and $w$ is the width of the fuse. This fuse may be posed as the boundary value problem of Figure 8.3.1A and solved. The left and right boundaries are at a constant but unknown potential, because they are far away from the constriction. In addition, the left boundary has eqn. (8.3.11). We also need to specify $\phi$ at least at one point, as proved in Appendix B. We shall choose this to be on our right boundary and thereby make the entire boundary be at zero potential. As the current $I$ is increased,

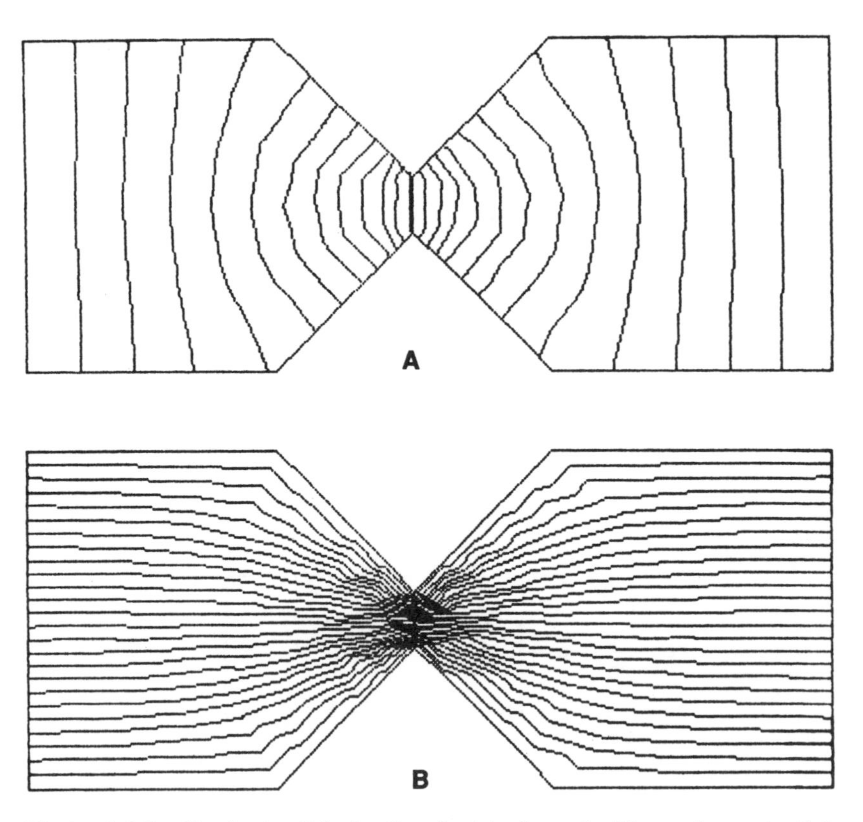


**Figure 8.3.2.** Equipotentials in the electric fuse. **A.** The scalar potential formulation. **B.** The vector potential formulation.

the potential on the left boundary will rise to accommodate it. The Neumann conditions imply no current egress along the other boundaries. Moreover, since this is a linear problem in a homogeneous medium, the conductivity may be taken as 1, without affecting our solution for $\phi$.

The formation of the finite element matrices will be as in chapter 7, except that the boundary elements, such as 234 for example, will have the term

$$\mathbf{v} = \frac{1}{2}\left(\frac{IL_{24}}{w}\right)[0 \quad 1 \quad 1]^t \tag{8.3.12}$$

added to the usual right-hand-side vector $\mathbf{q}$ of eqn. (7.6.3), as required by eqn. (8.3.9). However, if by chance the boundary element 234 had been numbered 342, with the first and second nodes being on the boundary, it is easy to verify that what we need to add is

$$\mathbf{v} = \frac{1}{2}\left(\frac{IL_{24}}{w}\right)[1 \quad 1 \quad 0]^t \tag{8.3.13}$$

The solution to this problem is shown plotted in Figure 8.3.2A.

### 8.3.3 A Complementary Formulation for the Fuse

An easier approach to the fuse problem, when it is the current rather than the voltage that is specified, is through a complementary formulation. If we return to the governing equations for the fuse in section 3.3.2, we will notice that the Poisson equation was derived from the governing equation pair

$$\nabla \times \mathbf{E} = 0, \tag{1.2.19}$$

$$\nabla \cdot \mathbf{J} = 0, \tag{3.3.6}$$

with the constitutive relationship

$$\mathbf{J} = \sigma \mathbf{E}. \tag{1.2.2}$$

There we chose to start with the divergence eqn. (1.2.29), and using the identity (A4), we set

$$\mathbf{E} = -\nabla \phi. \tag{1.2.21}$$

For this problem in two dimensions without a curl source, the complementary approach is simpler. Here we start with the curl eqn. (3.3.6) and, using the identity eqn. (A5), introduce the single component electric vector potential $\mathbf{T} = \mathbf{u}_z T$:

$$\mathbf{J} = \nabla \times \mathbf{u}_z T, \tag{3.3.14}$$

analogous to the magnetic vector potential of eqn. (1.2.31). Now, putting this into the divergence equation, we obtain

$$\sigma^{-1}\nabla T = 0. \tag{3.3.15}$$

We have already seen in section 1.2.4 that flow is along lines of equal vector potential so that the upper and lower boundaries become equi-$T$ lines; and the inlet and outlet become zero-gradient Neumann boundaries. It was also shown in section 4.4.1 that the total flow between two points is the difference in vector potential between those two points. Therefore, the constant potentials at which the upper and lower boundaries lie must differ by the specified current flow through the fuse. We also must specify the potential at one point on the boundary for reference. Taking this point on the lower boundary, we have the entire lower boundary at $T = 0$ and the entire upper boundary at $T = I$. The reader is left to establish the direction of flow, whether it is to the right or to the left, using eqn. (1.2.34). The corresponding boundary value problem shown in Figure 8.3.1B, it is seen, has no complex boundary conditions. The corresponding solution is shown plotted in Fig. 8.3.2B.

### 8.3.3 Handling Interior Line Charges in Finite Elements

Particularly in electron devices, it is common to encounter a specified line charge distribution interior to the solution region. By considering a small cylindrical pillbox with lids on either side of the boundary, it can be shown that

$$D_{n1} = -\varepsilon_1 \nabla\phi \cdot \mathbf{u}_{n1} = -D_{n2} + \mathbf{q} = \varepsilon_2 \nabla\phi \cdot \mathbf{u}_{n2} - \mathbf{q}, \tag{8.3.16}$$

where $\mathbf{q}$ is the line charge density. Observe that the clash in signs with eqn. (1.2.18) is on account of the normals $n1$ and $n2$ along $S$ being oppositely directed. We know how to specify the gradient at a boundary using eqn. (8.3.10), but how do we specify the condition interior to the solution region? To discover this, let the line containing the charge be $C$. We shall extend it along either end with sections $C_1$ and $C_2$ without charge $Q$, to cut the solution region $R$ and boundary $S$ into two parts $R_1$ and $R_2$, and $S_1$ and $S_2$, as shown in Figure 8.3.3. Let $g$ be $\nabla\phi \cdot \mathbf{u}_n$ along this extended line

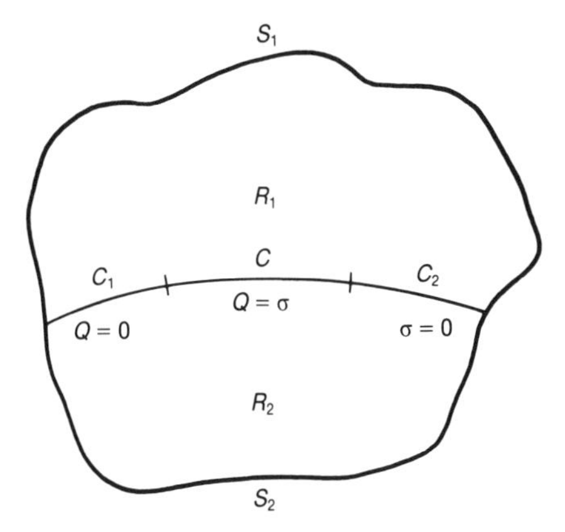


**Figure 8.3.3.** Modeling interior line charges.

with values $g_1$ and $g_2$ on either side of this line $C_1 + C + C_2$. Applying eqn. (8.3.10) to the two parts separately and imposing these unknown gradients $g_1$ and $g_2$ along the common boundary of the two regions,

$$\mathscr{L}_1(\phi) = \tfrac{1}{2}\iint_{R_1} \{\varepsilon_1[\nabla\phi]^2 - 2\phi\rho\}\, dR + \int_{C_1+C+C_2} \varepsilon_1 g_1 \phi\, dC, \quad (8.3.17)$$

$$\mathscr{L}_2(\phi) = \tfrac{1}{2}\iint_{R_2} \{\varepsilon[\nabla\phi]^2 - 2\phi\rho\}\, dR + \int_{C_1+C+C_2} \varepsilon_2 g_2 \phi\, dC. \quad (8.3.18)$$

Observe from eqn. (8.3.16) that $\varepsilon_1 g_1 + \varepsilon_2 g_2$ is zero along $C_1$ and $C_2$ because $Q$ is zero there, and that it is $-Q$ along $C$. Moreover, note that a simple Neumann or Dirichlet condition has been assumed along $S$. Adding the above two equations, we have

$$\mathscr{L}(\phi) = \tfrac{1}{2}\iint_{R} \{\varepsilon[\nabla\phi]^2 - 2\phi\rho\}\, dR - \int_C Q\phi\, dC. \quad (8.3.20)$$

This was to be expected because it corresponds to the integral $\phi\rho$ over $R$ as in the second term, representing the energy expended in bringing $Q$ to its present location.

To implement the functional of eqn. (8.3.20), consider the addition to the energy $\Delta\mathscr{L}$ from a segment of the line charge between two adjacent nodes 1 and 2. Assuming a linear interpolation, we have

$$\begin{aligned}\Delta\mathscr{L} &= \int_1^2 Q\left\{\phi_1 + \frac{[C - L_1][\phi_2 - \phi_1]}{L_2 - L_1}\right\} dC \\ &= \tfrac{1}{2} Q L_{12}[\phi_1 + \phi_2],\end{aligned} \quad (8.3.21)$$

where $L_{12}$ is the length between the two nodes, and $L_1$ and $L_2$ are the coordinates $C$ of the nodes along some coordinate system down the line charge. When the functional is extremized with respect to the potentials, the $\phi$ terms drop off so that the terms $\frac{1}{2}QL_{12}$ contribute to the first and second rows of the right-hand side of the global finite element matrix equation $\underset{\sim}{q}$ of eqn. (7.6.13). For example, if 3, 7, and 5 are adjacent nodes along the line charge $Q$, then node 7 is an interior node and row 7 of $\underset{\sim}{q}$ will have the terms $\frac{1}{2}QL_{37}$ and $\frac{1}{2}QL_{75}$ added to it, on account of the line charge contribution $Q$. However, if node 7 had been at an end of the line charge distribution, with, say, line 57 to its right but no charge to its left, then only the term $\frac{1}{2}QL_{75}$ will be added to the seventh row of $\underset{\sim}{q}$.

### 8.3.5 A Charge-Specified Capacitor

The parallel plate capacitor of Figure 8.3.4 with a specified plate charge was solved by this method to demonstrate the handling of interior line charges. An additional data set defines the line segments 12, 23, 34, and 45, etc., containing the line charges. After forming the usual finite element matrix by going through this list of

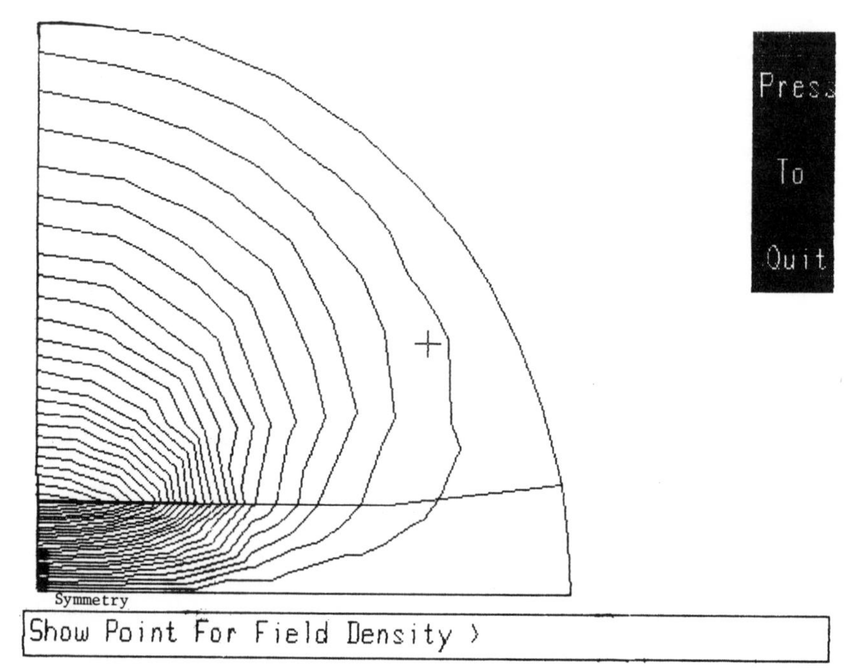


**Figure 8.3.4.** The charged parallel plate capacitor.

line charge elements, for every line element $ij$, the right-hand side vector has the term $\frac{1}{2}QL_{ij}$ added to the rows $i$ and $j$ of the right-hand side vector $\underset{\sim}{q}$. The solution in the symmetric quarter of the problem is depicted in Figure 8.3.4. Figure 10.7.2B shows the kind of solution that would have resulted if we had taken the potential to be specified on the plate.

Another example of interior line charge specification is found in section 10.4.2, where silicon-on-sapphire is analyzed.

### 8.3.6 Natural Impedance Boundary Conditions

In section 5.2.3 we encountered the impedance boundary condition which permits us to block off from our solution region those parts experiencing eddy currents, and in which we are not interested. This significantly reduces computational cost. The impedance boundary condition of eqn. (5.2.31) may in general be expressed as

$$\frac{\partial A}{\partial n} = kA. \tag{8.3.22}$$

Such a boundary condition is encountered in the escape of heat from a surface in heat flow studies and can be easily imposed naturally (Visser 1965). Applying the Galerkin principle as in eqn. (8.3.1)

$$\iint_R \psi[-\nu\nabla^2 A - J]\, dR + \int_\Gamma \nu\psi\left[\frac{\partial A}{\partial n} - kA\right] dS = 0, \tag{8.3.23}$$

where $R$ is the solution region with boundary $S$. $\Gamma$ is that part of $S$ on which the impedance boundary condition obtains. Manipulating as before, using the divergence theorem:

$$\iint_R [\nu\nabla\psi\,.\,\nabla A - \psi J]\, dR - \int_S \nu\psi\frac{\partial A}{\partial n}\, dS + \int_\Gamma \nu\psi\left[\frac{\partial A}{\partial n} - kA\right] dS = 0. \tag{8.3.24}$$

Since the first surface integral vanishes on account of the trial functions along those parts of the surface where $A$ is fixed, we obtain the local matrix equation from

$$\iint_A [\nu\nabla\psi\,.\,\nabla A - \psi J]\, dR - \int_S \nu k\psi A\, dR = 0. \tag{8.3.25}$$

When the finite element trial functions are put in, the surface integral term gives the contribution

$$\Delta \mathbf{P} = \tfrac{1}{6}L\begin{bmatrix} 2 & 1 \\ 1 & 2 \end{bmatrix} \tag{8.3.26}$$

to the local matrix **P**. $L$ is the length of the element edge along the surface, and the associated finite element 2-vector contains the vector potentials at the two vertices of the edge.

## 8.4 A SIMPLE HAND-WORKED EXAMPLE

### 8.4.1 A Test Problem with an Analytical Solution

As a means of explicating the approximation theory of the preceding sections, we shall here take up a simple numerical example. We shall try to approximate, by the various methods detailed, the solution to a problem that we know exactly by analytical methods.

Consider the function $\phi$ defined by:

$$\frac{d^2}{dx^2}\phi = x \qquad 0 \le x \le 1, \tag{8.4.1}$$

$$\frac{d}{dx}\phi = 0 \qquad x = 1, \tag{8.4.2}$$

$$\phi = 0 \qquad x = 0. \tag{8.4.3}$$

This is basically the Poisson equation in one dimension with $\rho(x) = -x$; the region of solution $R$ is the closed interval $[0, 1]$ with the two points $x = 0$ and $x = 1$ as boundary. At the left boundary we have a Dirichlet condition, eqn. (8.4.3), and at the right boundary a simple Neumann condition, eqn. (8.4.2).

To obtain the analytical solution, integrating eqn. (8.4.1), we get

$$\frac{d}{dx}\phi = \tfrac{1}{2}x^2 + c_1. \tag{8.4.4}$$

Using the condition of eqn. (8.4.2) we have $c_1 = -\tfrac{1}{2}$. Further integration gives

$$\phi = \frac{x^3}{6} - \tfrac{1}{2}x + c_2. \tag{8.4.5}$$

Finally, using the Dirichlet boundary condition of eqn. (8.4.3) we obtain $c_2 = 0$, so that

$$\phi = \frac{x^3}{6} - \frac{x}{2}. \qquad (8.4.6)$$

With this analytical solution available for purposes of comparison, let us try to obtain the same solution by approximate methods. Since the exact solution is of order 3 in $x$, so as not to arrive at the exact solution by approximate methods, we will always restrict our trial functions to order 2 or lower.

## 8.4.2 Galerkin—Strong Neumann, One Second-Order Element

Let our trial function be

$$\phi = a_0 + a_1 x + a_2 x^2 \qquad (8.4.7)$$

over the whole region of solution. The Dirichlet boundary condition of eqn. (8.4.3) is always explicitly imposed, so that $a_0$ must be zero. Moreover, since we are also explicitly imposing the Neumann condition here:

$$\frac{d}{dx}\phi = a_1 + 2a_2 x. \qquad (8.4.8)$$

Putting in eqn. (8.4.2), we have $a_1 + 2a_2 = 0$, so that

$$\phi = (-2x + x^2)a_2. \qquad (8.4.9)$$

We see that by explicitly imposing the Neumann condition, the trial function has been reduced to one free parameter $a_2$, which may vary to satisfy the governing eqn. (8.4.1).

Therefore, the approximation space is spanned by the function $(-2x + x^2)$, since any solution must be a multiple of it. That is, the dimension of the approximation space, $n$, is 1 and the only element of this basis, $\alpha_1$, is $(-2x + x^2)$. What must be determined then is $a_2$ of eqn. (8.4.8).

Applying the Galerkin formulation of eqn. (8.2.4), with $\gamma = \alpha$, $n = 1$, and $A = d^2/dx^2$, we get

$$\int_0^1 (-2x + x^2)\frac{d^2}{dx^2}(-2x + x^2)a_2\,dx = \int_0^1 (2x + x_2)x\,dx. \qquad (8.4.10)$$

Solving,

$$a_2 = \tfrac{5}{16}, \qquad (8.4.11)$$

$$\phi = \frac{5(-2x + x^2)}{16}. \qquad (8.4.12)$$

## 8.4.3 Collocation: Explicit Neumann, One Second-Order Element

When we explicitly impose the Neumann condition, with a second-order element encompassing the whole region of solution, the trial function would be exactly the same as eqn. (8.4.8). The residual for this trial function at any point, in accordance with eqn. (8.2.11), is

$$\mathcal{R} = \left[\frac{d^2}{dx^2}(-2x + x^2)a_2 - x\right] \tag{8.4.13}$$

and should be zero. To determine $a_2$ through collocation, we should determine one collocation point at which we may set the residual to zero. For the region of solution [0, 1], $x = 0.5$ is a representative point, so that, expanding eqn. (8.4.13), we get

$$\mathcal{R} = 2a_2 - x = 2a_2 - \tfrac{1}{2} = 0, \tag{8.4.14}$$

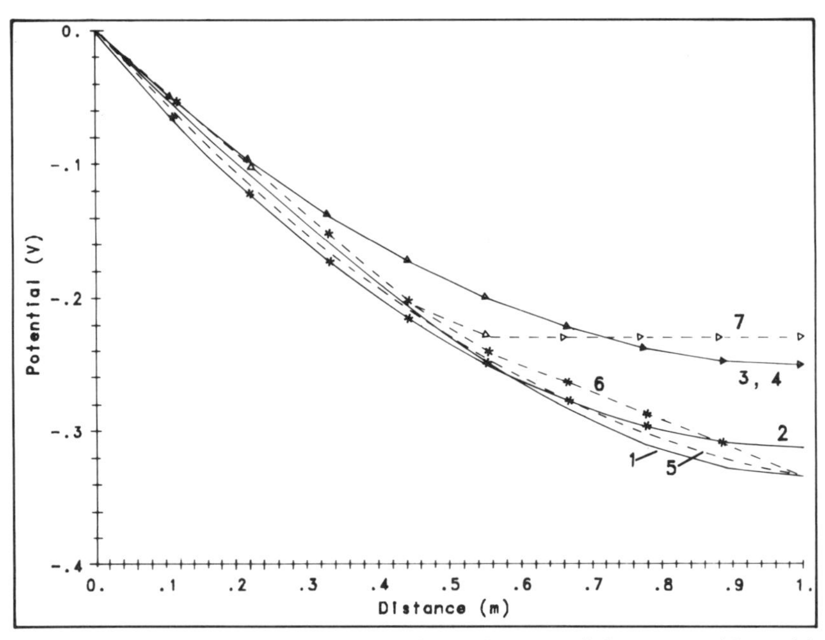


**Figure 8.4.1.** A plot of various Galerkin solutions of the test problem. (1) Exact. (2) Galerkin, second order, explicit Neumann, one element. (3) Collocation, second order, explicit Neumann, one element. (4) Least squares, second order, explicit Neumann, one element. (5) Galerkin, second order, weak Neumann, one element. (6) Galerkin, first order, weak Neumann, two elements. (7) Galerkin, first order, explicit Neumann, two elements.

from which we get

$$a_2 = \tfrac{1}{4}, \tag{8.4.15}$$

$$\phi = \frac{-2x + x^2}{4}, \tag{8.4.16}$$

which is reasonably close to the exact solution of eqn. (8.4.3) as seen from Figure 8.4.1. The importance of the choice of the collocation point to accuracy is seen from the fact that, had we chosen our collocation point as $x = 0$, we would have obtained $a_2 = 0$, and for the point $x = 1$, $a_2 = \frac{1}{2}$—both answers widely differing from the exact.

### 8.4.4 Least Squares: Strong Neumann, One Second-Order Element

Again we have the same trial function of eqn. (8.4.8), and, according to eqn. (8.2.12), the least-squares technique minimizes the integral of the residual $\mathcal{R}$ of eqn. (8.4.13) with respect to the free coefficient $a_2$:

$$\frac{\partial}{\partial a_2}\int_0^1 \mathcal{R}^2\,dx = \frac{\partial}{\partial a_2}\int_0^1 [2a_2 - x]^2\,dx = \frac{\partial}{\partial a_2}[4a_2^2 - 2a_2 + \tfrac{1}{3}] = 0, \tag{8.4.17}$$

giving the same results as eqns. (8.4.14) and (8.4.15). A way of obtaining the same answer through a direct application of eqn. (8.2.12) is to note first that $A\alpha_1 = (d^2/dx^2)(-2x + x^2) = 2$, so that

$$\int_0^1 2^*2a_2\,dx = \int_0^1 2^*x\,dx, \tag{8.4.18}$$

which again is the same result.

### 8.4.5 Galerkin: Weak Neumann, One Second-Order Element

Here we do not explicitly impose the Neumann condition, but use the weak formulation of eqn. (8.3.6). The trial function is eqn. (8.4.7), with $a_0$ zero in view of the Dirichlet condition at $x = 0$.

$$\phi = a_1x + a_2x^2. \tag{8.4.19}$$

The dimension $n$ of our approximation space has, as a result of our relaxing the Neumann condition, gone up by one, to two. The two functions $\zeta_1$ and $\zeta_2$, which span the approximation space, are $x$ and $x^2$, with the coefficients $a_1$ and $a_2$ to be determined. The right-hand-side term, $\rho = -x$, may be taken to be spanned by the function $\beta_1 = x$, belonging to the first-order space with $m = 1$. The operator $\nabla$ being $d/dx$, we have $\nabla\zeta_1 = 1$ and $\nabla\zeta_2 = 2x$; applying eqn.(8.3.6) for $k = 1$,

$$\int_0^1 1 * 1\, dx a_1 + \int_0^1 1 * 2x\, dx a_2 - \int_0^1 x * (-x)\, dx = 0. \quad (8.4.20)$$

This reduces to

$$a_1 + a_2 = -\tfrac{1}{3}. \quad (8.4.21)$$

Similarly, applying for $k = 2$,

$$\int_0^1 2x * 1\, dx a_1 + \int_0^1 2x * 2x\, dx a_2 - \int_0^1 x^2 * (-x)\, dx = 0, \quad (8.4.22)$$

giving

$$a_1 + \frac{4a_2}{3} = -\tfrac{1}{4}, \quad (8.4.23)$$

$$a_1 = -\tfrac{7}{12}; \qquad a_2 = \tfrac{1}{4}, \quad (8.4.24)$$

$$\phi = -\frac{7x}{12} + \frac{x^2}{4}. \quad (8.4.25)$$

### 8.4.6 Galerkin: Weak Neumann, Two First-Order Elements

This handworked example illustrates again, as in section 7.8, the use of general first-order linear interpolation and the assembly of contributions coming from the various elements of our mesh.

Let us divide our region of solution, [0, 1], into two elements, [0, 0.5] and [0.5, 1], with interpolation nodes 1, 2, and 3, respectively, at $x = 0$, $x = 0.5$, and $x = 1$. Thus, the two elements are described in terms of nodal connections as 1, 2 and 2, 3. In general, consider a first-order element with nodes 1 and 2 at $x = x_1$ and $x = x_2$. The first-order interpolation for this element is, by rearrangement of eqn. (7.3.7),

$$\phi(x) = \frac{(x_2 - x)}{(x_2 - x_1)}\phi_1 + \frac{(x - x_2)}{(x_2 - x_1)}\phi_2. \quad (8.4.26)$$

This provides us with the line coordinates $\zeta_1$ and $\zeta_2$, corresponding to our triangular coordinates earlier, defined by

$$\zeta_1 = \frac{(x_2 - x)}{(x_2 - x_1)} \qquad \zeta_2 = \frac{(x - x_2)}{(x_2 - x_1)}. \tag{8.4.27}$$

It may be easily shown that

$$\zeta_1 + \zeta_2 = 1, \tag{8.4.28}$$

$$\int \zeta_1^i \zeta_2^j \, dx = \frac{i!\, j!\, 1!\, L}{(i + j + 1)!}, \tag{8.4.29}$$

where $L = x_2 - x_1$, the length of the element, and

$$\frac{\partial \zeta_1}{\partial x} = -\frac{1}{L}; \qquad \frac{\partial \zeta_1}{\partial x} = -\frac{1}{L}. \tag{8.4.30}$$

Attention is drawn to the similarity of eqn. (7.5.9) to eqn. (8.4.28) and of eqn. (D1) to eqn. (8.4.29).

In general, we have two coefficients $\phi_1$ and $\phi_2$ per element to determine. The first-order Poissonian source $\rho = -x$ may be exactly expanded within the approximation space with the same two basis functions $\zeta_1$ and $\zeta_2$ using

$$x = x_1\zeta_1 + x_2\zeta_2, \tag{8.4.31}$$

so that $\beta = \alpha = \zeta$. Applying eqn. (8.3.6), with $k = 1$,

$$\int_{x=x_1}^{x_2} \frac{-1}{L}\frac{-1}{L}\, dx\, \phi_1 + \int_{x=x_1}^{x_2} \frac{-1}{L}\frac{1}{L}\, dx\, \phi_2 - \int_{x=x_1}^{x_2} \zeta_1(-\zeta_1 x_1 - \zeta_2 x_2)\, dx \equiv 0, \tag{8.4.32}$$

and for $k = 2$,

$$\int_{x=x_1}^{x_2} \frac{1}{L}\frac{-1}{L}\, dx\, \phi_1 + \int_{x=x_1}^{x_2} \frac{1}{L}\frac{1}{L}\, dx\, \phi_2 - \int_{x=x_1}^{x_2} \zeta_2(-\zeta_1 x_1 - \zeta_2 x_2)\, dx \equiv 0. \tag{8.4.33}$$

The congruence sign instead of the equals sign is used because the integrals are evaluated over an element. Equality will hold only when the integration is over the whole solution region or, equivalently, when the integrals over the elements are added together. The above two equations reduce, with the help of eqn. (8.4.29), to

$$\frac{\phi_1}{L} - \frac{\phi_2}{L} \equiv \frac{-L(2x_1 + x_2)}{6}, \tag{8.4.34}$$

$$\frac{-\phi_1}{L} + \frac{\phi_2}{L} \equiv \frac{-L(x_1 + 2x_2)}{6}. \tag{8.4.35}$$

It is to be noted that since the first point at $x = 0$ is a Dirichlet point, eqn. (8.4.34) will not apply when considering the first element. We get, for the first element, with $x_1 = 0$, $x_2 = 0.5$, and $\phi_1 = 0$:

$$2\phi_2 \equiv -\tfrac{1}{12}, \tag{8.4.36}$$

and for the second element, with $x_1 = 0.5$ and $x_2 = 1$:

$$2\phi_2 - 2\phi_3 \equiv -\tfrac{1}{6}, \tag{8.4.37}$$

$$-2\phi_2 + 2\phi_3 \equiv -\tfrac{5}{24}. \tag{8.4.38}$$

Combining the contributions from the two elements, as given by eqns. (8.4.36)–(8.4.38), according to a procedure similar to that in section 7.8:

$$\begin{bmatrix} 2+2 & -2 \\ -2 & 2 \end{bmatrix} \begin{bmatrix} \phi_2 \\ \phi_3 \end{bmatrix} = \begin{bmatrix} -\tfrac{1}{6} - \tfrac{1}{12} \\ -\tfrac{4}{24} \end{bmatrix}. \tag{8.4.39}$$

Solving this matrix equation, we get $\phi_2 = -\tfrac{11}{48}$ and $\phi_3 = -\tfrac{1}{3}$, so that according to eqn. (8.4.26),

$$\begin{aligned} \phi = & -\frac{11x}{24} && 0 \le x \le 0.5 \\ & -\tfrac{11}{48} + (-\tfrac{1}{3} + \tfrac{11}{48})(2x - 1) && 0.5 \le x \le 1. \end{aligned} \tag{8.4.40}$$

### 8.4.7 Galerkin: Explicit Neumann, Two First-Order Elements

Here, the interpolation will be exactly as before, but for $\phi_2 = \phi_3$ in view of the Neumann condition having to apply in the last element. That is, the field has only one degree of freedom. The equation corresponding to eqn. (8.4.36) will remain the same. But in the second element, since $\phi_3 = \phi_2$, the energy contribution will yield

$$2\phi_2 - 2\phi_2 \equiv -\tfrac{1}{6}, \tag{8.4.41}$$

$$-2\phi_2 + 2\phi_2 \equiv -\tfrac{5}{24}. \tag{8.4.42}$$

That is, there is no stored energy. Perhaps this is an example that vividly demonstrates why we cannot solve the element equations separately—for then, the above two equations become contradictions. Adding the equations together, we obtain:

$$2\phi_2 = -\tfrac{1}{12} - \tfrac{1}{6} - \tfrac{5}{24}, \tag{8.4.43}$$

giving $\phi_2 = -11/48$, so that the solution is

$$\phi = \begin{cases} -11x/24 & 0 \le x \le 0.5 \\ -\tfrac{11}{48} & 0.5 \le x \le 1 \end{cases}. \tag{8.4.44}$$

### 8.4.8 Some Observations

Having obtained by hand these various approximations, which are plotted in Figure 8.4.1, some observations are in order:

1. The Galerkin method with a weak Neumann solution gives the best overall solution, although the Neumann condition is not satisfied exactly. The explicit Neumann imposition results in exact satisfaction of the Neumann condition, but does not give a good overall fit of the solution.
2. The first-order two-element system and second-order one-element system both have two degrees of freedom when the weak Neumann formulation is used. And yet, the second-order system gives a better answer, because the higher polynomial order of the system allows it to satisfy the equation better.
3. The Galerkin method with the explicit Neumann condition cannot be employed with first-order elements, because the operator $d^2/dx^2$ will return a zero when differentiating an approximation function of order one.

## 8.5 HIGHER-ORDER FINITE ELEMENTS

### 8.5.1 Higher-Order Interpolations

We have so far worked with first-order interpolations— or straight-line variations of the trial function. In general, we have seen that the finite element solution is as good as the trial function we have assumed. An improved solution is achieved in one of two ways. Either we resort to a finer subdivision of the finite element mesh keeping the same shape functions, or, as will be described in this section, we resort to improved shape functions while keeping the same elements.

Two approaches are commonly taken in improving the shape function. The first approach is problem dependent, and here we know the kind of solutions to expect, such as exponential variations in eddy current problems (Keran and Lavers 1985; Weeber, Vidyasagar, and Hoole 1988) and Bessel functions in cylindrical waveguides. In these cases we would allow the unknown to vary as these special functions do and apply the Galerkin method to determine the numerical weights of these shape functions. In this chapter, we shall learn of the more general high-order polynomial shape functions, which may be readily applied without any expectation of the solution. As discussed in section 7.5, the similarity of the Pascal triangle will be used to work with suitably placed interpolation nodes in the high-order triangles shown in Figure 7.5.2. As demonstrated in that figure, a first-order triangle takes $1 + 2 = 3$ nodes (each node corresponding to a degree of freedom), a second-order triangle $1 + 2 + 3 = 6$ nodes, and in general, an $n$th-order triangle will contain nodes numbering

$$N(n) = 1 + 2 + \ldots + n + 1 = \tfrac{1}{2}(n + 1)(n + 2). \qquad (8.5.1)$$

As defined by Silvester (1969c, 1978), the $n$th-order triangle of Figure 8.5.1 has the trial function defined by the interpolation

$$\phi = \sum_{i+j+k=n} \alpha^n_{ijk}\phi_{ijk} = \tilde{\boldsymbol{\alpha}}^n \underset{\sim}{\boldsymbol{\phi}}, \qquad (8.5.2)$$

where each node is uniquely defined by the suffixes $i$, $j$, and $k$ totaling $n$, and Figures 8.5.1A and 8.5.1B, respectively, give the three suffix $ijk$ numbering of the nodes and the corresponding regular numbering adopted as convention. The summation in eqn. (8.5.2) is over all possible combinations of $i$, $j$, and $k$. $\tilde{\boldsymbol{\alpha}}^n$ is a row vector whose elements are $\alpha^n_{ijk}$. $\underset{\sim}{\boldsymbol{\phi}}$ is a column vector with $\phi_{ijk}$ as elements. The coordinates of the nodes are defined by

$$\zeta_1 = \frac{i}{n} \qquad \zeta_2 = \frac{j}{n} \qquad \zeta_3 = \frac{k}{n} \qquad (8.5.3)$$

and

$$\alpha^n_{ijk} = P^n_i(\zeta_1)P^n_j(\zeta_2)P^n_k(\zeta_3), \qquad (8.5.4)$$

where the $n$th-order polynomials $P^n_i(\zeta)$ are defined by

$$P^n_i(\zeta) = \begin{cases} 1 & \text{if } i = 0 \\ \displaystyle\prod_{j=1}^{i} \frac{n\zeta - j + 1}{j} & \text{otherwise.} \end{cases} \qquad (8.5.5)$$

To illustrate, for a first-order triangle,

$$\begin{aligned} \phi &= \alpha_{100}\phi_{100} + \alpha_{010}\phi_{010} + \alpha_{001}\phi_{001} \\ &= P^1_1(\zeta_1)P^1_0(\zeta_2)P^1_0\zeta_3\phi_{100} \\ &\quad + P^1_0(\zeta_1)P^1_1(\zeta_2)P^1_0(\zeta_3)\phi_{010} \\ &\quad + P^1_0(\zeta_1)P^1_0(\zeta_2)P^1_1(\zeta_3)\phi_{001} \\ &= \zeta_1\phi_1 + \zeta_2\phi_2 + \zeta_3\phi_3 \end{aligned} \qquad (8.5.6)$$

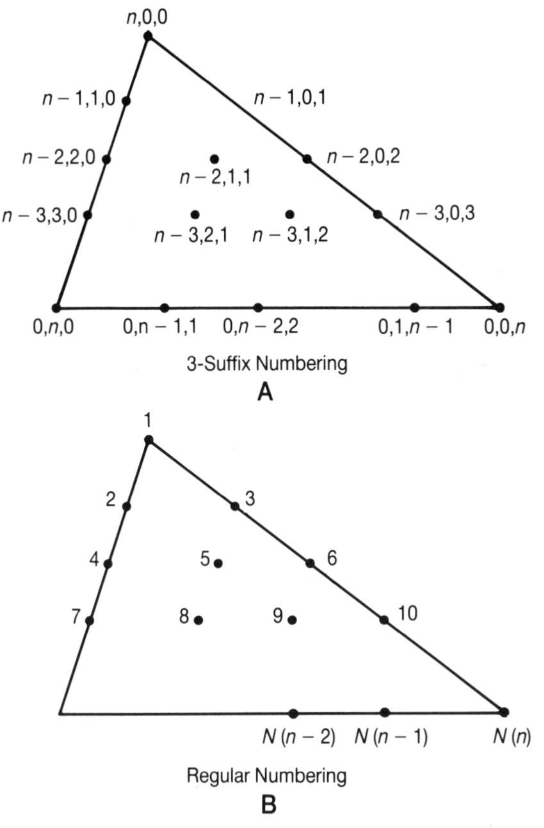


**Figure 8.5.1.** An *n*th-order triangle and node numbering convention. **A.** Three-suffix numbering. **B.** Regular one-digit numbering.

since

$$P_1^1(\zeta) = \zeta \qquad P_0^1 = 1, \tag{8.5.7}$$

according to eqn. (8.5.5), and $\phi_{100}$, $\phi_{010}$, and $\phi_{001}$ correspond to $\phi_1$, $\phi_2$, and $\phi_3$ at the three vertices as determined from the triangular coordinates defined in eqn. (8.5.3). As another example, consider the second-order triangle of Figure 8.5.2, where the trial function becomes

$$\begin{aligned}\phi = \alpha_{200}\phi_{200} + \alpha_{110}\phi_{110} + \alpha_{101}\phi_{101} + \alpha_{020}\phi_{020} \\ + \alpha_{011}\phi_{011} + \alpha_{002}\phi_{002}.\end{aligned} \tag{8.5.8}$$

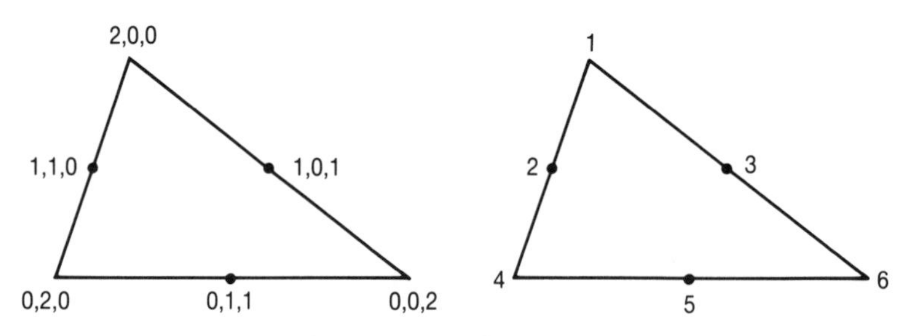


**Figure 8.5.2.** A second-order triangle.

Using eqn. (8.5.5),

$$P_2^2(\zeta) = \left[\frac{2\zeta - 1 + 1}{1}\right]\left[\frac{2\zeta - 2 + 1}{2}\right] = \zeta(2\zeta - 1), \quad (8.5.9a)$$

$$P_1^2(\zeta) = \frac{2\zeta - 1 + 1}{1} = 2\zeta, \quad (8.5.9b)$$

$$P_0^2 = 1. \quad (8.5.9c)$$

Putting eqns. (8.5.4) and (8.5.9) into eqn. (8.5.8) and simplifying the node numbers as indicated in Figure 8.5.1, we have

$$\begin{aligned}\phi = {} & \zeta_1(2\zeta_1 - 1)\phi_1 + 4\zeta_1\zeta_2\phi_2 + 4\zeta_1\zeta_3\phi_3 + \zeta_2(2\zeta_2 - 1)\phi_4 \\ & + 4\zeta_2\zeta_3\phi_5 + \zeta_3(2\zeta_3 - 1)\phi_6.\end{aligned} \quad (8.5.10)$$

The correctness of these interpolations is easily verified. For example, at vertex 1 of triangular coordinates $(\zeta_1, \zeta_2, \zeta_3) = (1, 0, 0)$, $\phi$ becomes $\phi_1$; at vertex 2 with coordinates $(\frac{1}{2}, \frac{1}{2}, 0)$, $\phi$ becomes $\phi_2$, and so on. Using the polynomials defined in this section, interpolations of any order may be defined.

## 8.5.2 Differentiation and Universal Matrices

For purposes of using high-order finite elements, we have already defined the interpolatory trial functions. Now to use them, we need to put them into the Galerkin or variational formulations of eqn. (7.5.18) or eqn. (8.2.4). In this section we shall demonstrate the implementation on the variational formulation, extensions to the Galerkin method being straightforward. We shall also describe precomputed universal matrices (Silvester 1978), which allow us to deal with higher-order finite elements using trivial computations that compare in difficulty only with those required for first-order finite elements.

To implement the variational formulation in two dimensions, we need to put the trial function eqn. (8.5.2) into the functional of eqn. (7.5.18) and minimize the functional with respect to the free parameters, which are the nodal potentials that make up $\phi$. Using an approximation for the known quantity $\rho$ similar to eqn. (8.5.2), we have

$$\begin{aligned}\mathscr{L}[\phi] &= \iint \{\tfrac{1}{2}[\nabla\phi]^2 - \rho\phi\}\, dR \\ &= \iint \{\tfrac{1}{2}[\nabla\tilde{\boldsymbol{\alpha}}^n\underset{\sim}{\boldsymbol{\phi}}] \cdot [\nabla\tilde{\boldsymbol{\alpha}}^n\underset{\sim}{\boldsymbol{\phi}}] - [\tilde{\boldsymbol{\alpha}}^n\underset{\sim}{\boldsymbol{\phi}}][\tilde{\boldsymbol{\alpha}}^n\underset{\sim}{\boldsymbol{\rho}}]\}\, dR.\end{aligned} \tag{8.5.11}$$

The terms $\nabla\tilde{\boldsymbol{\alpha}}$ involve differentiations with respect to $x$ and $y$, and the results of the differentiations must be integrated over $R$. Consider for example:

$$\begin{aligned}\frac{\partial\tilde{\boldsymbol{\alpha}}^n}{\partial x} &= \sum_{i=1}^{3} \frac{\partial\tilde{\boldsymbol{\alpha}}^n}{\partial\zeta_i}\frac{\partial\zeta_i}{\partial x} = \sum_{i=1}^{3} \frac{b_i}{\Delta}\frac{\partial\tilde{\boldsymbol{\alpha}}^n}{\partial\zeta_i} \\ &= \tilde{\boldsymbol{\alpha}}^{n-1}\sum_{i=1}^{3}\frac{b_i}{\Delta}\mathbf{G}_i^n = \tilde{\boldsymbol{\alpha}}^{n-1}\mathbf{D}_x^n,\end{aligned} \tag{8.5.12}$$

which defines the $n$th-order differentiation matrix $\mathbf{D}_x^n$ and the matrices $\mathbf{G}_i^n$ that make up $\mathbf{D}_x^n$. The derivative of $\zeta_i$ with respect to $x$ is obtained using eqn. (7.5.13). Note that $\tilde{\boldsymbol{\alpha}}^n$ being an $n$th-order polynomial, its derivative is of order $n - 1$ and so must be expressible in terms of $\tilde{\boldsymbol{\alpha}}^{n-1}$. And since $\tilde{\boldsymbol{\alpha}}^n$ is a row vector of length $N(n)$ as defined by eqn. (8.5.1), for compatibility $\mathbf{D}_x^n$ must be a matrix of size $N(n-1) \times N(n)$.

As an illustration of the foregoing, consider the second-order interpolation of eqn. (8.5.10):

$$\begin{aligned}\frac{\partial\tilde{\boldsymbol{\alpha}}^2}{\partial\zeta_1} &= \frac{\partial}{\partial\zeta_1}[\zeta_1(2\zeta_1 - 1)\;\; 4\zeta_1\zeta_2\;\; 4\zeta_1\zeta_3\;\; \zeta_2(2\zeta_2 - 1)\;\; 4\zeta_2\zeta_3\;\; \zeta_3(2\zeta_3 - 1)] \\ &= [4\zeta_1 - 1\;\; 4\zeta_2\;\; 4\zeta_3\;\; 0\;\; 0\;\; 0] \\ &= [\zeta_1\;\; \zeta_2\;\; \zeta_3]\begin{bmatrix} 3 & 0 & 0 & 0 & 0 & 0 \\ -1 & 4 & 0 & 0 & 0 & 0 \\ -1 & 0 & 4 & 0 & 0 & 0 \end{bmatrix},\end{aligned} \tag{8.5.13}$$

comparing with eqn. (8.5.12), we have

$$\mathbf{G}_1^2 = \begin{bmatrix} 3 & 0 & 0 & 0 & 0 & 0 \\ -1 & 4 & 0 & 0 & 0 & 0 \\ -1 & 0 & 4 & 0 & 0 & 0 \end{bmatrix}, \tag{8.5.14}$$

where the term $4\zeta_1 - 1$ has been rewritten as $4\zeta_1 - (\zeta_1 + \zeta_2 + \zeta_3)$ using eqn. (7.5.9). Similarly, we may work out $\mathbf{G}_2^2$ and $\mathbf{G}_3^2$.

An important observation is that these $\mathbf{G}$ matrices are independent of the shape of the triangle and so are called *universal* matrices. Although the matrix $\mathbf{D}_x^n$ does depend on shape, according to eqn. (8.5.12), it is derivable from the universal $\mathbf{G}$ matrices:

$$\mathbf{D}_x^n = \sum_{i=1}^{3} \frac{b_i}{\Delta} \mathbf{G}_i^n \tag{8.5.15a}$$

and similarly

$$\mathbf{D}_y^n = \sum_{i=1}^{3} \frac{c_i}{\Delta} \mathbf{G}_i^n. \tag{8.5.15b}$$

Consequently, although we are dealing with higher-order elements, if we compute the $\mathbf{G}$ matrices once and store them, the finite element computation would require only the calculation of the coefficients $b$ and $c$, as with first order finite elements.

A further simplification arises from the symmetry of the triangles about the three vertices (Silvester 1982b). That is, referring to Figure 8.5.3, $\mathbf{G}_2^2$ may be obtained by rotating the triangle of Figure 8.5.3A so that vertex 2 occupies the location of vertex 1 as in Figure 8.5.3B, and mapping the values of $\mathbf{G}_1^2$. That is, we are saying that $\mathbf{G}_2^2$ will be the same as $\mathbf{G}_1$ if vertex 2 is replaced by vertex 1. From Figures 8.5.3A and 8.5.3B, the mapping for the three finite element nodes, corresponding to the three vertices, is:

$$R^1[1 \quad 2 \quad 3] = [2 \quad 3 \quad 1]. \tag{8.5.16}$$

For the second-order example, $\mathbf{G}_2^2$ has six columns corresponding to a second-order triangle and three rows corresponding to a first-order triangle. The nodal mapping $R^2$ for the second-order triangles is extracted from Figures 8.5.3C and D:

$$R^2[1 \quad 2 \quad 3 \quad 4 \quad 5 \quad 6] = [6 \quad 3 \quad 5 \quad 1 \quad 2 \quad 4]. \tag{8.5.17}$$

The mapping for the $\mathbf{G}$ matrices, therefore, is in general

$$\mathbf{G}_2^n(i,j) = \mathbf{G}_1^n[R^{n-1}(i), R^n(j)]. \tag{8.5.18}$$

And similarly,

$$\mathbf{G}_3^n(i,j) = \mathbf{G}_2^n[R^{n-1}(i), R^n(j)] = \mathbf{G}_1^n[R^{n-1}R^{n-1}(i), R^nR^n(j)], \tag{8.5.19}$$

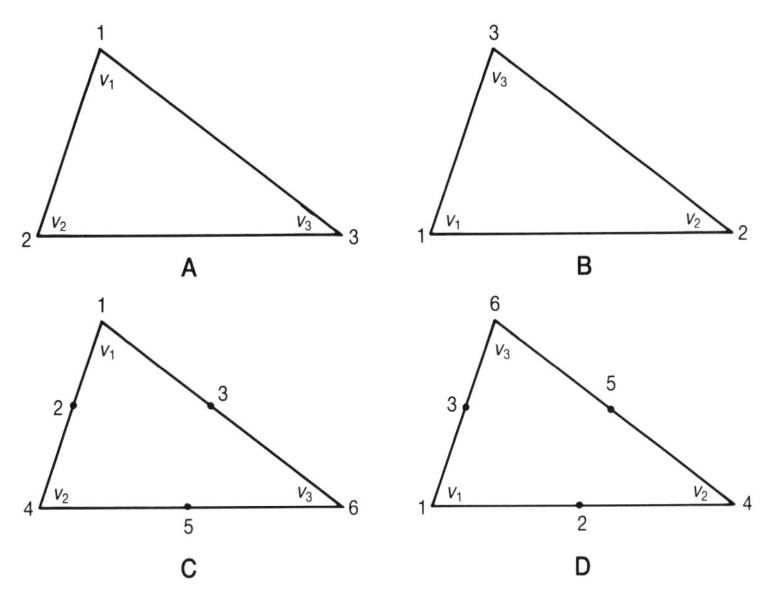


**Figure 8.5.3.** Triangle rotations and vertex symmetry.

so that the application of the same rotation defines $\mathbf{G}_3$. For example,

$$\mathbf{G}_2^2(1,2) = \mathbf{G}_1^2[R^1(1), R^2(2)] = \mathbf{G}_1^2(3,2) = 4, \tag{8.5.20}$$

$$\mathbf{G}_3^2(1,2) = \mathbf{G}_1^2[R^1R^1(1), R^2R^2(2)] = \mathbf{G}_1^2[R^1(3), R^2(2)] = \mathbf{G}_1^2(2,3) = 0. \tag{8.5.21}$$

Building thus, we get

$$\mathbf{G}_2^2 = \begin{bmatrix} \mathbf{G}_1^2[R^1(1), R^2(1)] & \mathbf{G}_1^2[R^1(1), R^2(2)] & \mathbf{G}_1^2[R^1(1), R^2(3)] \\ \mathbf{G}_1^2[R^1(2), R^2(1)] & \mathbf{G}_1^2[R^1(2), R^2(2)] & \mathbf{G}_1^2[R^1(2), R^2(3)] \\ \mathbf{G}_1^2[R^1(3), R^2(1)] & \mathbf{G}_1^2[R^1(3), R^2(2)] & \mathbf{G}_1^2[R^1(3), R^2(3)] \end{bmatrix.$$

$$\begin{matrix} \mathbf{G}_1^2[R^1(1), R^2(4)] & \mathbf{G}_1^2[R^1(1), R^2(5)] & \mathbf{G}_1^2[R^1(1), R^2(6)] \\ \mathbf{G}_1^2[R^1(2), R^2(4)] & \mathbf{G}_1^2[R^1(2), R^2(5)] & \mathbf{G}_1^2[R^1(2), R^2(6)] \\ \mathbf{G}_1^2[R^1(3), R^2(4)] & \mathbf{G}_1^2[R^1(3), R^2(5)] & \mathbf{G}_1^2[R^1(3), R^2(6)] \end{matrix}\right]$$

$$= \begin{bmatrix} \mathbf{G}_1^2(3,6) & \mathbf{G}_1^2(3,3) & \mathbf{G}_1^2(3,5) & \mathbf{G}_1^2(3,1) & \mathbf{G}_1^2(3,2) & \mathbf{G}_1^2(3,4) \\ \mathbf{G}_1^2(1,6) & \mathbf{G}_1^2(1,3) & \mathbf{G}_1^2(1,5) & \mathbf{G}_1^2(1,1) & \mathbf{G}_1^2(1,2) & \mathbf{G}_1^2(1,4) \\ \mathbf{G}_1^2(2,6) & \mathbf{G}_1^2(2,3) & \mathbf{G}_1^2(2,5) & \mathbf{G}_1^2(2,1) & \mathbf{G}_1^2(2,2) & \mathbf{G}_1^2(2,4) \end{bmatrix}$$

$$= \begin{bmatrix} 0 & 4 & 0 & -1 & 0 & 0 \\ 0 & 0 & 0 & 3 & 0 & 0 \\ 0 & 0 & 0 & -1 & 4 & 0 \end{bmatrix}, \tag{8.5.22}$$

and similarly:

$$\mathbf{G}_3^2 = \begin{bmatrix} 0 & 0 & 4 & 0 & 0 & -1 \\ 0 & 0 & 0 & 0 & 4 & -1 \\ 0 & 0 & 0 & 0 & 0 & 3 \end{bmatrix}. \qquad (8.5.23)$$

By working out $\mathbf{G}_2^2$ and $\mathbf{G}_3^2$ just the way $\mathbf{G}_1^2$ was in eqn. (8.5.13), the

**Order 1**

$R^1\,[1\quad 2\quad 3] = [3\quad 1\quad 2]$

$G_1{}^1 = [1\quad 0\quad 0]$

$$T^{1,1} = (\tfrac{1}{12}) \begin{bmatrix} 2 & 1 & 1 \\ 1 & 2 & 1 \\ 1 & 1 & 2 \end{bmatrix}$$

**Order 2**

$R^2 = [1\quad 2\quad 3\quad 4\quad 5\quad 6\;] = [6\quad 3\quad 5\quad 1\quad 2\quad 4]$

$$G_1^2 = \begin{bmatrix} 3 & 0 & 0 & 0 & 0 & 0 \\ -1 & 4 & 0 & 0 & 0 & 0 \\ -1 & 0 & 4 & 0 & 0 & 0 \end{bmatrix}$$

$$T^{2,2} = (\tfrac{1}{180}) \begin{bmatrix} 6 & 0 & 0 & -1 & -4 & -1 \\ 0 & 32 & 16 & 0 & 16 & -4 \\ 0 & 16 & 32 & -4 & 16 & 0 \\ -1 & 0 & -4 & 6 & 0 & -1 \\ -4 & 16 & 16 & 0 & 32 & 0 \\ -1 & -4 & 0 & -1 & 0 & 6 \end{bmatrix}$$

**Order 3**

$R^3[1\quad 2\quad 3\quad 4\quad 5\quad 6\quad 7\quad 8\quad 9\quad 10] = [10\quad 6\quad 9\quad 3\quad 5\quad 8\quad 1\quad 2\quad 4\quad 7]$

$$G_1^3 = (\tfrac{1}{8}) \begin{bmatrix} 44 & 0 & 0 & 0 & 0 & 0 & 0 & 0 & 0 & 0 \\ -1 & 36 & 0 & 9 & 0 & 0 & 0 & 0 & 0 & 0 \\ -1 & 0 & 36 & 0 & 9 & 0 & 0 & 0 & 0 & 0 \\ 8 & -36 & 0 & 72 & 0 & 0 & 0 & 0 & 0 & 0 \\ 8 & -18 & -18 & 9 & 54 & 9 & 0 & 0 & 0 & 0 \\ 8 & 0 & -36 & 0 & 0 & 72 & 0 & 0 & 0 & 0 \end{bmatrix}$$

$$T^{3,3} = (\tfrac{1}{6720}) \begin{bmatrix} 76 & 18 & 18 & 0 & 36 & 0 & 11 & 27 & 27 & 11 \\ 18 & 540 & -135 & 162 & -189 & 27 & -54 & -135 & -54 & 27 \\ 18 & -135 & 540 & -135 & 162 & -189 & 27 & -54 & -135 & 0 \\ 0 & 162 & -135 & 540 & 162 & -54 & 18 & 270 & -135 & 27 \\ 36 & -189 & 162 & 162 & 1944 & 162 & 36 & 162 & 162 & 36 \\ 0 & -135 & -189 & -54 & 162 & 540 & 27 & -135 & 270 & 18 \\ 11 & 0 & 27 & 18 & 36 & 27 & 76 & 18 & 0 & 11 \\ 27 & -135 & -54 & 270 & 162 & -135 & 18 & 540 & -189 & 0 \\ 27 & -54 & -135 & -135 & 162 & 270 & 0 & -189 & 540 & 18 \\ 11 & 27 & 0 & 27 & 36 & 18 & 11 & 0 & 18 & 76 \end{bmatrix}$$

**Figure 8.5.4.** Universal matrices.

correctness of the above rotations may be verified. What this means is that it is sufficient to store only one of the **G** matrices for each order of element, and the other two may be computed provided the rotations are also stored.

At this point we shall define another universal matrix, which we shall need in the next section. This is called the *metric tensor* and is defined by:

$$\mathbf{T}^{n,n} = A^{-1} \iint \tilde{\boldsymbol{\alpha}}^{nt} \tilde{\boldsymbol{\alpha}}^{n} \, dR, \tag{8.5.24a}$$

where $A$ is the area. Alternatively,

$$\mathbf{T}^{n,n}[i,j] = A^{-1} \iint \alpha_i \alpha_j \, dR, \tag{8.5.24b}$$

so that the matrix **T** is seen to be symmetric. $\mathbf{T}^{1,1}$ has already been evaluated in eqn. (7.6.12). It is seen that dividing the integral by the area again makes the matrix independent of shape, and therefore it is also deservingly given the adjective *universal*. These matrices also may be precomputed and stored for all orders of interpolation, as may be required using the integral relationship (D1). The rotations and **G** and **T** matrices are presented in Figure 8.5.4 for orders up to three and may easily be verified by the reader. For higher orders, these have already been evaluated and presented by Silvester (1978) and may be obtained from his classic paper. These matrices having been already computed by other workers, we are spared the burden of computing them and may use their results. However, in using their results, it is important to understand the principles that underlie them.

## 8.5.3 Functional Minimization

Having defined the universal matrices, we are now ready to return to the functional (8.5.11) and employ these matrices:

$$\begin{aligned}
\mathscr{L}[\phi] &= \sum_{\Delta} \iint \{\tfrac{1}{2}\varepsilon[\nabla\tilde{\boldsymbol{\alpha}}^n \underset{\sim}{\boldsymbol{\phi}}] \cdot [\nabla\tilde{\boldsymbol{\alpha}}^n \underset{\sim}{\boldsymbol{\phi}}] - [\tilde{\boldsymbol{\alpha}}^n \underset{\sim}{\boldsymbol{\phi}}][\tilde{\boldsymbol{\alpha}}^n \underset{\sim}{\boldsymbol{\rho}}]\} \, dR \\
&= \sum_{\Delta} \iint \left\{\tfrac{1}{2}\varepsilon\left[\left(\frac{\partial \tilde{\boldsymbol{\alpha}}^n \underset{\sim}{\boldsymbol{\phi}}}{\partial x}\right)^2 + \left(\frac{\partial \tilde{\boldsymbol{\alpha}}^n \underset{\sim}{\boldsymbol{\phi}}}{\partial y}\right)^2\right] - [\tilde{\boldsymbol{\alpha}}^n \underset{\sim}{\boldsymbol{\phi}}][\tilde{\boldsymbol{\alpha}}^n \underset{\sim}{\boldsymbol{\rho}}]\right\} dR \\
&= \sum_{\Delta} \iint \{\tfrac{1}{2}\varepsilon[(\tilde{\boldsymbol{\alpha}}^{n-1}\mathbf{D}_x \underset{\sim}{\boldsymbol{\phi}})^2 + (\tilde{\boldsymbol{\alpha}}^{n-1}\mathbf{D}_y \underset{\sim}{\boldsymbol{\phi}})^2] - [\tilde{\boldsymbol{\alpha}}^n \underset{\sim}{\boldsymbol{\phi}}][\tilde{\boldsymbol{\alpha}}^n \underset{\sim}{\boldsymbol{\rho}}]\} \, dR
\end{aligned}$$

$$= \sum_{\Delta} \left\{ \tfrac{1}{2}\varepsilon \left[ \underset{\sim}{\phi}^t \mathbf{D}_x^t \iint \tilde{\alpha}^{n-1t}\tilde{\alpha}^{n-1}\, dR\, \mathbf{D}_x \underset{\sim}{\phi} + \underset{\sim}{\phi}^t \mathbf{D}_y^t \iint \tilde{\alpha}^{n-1t}\tilde{\alpha}^{n-1}\, dR\, \mathbf{D}_y \underset{\sim}{\phi} \right] - \underset{\sim}{\phi}^t \iint \tilde{\alpha}^{nt}\tilde{\alpha}^{n}\, dR \right\} \underset{\sim}{\rho}$$

$$= \sum_{\Delta} \{ \tfrac{1}{2}\varepsilon A \underset{\sim}{\phi}^t [\mathbf{D}_x^t \mathbf{T}^{n-1,n-1}\mathbf{D}_x + \mathbf{D}_y^t \mathbf{T}^{n-1,n-1}\mathbf{D}_y] \underset{\sim}{\phi} - \underset{\sim}{\phi}^t A \mathbf{T}^{n,n} \underset{\sim}{\rho} \}. \tag{8.5.25}$$

Minimizing now according to the rules enunciated in eqns. (7.6.18) and (7.6.19):

$$\frac{\partial \mathscr{L}}{\partial \underset{\sim}{\phi}} = \sum_{\Delta} \{ \varepsilon A [\mathbf{D}_x^t \mathbf{T}^{n-1,n-1}\mathbf{D}_x + \mathbf{D}_y^t \mathbf{T}^{n-1,n-1}\mathbf{D}_y] \underset{\sim}{\phi} - A\mathbf{T}^{n,n}\underset{\sim}{\rho} \} = 0. \tag{8.5.26}$$

Thus, the local right-hand-side element vector is

$$\underset{\sim}{q} = A\mathbf{T}^{n,n}\underset{\sim}{\rho}, \tag{8.5.27}$$

and the local matrix is

$$\mathbf{P} = \varepsilon A [\mathbf{D}_x^t \mathbf{T}^{n-1,n-1}\mathbf{D}_x + \mathbf{D}_y^t \mathbf{T}^{n-1,n-1}\mathbf{D}_y]. \tag{8.5.28}$$

In Alg. 7.6.2 we have a scheme for forming the local matrices for first-order triangles. It uses the subroutine Triangle of Alg. 7.6.1 to compute the $b$ and $c$ terms and thence forms the local finite element matrix **P** and right-hand-side vector **q**. Here, using eqn. (8.5.26), we extend Alg. 7.6.2 to form the local matrix for an $n$th-order triangle, as shown in Alg. 8.5.2. Alg. 8.5.2 first forms the differentiation matrices using Alg. 8.5.1 based on eqn. (8.5.15) and then uses the differentiation matrices to form the local matrices. The rules for adding the local matrix contributions to form a global equation have already been described in Alg. 7.6.3. The difference here is that the local matrices of size $N \times N$ may be larger. For second order, for example, the matrix becomes $6 \times 6$, for third order, $10 \times 10$, and so on. The only modification required to the algorithm is in redefining the size of the local matrix from 3 to its present size.

For a given set of points connected as, say, second-order triangles, the accuracy will be greater than if they had been connected as first-order triangles. However, the solution of the matrix equation will require more work and storage for the higher-order matrix, because the sparsity of the mesh has been reduced although the matrix size remains the same. For example, as seen in Figure 8.5.5A, if we have six equilateral triangles meeting at a vertex, the vertex node will appear with the six surrounding nodes if it is a first-order mesh. However, if it is a second-order mesh as in Figure 8.5.5B, the vertex node will appear with 18 other

**Algorithm 8.5.1. Formation of Differentiation Matrices**

**Procedure DiffMats(Dx, Dy,** *A*, *Nn*, *Nn*_1, **x, y**, *Rn*, *Rn*_1, **Gn1, Tn_1n_1)**
{Function: Computes the local matrix **P** and right-hand-side vector for an *n*th-order triangle
Outputs: **Dx, Dy** = *Nn*_1 × *Nn* differentiation matrices of the triangle; *A* = area of the triangle
Inputs: *Nn* = *N*(*n*) and *Nn*_1 = *N*(*n* − 1), defined in eqn. (8.5.1); **x, y** = 3 × 1 arrays containing the coordinates of the three vertices; *Rn*, *Rn*_1 = Rotations on triangles of order *n* and *n* − 1; **Gn1** = *Nn*_1 × *Nn* matrix $G_1^n$—eqn. (8.5.12); **Tn_1n_1** = *N*(*n* − 1) × *N*(*n* − 1) metric tensor—eqn. (8.5.24)
Required: **Procedure Triangle** of Alg. 7.6.1}
**Begin**
  **Triangle(b, c,** *A*, **x, y)**
      {**c** is already divided by Δ as it comes from Triangle}
  **For** *i* ← 1 **To** *Nn*_1 **Do**
    **For** *j* ← 1 **To** *Nn* **Do**
      *i*1 ← *Rn*_1[*i*]
      *i*2 ← *Rn*_1[*i*1]
      *j*1 ← *Rn*[*j*]
      *j*2 ← *Rn*[*j*1]
      **Dx**[*i*, *j*] ← **b**[1]***Gn1**[*i*, *j*] + **b**[2]***Gn1**[*i*1, *j*1] + **b**[3]***Gn1**[*i*2, *j*2]
      **Dy**[*i*, *j*] ← **c**[1]***Gn1**[*i*, *j*] + **c**[2]***Gn1**[*i*1, *j*1] + **c**[3]***Gn1**[*i*2, *j*2]   {eqn. (8.5.15)}
**End**

**Algorithm 8.5.2. Formation of Local High-Order Matrices**

**Procedure LocalMatrix(P, q,** *Nn*, *Nn*_1, **x, y**, Eps, **Rho**, *Rn*, *Rn*_1, **Gn1, Tnn, Tn_1n_1)**
{Function: computes the local matrix **P** and right-hand-side vector for an *n*th-order triangle
Outputs: **P** = *Nn* × *Nn* local matrix; **q** = *Nn* × 1 right-hand-side local vector
Inputs: *Nn* = *N*(*n*) and *Nn*_1 = *N*(*n* − 1), defined in eqn. (8.5.1); **x, y** = 3 × 1 arrays containing the coordinates of the three vertices; Eps = material value ε in triangle; **Rho** = *Nn* × 1 vector containing ρ at the interpolation nodes; *Rn*, *Rn*_1 = rotations on triangles of order *n* and *n* − 1; **Gn1** = *Nn*_1 × *Nn* matrix $G_1^n$—eqn. (8.5.12); **Tnn, Tn_1n_1** = *Nn* × *Nn* and

$N(n-1) \times N(n-1)$ metric tensors—eqn. (8.5.24)
Required: **Procedure Diffmats** of Alg. 7.6.1}

```
Begin
  DiffMats(Dx, Dy, A, Nn, Nn_1, x, y, Rn, Rn_1, Gn1,
    Tn_1n_1)
  For i ← 1 To Nn Do
    q[i] ← 0.0
    For j ← 1 To Nn Do
      q[i] ← q[i] + Tnn[i, j]*Rho[j] {eqn. (8.5.27)}
      P[i, j] ← 0.0
      For k ← 1 To Nn_1 Do
        For m ← 1 To Nn_1 Do
          P[i, j] ← P[i, j] + Dx[k, i]*Tn_1n_1[k, m]*Dx[m, j]
             + Dy[k, i]*Tn_1n_1[k, m]*Dy[m, j] {eqn. (8.5.28)}
      P[i, j] ← A*Eps*P[i, j]
    q[i] ← A*q[i]
End
```

nodes so that the row of the matrix equation corresponding to that node will have 18 off-diagonal terms instead of six. Nodes of the second-order mesh such as node 2, which lie at the middle of element edges, will result only in eight off-diagonal terms.

In the plotting algorithms we have employed, straight lines were employed to draw equipotentials. Such plots would have exactly matched the first-order finite element solution, which assumes a linear variation. With higher orders, the plots would

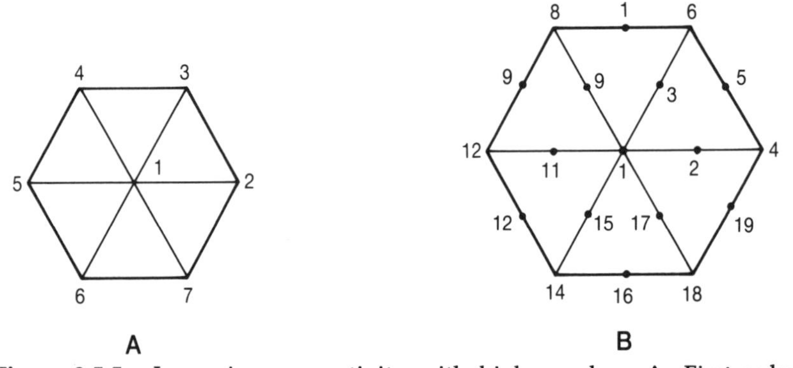


**Figure 8.5.5.** Increasing connectivity with higher orders. **A.** First-order mesh. **B.** Second-order mesh.

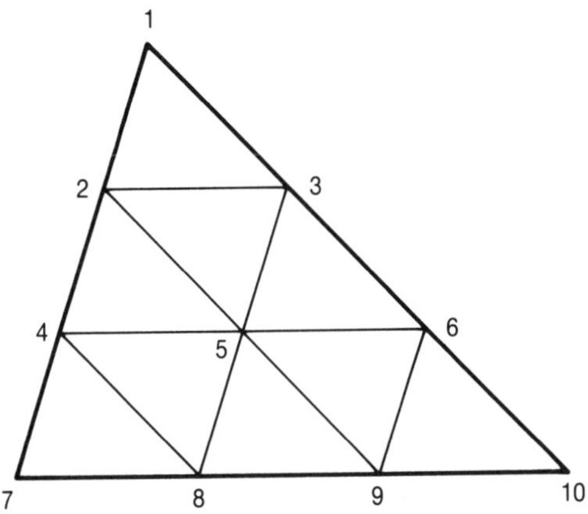


**Figure 8.5.6.** Breaking up a third-order triangle for plotting purposes.

really be smooth curves, except at element edges at which a match of the gradients in two adjacent triangles is not achieved (Irons and Draper 1965). However, since plots are always for eye evaluation of fields rather than any design task, for most purposes, reasonably good plots are achieved by the same algorithm applied to smaller triangles obtained by breaking up higher-order triangles and assuming a first-order variation in each of the smaller triangles. For this purpose, the breaking up of third-order triangles into smaller first-order triangles is shown in Figure 8.5.6 so that the same algorithm for plotting in first-order triangles may be utilized—for plotting in this third-order triangle, we need to call it nine times.

## 8.6 NUMERICAL INTEGRATION: QUADRATURE FORMULAE

Several physical quantities, such as energy and force, are expressed as integrals. Also, when the function being integrated is complex, it is convenient to perform the integration numerically. If we wish to integrate a function $f$ over a region $R$, we divide the region into triangular finite elements and perform the integrations over all the triangles and sum the results.

To this end, let us approximate the function $f$ as a polynomial of order $n$ and evaluate $f$ at the interpolation nodes numbering $N$, as defined in eqn. (8.5.1):

$$f = \tilde{\boldsymbol{\alpha}}^n \underset{\sim}{\mathbf{f}} = \sum_{i=1}^{N} \alpha_i f_1. \tag{8.6.1}$$

Observe that by the way the interpolation functions $\alpha_i$ are defined (1 at node $i$ and 0 at all other nodes), the approximation will match itself at all the interpolation nodes. The more nodes we have, the more accurate this will be. Now performing the integral,

$$\iint f\,dR = \iint \sum \alpha_i f_i = \sum f_i \left\{ \iint \alpha_i\,dR \right\} = A^r \sum w_i f_i, \tag{8.6.2}$$

where $A^r$ is the area of the triangle and $w_i$, called a quadrature weight at node $i$, is defined by

$$w_i = \frac{1}{A^r} \iint \alpha_i\,dR \tag{8.6.3}$$

and needs to be evaluated only once and stored using eqn. (D1), to be recalled at will. For example, for first-order interpolation, as seen from eqn. (7.5.3), $\alpha_1$, $\alpha_2$, and $\alpha_3$ are, respectively, $\zeta_1$, $\zeta_2$, and $\zeta_3$, so that, from eqn. (D1):

$$w_1 = \frac{1}{A^r} \iint \zeta_1\,dR = \frac{2!}{3!} = \tfrac{1}{3}. \tag{8.6.4}$$

Similarly,

$$w_2 = w_3 = \tfrac{1}{3}, \tag{8.6.5}$$

and we obtain

$$\iint f\,dR = A^r[\tfrac{1}{3}f_1 + \tfrac{1}{3}f_2 + \tfrac{1}{3}f_3]. \tag{8.6.6}$$

If we integrate $f = 1$ over a triangle, $f_1$, $f_2$, and $f_3$ are all 1 and we get the area of the triangle. Similarly, if we integrate $x$, the result is the area times the value of $x$ at the centroid, as expected. In both these instances we get exact answers because 1 and $x$ are exactly fitted by a first-order fit (eqn. (8.6.1)) and no gain in accuracy will be realized by using higher orders. However, as more commonly we integrate general functions not necessarily fitted properly by low-order polynomials, the higher the order we choose in eqn. (8.6.1), the better.

The quadrature weights for various orders have been already worked out and tabulated by Silvester (1970). We may conveniently use his results.

An example of using the integral formulas is in error computation. If we knew the exact solution $\phi$ and we wished to find the error in the finite element solution $\phi^f$, let us define the error function as $e = \phi - \phi^f$, computable at every node, and the total percentage error

$$\text{Error} = 100 \times \sqrt{\frac{\iint e^2 \, dR}{\iint \phi^2 \, dR}}. \tag{8.6.7}$$

These terms may be easily evaluated. For example,

$$\iint \phi^2 \, dR = \sum_{\Delta} \iint [(\tilde{\alpha}^n \underset{\sim}{\phi})(\tilde{\alpha}^n \underset{\sim}{\phi}) \, dR = \sum_{\Delta} \underset{\sim}{\phi}^t \left[ \iint \tilde{\alpha}^{nt} \tilde{\alpha}^n \, dR \right] \underset{\sim}{\phi} \, dR \tag{8.6.8}$$

$$= \sum_{\Delta} A^r \underset{\sim}{\phi}^r \mathbf{T}^{n,n} \underset{\sim}{\phi}$$

where the integral over the domain has been split over the elements and eqn. (8.5.24) has been used to express it in terms of the precomputed universal **T** matrix. Thus, the quantity $\phi$ must be evaluated at all the interpolation nodes and eqn. (8.6.8) evaluated. For first order, for example, $\mathbf{T}^{1,1}$ is defined in eqn. (7.6.12), and if $\phi$ is $\phi_1$, $\phi_2$, and $\phi_3$ at the three vertices, over the element of area $A^r$, eqn. (8.6.8) will work out to $A^r[\phi_1^2 + \phi_2^2 + \phi_3^2 + \phi_1\phi_2 + \phi_2\phi_3 + \phi_3\phi_1]/12$. An application of this follows in the next section.

## 8.7 FINITE ELEMENTS AND FINITE DIFFERENCES

It may be shown in a rigidly mathematical sense that the finite difference method is a part of the generalized finite element method in the sense of the Galerkin approach (Zienkiewicz 1977). The same may be seen, albeit with less generality, from a simple square finite element mesh (Zienkiewicz and Cheung 1965). Consider the finite element mesh of Figure 8.7.1. For the typical right-angled element 302 (Figure 8.7.1a) of base $h$ and height $h$, say, it may be shown, corresponding to eqns. (7.6.3) and (7.6.4), that the differentiation matrices $\tilde{\mathbf{b}}$ and $\tilde{\mathbf{c}}$ are $h^{-1}[-1 \ 1 \ 0]$ and $h^{-1}[0 \ -1 \ 1]$. The local

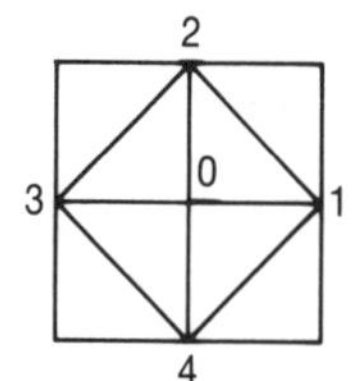


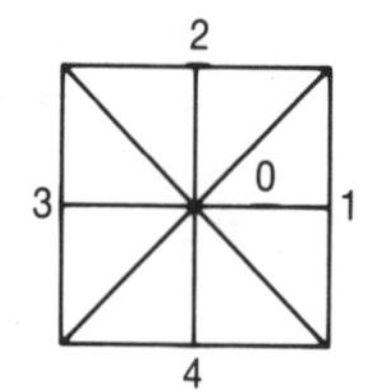


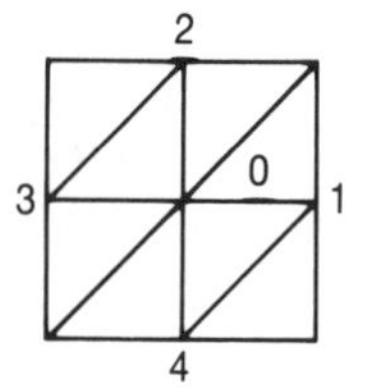


**Figure 8.7.1.** Different finite element discretizations of a finite difference mesh.

matrices corresponding to eqn. (7.6.17), therefore, are:

$$\mathbf{P} = \tfrac{1}{2}\varepsilon \begin{bmatrix} 1 & -1 & 0 \\ -1 & 2 & -1 \\ 0 & -1 & 1 \end{bmatrix} \tag{8.7.1}$$

and

$$\underset{\sim}{q} = \tfrac{1}{3}\tfrac{1}{2}h^2\rho_0[1 \quad 1 \quad 1]^t, \tag{8.7.2}$$

where a constant $\rho$ has been assumed.

Considering Figure 8.7.1A now and adding the contributions of the four triangles 302, 304, 401, and 102 to the row corresponding to the node 0, we will get the equation:

$$\varepsilon[\phi_1 + \phi_2 + \phi_3 + \phi_4 - 4\phi_0] = \frac{2h^2\rho_0}{3}. \tag{8.7.3}$$

In this, all the contributions to the row corresponding to node 0 come from the second row of the matrix **P**. Similarly, for the mesh of Figure 8.7.1B, the contributions to the global matrix come from the eight triangles 015, 025, 026, 036, 037, 047, 048, and 018, all suitably numbered to have the same local matrix **P**. The contributions will, therefore, come from row 1 of **P**. The equation will be

$$\varepsilon[\phi_1 + \phi_2 + \phi_3 + \phi_4 - 4\phi_0] = \frac{4h^2\rho_0}{3}. \tag{8.7.4}$$

Similarly, for the mesh of Figure 8.7.1C, the nodal equation will be

$$\varepsilon[\phi_1 + \phi_2 + \phi_3 + \phi_4 - 4\phi_0] = h^2\rho_0, \tag{8.7.5}$$

which is the standard finite difference approximation. Thus, we see that a particular finite element discretization gives us the finite difference approximation. We further see that, depending on the discretization, the answers will be different. For instance, two similar nodes give different equations differing by terms of order

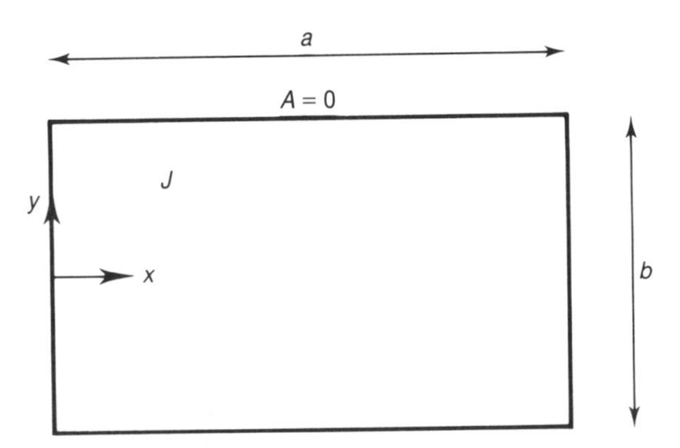


**Figure 8.7.2.** A stranded rectangular AC conductor surrounded by copper.

$h^2$, depending on the discretization in their locality. We have already seen, however, that the finite difference approximations are at least of accuracy $h^2$.

To confirm these observations, consider the boundary value problem of Figure 8.7.2, where a rectangular, stranded AC conductor is surrounded by a copper sheet so that the boundary of the conductor is a flux line and the current density is a constant inside. The exact solution of the potential is (Timoshenko and Woinowsky-Krieger 1959):

$$A = \frac{4a^2J}{\pi^3} \sum_{m=1,3,5}^{\alpha} \frac{\sin m\pi x/a}{m^3} \left[1 - \frac{\cosh m\pi y/a}{\cosh m\pi b/2a}\right]. \tag{8.6.9}$$

This problem was divided into a rectangular mesh as shown in Figure 8.7.3 and solved by five-point and nine-point finite differences (Van de Vooren and Vliegenthart 1967). The nine-point

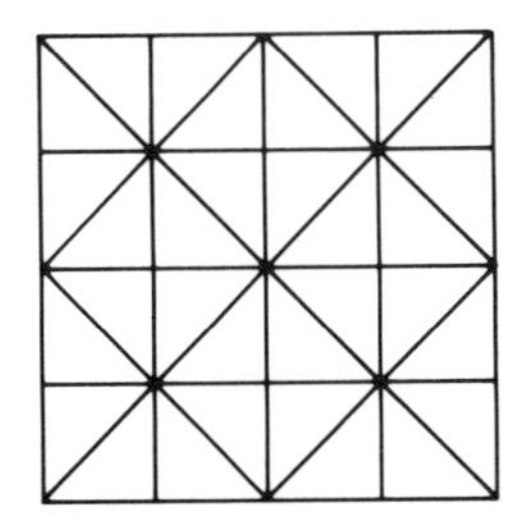

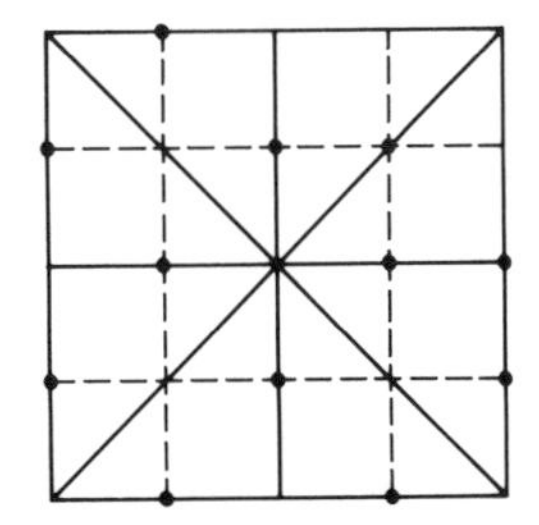

**Figure 8.7.3.** Common mesh for five- and nine-point finite difference and first- and second-order finite elements.

finite difference formula is a more accurate finite difference approximation that relates the potential at a node 0 to the eight that surround it:

$$4[\phi_1 + \phi_2 + \phi_3 + \phi_4] + \phi_5 + \phi_6 + \phi_7 + \phi_8 - 20\phi_0 = 6h^2\varepsilon^{-1}\rho_0. \tag{8.6.10}$$

Thereafter, cutting each rectangle into two triangles, the problem was solved by first- and second-order finite elements. The cutting of the rectangles was such as to have nodes of the type of Figures 8.7.1A and 8.7.1B alternating in the mesh. Indeed, there is no point in cutting them to yield a mesh of nodes, consistently of the type depicted in Figure 8.7.1 since they would give the same answers as the finite difference approximations. The errors were computed using eqn. (8.6.7) and are presented in Figure 8.7.4. This confirms the efficacy of higher-order methods and that they ought to be preferred to lower-order treatment. It further confirms that the term $\nabla^2$ in a rectangular region is better modeled by the finite difference approximation. It is added that many will quarrel with this last statement. Indeed, the superiority of the finite element method comes with our ability to (1) have finer meshes in regions of importance; (2) use higher orders when necessary; (3) model complex geometries; and (4) model the source term more accurately.

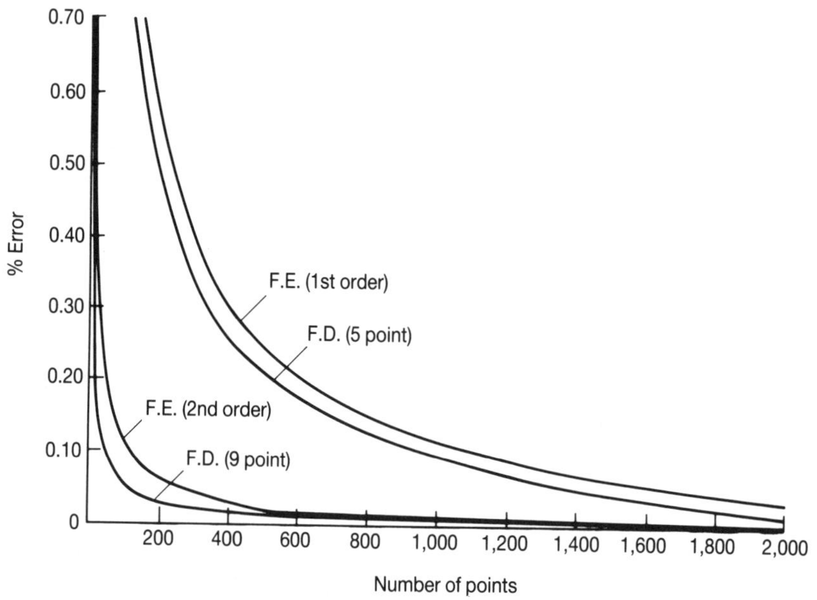


**Figure 8.7.4.** Errors by the four methods.

## 8.8 MIXED FINITE ELEMENTS

Two kinds of mixing of finite elements are possible. First, we may mix elements of different order, as shown in Figure 8.8.1A. These are referred to as *nonconforming* elements, which are useful although theoretically difficult to justify. The term *nonconforming* is employed because, when, say, a second-order element lies on one side of a line while two first-order elements lie on the other side, we have a contradiction. Each postulates a different kind of variation along the common line. Experience, however, is that such mixing gives good results. For instance, when we have a large homogeneous region abutting a small detailed region, it is convenient, in terms of both data preparation and accuracy, to have a few large high-order elements covering the expanse and several small first-order elements modeling the smaller regions involving detailed geometry (Csendes 1980). Another use of nonconforming elements is as in Figure 8.8.1B, where we have two first-order triangles sharing the edge of a larger first-order triangle. Thus, in the larger triangle the potential must vary linearly down the edge, whereas on the other side of the edge, although the variation is linear, it is in two straight lines rather than one. That is, an additional node is present giving the trial function a greater ability to fit the governing equation. The use of such element groups is natural in modeling open-boundary problems where, as we go out to far regions of low field, we are not too particular about accuracy, which suffers as a result in this case, and we may wish to use elements of larger size.

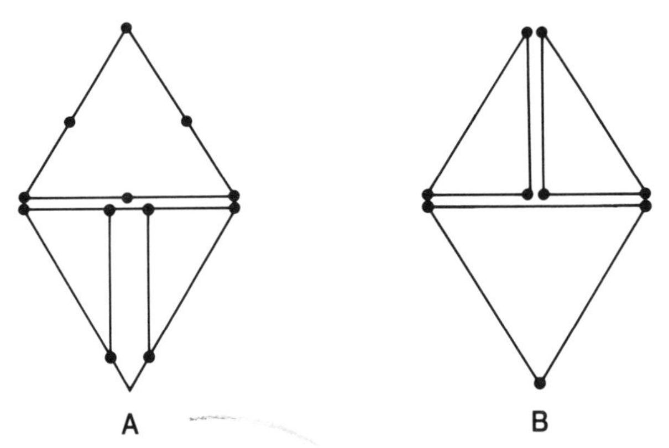


**Figure 8.8.1.** Nonconforming finite elements. **A.** Higher- and lower-order triangles without skipping nodes. **B.** First-order triangles skipping nodes.

The second kind of mixing involves the simultaneous use of elements of different dimensions. No contradiction (or nonconformity) need be inherent to this kind of mixing. Often this is required when we have thin objects in the presence of wider (or thicker) ones. Examples are in modeling a thin bolt of lightning, where, in the axisymmetric system, the lightning channel is conveniently modeled by one-dimensional line elements and the surrounding air by two-dimensional triangular elements (Hoole and Hoole 1986c, 1987). Zienkiewicz, Bahrani, and Arlett (1967), in studying the flow from a thin electrode in an electrolytic tank, model the electrode using one-dimensional elements and the electrolyte in three dimensions.

As another example, in studying the earthing arrangement of Figure 3.3.5, let the earthing rod be so thin that any variation in potential can only be down its length. We will show that using triangular elements is the same as using line elements under these conditions. Let us first model the rod as two-dimensional triangular elements as shown in Figure 8.8.2. The element 123, of thickness $t$ and height $h$, gives us, corresponding to eqns. (7.6.3) and (7.6.4), the differentiation matrices $\tilde{\mathbf{b}}$ and $\tilde{\mathbf{c}}$: $t^{-1}[-1\ 1\ 0]$ and $h^{-1}[0\ -1\ 1]$. The local matrices corresponding to eqn. (7.6.17) therefore are

$$\mathbf{P} = \tfrac{1}{2}\sigma ht \begin{bmatrix} t^{-2} & -t^{-2} & 0 \\ -t^{-2} & t^{-2} + h^{-2} & -h^{-2} \\ 0 & -h^{-2} & h^{-2} \end{bmatrix} \tag{8.8.1}$$

and

$$\underset{\sim}{q} = \tfrac{1}{3}\tfrac{1}{2}ht\rho_0[1 \quad 1 \quad 1]^t, \tag{8.8.2}$$

where $\sigma$ is the conductivity. The corresponding local finite element

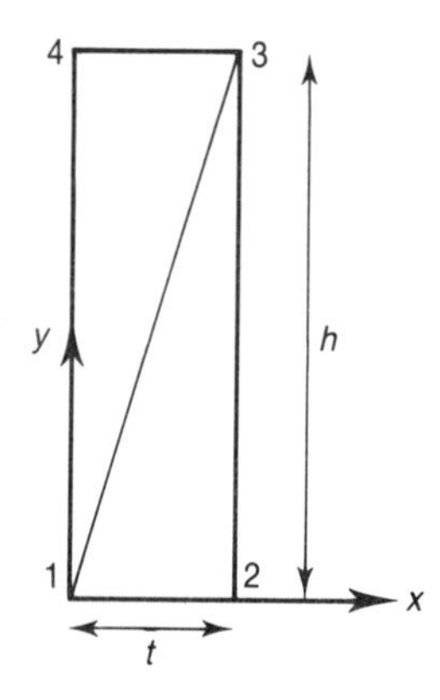


**Figure 8.8.2.** Condensation—thin triangles into line elements.

vector $\Phi$ is $[\phi_1 \quad \phi_2 \quad \phi_3]^t$. The energy of the system is given by $\frac{1}{2}\Phi^t P \Phi - \Phi^t q$. If we now set that no variations shall occur along the thickness, then $\phi_1 = \phi_2$. Expressing $\Phi$ as

$$\Phi = \begin{bmatrix} \phi_1 \\ \phi_2 \\ \phi_3 \end{bmatrix} = \begin{bmatrix} 1 & 0 \\ 1 & 0 \\ 0 & 1 \end{bmatrix} \begin{bmatrix} \phi_1 \\ \phi_3 \end{bmatrix}, \tag{8.8.3}$$

in terms of the new one-dimensional finite element vector $\Phi^* = [\phi_1 \quad \phi_3]^t$, and substituting in the energy, we obtain the energy in the form $\frac{1}{2}\Phi^{*t}P^*\Phi^* - \Phi^{*t}q^*$. The new local matrices must therefore be, upon evaluation,

$$P^* = \tfrac{1}{2}\sigma h^{-1} t \begin{bmatrix} 1 & -1 \\ -1 & 1 \end{bmatrix}, \tag{8.8.4}$$

and

$$q^* = \tfrac{1}{3}\tfrac{1}{2} h t \rho_0 [2 \quad 1]^t. \tag{8.8.5}$$

Similarly, for the element 341 of the same shape, we will have the same local matrices, but a different local finite element vector $[\phi_3 \quad \phi_1]$. Adding the contributions of the two elements together by the rules of eqn. (7.6.23) for the long (or thin) rectangular region 1234:

$$P = \tfrac{1}{2}\sigma h^{-1} t \begin{bmatrix} 1+1 & -1-1 \\ -1-1 & 1+1 \end{bmatrix} = \sigma h^{-1} t \begin{bmatrix} 1 & -1 \\ -1 & 1 \end{bmatrix}, \tag{8.8.6}$$

$$q^* = \tfrac{1}{3}\tfrac{1}{2} h t \rho_0 [2+1 \quad 1+2]^t = \tfrac{1}{2} h t \rho_0 [1 \quad 1], \tag{8.8.7}$$

which is about what we will get by assuming a linear variation down the rod, from node 1 to node 3. This may be verified by going through steps similar to those of section 7.3.1. Observe the presence of the additional factor $t$, which relates the dimensions to those of the exterior sections.

## 8.9 SPARSITY PATTERN COMPUTATION

By now, particularly after the previous examples of section 8.4 where we solved for various matrix sizes, it must have been impressed upon the reader that full storage cannot be used indefinitely with impunity. In the solution of practical problems, large matrices are encountered and the sparsity and symmetry of the equations must be exploited, as detailed in section 6.4. We saw that the coefficients should be ideally stored in a one-dimensional array and that they may be retrieved using the procedure Locate of Alg. 6.4.2, employing the data structures **Diag** and **Col**. Alternatively, with profile storage, we will replace the vector **Col** by **FC**. In this section we will deal exclusively with sparse symmetric storage, because the principles extend trivially to profile storage.

To effect sparse storage we must be in possession of the arrays **Diag** and **Col**. While these may be manually created by inspection of the mesh, such an approach is neither economic nor feasible for large problems. Much is to be gained by using the computer to create them automatically. In finite element analysis, we may easily devise algorithms to create the relevant arrays by inspection of the mesh, as shown in Algs. 8.9.1 and 8.9.2. In these algorithms we initially assume that the matrix is diagonal. So, for an $n \times n$ matrix we take **Diag**[$i$] as $i$ and **Col**[$i$] as $i$, since only the diagonal is occupied. NElems is $n$, since there are $n$ coefficients, one on each row. It is for us now to allocate storage for all off-diagonal elements,

**Algorithm 8.9.1. Procedures Required in Creating Sparse Data Structures**

**Procedure Modify Diag(Diag,** RowN, NUnk)
{Function: Modifies **Diag**, when an element is added to row number RowN}
**Begin**
  **For** $i \leftarrow$ RowN **To** NUnk **Do**
    **Diag**[$i$] $\leftarrow$ **Diag**[$i$] + 1
**End**

**Procedure ModifyCol(Diag,** ColN, $ij$, NElems)
{Function: Modifies **Col** when an element is added to location RowN, ColN after the item $ij$ in the one-dimensionally stored matrix. Observe that when Locate returns Found as False, $ij$ is the location of the previous item before the missing one}
**Begin**
  **For** $i \leftarrow ij + 1$ **To** NElems **Do**
    $j \leftarrow ij + 1 +$ NElems $- i$ {$j$ goes from NElems down to $ij + 1$}
    **Col**[$j + 1$] $\leftarrow$ **Col**[$j$]
  **Col**[$ij + 1$] $\leftarrow$ ColN
  NElems $\leftarrow$ NElems + 1
**End**

**Algorithm 8.9.2. Creating Sparsity Data Structures**

```
Procedure Sparsity(Diag, Col, NElems, TriaNum, NUnk, N)
{Function: Creates the sparsity data Diag, Col, NElems for a
finite element mesh.
Input: Number of triangles TriaNum; Number of Unknowns
NUnk, and the number of interpolation nodes N in the
nth-order triangle.
Required: Procedures Locate of Alg. 6.4.2 and ModifyDiag and
ModifyCol above}
Begin
  For i ← 1 To NUnk Do
    Diag[i] ← i
    Col[i] ← i
  NElems ← NUnk
  For i ← 1 To TriaNum Do
    Get v[1], v[2], v[3], . . . , v[N], the N vertices of nth-order
    triangle i
    For j ← 1 To N − 1 Do
      For k ← j + 1 To N Do
        If (v[j] ≤ NUnk) And (v[k] ≤ NUnk)
          Then
            If v[j] ≥ v[k]
              Then
                RowN ← v[j]
                ColN ← v[k]
              Else
                RowN ← v[k]
                ColN ← v[j]
            Locate(ij, Found, Diag, Col, NElems, RowN, ColN)
            If Not Found
              Then
                ModifyDiag(Diag, RowN, NUnk)
                ModifyCol(Diag, ColN, ij, NElems)
End
```

using the fact that if two unknown nodes $p$ and $q$ are connected in the mesh (that is, if they are vertices in the same triangle), then location $(p, q)$ will be occupied in the matrix so that storage must be assigned to it.

To this end, we read each triangle in Alg. 8.9.2 and, for each unknown vertex, find out using Locate (after ensuring that we are indeed in the lower half of the matrix) if space has been assigned for the column ColN and row RowN given by the vertices (if they are both unknown). If no space has been assigned, as indicated by Found from Locate being returned False, a study of Alg. 6.4.2 will show that for this case the value *ij* returned by Locate contains the item after which location (RowN, ColN) should be, but is missing. If we assign space for (RowN, ColN), row RowN, and all rows after, it will terminate one item later, so that **Diag** must be incremented by 1 as in the procedure ModifyDia of Alg. 8.9.1. Similarly, **Col** should contain an extra item after item *ij* returned by Locate. This item reflects column position ColN. Thus, all items after position *ij* are shifted down an item and the vacancy created at position $ij + 1$ is assigned the value ColN in the procedure ModifyCol of Alg. 8.9.1. Once the sparsity data are thus automatically created, it will be found that much larger problems may be solved and solved quickly.

# 9 POSTPROCESSING

## 9.1 WHY POSTPROCESSING?

So far in this book we have learned the fundamental esoterics of field computation. In analyzing a device on the computer, we now know how to find the electromagnetic fields that flow in the device. What the computer returns, whether by the finite element, finite difference, or boundary integral method, is a clutter of numbers that do not immediately and easily speak their meaning to us. The purpose of postprocessing, therefore, is to use the computer to interpret the results of our computer-aided analysis of the electromagnetic fields. Indeed, the whole purpose of our analysis is to obtain some physically meaningful quantity, such as energy, flux density, or force. How to interpret our numerical results easily and accomplish the computation of such quantities from them is the stuff of this chapter.

## 9.2 COMPUTER-DRAWN EQUIPOTENTIAL PLOTS

### 9.2.1 Why Equipotential Plots?

We have already seen that equipotentials provide a quick means of getting an idea of the solution of the electromagnetic fields. As discussed in section 1.2.4, if we define a local coordinate system with the $x$ axis along the tangent to an equipotential as shown in Figure 9.2.1, the term $\partial\phi/\partial x$ will be zero, since $\phi$ is the same along the contour. The normal derivative $\partial\phi/\partial y$ will be $\delta\phi/\delta n$, the

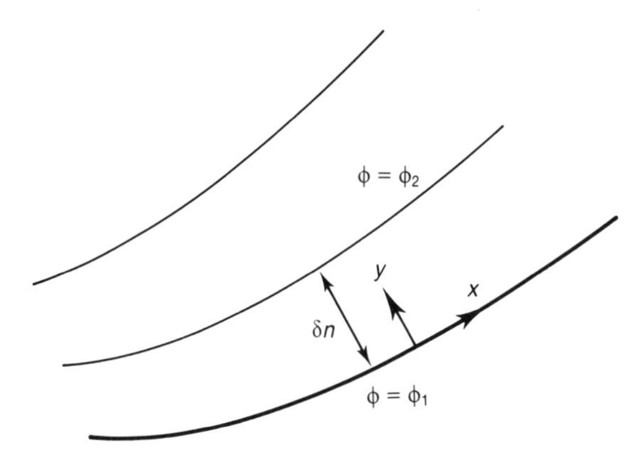


**Figure 9.2.1.** Interpreting equipotentials.

potential difference between two adjacent contours $\delta\phi$ divided by the distance between them $\delta n$. In general, if the equipotential lines are drawn in equal steps of potential so that $\delta\phi$ is the same everywhere, where the lines are crowded is where the fields are higher, since $\delta n$ decreases and thereby increases the gradient. Moreover, according to eqns. (1.2.33) and (1.2.34), electric fields, with no $x$ component, will flow normally to the equipotentials, and magnetic fields, with no $y$ component, will flow along the equipotentials.

Not only do equipotentials tell us the solution, but they are also means for judging whether or not our programming has been error free—for surely with a little experience we should be able to judge from the equipotentials whether the answers look physically reasonable or not. For this latter purpose of judging the correctness of solution the following three general rules apply:

1. The contours should get crowded in parts of higher permeability (which allow magnetic flux to permeate more) and sparser in regions of higher permittivity (since in electric field problems the permittivity is analogous to inverse permeability, as may be observed from eqns. (1.2.22) and (1.2.32)).
2. Where we expect higher fields, the lines should be crowded. Therefore, as we move away from current or charge-carrying parts, the density of contours should drop, corresponding to the drop in electric and magnetic fluxes.
3. When magnetic flux enters a more permeable region, it should bend so as to hug the surface more inside the permeable region. Similarly, electric fields should move towards the normal to the

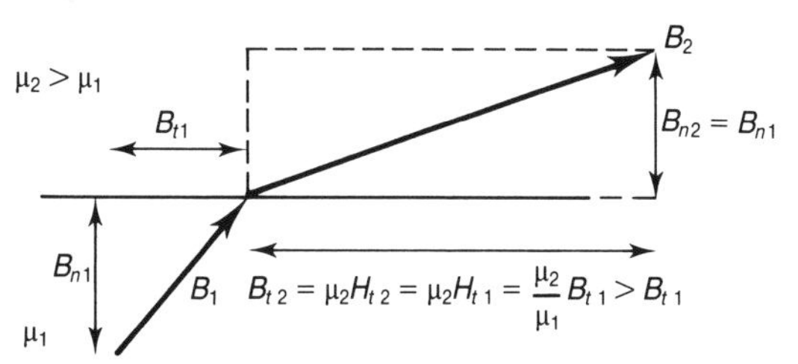


**Figure 9.2.2.** Bending of flux at a material inhomogeneity.

surface upon entering a high-permittivity region. This may be established from the continuity conditions, eqns. (1.2.15)–(1.2.18), and is illustrated in Figure 9.2.2.

## 9.2.2 World and Screen Coordinates

Before we attempt plotting equipotentials on a graphics device such as a computer screen, graphics printer, or plotter, we need to determine the scaling of the drawing, as the device being analyzed will not have its coordinates directly translating onto the coordinates of the graphics output device. As an example, a Hewlett–Packard graphics screen has coordinates from $(-1, -1)$ to $(1, 1)$, whereas a device measuring, say, 10 ft by 50 ft, with respect to some world coordinate system, may lie in a rectangle with $(0, 0)$ at the lower left corner and $(10, 50)$ at the upper right corner. Thus, without some scaling we shall not be able to draw equipotentials in the region $(10, 50)$ of the device, which is outside the graphics screen. The screen (or plotter or printer) coordinates, say, (*x*_*L*_*S*, *y*_*L*_*S*) at the lower left corner of the screen and (*x*_*U*_*S*, *y*_*U*_*S*) at the upper right corner, may again be obtained by referring to the manual of the graphics output device. The corresponding limits of world coordinates of the device (*x*_*L*_*W*, *y*_*L*_*W*) and (*x*_*U*_*W*, *y*_*U*_*W*) of course will necessarily follow from the dimensions of the device. Therefore, in translating from the world coordinates $(x_w, y_w)$ to the screen coordinates $(x_s, y_s)$, we need to use a linear mapping such as

$$x_s = m_x x_w + c_x, \qquad (9.2.1a)$$

$$y_s = m_y y_w + c_y, \qquad (9.2.2a)$$

where $m_x$ and $m_y$ are the slopes of the mapping and $c_x$ and $c_y$ are the offsets. Substituting the mappings $x\_L\_W \rightarrow x\_L\_S$ and $x\_U\_W \rightarrow x\_U\_S$, we have

$$x\_L\_S = m_x x\_L\_W + c_x, \tag{9.2.3}$$

$$x\_U\_S = m_x x\_U\_W + c_x, \tag{9.2.4}$$

and solving for $m_x$ and $c_x$ and similarly substituting and solving for $m_y$ and $c_y$, there obtains:

$$x_s = \frac{x\_U\_S - x\_L\_S}{x\_U\_W - x\_L\_W} x_w + \frac{x\_L\_S^* x\_U\_W - x\_U\_S^* x\_L\_W}{x\_U\_W - x\_L\_W}, \tag{9.2.1b}$$

$$y_s = \frac{y\_U\_S - y\_L\_S}{y\_U\_W - y\_L\_W} y_w + \frac{y\_L\_S^* y\_U\_W - y\_U\_S^* y\_L\_W}{y\_U\_W - y\_L\_W}. \tag{9.2.1b}$$

This translation from world to screen coordinates may be made into a procedure as depicted in Alg. 9.2.1. While the world coordinate

**Algorithm 9.2.1. Mapping World to Screen Coordinates**

**Procedure World_To_Screen**($x_s, y_s, x_w, y_w,$
$x\_L\_W, y\_L\_W, x\_U\_W, y\_U\_W,$
$x\_L\_S, y\_L\_S, x\_U\_S, y\_U\_S$)

{Inputs: $x_w, y_w$ = world coordinates to be mapped; $x\_L\_W$, $y\_L\_W$, $x\_U\_W$, $y\_U\_W$ = limits of world coordinates; $x\_L\_S$, $y\_L\_S$, $x\_U\_S$, $y\_U\_S$ = limits of device coordinates
Outputs: $x_s, y_s$ = Screen coordinates corresponding to $x_w, y_w$}

**Begin**

$$x_s \leftarrow (x\_U\_S - x\_L\_S)^* x_w + x\_L\_S^* x\_U\_W - x\_U\_S^* x\_L\_W$$

$$x_s \leftarrow \frac{x_s}{(x\_U\_W - x\_L\_W)}$$

$$y_s \leftarrow (y\_U\_S - y\_L\_S)^* y_w + y\_L\_S^* y\_U\_W - y\_U\_S^* y\_L\_W$$

$$y_s \leftarrow \frac{y_s}{(y\_U\_W - y\_L\_W)}$$

**End**

system may always be represented by real numbers, in some computer systems, the screen coordinate system is sometimes described by integers—in such an event, the algorithm may be slightly modified so that using the function for rounding a real number to an integer (customarily called Round) we round the computed values of $x_s$ and $y_s$ before putting them into $x_s$ and $y_s$.

Most mapping between the screen and the world is accomplished through the primitive graphics commands Move$(x, y)$, which moves the graphics cursor to the screen coordinates $(x, y)$, and Line$(x, y)$, which draws a line from wherever the cursor is to the point $(x, y)$ and leaves the cursor there. That is, the line from $(x_1, y_1)$ to $(x_2, y_2)$ is accomplished by the successive application of the two commands Move$(x_1, y_1)$ and Line$(x_2, y_2)$; now since the cursor is at $(x_2, y_2)$, the command Line$(x_3, y_3)$ will draw a line from $(x_2, y_2)$ to $(x_3, y_3)$. These commands Move and Line are invariably available in all graphics systems, but perhaps by different names, and the corresponding names may be found in the manuals. For our purposes, therefore, it is convenient to redefine these commands through Alg. 9.2.2. OurMove and, similarly, OurLine, so that we do not have to do any scaling every time we wish to draw lines. Whenever the coordinates are in world coordinates, these procedures scale them to the screen system and then perform the Move and Line operations in the screen coordinates.

**Algorithm 9.2.2. Scaling Move Command to Map World onto the Screen**

```
Procedure OurMove(x, y)
{Input: x, y = world coordinates
Function: moves graphics cursor to corresponding point on the
screen
Predefined constants: x_L_W, y_L_W, x_U_W, y_U_
W, x_L_S, y_L_S, x_U_S, y_U_S = defined in Alg.
9.2.1
Uses: System Function Move(x, y); World_To_Screen}
Begin
  World_To_Screen(x, y, x, y,
      x_L_W, y_L_W, x_U_W, y_U_W, x_L_S,
      y_L_S, x_U_S, y_U_S)
  Move(x, y)
End
```

### 9.2.3 An Algorithm for Plotting Equipotentials

We have already seen that in plotting solutions from a finite difference or boundary integral simulation (section 3.4), we may conveniently plot equipotentials rectangle by rectangle, and from a finite element solution (section 7.8) we need to plot equipotentials triangle by triangle. While a triangle is not the same as a rectangle, the algorithms are in essence the same, for we plot from one line to another, using straight-line movements. The difference lies in our having to draw between six pairs of lines in a rectangle (from sides $1 \to 2$, $1 \to 3$, $1 \to 4$; $2 \to 3$, $2 \to 4$; and $3 \to 4$) and three pairs of lines in a triangle (from sides $1 \to 2$, $1 \to 3$; and $2 \to 3$). Therefore we shall generalize our algorithm for drawing equipotentials in an $n$-sided polygon of given nodal coordinates and precomputed nodal potentials. Thus, by using $n = 3$, the algorithm may be used for plotting a finite element solution from a triangular mesh, and, with $n = 4$, for a rectangular finite difference mesh.

Let the algorithms for drawing these lines on the computer be considered. Any graphics system must be initialized before use, i.e., have its screen wiped clean and made ready to accept commands. The initializing commands naturally vary from system to system and must be looked up by the programmer from the manuals.

As evidenced by the examples of sections 3.4 and 7.8, the plotting procedure involves taking edge 1 of the polygon and plotting all the equipotentials that cross the edge and lead to edges $2, 3, \ldots, n$; subsequently, we take edge 2 and plot all lines that go from there to edges $3, 4, \ldots, n$. And so on we proceed, until we plot the lines from edge $n - 1$ to $n$. Edge $i$, in general, goes from node $i$ to node $i + 1$, except for the $n$th edge, which goes from node $n$ to node 1. This may be defined as going from node $i$ to node $i$ Mod $n + 1$, where the mathematical operation $i$ Mod $n$ gives the remainder after division by $n$, and is under the same nomenclature in PASCAL and is denoted by Mod$(i, n)$ in FORTRAN. When dealing with a finite difference solution, the triangle may be read straight off the data files since triangles are directly stored. However, while dealing with finite differences where the data do not include rectangles (unless we choose to do so at greater effort in data preparation), we may identify a rectangle by taking a node, finding the node to its east, the node to the north of the latter node, and the node to the west of the latter.

A possible scheme for the algorithm for plotting in a polygon is presented in Alg. 9.2.3. The main Plot_Polygon algorithm, Alg.

**Algorithm 9.2.3A. Breaking up Plotting inside a Polygon into Plotting from Edge to Edge**

```
Procedure PlotPolygon(n, Phi, x, y, Lwst, Hghst, Step)
{Inputs: n = number of sides of polygon; Phi = n × 1 array
containing the n vertex potentials; x, y = n × 1 arrays contain-
ing the vertex coordinates; Lwst, Hghst = lowest and highest
potential contours for plotting; Step = potential difference
between two adjacent contours}
Begin
  For FirstEdge ← 1 To n − 1 Do
    xFE[1] ← x[FirstEdge]
    yFE[1] ← y[FirstEdge]
    PhiFE[1] ← Phi[FirstEdge]
    xFE[2] ← x[FirstEdge + 1]
    yFE[2] ← y[FirstEdge + 1]
    PhiFE[2] ← Phi[FirstEdge + 1]
    For SecondEdge := FirstEdge + 1 To n Do
      xSE[1] ← x[SecondEdge]
      ySE[1] ← y[SecondEdge]
      PhiSE[1] ← Phi[SecondEdge]
      xSE[2] ← x[SecondEdge Mod n + 1]
      ySE[2] ← y[SecondEdge Mod n + 1]
      PhiSE[2] ← Phi[(SecondEdge) Mod n + 1]
      Plot_Edge_To_Edge(xFE, yFE, xSE, ySE, PhiFE,
        PhiSE, Lwst, Hghst, Step)
End
```

**Algorithm 9.2.3B. Plotting Contours from Line to Line**

```
Procedure Plot_Edge_To_Edge(xFE, yFe, xSE, ySE, PhiFE,
PhiSE, Lwst, Hghst, Step)
{Function: Plots equipotentials from one edge (First Edge) to
another (Second Edge)
Inputs: PhiFE = 2 × 1 array with first edge vertex potentials;
xFE, yFE = 2 × 1 arrays with first edge vertex coordinates;
PhiSE = 2 × 1 array with second edge vertex potentials; xSE,
ySE = 2 × 1 arrays with second edge vertex coordinates;
Lwst, Hghst = lowest and highest potential contours to plot;
Step = potential difference between two adjacent contours
Uses: Procedures Inc_Order, Interpolate, OurMove, OurLine}
Begin
  Inc_Order(PhiFE, xFE, yFE) {Make PhiFE[1] ≤ PhiFE[2]}
  Inc_Order(PhiSE, xSE, ySE) {Make PhiSE[1] ≤ PhiSE[2]}
```

```
    Pot ← Lwst;
    While Pot < PhiFE[1] Do
      Pot ← Pot + Step;
    While Pot < Hghst Do
        If (Pot ≥ PhiSE[1]) And (Pot ≤ PhiSE[2])
          Then
            Interpolate(xx, yy, xFE, yFE, PhiFE, Pot)
            OurMove(xx, yy)
            Interpolate(xx, yy, xSE, ySE, PhiSE, Pot)
            OurLine(xx, yy)
            Pot ← Pot + Step
  End
```

**Algorithm 9.2.3C. Renumbering an Edge in Order of Increasing Potential**

```
  Procedure Inc_Order(Phi, x, y)
  {Inputs: Phi = 2 × 1 array containing potentials at the two
  nodes of an edge; x, y = 2 × 1 arrays containing the coor-
  dinates of the two nodes
  Function: If Phi[2] is less than Phi[1], the two nodes are
  renumbered such that Phi[1] ≤ Phi[2]
  Uses: Procedure Swap}
  Begin
    If Phi[2] < Phi[1]
      Then
        Swap(Phi[1], Phi[2])
        Swap(x[1], x[2])
        Swap(y[1], y[2])
  End
```

**Algorithm 9.2.3D. Swapping Two Numbers**

```
  Procedure Swap(x, y)
  {Function: Swaps the values of x and y}
  Begin
    Temp ← x
    x ← y
    y ← Temp
  End
```

**Algorithm 9.2.3E. Locating a Contour on an Edge by Interpolation**

**Procedure Interpolate**($xx, yy$, **xE**, **yE**, **PhiE**, Pot)
{Inputs: **xE**, **yE**: $2 \times 1$ Arrays with the coordinates of the two nodes of an edge; **PhiE**: $2 \times 1$ Array with the nodal potentials of an edge; Pot: Potential level of contour crossing edge;
Output: $xx, yy$: Coordinate location of potential contour Pot crossing edge}
**Begin**
    $xx$ ← **xE**[1] + (**xE**[2] − **xE**[1])*(Pot − **PhiE**[1])/(**PhiE**[2] − **PhiE**[1])
    $yy$ ← **yE**[1] + (**yE**[2] − **yE**[1])*(Pot − **PhiE**[1])/(**PhiE**[2] − **PhiE**[1])
**End**

9.2.3A, requires as input the coordinates and precomputed potentials of the $n$ nodes, the contour levels to be plotted given by the lowest and highest contours, and the step from contour to contour (such as, for example, from 0 to 100 in steps of two). The algorithm goes through the polygon and, after identifying two sides, First Edge and Second Edge, through which potential contours are to be plotted, passes these on to Plot_Edge_To_Edge (Alg. 9.2.3B), which draws the various contours that pass through both. This algorithm requires the potential of the first node of each edge to be at least lower than the potential of the second. This is ensured by the procedure Inc_Order (Alg. 9.2.3C), which checks for this and, if it does not hold, calls the procedure Swap (Alg. 9.2.3D) to rearrange the ordering of the nodes of the edge. Algorithm Plot_Edge_To_Edge thereafter identifies the first contour level Pot that passes through the first edge by setting Pot to the level of the lowest contour Lwst and then incrementing Pot by Step until Pot is higher than the potential of the first node. Thereafter, if the contour level Pot lies between the potential levels of the Second Edge, it means that the contour runs from the First Edge to the Second Edge, so that it must be drawn. To draw the contour of potential level Pot, through linear interpolation, the location where it intersects First Edge is identified and the cursor is moved there using the command OurMove (Alg. 9.2.2), and then, after identifying where it leads to on the Second Edge, using OurLine, a straight line is drawn. Having dealt with the contour level Pot leading from First Edge to Second Edge, we increment Pot by Step for the next contour, and, if the new level is also on First Edge, then we deal with it as before.

In using these procedures in finite element analysis, therefore, we would sequentially read every triangle defined by the three vertices, get the vertex potentials and coordinates, and call PlotPolygon with $n = 3$. Although we are plotting polygon by polygon, the overall continuity of contours will be maintained because the interpolation to find the position of a contour on an edge uses only the potentials at the ends of the edges—so that a particular contour at potential Pot will be located at the same position on the edge, when drawing in the polygon on either side of the edge. Figure 9.2.3A gives the equipotential plot for a magnetic lifting device half-way through plotting from a finite element solution, and Figure 9.2.3B gives the final plot. The contour continuity may be noted.

When plotting from a finite difference solution, the identification of rectangles is more difficult. Each rectangle may be given the same number as the vertex on its lower left corner. Thus, referring to Figure 2.8.1, while rectangle 1 is 1, 2, 5, 3, there is no rectangle 34. Using the dummy notation of section 3.4, we may implement Get_Rectangle (Alg. 9.2.4) for extracting rectangles. The algorithm assumes an array **Nodes** containing the nodes to the east, north, west, and south of each node. In this algorithm for identifying a

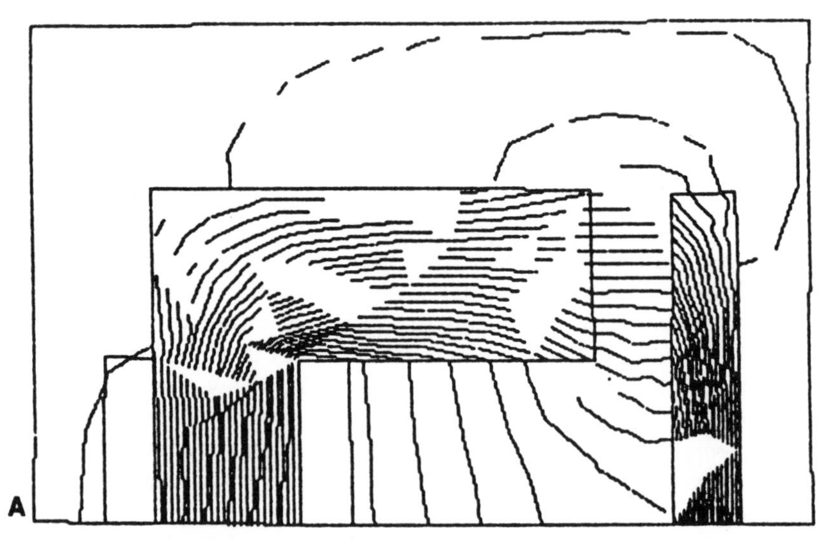


**Figure 9.2.3.** Plotting in a lifting magnet. **A.** Halfway through plotting. **B.** Final continuous plot.

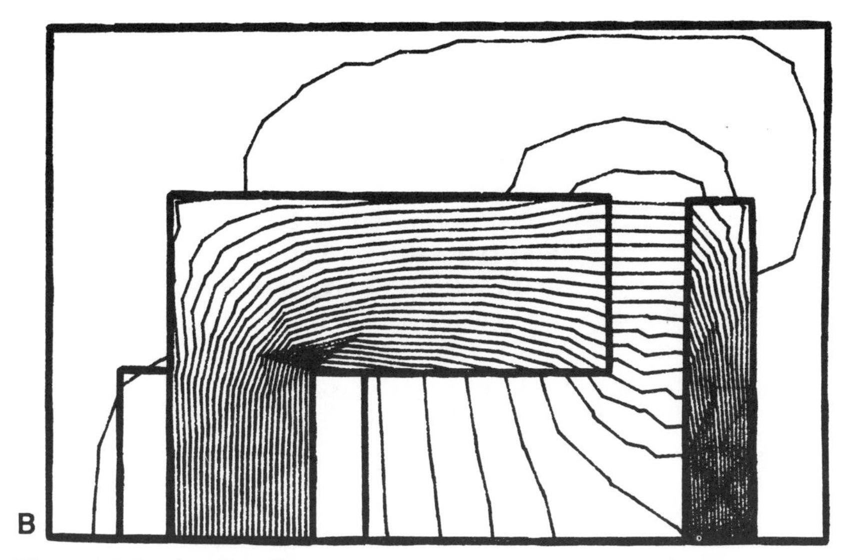


**Figure 9.2.3.** *(continued)*

## Algorithm 9.2.4. Finding a Rectangle in a Finite Difference Mesh

```
Procedure Get_Rectangle(i, Found, v)
{Input: i = vertex at lower left corner of rectangle
Outputs: Found = Boolean indicating whether such a rectangle
exists; v = if found, the numbering of the rectangle, anticlock-
wise from i}
Begin
  Found ← False
  For j ← 2 To 4 Do
    v[j] ← 0
  v[1] ← i
  If Nodes[v[1], 1]⟨ ⟩0
    Then {i Not on Dirichlet Boundary}
      If Nodes[v[1], 1] ⟨ ⟩ Nodes[v[1], 3]
        Then {v1 Not Neumann}
          v[2] ← Nodes[v[1], 1]
        Else {Find if E or W of i is the Dummy Node}
          If Nodes[Nodes[v[1], 1], 3] = Nodes[v[1], 1]
            Then
              v[2] ← Nodes[v[1], 1]
```

```
            If v[2] > 0
              Then
                If Nodes[v[2], 2] ⟨ ⟩0
                  Then {v2 Not Dirichlet}
                    If Nodes[v[2], 2] ⟨ ⟩ Nodes[v[2], 4]
                      Then
                        v[3] ← Nodes[v[2], 2]
                      Else {Find if N or S of v2 is Dummy}
                        If Nodes[Nodes[v[2], 2], 4] = Nodes[v[2], 2]
                          Then
                            v[3] ← Nodes[v[2], 2]
                If v[3] > 0
                  Then
                    If Nodes[v[3], 3] ⟨ ⟩ 0
                      Then
                        If Nodes[v[3], 1] ⟨ ⟩ Nodes[v[3], 3]
                          Then {v3 Not Neumann}
                            v[4] ← Nodes[v[3], 3]
                          Else
                            If Nodes[Nodes[v[3], 3], 1] =
                            Nodes[v[3], 3]
                              Then
                                v[4] ← Nodes[v[3], 3]
        If v[4] > 0 Then
          Found ← True
      End
```

rectangle $i$, we first read the node to the east of $i$, **Nodes**[$i$, 1]. If this is zero then no rectangle $i$ exists. Otherwise, if $i$ is a regular node, **Node**[$i$, 1] is the second node of the rectangle. However, if $i$ is on a vertical Neumann boundary then two possibilities exist; first, actually no node exists to its right, but the number to the west of $i$ is entered as a dummy node to account for Neumann conditions; second, no node exists to the left of $i$ and the number to the east has been entered as a dummy node in its place. If the latter is the case the rectangle may exist, and not otherwise. To check this, we say that if the node to the east, Nodes [**v**[1], 1] is a real node, then the node to the west of that node, **Nodes[Nodes[v**[1], 1], 3], should bring us back to the original node **Nodes[v**[1], 1]. By the application of the same principle, the other nodes are also identified.

Thus, for every node we will call this procedure Get_Rectangle and determine if Found is true. If it is true, then a rectangle corresponds to that node and so we will call Plot_Polygon (Alg. 9.2.3a) to draw all the potentials that pass through the rectangle. Thereby, a complete plot within the device may be obtained.

## 9.3 FIELD DENSITY

A quantity of practical interest is the field density. In nondestructive testing of insulators, it is the electric field density

$$\mathbf{E} = -\mathbf{u}_x \frac{\partial \phi}{\partial x} - \mathbf{u}_y \frac{\partial \phi}{\partial y} \tag{1.2.21}$$

that determines whether the insulator can or cannot withstand the applied electric field. In magnetic field problems, the flux density

$$\mathbf{B} = \mathbf{u}_x \frac{\partial A}{\partial y} - \mathbf{u}_y \frac{\partial A}{\partial x} \tag{1.2.34}$$

tells us, for example, whether the flux density is high in a transformer leg. If so, the ensuing saturation will result in leakage so that we will need to redesign the system with larger steel cross sections to reduce the flux density. In sections 4.5 and 4.6, in dealing with nonlinear magnetics we came across another application. We had to compute repeatedly the flux densities so as to compute the new permeability associated with each iteration. There we computed the flux densities for a finite difference mesh, and therefore here we shall treat only finite element meshes. In the next two sections, we shall see further applications in which the flux densities are required—namely, in energy and force computation.

Consider an $n$th-order finite element mesh in which we have already solved for the electric potential $\phi$:

$$\phi = \tilde{\boldsymbol{\alpha}}^n \underset{\sim}{\boldsymbol{\phi}}. \tag{8.5.2}$$

Putting this into eqn. (1.2.21) and using the concept of differentiation matrices defined in eqn. (8.5.12), we have

$$\mathbf{E} = -\mathbf{u}_x \tilde{\boldsymbol{\alpha}}^{n-1} \mathbf{D}_x^n \underset{\sim}{\boldsymbol{\phi}} - \mathbf{u}_y \tilde{\boldsymbol{\alpha}}^{n-1} \mathbf{D}_y^n \underset{\sim}{\boldsymbol{\phi}}, \tag{9.3.1}$$

in any triangle in which the differentiation matrices $\mathbf{D}_x^n$ and $\mathbf{D}_y^n$ may be computed using Alg. 8.5.1. Observe that for a first-order triangle $\mathbf{D}_x^1$ and $\mathbf{D}_y^1$ are $\tilde{\mathbf{b}}$ and $\tilde{\mathbf{c}}$ of eqns. (7.6.3) and (7.6.4) and $\tilde{\boldsymbol{\alpha}}^{n-1}$ is 1, so that flux density is a constant anywhere in the triangle. Therefore, if we want the flux density at a point we first need to locate the triangle in which it lies, and for first-order approximations, the constant value given by eqn. (9.3.1) will apply throughout the triangle. For higher orders of interpolation, using eqn. (7.5.13), we first must compute the triangular coordinate values $\zeta_1$, $\zeta_2$, and $\zeta_3$ from the rectangular coordinates of the point $(x, y)$, and then, substituting these values in the polynomial row vector $\tilde{\boldsymbol{\alpha}}^{n-1}$, we may compute the flux density at that point. The computation of magnetic flux density would follow along very similar lines.

## 9.4 STORED ENERGY COMPUTATION

The stored electric energy in a region is

$$\begin{aligned}\mathscr{L} &= \iint \tfrac{1}{2}\mathbf{D}\,.\,\mathbf{E}\,dR = \sum \tfrac{1}{2}\varepsilon \iint [E_x^2 + E_y^2]\,dR \\ &= \sum \tfrac{1}{2}\varepsilon \iint \left[\left(\frac{\partial\phi}{\partial x}\right)^2 + \left(\frac{\partial\phi}{\partial y}\right)^2\right] dR,\end{aligned} \tag{9.4.1}$$

where the integral over $R$ has been replaced by integrals over subdomains which may conveniently be taken as triangles. The expression for magnetic fields will have $\mu^{-1}$ taking the place of $\varepsilon$, and $A$ that of $\phi$. The computation of stored energy is very useful in force, capacitance, and inductance computations, as we shall see.

In a finite difference analysis, a rectangle 1234 in the $h \times k$ mesh may be divided into two right-angled triangles 124 and 234, as shown in Figure 9.4.1. In triangle 124, $\partial\phi/\partial x$ is $(\phi_2 - \phi_1)/h$ and $\partial\phi/\partial y$ is $(\phi_4 - \phi_1)/k$. The area of the triangle is $\frac{1}{2}hk$. With a similar expression for triangle 234, we have the contribution of the rectangle 1234 as

$$\begin{aligned}\mathscr{L}_{1234} = \tfrac{1}{2}hk\Bigg\{&\left[\frac{\phi_2 - \phi_1}{h}\right]^2 + \left[\frac{\phi_4 - \phi_1}{k}\right]^2 \\ &+ \left[\frac{\phi_3 - \phi_4}{h}\right]^2 + \left[\frac{\phi_3 - \phi_2}{k}\right]^2\Bigg\}.\end{aligned} \tag{9.4.2}$$

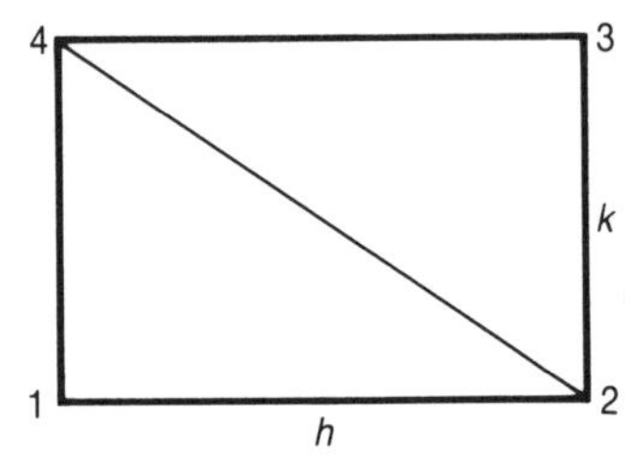


**Figure 9.4.1.** A rectangle as two triangles for numerical integration.

In finite element analysis, from eqns. (8.5.25) and (8.5.28) where we have evaluated the functional over a triangle, we see that the energy term is given by the finite element local matrix **P**.

$$\mathcal{L} = \tfrac{1}{2}\phi^t \mathbf{P} \phi. \tag{9.4.3}$$

Thus, to compute the energy computation of a triangle, we merely compute the local matrix **P** using Alg. 8.5.2 and then take a half of the multiple $\phi^t \mathbf{P} \phi$.

## 9.5 INDUCTANCE AND CAPACITANCE: THE THREE-CONDUCTOR CABLE

When computing the inductance $L$ or capacitance $C$ of a circuit, one may use the relationships:

$$W = \tfrac{1}{2}CV^2 \tag{9.5.1}$$

and

$$W = \tfrac{1}{2}LI^2. \tag{4.1.23}$$

That is, we set up a boundary value electric field problem sourced by the voltage $V$ when we wish to determine the capacitance or a magnetic field problem sourced by the current $I$. Once we solve the field problem, we may compute the stored energy by the schemes above and thence compute the capacitance or energy. The analysis of the three-conductor cable of Figure 3.5.1 for capacitance is shown in Figure 9.5.1. In solving this electric field problem, we have the conductor whose capacitance we want at a voltage $V$, and all other conductors have been set at rest. Symmetry has been used to solve in half the domain. For the magnetic field problem of finding inductance, a current density computed from a specified current $I$ divided by the cross sectional area of the conductor is used.

Alternatively, we may use the definitions

$$q = CV, \tag{9.5.2}$$

as in eqn. (2.4.8), and

$$\Psi = LI, \tag{9.5.3}$$

from eqn. (4.1.25), to compute the capacitance and inductance. The computation of the charge on the conductor is accomplished by taking a surface $S$, conveniently running along mesh lines, and

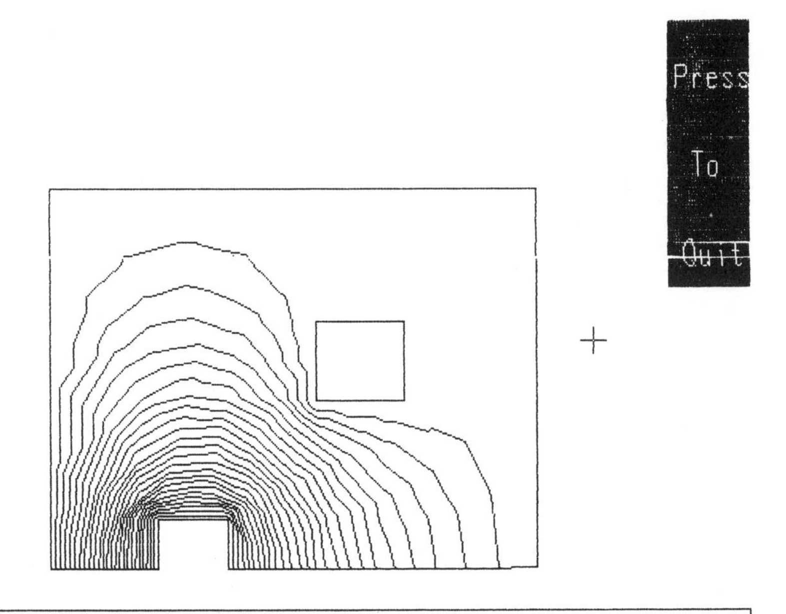


**Figure 9.5.1.** Capacitance analysis of a three-conductor cable.

using a form of Gauss's law in two dimensions:

$$q = \int \mathbf{D} \cdot \mathbf{dS} = -\int \varepsilon \nabla\phi \cdot \mathbf{u}_n \, dS = -\int \varepsilon \frac{\partial \phi}{\partial n} \, dS. \qquad (9.5.4)$$

While dealing with finite differences, the evaluation of the above contour integral (running anticlockwise) is straightforward. Since the contour is made up of vertical and horizontal pieces as shown in Figure 9.5.2A, at a typical vertical edge 12, with another parallel

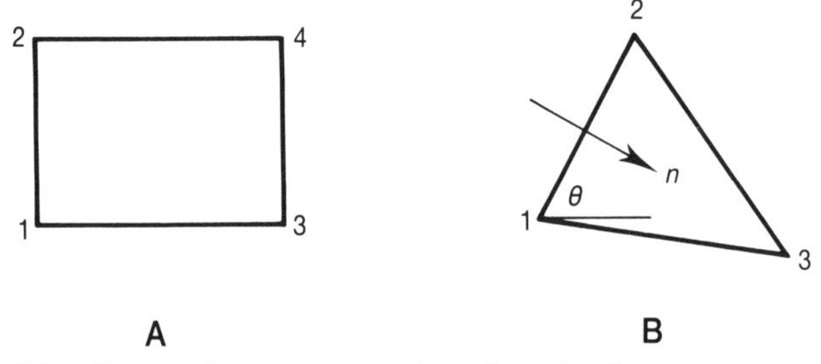


**Figure 9.5.2.** Integration over section $1 \rightarrow 2$ of a contour. **A.** Finite differences. **B.** Finite elements.

edge 34 to its right, we may take $\partial\phi/\partial n$ as $(\phi_3 - \phi_1)/h$ at 1 and $(\phi_4 - \phi_2)$ at 2, so that using an average, the integral over 12 becomes $\frac{1}{2}k(\phi_3 + \phi_4 - \phi_1 - \phi_2)/h$. A similar expression is possible for a horizontal edge. Notice that we could equally well have used differentials to the left of the edge 12 and the answers would have been different—the discrepancy being on account of the numerical approximations yielding step changes in the gradients rather than smooth changes.

While dealing with finite elements, the edges will run in arbitrary directions—unless we intentionally use a finite difference mesh with diagonals drawn and thereby lose the power of the finite element method. The normal gradient, therefore, must be obtained by resolving vectors. We know that the vector $\nabla\phi$ has components $\partial\phi/\partial x$ along the $x$ axis and $\partial\phi/\partial y$ along the $y$ axis. These we know how to compute from the differentiation matrices as in eqn. (9.3.1); for example, for first-order finite elements, they become $\mathbf{D}_x\underset{\sim}{\boldsymbol{\phi}}$ and $\mathbf{D}_y\underset{\sim}{\boldsymbol{\phi}}$ using $\phi$ from the neighboring triangle 123 outside the contour, as shown in Figure 9.5.2B. To obtain $\nabla\phi$ along the normal direction, then, we need only define direction cosines of the normal to edge 12, which makes an angle $\theta$ with the $x$ axis computable from

$$\cos\theta = \frac{x_2 - x_1}{L_{12}} \tag{9.5.5}$$

and

$$\sin\theta = \frac{y_2 - y_1}{L_{12}}, \tag{9.5.6}$$

where $L_{12}$ is the length of the edge $\sqrt{[(x_2 - x_1)^2 + (y_2 - y_1)^2]}$. The direction cosines of the normal, 90° behind $\theta$, are then obtained from

$$\cos\beta = \cos(\theta - 90) = \cos(90 - \theta) = \sin\theta, \tag{9.5.7}$$

and

$$\sin\beta = \sin(\theta - 90) = -\sin(90 - \theta) = -\cos\theta. \tag{9.5.8}$$

The component of $\nabla\phi$ along the outward normal, then, is

$$\begin{aligned}\frac{\partial\phi}{\partial n} &= \frac{\partial\phi}{\partial x}\cos\beta + \frac{\partial\phi}{\partial y}\sin\beta \\ &= \sin\theta\mathbf{D}_x\underset{\sim}{\boldsymbol{\phi}} - \cos\theta\mathbf{D}_y\underset{\sim}{\boldsymbol{\phi}}\end{aligned} \tag{9.5.9}$$

for first order. Since this is a constant, the integral over 12 is merely $L_{12}\,\partial\phi/\partial n$.

When performing inductance computation, the computation of the flux linkage $\psi$ is elementary. As already explained, the flux generated by the current $I$ is the vector potential difference between the point in the middle of the conductor (where it is at its highest or lowest depending on the direction of current) and the boundary at zero potential.

Be it noted that we have shown in section 8.1 that the energy computed from a variational finite element solution is of an order of accuracy superior to the potential solutions. And since the finite difference solution is a special case of the finite element method, the same holds for finite difference solutions. Therefore, capacitance and inductance computation using energy is the recommended course of action. It is left as a mental exercise to the reader to ponder on whether or not the energy from the exterior region is zero, when we impose an artificial Dirichlet boundary nearby for convenience. If not, then in computing the capacitance or inductance we will be computing only some of the energy. In thinking about this, a useful approach is to look at the flux densities in the elements just inside the boundary and thence, using the continuity conditions on the flux density, get the densities outside.

## 9.6 FORCE COMPUTATION

### 9.6.1 On Force Computation

In field analysis, we have seen that energy computation is of an order of accuracy higher than the potentials we compute. On the other hand, computed forces are of an order of accuracy lower (Hoole and Hoole 1985), particularly when we use the square of field density to evaluate force. As a result, it is important to use extremely fine meshes when force is of practical interest.

### 9.6.2 Force Density on Source-Carrying Parts

One of the easiest ways to compute force on charge-carrying parts is to use the expression

$$\begin{aligned}\mathbf{F} &= \iint \rho \mathbf{E}\, dR = -\iint \rho \nabla \phi\, dR \\ &= -\mathbf{u}_x \sum A^r \rho \frac{\partial \phi}{\partial x} - \mathbf{u}_y \sum A^r \rho \frac{\partial \phi}{\partial y},\end{aligned} \tag{9.6.1}$$

which arises from the definition of the electric field as the force on a unit point charge. $A^r$ is the area of each triangle (or rectangle when using finite differences) in the mesh, the summation is over all such elemental areas $A^r$ in the charge carrying part, and the slopes of $\phi$ are computed as described in section 9.3.

Similarly in magnetic field problems, we use the definition

$$\begin{aligned}\mathbf{F} &= \iint \mathbf{J} \times \mathbf{B}\, dR = \iint \mathbf{u}_z J \times \left[\mathbf{u}_x \frac{\partial A}{\partial y} - \mathbf{u}_y \frac{\partial A}{\partial x}\right] dR \\ &= \iint J\left[\mathbf{u}_x \frac{\partial A}{\partial x} + \mathbf{u}_y \frac{\partial A}{\partial y}\right] dR \\ &= \mathbf{u}_x \sum A^r J \frac{\partial A}{\partial x} + \mathbf{u}_y \sum A^r J \frac{\partial A}{\partial y},\end{aligned} \tag{9.6.2}$$

which may be evaluated just as with electric forces.

### 9.6.3 The Maxwell Stress Method

When computing the electrically caused force on a part that has no charge, the force density expression is not valid. It may be shown by writing (Panofsky and Phillips 1969):

$$\mathbf{F} = \iint \rho \mathbf{E}\, dR = \iint (\nabla \,.\, \mathbf{D})\mathbf{E}\, dR, \tag{9.6.3}$$

and converting this expression for force into a surface integral over a contour surrounding the body on which we want the force, that

$$\mathbf{F} = \iint \tfrac{1}{2}\varepsilon_0[E_n^2 - E_t^2]\mathbf{u}_n\, dS + \iint \varepsilon_0 E_n E_t \mathbf{u}_t\, dS. \tag{9.6.4a}$$

Force has been expressed as the integral of the so-called Maxwell stress tensor. In the above, $t$ and $n$ refer, respectively, to the tangential and normal directions. As we did with charge computation, we may conveniently take the contour to run along the mesh lines. Thus, we may compute **E** according to section 9.3, resolve into normal and tangential components as in eqn. (9.5.9), and perform the integrations required. The expression for magnetically caused force, similarly, is shown to be (Carpenter 1959):

$$\mathbf{F} = \iint \tfrac{1}{2}\mu_0[H_n^2 - H_t^2]\mathbf{u}_n\, dS + \iint \mu_0 H_n H_t \mathbf{u}_t\, dS. \tag{9.6.4b}$$

### 9.6.4 Force through Virtual Work

The concept of virtual work may be used to obtain the force acting on a part. If the part is moved by an infinitesimal distance $\delta x$, the work done is the component of force $F_x$ in that direction times $\delta x$. This work must come from the stored energy $W$ of the system:

$$F_x = \frac{-\delta W}{\delta x}. \tag{9.6.5}$$

Therefore, we must solve the field problem once, then repeat the problem with the part in question moved very slightly through a known distance $\delta x$ and solve again. The difference in stored energy $\delta W$ is then computed from the two solutions. The disadvantage there is that a third solution must be performed to compute $F_y$—unless we know the direction of the force from the field lines and move the part in that direction. Moreover, the numerical differentiation being performed on the stored energy requires $\delta x$ to be extremely small, and this causes implementational difficulties (particularly when using finite differences), as a result of which numerical error arises. The shift is accomplished more easily when using finite elements, however, since there are no rigid requirements on the shape of the mesh. A good way to perform the shift is to add a small number computationally, first to the $x$ coordinate of all nodes on the surface of the part on which force is being computed, and then to the $y$ coordinate.

In what may well prove a significant development in force computation, Coulomb (1983) has come forward with an elegant scheme for implementing the virtual work principle without the need for a second solution. It employs the fact that the finite element solutions are functions of space as given by the trial functions, and not just the explicit nodal values that are returned. Using this, the energy $W$ is evaluated as a function of the nodal values and spatial coordinates and differentiated mathematically, before the integration over the solution space is performed to eliminate the $x$ and $y$ coordinates. The answers obtained through this scheme are more accurate than those by the other methods, because it uses the explicit energy expression, which is highly accurate in finite element analysis. The more interested reader is referred to the works by Coulomb (1983) and Coulomb and Meunier (1984).

# 10 FINITE ELEMENTS: SPECIAL PROBLEMS

## 10.1 ANISOTROPIC FIELD PROBLEMS

In a large class of field problems, unlike the isotropic situation we have assumed up to now under which flux has no preferred direction of flow within a material, flux finds it easier to flow along a particular direction. That is, we need to deal with anisotropy. In magnetics this is particularly common, on account of the manufacturing process of steel. As steel is rolled, the grains are oriented depending on the direction of rolling and the permeability is increased in a certain direction. Moreover, the use of laminations to avoid flux leakage also gives the fluxes a preferred direction of flow. Laminations are several thin plates of steel, magnetically insulated from each other by the application of some kind of lacquer and stacked together. Thus, while flux can easily flow along the two larger dimensions of a plate, it cannot do so in the third on account of the need to cross the lacquer and the trapped air between plates. By these means, the flux may be directed by the designer of electric machinery and leakage into air minimized. Furthermore, since eddy currents flow transverse to the fluxes, even they must cross the plates and thus are minimized.

We shall restrict our discussion to magnetics here, on the understanding that extension to electric field problems is trivial, by mathematical analogy. In isotropic problems, since the permeability $\mu$ and, therefore, the reluctivity $\nu$ are the same in all directions, we have $H_x = \nu B_x$ and $H_y = \nu B_y$, and it was combining these that we got

$$\mathbf{H} = \nu \mathbf{B}, \tag{10.1.1}$$

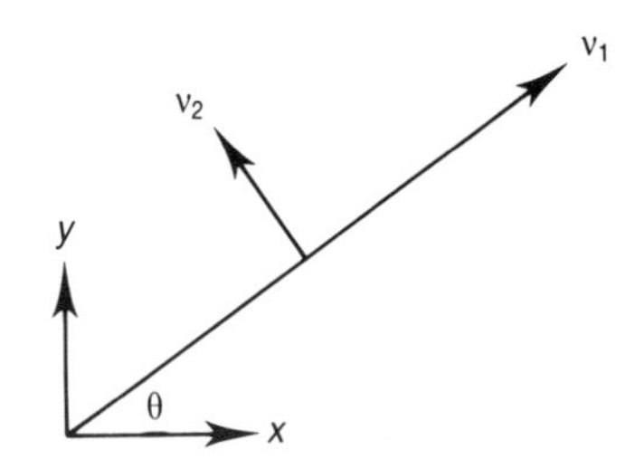


**Figure 10.1.1.** Anisotropy.

as in eqn. (1.2.7). However, in anisotropic problems, flux has a preferred direction of flow. Let us say that this preferred direction makes an angle $\theta$ with the $x$ axis and has a low reluctivity $\nu_1$, and normal to this direction the reluctivity is higher and is $\nu_2$. Therefore, referring to Figure 10.1.1, resolving in these two orthogonal directions,

$$H_x \cos\theta + H_y \sin\theta = \nu_1[B_x \cos\theta + B_y \sin\theta], \tag{10.1.2}$$

and

$$H_y \cos\theta - H_y \sin\theta = \nu_2[B_y \cos\theta - B_x \sin\theta]. \tag{10.1.3}$$

Solving for $H_x$ and $H_y$, we have

$$\begin{bmatrix} H_x \\ H_y \end{bmatrix} = \begin{bmatrix} \nu_1 \cos^2\theta + \nu_2 \sin^2\theta & -(\nu_2 - \nu_1)\sin\theta\cos\theta \\ -(\nu_2 - \nu_1)\sin\theta\cos\theta & \nu_1 \sin^2\theta + \nu_2 \cos^2\theta \end{bmatrix} \begin{bmatrix} B_x \\ B_y \end{bmatrix} \tag{10.1.4}$$

A quick verification is provided at $\theta = 0$ or $\frac{1}{2}\pi$, where the preferred direction is seen to be, respectively, along the $x$ and $y$ axes. For anisotropic materials, therefore, $H$ and $B$ are related by the symmetric reluctivity matrix (or the reluctivity tensor as it is often grandiosely described):

$$\begin{bmatrix} H_x \\ H_y \end{bmatrix} = [\nu]\begin{bmatrix} B_x \\ B_y \end{bmatrix} = \begin{bmatrix} \nu_{11} & \nu_{12} \\ \nu_{21} & \nu_{22} \end{bmatrix}\begin{bmatrix} B_x \\ B_y \end{bmatrix} \tag{10.1.5}$$

where the coefficients $\nu_{ij}$ of eqn. (10.1.5) may be obtained through direct correspondence with eqn. (10.1.4). For any part of the device, then, the coefficients of $[\boldsymbol{\nu}]$ may be computed with respect to the coordinate system we are working with.

Now to use these relationships in our finite element analysis, first notice that:

$$\mathbf{B}\,.\,\mathbf{H} = B_xH_x + B_yH_y = \mathbf{B}^t\mathbf{H} = \mathbf{B}^t[\nu]\mathbf{B}. \tag{10.1.6}$$

Using an $n$th-order finite element approximation for the vector potential $A$ over a triangular mesh spanning the solution region, from eqns. (1.2.34) and (8.5.15):

$$\mathbf{B} = \mathbf{u}_x \frac{\partial A}{\partial y} - \mathbf{u}_y \frac{\partial A}{\partial x} = \begin{bmatrix} \dfrac{\partial A}{\partial y} \\ -\dfrac{\partial A}{\partial x} \end{bmatrix} = \begin{bmatrix} \tilde{\boldsymbol{\alpha}}^{n-1}\mathbf{D}_y\underset{\sim}{\mathbf{A}} \\ -\tilde{\boldsymbol{\alpha}}^{n-1}\mathbf{D}_x\underset{\sim}{\mathbf{A}} \end{bmatrix} = \begin{bmatrix} \tilde{\boldsymbol{\alpha}}^{n-1}\mathbf{D}_y \\ -\boldsymbol{\alpha}^{n-1}\mathbf{D}_x \end{bmatrix}\mathbf{A}. \tag{10.1.7}$$

Putting these into our energy functional:

$$\mathscr{L} = \iint \{\tfrac{1}{2}\mathbf{B}\,.\,\mathbf{H} - JA\}\, dR = \iint \{\tfrac{1}{2}\mathbf{B}^t[\nu]\mathbf{B} - JA\}\, dR$$

$$= \sum \iint \{\tfrac{1}{2}\underset{\sim}{\mathbf{A}}^t[\mathbf{D}_y^t\tilde{\boldsymbol{\alpha}}^{n-1t} \quad -\tilde{\boldsymbol{\alpha}}^{n-1}\mathbf{D}_x^t][\nu]\begin{bmatrix} \tilde{\boldsymbol{\alpha}}^{n-1}\mathbf{D}_y \\ -\tilde{\boldsymbol{\alpha}}^{n-1}\mathbf{D}_x \end{bmatrix}\underset{\sim}{\mathbf{A}} - \underset{\sim}{\mathbf{A}}^t\tilde{\boldsymbol{\alpha}}^{nt}\tilde{\boldsymbol{\alpha}}^n\underset{\sim}{\mathbf{J}}\}\, dR. \tag{10.1.8}$$

The local coefficient matrix, upon expansion, is

$$\mathbf{P} = \iint \left\{[\mathbf{D}_y^t\tilde{\boldsymbol{\alpha}}^{n-1t} \quad -\tilde{\boldsymbol{\alpha}}^{n-1}\mathbf{D}_x^t]\begin{bmatrix} \nu_{11} & \nu_{12} \\ \nu_{21} & \nu_{22} \end{bmatrix}\begin{bmatrix} \tilde{\boldsymbol{\alpha}}^{n-1}\mathbf{D}_y \\ -\tilde{\boldsymbol{\alpha}}^{n-1}\mathbf{D}_x \end{bmatrix}\right\} dR$$

$$= \iint \left\{[\mathbf{D}_y^t\tilde{\boldsymbol{\alpha}}^{n-1t} \quad -\tilde{\boldsymbol{\alpha}}^{n-1}\mathbf{D}_x^t]\begin{bmatrix} \nu_{11}\tilde{\boldsymbol{\alpha}}^{n-1}\mathbf{D}_y - \nu_{12}\tilde{\boldsymbol{\alpha}}^{n-1}\mathbf{D}_x \\ \nu_{21}\tilde{\boldsymbol{\alpha}}^{n-1}\mathbf{D}_y - \nu_{22}\tilde{\boldsymbol{\alpha}}^{n-1}\mathbf{D}_x \end{bmatrix}\right\} dR \tag{10.1.9}$$

$$= \iint \{\nu_{11}\mathbf{D}_y^t\tilde{\boldsymbol{\alpha}}^{n-1t}\tilde{\boldsymbol{\alpha}}^{n-1}\mathbf{D}_y + \nu_{22}\mathbf{D}_x^t\tilde{\boldsymbol{\alpha}}^{n-1t}\tilde{\boldsymbol{\alpha}}^{n-1}\mathbf{D}_x - 2\nu_{12}\mathbf{D}_y^t\tilde{\boldsymbol{\alpha}}^{n-1t}\tilde{\boldsymbol{\alpha}}^{n-1}\mathbf{D}_x\}\, dR$$

$$= A^r\{\nu_{11}\mathbf{D}_y^t\mathbf{T}^{n-1,n-1}\mathbf{D}_y + \nu_{22}\mathbf{D}_x^t\mathbf{T}^{n-1,n-1}\mathbf{D}_x - 2\nu_{12}\mathbf{D}_y^t\mathbf{T}^{n-1,n-1}\mathbf{D}_x\},$$

where we have used $\nu_{12} = \nu_{21}$ and the fact that $\mathbf{D}_x^t\tilde{\boldsymbol{\alpha}}^{n-1t}\tilde{\boldsymbol{\alpha}}^{n-1}\mathbf{D}_y$ is a scalar to transpose it and write it as $\mathbf{D}_y^t\tilde{\boldsymbol{\alpha}}^{n-1t}\tilde{\boldsymbol{\alpha}}^{n-1}\mathbf{D}_x$. The matrices $\mathbf{T}^{n-1,n-1}$ are independent of element, are defined in eqn. (8.5.24), and are 1 for first-order interpolation. $A^r$ is the area of an element. Observe that for isotropic situations, with $\nu_{11} = \nu_{22} = \nu$ and $\nu_{12} = 0$ from eqn. (10.1.9), the local matrix reduces to that derived in eqn. (8.5.25).

The source term in eqn. (10.1.8) is simply the usual $A^r\underset{\sim}{\mathbf{A}}^t\mathbf{T}^{n,n}\underset{\sim}{\mathbf{J}}$. The single phase transformer of Figure 10.1.2, with anisotropy, was analyzed by this scheme using first-order finite elements. The limbs have a permeability of $10\mu_0$ along the limb and $\mu_0$ normal to it. So, for example, in the vertical limb, $\theta = 90$, $\nu_1 = \nu_0$, and $\nu_2 = 0.1\nu_0$. Working out, we have $\nu_{11} = \nu_0$, $\nu_{22} = 0.1\nu_0$, and $\nu_{12} = \nu_{21} = 0$. The first-order mesh and solution to this field problem are shown, respectively, in Figures 10.1.2A and 10.1.2B.

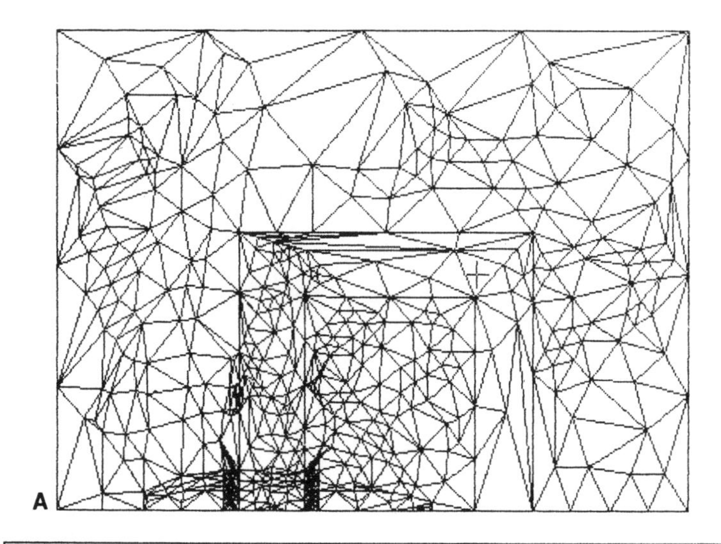


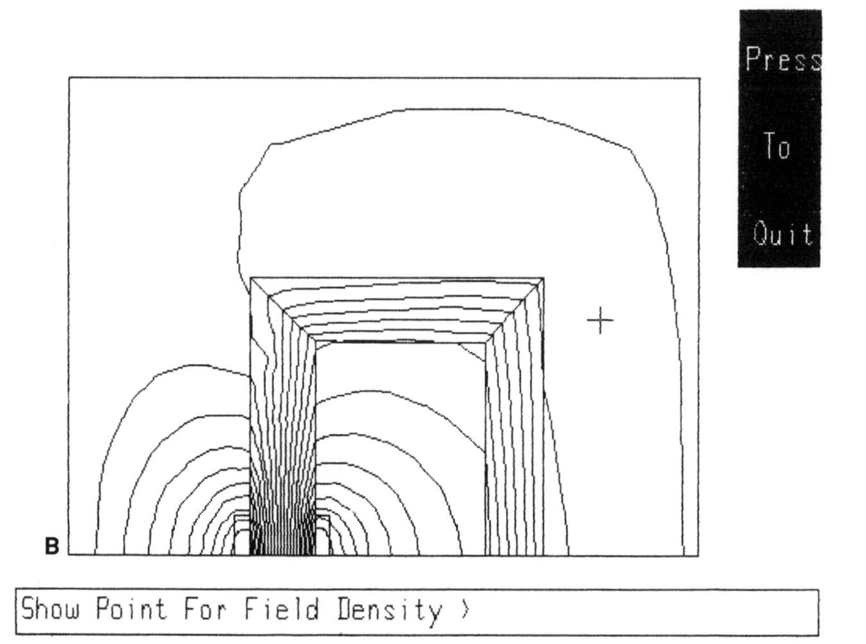


**Figure 10.1.2.** The anisotropic single phase transformer. **A**. The mesh. **B**. The solution.

# 10.2 AXISYMMETRIC SYSTEMS

## 10.2.1 Axisymmetric Electric Field Problems

Axisymmetric electric field problems are those in which, in cylindrical coordinates $r$, $\theta$, $z$, with line elements $dr$, $r\,d\theta$, and $dz$, no changes occur in the coordinate direction $\theta$, as discussed and developed in section 1.3.

For electric field problems, by extension of eqn. (7.5.8) to three dimensions, we have a general functional:

$$\mathscr{L}[\phi] = \iiint [\tfrac{1}{2}\varepsilon(\nabla\phi)^2 - \rho\phi]\,dV. \tag{10.2.1}$$

Writing the volume element $dV$ as $(dr)(r\,d\theta)\,dz$, we have

$$\begin{aligned}\mathscr{L}[\phi] &= \iiint \left\{\tfrac{1}{2}\varepsilon\left[\left(\frac{\partial\phi}{\partial r}\right)^2 + \left(\frac{\partial\phi}{\partial z}\right)^2\right] - \rho\phi\right\} r\,dr\,d\theta\,dz \\ &= \iint 2\pi r\left\{\tfrac{1}{2}\varepsilon\left[\left(\frac{\partial\phi}{\partial r}\right)^2 + \left(\frac{\partial\phi}{\partial z}\right)^2\right] - \rho\phi\right\} dr\,dz,\end{aligned} \tag{10.2.2}$$

performing the integral of $d\theta$ from 0 to $2\pi$. In the first paper on axisymmetric finite elements, Zienkiewicz and Cheung (1965) assumed $r$ to be constant in each triangle of the mesh so that it was pulled out of the integral sign. The average value of $r$ over the triangle was used towards this end. Subsequently, the rest of the functional integration becomes just as in two dimensions and is very straightforward. We shall work out here the functional for more general finite elements and show that the choice of average $r$ taken by Zienkiewicz and Cheung (1965) is in fact the same as that yielded by an exact evaluation.

Using the usual finite element approximation $\tilde{\boldsymbol{\alpha}}^n \underset{\sim}{\boldsymbol{\phi}}$ over the coordinate plane $r$–$z$, all the approximations and differential relations of section 9.5 remain applicable except for the symbolic replacement of $x$ and $y$ by $r$ and $z$. Also observe that $r$ is a first-order function of space, so that

$$r = \tilde{\boldsymbol{\alpha}}^1 \underset{\sim}{\mathbf{r}} \tag{10.2.3}$$

is an exact relationship, where in a triangle, $r$ contains the values of

$r$ at the three vertices. Putting these into our functional:

$$
\begin{aligned}
\mathscr{L}[\phi] &= \sum \iint 2\pi(\tilde{\boldsymbol{\alpha}}^1\underset{\sim}{\mathbf{r}})\{\tfrac{1}{2}\varepsilon[(\tilde{\boldsymbol{\alpha}}^{n-1}\mathbf{D}_r\underset{\sim}{\boldsymbol{\phi}})^2 + (\tilde{\boldsymbol{\alpha}}^{n-1}\mathbf{D}_z\underset{\sim}{\boldsymbol{\phi}})^2] \\
&\quad - (\tilde{\boldsymbol{\alpha}}^n\underset{\sim}{\rho})(\tilde{\boldsymbol{\alpha}}^n\underset{\sim}{\boldsymbol{\phi}})\}\, dr\, dz \\
&= \sum \iint \{\pi\varepsilon(\tilde{\boldsymbol{\alpha}}^1\underset{\sim}{\mathbf{r}})[(\underset{\sim}{\boldsymbol{\phi}}^t\mathbf{D}_r^t\tilde{\boldsymbol{\alpha}}^{n-1t})(\tilde{\boldsymbol{\alpha}}^{n-1}\mathbf{D}_r\underset{\sim}{\boldsymbol{\phi}}) \\
&\quad + (\underset{\sim}{\boldsymbol{\phi}}^t\mathbf{D}_z^t\tilde{\boldsymbol{\alpha}}^{n-1t})(\tilde{\boldsymbol{\alpha}}^{n-1}\mathbf{D}_z\underset{\sim}{\boldsymbol{\phi}})]\, dr\, dz \\
&\quad - \sum \iint 2\pi(\tilde{\boldsymbol{\alpha}}^1\underset{\sim}{\mathbf{r}})(\underset{\sim}{\boldsymbol{\phi}}^t\tilde{\boldsymbol{\alpha}}^{nt})(\tilde{\boldsymbol{\alpha}}^n\underset{\sim}{\boldsymbol{\rho}})\, dr\, dz \\
&= \sum \pi\varepsilon A^r[\underset{\sim}{\boldsymbol{\phi}}^t D_r^t\mathbf{M}^n\mathbf{D}_r\underset{\sim}{\boldsymbol{\phi}} + \underset{\sim}{\boldsymbol{\phi}}^t\mathbf{D}_z^t\mathbf{M}^n\mathbf{D}_z\underset{\sim}{\boldsymbol{\phi}}] - 2\pi A^r\underset{\sim}{\boldsymbol{\phi}}^t\mathbf{M}^{n+1}\underset{\sim}{\boldsymbol{\rho}},
\end{aligned}
\tag{10.2.4}
$$

where $A^r$ is the area, $\mathbf{D}_r$ and $\mathbf{D}_z$ are the differentiation matrices of eqn. (8.5.15), $\mathbf{T}^{n,n}$ is the universal matrix of eqn. (8.5.24), and $\mathbf{M}^n$ is the symmetric matrix

$$
\mathbf{M}^n = \frac{1}{A^r}\iint (\tilde{\boldsymbol{\alpha}}^1\underset{\sim}{\mathbf{r}})\tilde{\boldsymbol{\alpha}}^{n-1t}\tilde{\boldsymbol{\alpha}}^{n-1}\, dr\, dz, \tag{10.2.5}
$$

of size $N(n-1) \times N(n-1)$. The term at row $i$ and column $j$ of $\mathbf{M}$ is, therefore,

$$
\begin{aligned}
M_{ij}^n &= \frac{1}{A^r}\iint (\tilde{\boldsymbol{\alpha}}^1\underset{\sim}{\mathbf{r}})\alpha_i^{n-1}\alpha_j^{n-1}\, dr\, dz \\
&= \frac{1}{A^r}\iint (\zeta_1 r_1 + \zeta_2 r_2 + \zeta_3 r_3)\alpha_i^{n-1}\alpha_j^{n-1}\, dr\, dz
\end{aligned}
\tag{10.2.6}
$$

and may be evaluated from the three-suffix universal parameters

$$
p_{ijk}^n = \frac{1}{A^r}\iint \alpha_i^{n-1}\alpha_j^{n-1}\zeta_k\, dr\, dz \qquad i,j = 1,\ldots,N(n^{-1}); \quad k = 1,2,3, \tag{10.2.7}
$$

which are independent of element shape and need to be evaluated only once for each order. With their aid, eqn. (10.2.6) reduces to

$$
M_{ij}^n = r_1 p_{ij1} + r_2 p_{ij2} + r_3 p_{ij3}. \tag{10.2.8}
$$

For the particular and common example of first-order interpolation, $n$ is 1 and the zeroth-order polynomials $\tilde{\boldsymbol{\alpha}}^{n-1}$ are 1, representing the result of differentiation of a first-order function. For this case, $\mathbf{M}^1$ is a constant and we have:

$$\mathbf{M}^1 = \left(\frac{1}{A^r}\right)\iint (\tilde{\boldsymbol{\alpha}}^1 \underline{\mathbf{r}})\, dr\, dz = \frac{1}{A^r}\iint [\zeta_1 r_1 + \zeta_2 r_2 + \zeta_3 r_3]\, dr\, dz \tag{10.2.9}$$
$$= \tfrac{1}{3}[r_1 + r_2 + r_3],$$

using eqn. (D1). This is exactly the same as taking $r$ to be the constant value at the centroid in each triangle. As a further example, if we are using second-order finite elements, we have from eqn. (8.5.4):

$$\tilde{\boldsymbol{\alpha}}^{n-1} = \tilde{\boldsymbol{\alpha}}^1 = [\zeta_1 \quad \zeta_2 \quad \zeta_3], \tag{10.2.10}$$

so that we need to evaluate 27 coefficients, which in this case only correspond to the coefficients of eqns. (10.4.36)–(10.4.41). $\mathbf{M}^2$ is a $3 \times 3$ matrix given by:

$$\mathrm{A}^r\mathbf{M}^2 = \iint \begin{bmatrix} (\zeta_1 r_1 + \zeta_2 r_2 + \zeta_3 r_1)\zeta_1^2 & (\zeta_1 r_1 + \zeta_2 r_2 + \zeta_3 r_3)\zeta_1\zeta_2 & (\zeta_1 r_1 + \zeta_2 r_2 + \zeta_3 r_2)\zeta_1\zeta_3 \\ (\zeta_1 r_1 + \zeta_2 r_2 + \zeta_3 r_3)\zeta_2\zeta_1 & (\zeta_1 r_1 + \zeta_2 r_2 + \zeta_3 r_3)\zeta_2^2 & (\zeta_1 r_1 + \zeta_2 r_2 + \zeta_3 r_3)\zeta_2\zeta_3 \\ (\zeta_1 r_1 + \zeta_2 r_2 + \zeta_3 r_3)\zeta_3\zeta_1 & (\zeta_1 r_1 + \zeta_2 r_2 + \zeta_3 r_3)\zeta_3\zeta_2 & (\zeta_1 r_1 + \zeta_2 r_2 + \zeta_3 r_3)\zeta_3^2 \end{bmatrix} dR, \tag{10.2.11}$$

giving us

$$\mathbf{M}^2 = \begin{bmatrix} \frac{r_1}{10} + \frac{r_2}{30} + \frac{r_3}{30} & \frac{r_1}{30} + \frac{r_2}{30} + \frac{r_3}{60} & \frac{r_1}{30} + \frac{r_2}{60} + \frac{r_3}{30} \\ \frac{r_1}{30} + \frac{r_2}{30} + \frac{r_3}{60} & \frac{r_1}{30} + \frac{r_2}{10} + \frac{r_3}{30} & \frac{r_1}{60} + \frac{r_2}{30} + \frac{r_3}{30} \\ \frac{r_1}{30} + \frac{r_2}{60} + \frac{r_3}{30} & \frac{r_1}{60} + \frac{r_2}{30} + \frac{r_3}{30} & \frac{r_1}{30} + \frac{r_2}{30} + \frac{r_3}{10} \end{bmatrix} \tag{10.2.12}$$

where we have employed the relationships

$$\iint \zeta_i^3\, dR = A^r \frac{1}{10}, \tag{10.2.13}$$

$$\iint \zeta_i \zeta_j^2\, dR = A^r \frac{1}{30}, \tag{10.2.14}$$

$$\iint \zeta_i \zeta_j \zeta_k\, dR = A^r \frac{1}{60}, \tag{10.2.15}$$

for distinct values $i$, $j$, and $k$. It is seen that $\mathbf{M}^n$ may be expressed as a sum of three universal matrices $\mathbf{M}_i^n$ weighted by the vertex coordinate radii

$$\mathbf{M}^n = r_1\mathbf{M}_1^n + r^2\mathbf{M}_2^n + r_3\mathbf{M}_3^n, \tag{10.2.16}$$

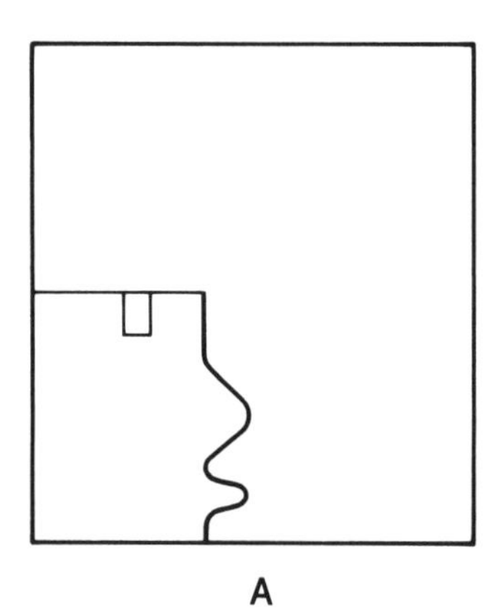


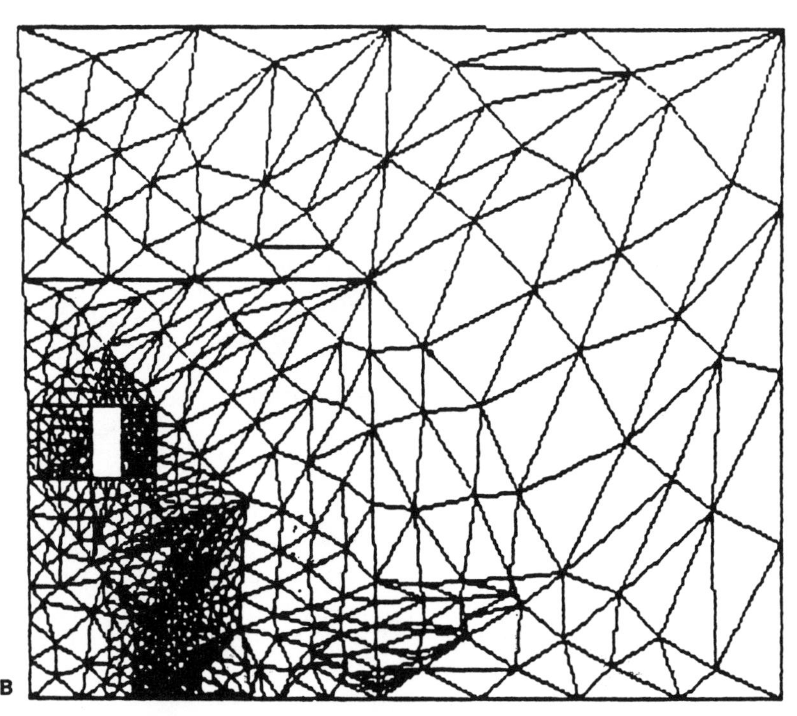


**Figure 10.2.1.** The axisymmetric insulator string. **A**. The problem further reduced. **B**. The mesh. **C**. The solution.

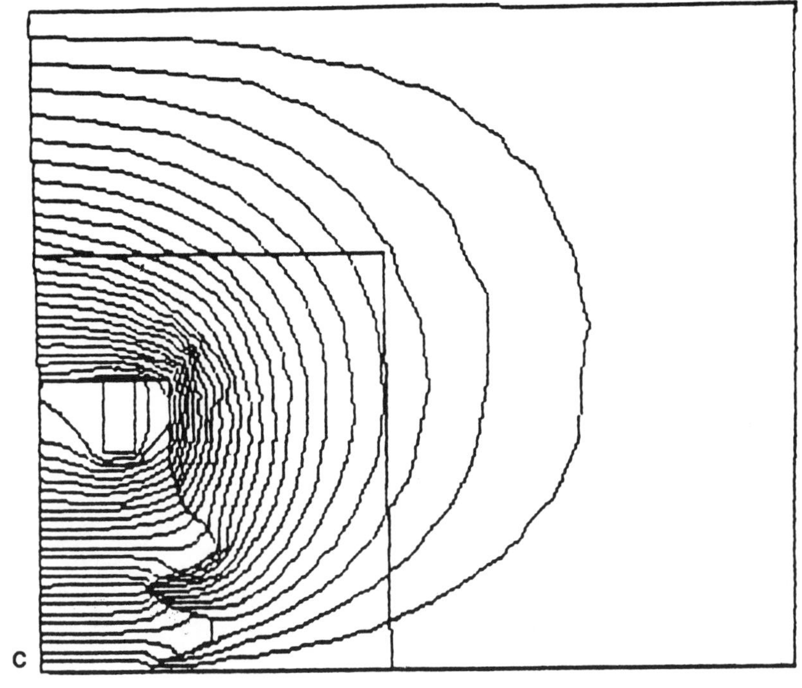


**Figure 10.2.1.** *(continued)*

where, for example,

$$\mathbf{M}_1^2 = \begin{bmatrix} \frac{1}{10} & \frac{1}{30} & \frac{1}{30} \\ \frac{1}{30} & \frac{1}{30} & \frac{1}{60} \\ \frac{1}{30} & \frac{1}{60} & \frac{1}{30} \end{bmatrix}. \qquad (10.2.17)$$

The other two matrices, $\mathbf{M}_2^n$ and $\mathbf{M}_3^n$, must clearly follow from rotation of $\mathbf{M}_1^n$ as defined in eqn. (8.5.16).

Extremizing eqn. (10.2.4) with respect to the unknowns $\phi$ by the rules of eqns. (7.6.18) and (7.6.19), and scaling by the common factor $2\pi$, we have

$$\frac{1}{2\pi}\frac{\partial \mathscr{L}}{\partial \underset{\sim}{\boldsymbol{\phi}}} = \sum \varepsilon A^r[\mathbf{D}_r^t \mathbf{M}^n \mathbf{D}_r \underset{\sim}{\boldsymbol{\phi}} + \mathbf{D}_z^t \mathbf{M}^n \mathbf{D}_z \underset{\sim}{\boldsymbol{\phi}}] - A^r \mathbf{M}^{n+1} \underset{\sim}{\boldsymbol{\rho}} = 0. \quad (10.2.18)$$

The axisymmetric insulator string of Figure 1.3.1 was solved by this scheme. Provided the voltages were equal and opposite, the problem may be reduced further to that of Figure 10.2.1A. The mesh and solution are given in Figures 10.2.1B and 10.2.1C. Figure 10.2.2 gives the mesh and solution of a more complexly shaped insulator, taken from Konrad (1985b).

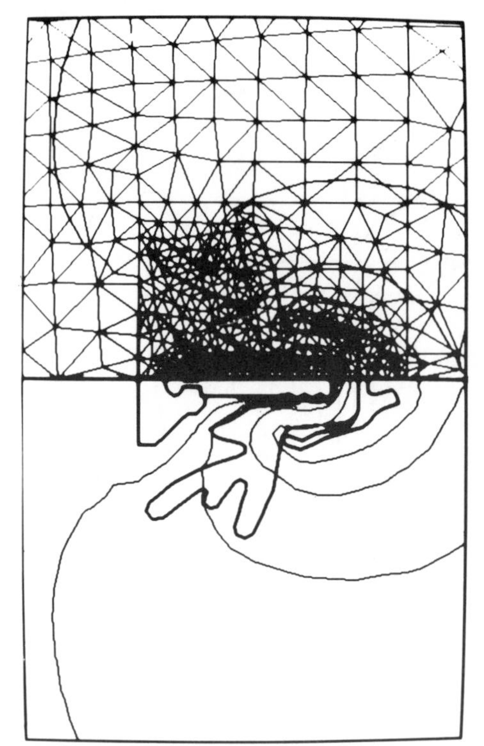

**Figure 10.2.2.** Another axisymmetric insulator string.

## 10.2.2 Axisymmetric Magnetic Field Problems

While dealing with axisymmetric magnetic fields, the treatment is unfortunately not as straightforward as with electric fields. We have seen in section 1.3 that the magnetic field is given by

$$\begin{aligned} \mathbf{B} &= -\mathbf{u}_r \frac{\partial A}{\partial z} + \mathbf{u}_z r^{-1} \frac{\partial rA}{\partial r} \\ &= -\mathbf{u}_r \frac{\partial A}{\partial z} + \mathbf{u}_z r^{-1} \left[ A + r \frac{\partial A}{\partial r} \right] \end{aligned} \tag{10.2.19}$$

where $A$ is the component of the vector potential in the $\theta$ direction. Putting this into the energy functional that we have already derived,

$$\mathscr{L} = \iiint \{\tfrac{1}{2}\mathbf{B} \,.\, \mathbf{H} - \mathbf{A} \,.\, \mathbf{J}\}\, dV = \iiint \{\tfrac{1}{2}\nu B^2 - \mathbf{A} \,.\, \mathbf{J}\}\, dV$$

$$= \iiint \left\{ \tfrac{1}{2}\nu\left[\left(\frac{\partial A}{\partial z}\right)^2 + \left(r^{-1}A + \frac{\partial A}{\partial r}\right)^2\right] - AJ \right\} dr(r\,d\theta)\,dz \tag{10.2.20}$$
$$= \iint 2\pi r \left\{ \tfrac{1}{2}\nu\left[\left(\frac{\partial A}{\partial z}\right)^2 + \left(\frac{\partial A}{\partial r}\right)^2 + r^{-2}A^2 + 2r^{-1}A\frac{\partial A}{\partial r}\right] - AJ \right\} dr\,dz.$$

It is seen that we have, in addition to the terms of the functional of the electric scalar potential (10.2.2), the terms $r^{-1}A$ and $2A\,\partial A/\partial r$. This poses two difficulties. First, along the axis of symmetry, $r$ is zero. Second, the term $r^{-1}$ introduces triangular coordinates in the denominator, which makes integration difficult. Also notice that the term $A\,\partial A/\partial r$, after introducing the appropriate approximations and transposing $A$, appears to give us the asymmetric term $2\underset{\sim}{\mathbf{A}}^t\tilde{\boldsymbol{\alpha}}^{nt}\tilde{\boldsymbol{\alpha}}^{n-1}\mathbf{D}_r\underset{\sim}{\mathbf{A}}$. But this in fact contributes a symmetric term as in eqn. (7.6.19), and is quickly realized to be such, upon writing it as $A^t(\partial A/\partial r) + (\partial A/\partial r)^t A = \underset{\sim}{\mathbf{A}}^t[\tilde{\boldsymbol{\alpha}}^{nt}\tilde{\boldsymbol{\alpha}}^{n-1}\mathbf{D}_r + \mathbf{D}_r^t\tilde{\boldsymbol{\alpha}}^{n-1t}\tilde{\boldsymbol{\alpha}}^n]\underset{\sim}{\mathbf{A}}$.

To overcome the difficulty, following the work of Silvester and Konrad (1973), we shall introduce the modified vector potential

$$A^m = r^{-1/2}A, \tag{10.2.21}$$

and the modified current density

$$J^m = r^{-1/2}J, \tag{10.2.22}$$

and solve for this modified potential everywhere. Observe that the modified source $J^m$ may be trivially evaluated everywhere except at the axis of symmetry, where by l'Hopital's rule:

$$\lim_{r\to 0} J^m = \lim_{r\to 0}\frac{\dfrac{dJ}{dr}}{\dfrac{d\sqrt{r}}{dr}} = 0. \tag{10.2.23}$$

With this, the functional eqn. (10.2.20), after scaling by $\pi$, becomes:

$$\mathscr{L} = \iint \left\{ \nu r\left(\frac{\partial r^{1/2}A^m}{\partial z}\right)^2 + \nu r\left(\frac{\partial r^{1/2}A^m}{\partial r}\right)^2 + \nu r^{-1}(r^{1/2}A^m)^2 \right.$$
$$\left. + 2\nu r^{1/2}A^m\left(\frac{\partial r^{1/2}A^m}{\partial r}\right) - 2rr^{1/2}A^m r^{1/2}J^m \right\} dr\,dz$$
$$= \iint \left\{ \nu r^2\left(\frac{\partial A^m}{\partial z}\right)^2 + \nu r\left[r\left(\frac{\partial A^m}{\partial r}\right)^2 + A^m\frac{\partial A^m}{\partial r} + \tfrac{1}{4}r^{-1}(A^m)^2\right] \right. \tag{10.2.24}$$
$$\left. + \nu(A^m)^2 + 2\nu r^{1/2}A^m\left(\frac{r^{1/2}\partial A^m}{\partial r} + \tfrac{1}{2}r^{-1/2}A^m\right) - 2r^2A^mJ^m \right\} dr\,dz$$
$$= \iint \left\{ \nu r^2\left(\frac{\partial A^m}{\partial z}\right)^2 + \nu r^2\left(\frac{\partial A^m}{\partial r}\right)^2 + \frac{9\nu(A^m)^2}{4} \right.$$
$$\left. + 3\nu r^2 A\frac{\partial A}{\partial r} - 2r^2A^mJ^m \right\} dr\,dz,$$

which, it is pointed out, contains an important correction to the functional for axisymmetric fields provided by Silvester and Ferrari (1983) and Silvester and Konrad (1973). In their equation, the term $3\nu r^2 A\, \partial A/\partial r$ does not have the coefficient 3.

Dividing the solution region into triangles with, say, $n$th-order variations in $A^m$ and $J^m$, we have

$$\mathscr{L} = \sum \Big\{ \nu \underset{\sim}{\mathbf{A}}^{mt}\mathbf{D}_z^t \Big[\iint r^2 \tilde{\boldsymbol{\alpha}}^{n-1t}\tilde{\boldsymbol{\alpha}}^{n-1}\, dR\Big]\mathbf{D}_z \underset{\sim}{\mathbf{A}}^m$$

$$+ \nu \underset{\sim}{\mathbf{A}}^{mt}\mathbf{D}_r^t \Big[\iint r^2 \tilde{\boldsymbol{\alpha}}^{n-1t}\tilde{\boldsymbol{\alpha}}^{n-1}\, dR\Big]\mathbf{D}_r^t \underset{\sim}{\mathbf{A}}^m + \tfrac{9}{4}\nu \underset{\sim}{\mathbf{A}}^{mt}\Big[\iint \tilde{\boldsymbol{\alpha}}^{nt}\tilde{\boldsymbol{\alpha}}^{n}\, dR\Big]\underset{\sim}{\mathbf{A}}^m \tag{10.2.25}$$

$$+ 3\nu \underset{\sim}{\mathbf{A}}^{mt}\Big[\iint r^2 \tilde{\boldsymbol{\alpha}}^{nt}\tilde{\boldsymbol{\alpha}}^{n-1}\mathbf{D}_r\, dR\Big]\underset{\sim}{\mathbf{A}}^m - 2\underset{\sim}{\mathbf{A}}^{mt}\Big[\iint r^2 \tilde{\boldsymbol{\alpha}}^{nt}\tilde{\boldsymbol{\alpha}}^{n}\, dR\Big]\underset{\sim}{\mathbf{J}}^m \Big\}$$

$$= \sum A^r\{\nu \underset{\sim}{\mathbf{A}}^{mt}[\mathbf{D}_z^t\mathbf{M}_1^n\mathbf{D}_z + \mathbf{D}_r^t\mathbf{M}_1^n\mathbf{D}_r + \tfrac{9}{2}\mathbf{T}^{n,n} + 3\mathbf{M}_2^n\mathbf{D}_r + 3\mathbf{D}_r^t\mathbf{M}_2^{nt}]\underset{\sim}{\mathbf{A}}^m$$

$$- 2\underset{\sim}{\mathbf{A}}^{mt}\mathbf{M}_1^{n+1}\underset{\sim}{\mathbf{J}}^m\},$$

where $A^r$ is the area of a triangle, $\mathbf{D}_z$ and $\mathbf{D}_r$ are the differentiation matrices defined in section 8.5, and $\mathbf{M}_1^n$, $\mathbf{M}_2^n$ are matrices that may be expressed in terms of the nodal radii and universal matrices independent of element shape:

$$\mathbf{M}_1^n = \frac{1}{A^r}\iint r^2 \tilde{\boldsymbol{\alpha}}^{n-1t}\tilde{\boldsymbol{\alpha}}^{n-1}\, dR, \tag{10.2.26}$$

and

$$\mathbf{M}_2^n = \frac{1}{A^r}\iint r^2 \tilde{\boldsymbol{\alpha}}^{n-1t}\tilde{\boldsymbol{\alpha}}^{n}\, dR. \tag{10.2.27}$$

For example, for first-order finite elements, $\tilde{\boldsymbol{\alpha}}^{n-1}$ is the zeroth-order function 1, so that using an exact first-order expansion for $r$ as in eqn. (10.2.3), we have

$$\begin{aligned}
\mathbf{M}_1^1 &= \frac{1}{A^r}\iint r^2\, dR = \frac{1}{A^r}\iint [\zeta_1 r_1 + \zeta_2 r_2 + \zeta_3 r_3]^2\, dR \\
&= \frac{1}{A^r}\iint (\zeta_1^2 r_1^2 \zeta_2^2 r_2^2 + \zeta_3^2 r_3^2 + 2\zeta_1\zeta_2 r_1 r_2 + 2\zeta_2\zeta_3 r_2 r_3 + \zeta_1\zeta_3 r_1 r_3)\, dR \\
&= \frac{r_1^2}{6} + \frac{r_2^2}{6} + \frac{r_3^2}{6} + \frac{r_1 r_2}{6} + \frac{r_1 r_3}{6} + \frac{r_2 r_3}{6} \qquad (10.2.28) \\
&= \tfrac{1}{12}\underset{\sim}{\mathbf{r}}^t \begin{bmatrix} 2 & 1 & 1 \\ 1 & 2 & 1 \\ 1 & 1 & 2 \end{bmatrix} \underset{\sim}{\mathbf{r}},
\end{aligned}$$

and:

$$
\begin{aligned}
\mathbf{M}_2^1 &= \frac{1}{A^r} \iint (\zeta_1 r_1 + \zeta_2 r_2 + \zeta_3 r_3)^2 [\zeta_1 \quad \zeta_2 \quad \zeta_3] \, dR \\
&= \frac{1}{A^r} \iint [\zeta_1(\zeta_1^2 r_1^2 + \zeta_2^2 r_2^2 + \zeta_3^2 r_3^2 + 2\zeta_1\zeta_2 r_1 r_2 + 2\zeta_2\zeta_3 r_2 r_3 + \zeta_1\zeta_3 r_1 r_3) \\
&\qquad \zeta_2(\zeta_1^2 r_1^2 + \zeta_2^2 r_2^2 + \zeta_3^2 r_3^2 + 2\zeta_1\zeta_2 r_1 r_2 + 2\zeta_2\zeta_3 r_2 r_3 + \zeta_1\zeta_3 r_1 r_3) \\
&\qquad \zeta_3(\zeta_1^2 r_1^2 + \zeta_2^2 r_2^2 + \zeta_3^2 r_3^2 + 2\zeta_1\zeta_2 r_1 r_2 + 2\zeta_2\zeta_3 r_2 r_3 + \zeta_1\zeta_3 r_1 r_3)] \, dR
\end{aligned}
\tag{10.2.29}
$$

$$
\begin{aligned}
&= \tfrac{1}{60}[(6r_1^2 + 4r_2^2 + 4r_3^2 + 4r_1 r_2 + 4r_1 r_3 + r_2 r_3) \\
&\qquad (4r_1^2 + 6r_2^2 + 4r_3^2 + 4r_1 r_2 + r_1 r_3 + 4r_2 r_3) \\
&\qquad (6r_1^2 + 4r_2^2 + 6r_3^2 + r_1 r_2 + 4r_1 r_3 + 4r_2 r_3)],
\end{aligned}
$$

where we have put the relationships of eqns. (10.2.13)–(10.2.15) to use.

The mesh and corresponding solution of the axisymmetric magnetic pot-core reactor of Figure 1.3.2 are shown in Figures 10.2.3A and 10.2.3B.

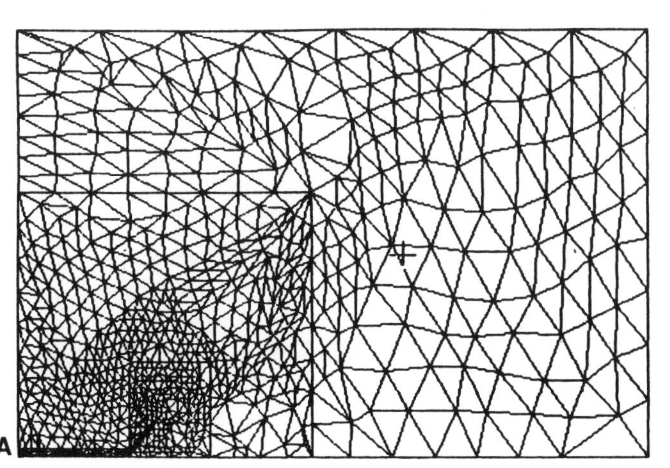


**Figure 10.2.3.** The pot-core reactor. **A.** The mesh. **B.** The solution.

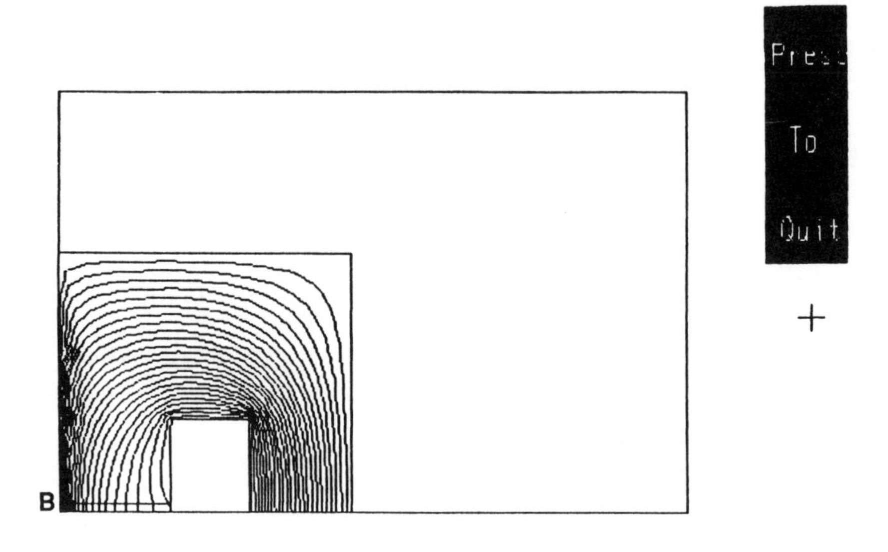


**Figure 10.2.3.** *(continued)*

## 10.3 THE DIFFUSION EQUATION IN EDDY CURRENT PROBLEMS

In steady-state low-frequency eddy current systems in magnetic structures, the single-component complex vector potential is governed by the diffusion equation

$$-\nu\nabla^2\hat{A} = \hat{J}_0 - j\omega\sigma\hat{A}, \tag{5.1.17}$$

where $\omega$ is the frequency and $\sigma$ is the conductivity. We have already encountered the solution of this problem in section 5.1 using finite differences, and we realized that as the depth of penetration $d = \sqrt{(\omega\mu\sigma)}$ decreased, the currents and magnetic fields became pronounced along the skin of conducting bodies. An accurate modeling of this requires a fine mesh close to conductor surfaces, and therefore the finite element method ideally fits the bill with its ability to accommodate more mesh points where required.

We shall follow in the footsteps of Chari and Silvester (1971) and Chari (1970, 1974) and identify a functional for this equation.

We already know from eqn. (8.1.5) that when we take differentials of

$$\mathscr{L}_1 = \iint \{\tfrac{1}{2}\nu[\nabla A]^2 - AJ_0\}\, dR, \tag{7.5.8}$$

what results under Neumann and Dirichlet boundary conditions is:

$$\delta\mathscr{L}_1 = \iint \delta A[-\nu\nabla^2 A - J_0]\, dR. \tag{10.3.1}$$

It is also simple to verify that

$$\mathscr{L}_2 = \iint \frac{j\omega\sigma}{2} A^2\, dR \tag{10.3.2}$$

has the differential

$$\delta\mathscr{L}_2 = \iint \delta A[j\omega\sigma A]\, dR. \tag{10.3.3}$$

Clearly, then, the functional we seek is:

$$\mathscr{L} = \mathscr{L}_1 + \mathscr{L}_2 = \mathscr{L}_1 = \iint \{\tfrac{1}{2}\nu[\nabla\hat{A}]^2 + \tfrac{1}{2}j\omega\sigma\hat{A}^2 - \hat{A}\hat{J}_0\}\, dR, \tag{10.3.4}$$

for we shall obtain, under the given boundary conditions,

$$\delta\mathscr{L} = \delta\mathscr{L}_1 + \delta\mathscr{L}_2 = \iint \widehat{\delta A}[-\nu\nabla^2\hat{A} - \hat{A}\hat{J}_0 + j\omega\sigma\hat{A}]\, dR, \tag{10.3.5}$$

so that at the extremum of this functional, we must have the diffusion equation satisfied.

To solve an eddy current problem, therefore, we shall assume the usual interpolation of the unknown complex values of $\hat{A}$ at the vertices so that the functional reduces to

$$\begin{aligned}\mathscr{L} &= \sum \iint \{\tfrac{1}{2}\nu\underset{\sim}{\hat{\mathbf{A}}}^t[\nabla\tilde{\boldsymbol{\alpha}}^{nt}\nabla\tilde{\boldsymbol{\alpha}}^n]\underset{\sim}{\hat{\mathbf{A}}} + \tfrac{1}{2}j\omega\sigma\underset{\sim}{\hat{\mathbf{A}}}^t\tilde{\boldsymbol{\alpha}}^{nt}\tilde{\boldsymbol{\alpha}}^n\underset{\sim}{\hat{\mathbf{A}}} - \underset{\sim}{\hat{\mathbf{A}}}^t\tilde{\boldsymbol{\alpha}}^{nt}\tilde{\boldsymbol{\alpha}}^n\underset{\sim}{\mathbf{J}_0}\}\, \mathrm{dR} \\ &= \sum \tfrac{1}{2}\underset{\sim}{\hat{\mathbf{A}}}^t\mathbf{P}\underset{\sim}{\hat{\mathbf{A}}} + \tfrac{1}{2}j\omega\sigma A^r\underset{\sim}{\hat{\mathbf{A}}}^t\mathbf{T}^{n,n}\underset{\sim}{\hat{\mathbf{A}}} - A^r\underset{\sim}{\hat{\mathbf{A}}}^t\mathbf{T}^{n,n}\underset{\sim}{\mathbf{J}_0},\end{aligned} \tag{10.3.6}$$

where the matrices $\mathbf{P}$ and $\mathbf{T}^{n,n}$ are already defined and are the same as those we encounter for the static problem. Differentiating with respect to $\underset{\sim}{\hat{\mathbf{A}}}$ by the rules of section 7.6 we have for solution the complex matrix equation

$$\sum [\mathbf{P} + j\omega\sigma A^r\mathbf{T}^{n,n}]\underset{\sim}{\hat{\mathbf{A}}} = \sum A^r\mathbf{T}^{n,n}\underset{\sim}{\hat{\mathbf{J}}_0}. \tag{10.3.7}$$

Notice that the sourcing function $\hat{J}_0$ may be taken as the reference phasor, as a result of which we have the simplification

$$\hat{J}_0 = J_{r0} + j0. \tag{10.3.8}$$

Eqn. (10.3.7) may be solved using some library on large computers or modifying our routine (Alg. 2.3.1 or Alg. 3.2.1) for complex arithmetic. Alternatively, if complex arithmetic is not supported by the operating system on the machine, we may split $\hat{A}$ into $A_r + jA_i$ and split the complex equation into its real and imaginary parts:

$$\sum [\mathbf{P}\underset{\sim}{\mathbf{A}}_r - \omega\sigma A^r \mathbf{T}^{n,n} \underset{\sim}{\mathbf{A}}_i] = \sum A^r \mathbf{T}^{n,n} \underset{\sim}{\mathbf{J}}_{r0}, \tag{10.3.9}$$

$$\sum [\omega\sigma A^r \mathbf{T}^{n,n} \underset{\sim}{\mathbf{A}}_r + \mathbf{P}\underset{\sim}{\mathbf{A}}_i] = 0. \tag{10.3.10}$$

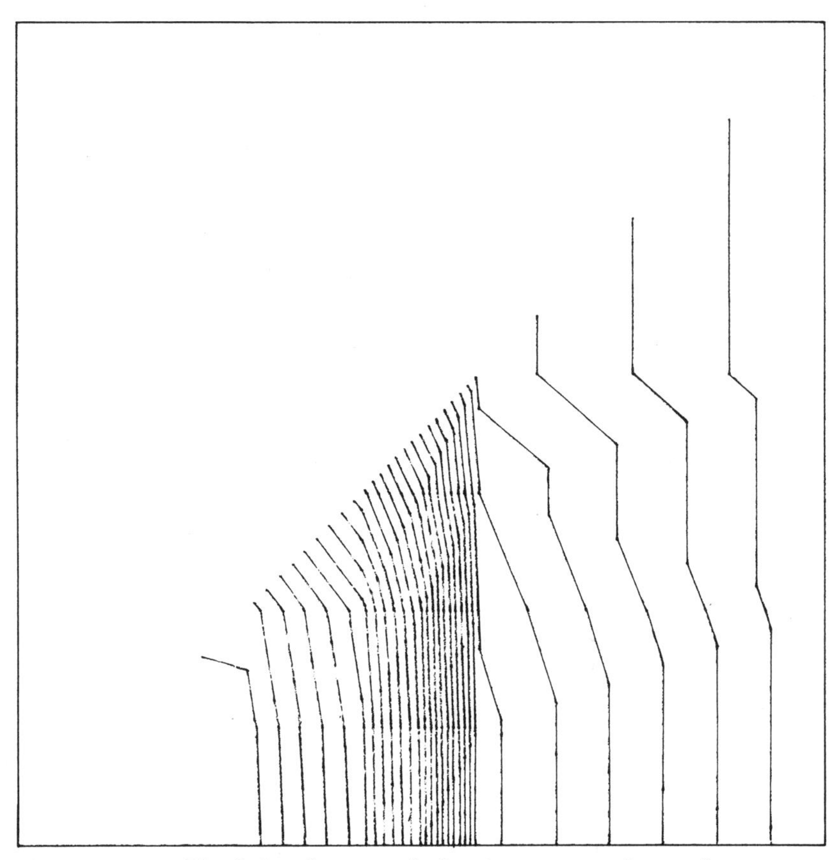

**Figure 10.3.1.** The finite element solution in a rectangular conductor.

This may be put into matrix form and solved for $A_r$ and $A_i$. The asymmetric nature of this matrix precludes the use of several powerful solution routines available to us. A way out is to premultiply by the transpose of the coefficient matrix and make the system symmetric. The disadvantage to such a course of action is that such squaring of the coefficient matrix reduces the sparsity of the system of equations. Since the coefficients of eqn. (10.3.7) are already in sparse, symmetric, and positive definite form, it is commonly convenient not to resort to splitting the matrix equation into real and imaginary parts, but rather to solve the equation straight using Liebmann's method described in Alg. 3.2.1. Where the computing environment does not support complex arithmetic, it is easy to define our own procedures implementing complex operations. Moreover, the coefficients of eqn. (10.3.7) are already in real and imaginary parts $\mathbf{P}$ and $A'\mathbf{T}^{n,n}$, so that we may easily assemble the complex coefficients in two separate matrices, one containing the real part and the other the imaginary part.

Figure 10.3.1 shows the solution of eddy currents in a symmetric eighth of a rectangular conductor by first-order finite elements. Observe the flux hugging the boundary of the conductor, as expected.

## 10.4 NONLINEAR PROBLEMS: SOLUTION BY NEWTON LINEARIZATION

### 10.4.1 Magnetic Field Problems

Under nonlinear conditions, the equation governing the sole component of the vector potential changes. Previously we had assumed, even in the chord method of section 4.6 for nonlinear problems, that $\nu$ is a constant. If we return to the origin of the equation for the vector potential, we will find that it is obtained by substituting $\mathbf{H} = \nu\nabla \times \mathbf{A}$ in eqn. (1.2.14). Thus in two dimensions, using eqn. (1.2.34) for flux density $\mathbf{B}$,

$$\nabla \times \nu\left[\mathbf{u}_x\frac{\partial A}{\partial y} - \mathbf{u}_y\frac{\partial A}{\partial x}\right] = \mathbf{u}_z J. \tag{10.4.1}$$

Expanding using eqns. (A1) and (A3), noting that in two dimensions $\partial/\partial z$ is zero and equating the single components in the $z$

direction that result, we obtain

$$-\frac{\partial}{\partial x}\nu\frac{\partial A}{\partial x}-\frac{\partial}{\partial y}\nu\frac{\partial A}{\partial y}=J, \tag{10.4.2}$$

which may be conveniently expressed as

$$-\nabla\,.\,\nu\nabla A=J, \tag{10.4.3}$$

which reduces to eqn. (1.2.32) when $\nu$ is constant.

We were introduced in section 3.6.3 to Newton's scheme for nonlinear problems, which has clear computational advantages over the chord scheme. However, to apply the scheme we must be able to obtain a linear equation for the error $\Delta A$ in the vector potential $A$. An obvious operator form to which the Poisson eqn. (10.4.3) lends itself is based on the residual

$$\mathcal{R}_1(A)=\nabla\,.\,\nu\nabla A+J=0. \tag{10.4.4}$$

Following Chari (1970) and Silvester and Chari (1970), let us try to linearize this for an error $\Delta A^i$ in a current estimate $A^i$ in the vector potential, following the rules of section 3.6.3. Since $A^i+\Delta A^i$ is the correct solution of (10.4.4),

$$\mathcal{R}_1(A^{i+1})=0. \tag{10.4.5}$$

Or:

$$\mathcal{R}_1(A^i+\Delta A^i)=\nabla\,.\,\nu(A^i+\Delta A^i)\nabla(A^i+\Delta A^i)+J=0. \tag{10.4.6}$$

We must now expand this and obtain a linear equation for $\Delta A^i$ in terms of the known quantities $A^i$. Before trying to do this, let us rewrite eqn. (10.2.19) as

$$B^2=B_x^2+B_y^2=\left(\frac{\partial A}{\partial y}\right)^2+\left(\frac{\partial A}{\partial x}\right)^2=\nabla A\,.\,\nabla A=[\nabla A]^2, \tag{10.4.7}$$

and consider

$$\begin{aligned}\frac{\partial\nu}{\partial A}\Delta A&=\frac{\partial\nu}{\partial B^2}\frac{\partial B^2}{\partial A}\Delta A=\kappa\Delta B^2=\kappa\{[\nabla(A^i+\Delta A^i)]^2-[\nabla A^i]^2\}\\&=\kappa\{[\nabla A^i]^2+2[\nabla A^i][\nabla\,\Delta A^i]+[\nabla\,\Delta A^i]^2-[\nabla A^i]^2\}\\&\approx 2\kappa[\nabla A^i][\nabla\,\Delta A^i],\end{aligned} \tag{10.4.8}$$

where the second-order terms in $\Delta A^i$ have been neglected and $\kappa$ is the derivative of $\nu$ with respect to $B^2$.

With this result, we are ready to expand eqn. (10.4.6):

$$\mathcal{R}_1(A^i + \Delta A^i) \approx \nabla . \{[\nu(A^i) + \frac{\partial \nu}{\partial A}(A^i)\Delta A^i][\nabla A^i + \nabla \Delta A^i]\} + J$$

by Taylor's series

$$= \nabla . \{[\nu^i + 2\kappa(\nabla A^i) . (\nabla \Delta A^i)][\nabla A^i + \nabla \Delta A^i]\} + J$$

using eqn. (10.4.8)

$$\approx \nabla . \nu^i \nabla A^i + \nabla . \nu^i \nabla \Delta A^i + \nabla . [2\kappa^i(\nabla A^i) . (\nabla \Delta A^i)\nabla A^i] + J = 0, \tag{10.4.9}$$

where again, we have neglected the higher order terms of $\Delta A^i$, and $\nu^i$ and $\kappa^i$ represent the values of $\nu$ and $\kappa$ evaluated when $A = A^i$. This must be solved in the solution region $R$ with either a Dirichlet condition imposed through the trial function or the Neumann condition

$$\nu \frac{\partial(A^i + \Delta A^i)}{\partial n} = \nu \nabla(A^i + \Delta A^i) . \mathbf{u}_n = \sigma, \tag{10.4.10}$$

which has a residual

$$\mathcal{R}_2 = \sigma - \nu \nabla(A^i + \Delta A^i) . \mathbf{u}_n \tag{10.4.11}$$

along the boundary $S$.

Resorting to a finite element solution, we shall divide the solution region into triangles and consider only first-order interpolation; a great simplification arises on account of this, as we shall see. Theoretically, higher-order interpolations may be used, as in section 8.5, but this will result in inordinate complexity. Therefore by the Galerkin principle, after putting in the approximations of eqn. (7.5.3) for the vector potential $A$ and the error $\Delta A$:

$$A = \tilde{\boldsymbol{\alpha}} \underset{\sim}{\mathbf{A}}, \tag{10.4.12}$$

$$\Delta A = \tilde{\boldsymbol{\alpha}} . \underset{\sim}{\Delta \mathbf{A}}, \tag{10.4.13}$$

we must have

$$\iint \tilde{\boldsymbol{\alpha}}^t \mathcal{R}_1 \, dR + \int \tilde{\boldsymbol{\alpha}}^t \mathcal{R}_2 \, dS = 0, \tag{10.4.14}$$

where eqn. (10.4.14) represents three equations where in each row of the equation, we are reducing the residual in the "direction" of each component of the interpolation vector $\tilde{\boldsymbol{\alpha}}$:

$$\iint \tilde{\boldsymbol{\alpha}}^t \{\nabla . \nu^i \nabla \tilde{\boldsymbol{\alpha}} \underset{\sim}{\mathbf{A}}^i + \nabla . \nu^i \nabla \tilde{\boldsymbol{\alpha}} \underset{\sim}{\Delta \mathbf{A}}^i$$
$$+ \nabla . [2\kappa^i(\nabla \tilde{\boldsymbol{\alpha}} \underset{\sim}{\mathbf{A}}^i) . (\nabla \tilde{\boldsymbol{\alpha}} \, \underset{\sim}{\Delta \mathbf{A}}^i) \nabla \tilde{\boldsymbol{\alpha}} \underset{\sim}{\mathbf{A}}^i] + \tilde{\boldsymbol{\alpha}} \underset{\sim}{\mathbf{J}}\} \, dR$$
$$+ \int \tilde{\boldsymbol{\alpha}}^t [\sigma - \nu \nabla(\tilde{\boldsymbol{\alpha}} \underset{\sim}{\mathbf{A}}^i + \tilde{\boldsymbol{\alpha}} \, \underset{\sim}{\Delta \mathbf{A}}^i) . \mathbf{u}_n] \, dS = 0. \tag{10.4.15}$$

Note that $\tilde{\boldsymbol{\alpha}}$ is a row vector of length 3 defined in eqn. (7.5.4), and $\underset{\sim}{\mathbf{A}}$ is a column vector of the same length, so that their compatible multiple is a scalar. $\nabla$ operating on a scalar gives a 2-vector, the components consisting of derivatives with respect to $x$ and $y$.

Now we have from the identity eqn. (A11) for any row vector $\mathbf{P}$ of compatible length 3:

$$\tilde{\boldsymbol{\alpha}}^t\nabla \,.\, \mathbf{P} = \nabla \,.\, \tilde{\boldsymbol{\alpha}}^t\mathbf{P} - [\nabla\tilde{\boldsymbol{\alpha}}^t] \,.\, \mathbf{P}, \tag{10.4.16}$$

and from eqn. (A13):

$$\iint \tilde{\boldsymbol{\alpha}}^t\nabla \,.\, \mathbf{P}\, dR = \int \tilde{\boldsymbol{\alpha}}^t\mathbf{P} \,.\, \mathbf{u}_n\, dS - \iint [\nabla\tilde{\boldsymbol{\alpha}}^t] \,.\, \mathbf{P}\, dR. \tag{10.4.17}$$

Applying this result to eqn. (10.4.15):

$$\begin{aligned}
&\int v^i\tilde{\boldsymbol{\alpha}}^t\nabla\tilde{\boldsymbol{\alpha}}\underset{\sim}{\mathbf{A}}^i \,.\, \mathbf{u}_n\, dS - \iint v^i\nabla\tilde{\boldsymbol{\alpha}}^t\nabla\tilde{\boldsymbol{\alpha}}\underset{\sim}{\mathbf{A}}^i\, dR \\
&\quad + \int v^i\tilde{\boldsymbol{\alpha}}^t\nabla\tilde{\boldsymbol{\alpha}}\, \underset{\sim}{\Delta\mathbf{A}}^i \,.\, \mathbf{u}_n\, dS - \iint v^i\, \nabla\tilde{\boldsymbol{\alpha}}^t\nabla\tilde{\boldsymbol{\alpha}}\, \underset{\sim}{\Delta\mathbf{A}}^i\, dR \\
&\quad + \int \alpha^t[2\kappa^i(\nabla\tilde{\boldsymbol{\alpha}}\underset{\sim}{\mathbf{A}}^i) \,.\, (\nabla\tilde{\boldsymbol{\alpha}}\, \underset{\sim}{\Delta\mathbf{A}}^i)\nabla\underset{\sim}{\mathbf{A}}^i] \,.\, \mathbf{u}_n\, dS \\
&\quad - \iint 2\kappa\nabla\tilde{\boldsymbol{\alpha}}^t[(\nabla\tilde{\boldsymbol{\alpha}}\underset{\sim}{\mathbf{A}}^i) \,.\, (\nabla\tilde{\boldsymbol{\alpha}}\, \underset{\sim}{\Delta\mathbf{A}}^i)\nabla\tilde{\boldsymbol{\alpha}}\underset{\sim}{\mathbf{A}}^i]\, dR \\
&\quad + \iint \tilde{\boldsymbol{\alpha}}^t\tilde{\boldsymbol{\alpha}}\underset{\sim}{\mathbf{J}}\, dR + \int \tilde{\boldsymbol{\alpha}}^t[\sigma - v^i\nabla(\tilde{\boldsymbol{\alpha}}\underset{\sim}{\mathbf{A}}^i + \tilde{\boldsymbol{\alpha}}\, \underset{\sim}{\Delta\mathbf{A}}^i) \,.\, \mathbf{u}_n]\, dS = 0.
\end{aligned} \tag{10.4.18}$$

All but one of the surface terms cancel off because of the boundary conditions and we are left with:

$$\begin{aligned}
&- \iint v^i\nabla\tilde{\boldsymbol{\alpha}}^t\nabla\tilde{\boldsymbol{\alpha}}\underset{\sim}{\mathbf{A}}^i\, dR - \iint v^i\nabla\tilde{\boldsymbol{\alpha}}^t\nabla\tilde{\boldsymbol{\alpha}}\, \underset{\sim}{\Delta\mathbf{A}}^i\, dR \\
&\qquad - \iint 2\kappa^i\nabla\tilde{\boldsymbol{\alpha}}^t[(\nabla\tilde{\boldsymbol{\alpha}}\underset{\sim}{\mathbf{A}}^i) \,.\, (\nabla\tilde{\boldsymbol{\alpha}}\, \underset{\sim}{\Delta\mathbf{A}}^i)\, \nabla\tilde{\boldsymbol{\alpha}}\underset{\sim}{\mathbf{A}}^i]\, dR \\
&\qquad\qquad + \iint \tilde{\boldsymbol{\alpha}}^t\tilde{\boldsymbol{\alpha}}\underset{\sim}{\mathbf{J}}\, dR + \int \tilde{\boldsymbol{\alpha}}^t\sigma\, dS = 0.
\end{aligned} \tag{10.4.19}$$

Let us look at the individual terms:

$$\begin{aligned}
\iint v^i\nabla\tilde{\boldsymbol{\alpha}}^t\nabla\tilde{\boldsymbol{\alpha}}^n\underset{\sim}{\mathbf{A}}^i\, dR &= \iint v^i\left[\left(\frac{\partial\tilde{\boldsymbol{\alpha}}}{\partial x}\right)^2 + \frac{\partial\tilde{\boldsymbol{\alpha}}}{\partial y}\right]\underset{\sim}{\mathbf{A}}^i\, dR \\
&= \iint v^i[(\mathbf{D}_x)^2 + (\mathbf{D}_y)^2]\underset{\sim}{\mathbf{A}}^i\, dR \qquad \text{from eqn. (8.5.12),}
\end{aligned} \tag{10.4.20}$$

$$= \nu^i \iint [(\mathbf{D}_x^t)(\mathbf{D}_x) + (\mathbf{D}_y^t)(\mathbf{D}_y)]\underset{\sim}{\mathbf{A}}^i \, dR$$

$$= \nu_i \mathbf{D}_x^t \left\{ \iint dR \right\} \mathbf{D}_x \underset{\sim}{\mathbf{A}}^i + \nu^i \mathbf{D}_y^t \left\{ \iint dR \right\} \mathbf{D}_y \underset{\sim}{\mathbf{A}}^i$$

$$= \nu^i A^r [\mathbf{D}_x^t \mathbf{D}_x + \mathbf{D}_y^t \mathbf{D}_y] \underset{\sim}{\mathbf{A}}^i$$

$$= \mathbf{P}^i \underset{\sim}{\mathbf{A}}^i,$$

where $A^r$ is the area of the triangle, the matrix $\mathbf{T}$ and $\mathbf{P}$ is defined in eqn. (8.5.28), and the first-order differentiation matrices $\mathbf{D}_x$ and $\mathbf{D}_y$ are the same as $\tilde{\mathbf{b}}$ and $\tilde{\mathbf{c}}$ defined in eqns. (7.6.3) and (7.6.4). Observe from section 8.5 that $\mathbf{P}$ is the usual finite element matrix arising in linear analysis. Notice that because we assumed first-order interpolation, the derivative of $\tilde{\boldsymbol{\alpha}}$ will be constant in each triangle, so that $\nu$ too will be constant; as a result we have been able to pull $\nu$ outside the integral sign. Had we chosen higher-order interpolations the integral would have been very complex and might have required the integration methods of section 8.6 to evaluate. In addition to this disadvantage, the third term considered next (if of higher order) poses the hazard of getting stuck in a mathematical bog.

The second term is very similar to the first and needs no special mention. The third term, however, would have been very complex had we decided on higher-order interpolations. It would have required careful expansion into a complicated matrix whose coefficients will be expressible in terms of $\zeta_1$, $\zeta_2$, $\zeta_3$, the known values $\underset{\sim}{\mathbf{A}}^i$, and the unknowns $\underset{\sim}{\mathbf{\Delta A}}^i$; upon integration using eqn. (D1), the term results in the form $\mathbf{M}\underset{\sim}{\mathbf{\Delta A}}^i$ where $\mathbf{M}$ is a known coefficient matrix. Instead of risking a mess, we have chosen to use first-order interpolation. Taking up the term under the third integral and using the scalar product of the two vectors:

$$\iint 2\kappa \nabla \tilde{\boldsymbol{\alpha}}^t [(\nabla \tilde{\boldsymbol{\alpha}} \underset{\sim}{\mathbf{A}}^i) . (\nabla \tilde{\boldsymbol{\alpha}} \, \underset{\sim}{\mathbf{\Delta A}}^i) \nabla \tilde{\boldsymbol{\alpha}} \underset{\sim}{\mathbf{A}}^i] \, dR$$

$$= \iint 2\kappa \nabla \tilde{\boldsymbol{\alpha}}^t [(\underset{\sim}{\mathbf{A}}^{it} \nabla \tilde{\boldsymbol{\alpha}}^t \nabla \tilde{\boldsymbol{\alpha}} \, \underset{\sim}{\mathbf{\Delta A}}^i) \nabla \tilde{\boldsymbol{\alpha}} \underset{\sim}{\mathbf{A}}^i] \, dR \tag{10.4.21}$$

$$= 2\kappa \iint \{\nabla \tilde{\boldsymbol{\alpha}}^t (\underset{\sim}{\mathbf{A}}^{it} \mathbf{S} \underset{\sim}{\mathbf{\Delta A}}^i) \nabla \tilde{\boldsymbol{\alpha}} \underset{\sim}{\mathbf{A}}^i\} \, dR,$$

where, from eqn. (7.6.9),

$$\mathbf{S} = \nabla \tilde{\boldsymbol{\alpha}}^t \nabla \tilde{\boldsymbol{\alpha}} = \frac{\mathbf{P}}{\nu^i A^r}, \tag{10.4.22}$$

a constant $3 \times 3$ matrix because $\tilde{\alpha}$ is of order 1. Now since $\underset{\sim}{\mathbf{A}}^i$ and $\underset{\sim}{\Delta \mathbf{A}}^i$ are both column vectors of length 3, $\underset{\sim}{\mathbf{A}}^{it}\mathbf{S}\underset{\sim}{\Delta\mathbf{A}}^i$ is a constant scalar and so may be moved to the right of $\nabla\tilde{\alpha}\underset{\sim}{\mathbf{A}}^i$. Doing so and dropping the brackets because of the size compatibility as shown by the suffixes, we have:

$$\iint 2\kappa \nabla\tilde{\alpha}^t[(\nabla\tilde{\alpha}\underset{\sim}{\mathbf{A}}^i)\,.\,(\nabla\tilde{\alpha}\ \underset{\sim}{\Delta\mathbf{A}}^i)\nabla\underset{\sim}{\mathbf{A}}^i]\ dR$$

$$= 2\kappa \iint \{\nabla\tilde{\alpha}^t(\underset{\sim}{\mathbf{A}}^{it}\mathbf{S}\ \underset{\sim}{\Delta\mathbf{A}}^i)\nabla\tilde{\alpha}\underset{\sim}{\mathbf{A}}^i\}\ dR$$

$$= 2\kappa \iint (\nabla\tilde{\alpha}^t)_{3\times 2}(\nabla\tilde{\alpha})_{2\times 3}\underset{\sim}{\mathbf{A}}^i_{3\times 1}(\underset{\sim}{\mathbf{A}}^{it}_{1\times 3}\mathbf{S}_{3\times 3}\ \underset{\sim}{\Delta\mathbf{A}}^i_{3\times 1})\ dR \qquad (10.4.23)$$

$$= 2\kappa[\nabla\tilde{\alpha}^t\nabla\tilde{\alpha}]\underset{\sim}{\mathbf{A}}^i\underset{\sim}{\mathbf{A}}^{it}\mathbf{S}\ \underset{\sim}{\Delta\mathbf{A}}^i \iint dR$$

$$= \frac{2\kappa\mathbf{P}\underset{\sim}{\mathbf{A}}^i\underset{\sim}{\mathbf{A}}^{it}\mathbf{P}\ \underset{\sim}{\Delta\mathbf{A}}^i}{v^{i2}A^r}.$$

The integral $\int \varepsilon\sigma\tilde{\alpha}^t\, dS$ has already been dealt with at the end of section 8.3. Thus, the finite element solution requires our finding $\underset{\sim}{\Delta\mathbf{A}}^i$ from eqn. (10.4.19):

$$-\mathbf{P}\underset{\sim}{\mathbf{A}}^i - \mathbf{P}\ \underset{\sim}{\Delta\mathbf{A}}^i - \frac{2\kappa}{v^{i2}A^r}\mathbf{P}\underset{\sim}{\mathbf{A}}^i\underset{\sim}{\mathbf{A}}^{it}\mathbf{P}\ \underset{\sim}{\Delta\mathbf{A}}^i + A^r\mathbf{T}^{11}\underset{\sim}{\mathbf{J}} + \mathbf{v} = 0, \qquad (10.4.24)$$

where the vector $\mathbf{v}$ is defined in eqn. (8.3.9). Thus, we repeatedly solve

$$\left[\mathbf{P} + \frac{2\kappa}{v^{i2}A^r}\mathbf{P}\underset{\sim}{\mathbf{A}}^i\underset{\sim}{\mathbf{A}}^{it}\mathbf{P}\right]\underset{\sim}{\Delta\mathbf{A}}^i = A^r\mathbf{T}^{11}\underset{\sim}{\mathbf{J}} + \mathbf{v} - \mathbf{P}\underset{\sim}{\mathbf{A}}^i \qquad (10.4.25)$$

for $\underset{\sim}{\Delta\mathbf{A}}^i$, add it to $\underset{\sim}{\mathbf{A}}^i$, and get the new value

$$\underset{\sim}{\mathbf{A}}^{i+1} = \underset{\sim}{\mathbf{A}}^i + \underset{\sim}{\Delta\mathbf{A}}^i, \qquad (10.4.26)$$

from which by differentiation according to eqn. (1.2.34) we get $B$ and the corresponding values of $v$ and $\kappa$, and then solve for $\underset{\sim}{\Delta\mathbf{A}}^i$ again, and so on, until $\underset{\sim}{\Delta\mathbf{A}}^i$ becomes acceptably small. The coefficient matrix of eqn. (10.4.33) is best formed by computing $\mathbf{P}$, the usual finite element local matrix, adding it to an initialized matrix, then computing

$$\underset{\sim}{\mathbf{w}} = \mathbf{P}\underset{\sim}{\mathbf{A}}^i, \qquad (10.4.27)$$

and then using the fact that $\mathbf{P}$ is symmetric to write:

$$\mathbf{P}\underset{\sim}{\mathbf{A}}^i\underset{\sim}{\mathbf{A}}^{it}\mathbf{P} = (\mathbf{P}\underset{\sim}{\mathbf{A}}^i)(\mathbf{P}\underset{\sim}{\mathbf{A}}^i)^t = \underset{\sim}{\mathbf{w}}\underset{\sim}{\mathbf{w}}^t, \qquad (10.4.28)$$

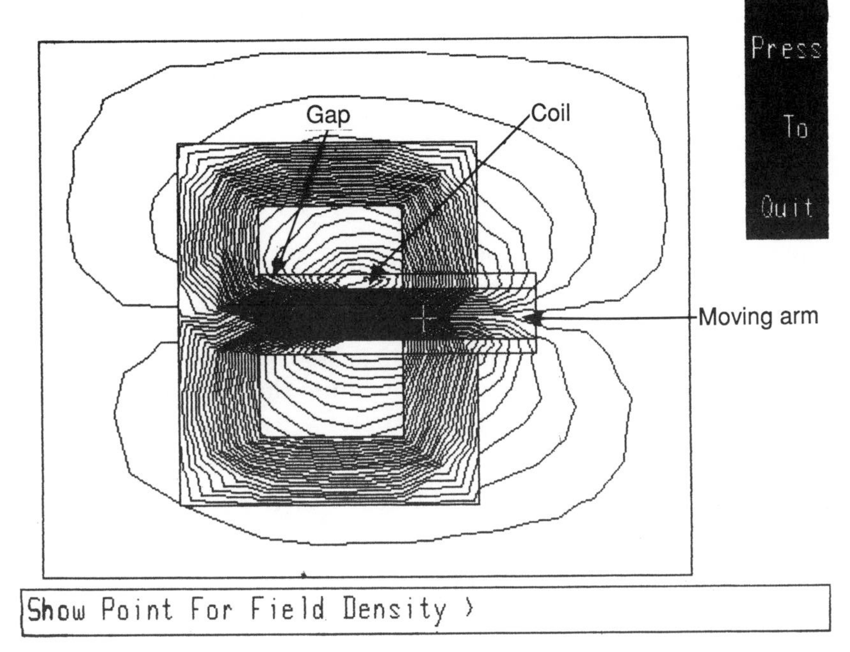


**Figure 10.4.1.** The nonlinear magnetic switch.

which may be appropriately scaled and added to the coefficient matrix to obtain its value.

A solution of an electric switch by this method is shown plotted in Figure 10.4.1. Here, as the coils are excited the flux flows through the air gap and the pole pieces across the gap act as opposite poles. The moving arm is thus attracted and as it moves, a spring-loaded switch is opened.

## 10.4.2 Electric Field Problems: Silicon-on-Sapphire

When a material of permittivity $\varepsilon$ is affected by an electric field, the governing equation from eqns. (1.2.11), (1.2.13), and (1.2.21) is:

$$-\nabla \cdot \varepsilon \nabla \phi = \rho, \qquad (10.4.29)$$

completely in analogy with eqn. (10.4.3). When $\varepsilon$ is constant, it reduces to eqn. (1.2.22), which we have employed up to now. Therefore by analogy, the Newton equation for solution is eqn. (10.4.25) with $\phi$ taking the place of $A$, $\rho$ the place of $J$, and $\varepsilon$ the place of $v$. All the considerations of section 10.4.1 will then apply.

In electric field problems, however, the nonlinearity more commonly arises from the sources, as we saw in section 3.6.5. There we had to solve the nonlinear Poisson eqn. (3.6.30) in an electron device, and we linearized this to the form

$$-\varepsilon\nabla^2\Delta\phi + f[\phi]\,\Delta\phi = \varepsilon\nabla^2\phi + g[\phi], \tag{3.6.30}$$

which we must repeatedly solve for $\Delta\phi$ until it reduces to acceptably small amounts. To solve this equation by finite elements, we will as usual discretize the solution region into elements, and, as with finite differences in section 3.7, associate two variables with each node—$\phi$ and $\Delta\phi$. Putting in the finite element approximations $\tilde{\boldsymbol{\alpha}}^n\underset{\sim}{\boldsymbol{\phi}}$ and $\tilde{\boldsymbol{\alpha}}^n\underset{\sim}{\boldsymbol{\Delta\phi}}$ for $\phi$ and $\Delta\phi$, and applying the Galerkin principle:

$$\iint \tilde{\boldsymbol{\alpha}}^{nt}\{-\varepsilon\nabla^2\tilde{\boldsymbol{\alpha}}^n\underset{\sim}{\boldsymbol{\Delta\phi}} + f[\tilde{\boldsymbol{\alpha}}^n\underset{\sim}{\boldsymbol{\phi}}]\tilde{\boldsymbol{\alpha}}^n\,\underset{\sim}{\boldsymbol{\Delta\phi}} - \varepsilon\nabla^2\tilde{\boldsymbol{\alpha}}^n\underset{\sim}{\boldsymbol{\phi}} - g[\tilde{\boldsymbol{\alpha}}^n\underset{\sim}{\boldsymbol{\phi}}]\}\,dR + \int_\Gamma \varepsilon\nabla\phi\,.\,\mathbf{u}_n\,dS = 0, \tag{10.4.30}$$

where we have assumed a zero-gradient Neumann boundary along a portion $\Gamma$ of the boundary $S$ to keep the equations simple. Generalizing this to more complex conditions along the lines of section 10.4.1, using Gauss's theorem (A13) as with the magnetic field problem, and cancelling the surface terms, we have:

$$\iint \varepsilon\{\nabla\tilde{\boldsymbol{\alpha}}^{nt}\,.\,\underset{\sim}{\nabla\tilde{\boldsymbol{\alpha}}^n}\}\,dR\,\underset{\sim}{\boldsymbol{\Delta\phi}} + \iint \{\tilde{\boldsymbol{\alpha}}^{nt}f[\tilde{\boldsymbol{\alpha}}^n\underset{\sim}{\boldsymbol{\phi}}]\tilde{\boldsymbol{\alpha}}^n\}\,dR\,\underset{\sim}{\boldsymbol{\Delta\phi}}$$
$$= \iint \{\tilde{\boldsymbol{\alpha}}^{nt}g[\tilde{\boldsymbol{\alpha}}^n\underset{\sim}{\boldsymbol{\phi}}]\tilde{\boldsymbol{\alpha}}^n\}\,dR\underset{\sim}{\boldsymbol{\phi}} - \iint \{\varepsilon\nabla\tilde{\boldsymbol{\alpha}}^{nt}\,.\,\nabla\tilde{\boldsymbol{\alpha}}^n\}\,dR\,\underset{\sim}{\boldsymbol{\phi}}. \tag{10.4.31}$$

The first and last integrals are very straightforward and are in fact the usual static linear finite element matrices **P** of eqn. (8.5.28). The other two integrals are horrendous to evaluate exactly, if at all possible, so we must resort to the approximate techniques of section 8.6. What we shall do is evaluate $f$ at the $N(n)$ interpolation nodes on each triangle where $\phi$ is known and make the $n$th-order approximation as in eqn. (8.6.1):

$$f = \tilde{\boldsymbol{\alpha}}^n\underset{\sim}{\mathbf{f}} = \sum \alpha_i f_i, \tag{10.4.32}$$

where the summation is understood to run from 1 to $N$. Now

consider the integral:

$$\iint \{\tilde{\boldsymbol{\alpha}}^{nt} f \tilde{\boldsymbol{\alpha}}^{n}\}\, dR = \iint \begin{bmatrix} f\alpha_1^2 & f\alpha_1\alpha_2 & f\alpha_1\alpha_3 \ldots f\alpha_1\alpha_N \\ f\alpha_2\alpha_1 & f\alpha_2^2 & f\alpha_2\alpha_3 \ldots f\alpha_2\alpha_N \\ \cdot & \cdot & \cdot \\ \cdot & \cdot & \cdot \\ f\alpha_N\alpha_1 & f\alpha_N\alpha_2 & f\alpha_N\alpha_3 \ldots f\alpha_N^2 \end{bmatrix} dR. \quad (10.4.33)$$

The result is obviously an $N \times N$ symmetric matrix $\mathbf{M}_f$ with the coefficient at row $i$, column $j$ being given by:

$$\mathbf{M}_{fij} = \iint f\alpha_i\alpha_j\, dR = \iint \alpha_i\alpha_j \sum_k \alpha_k f_k\, dR = \sum_k \left\{\iint \alpha_i\alpha_j\alpha_k\, dR\right\} f_k$$

$$= A^r \sum_k p_{ijk} f_k, \quad (10.4.34)$$

where the summation is over the suffix $k$, $A^r$ is the area of the element, and:

$$p_{ijk} = \frac{1}{A^r}\iint \alpha_i\alpha_j\alpha_k\, dR \qquad i, j, k = 1, 2, \ldots, N \quad (10.4.35)$$

is a three-suffix coefficient independent of area and may be evaluated using eqn. (D1). These coefficients $p_{ijk}$ are like the universal matrices of section 8.5.2 and the weights of section 8.6—for they need to be evaluated only once, tabulated, and stored, to be recalled when necessary. For example, for first-order elements, we have from eqn. (7.5.3) $N = 3$ $\alpha_1 = \zeta_1$, $\alpha_2 = \zeta_2$, and $\alpha_3 = \zeta_3$, so that we need to evaluate only 27 coefficients:

$$p_{111} = \frac{1}{A^r}\iint \zeta_1^3\, dR = \frac{3!\,2!}{5!} = \frac{1}{10}, \quad (10.4.36)$$

$$p_{222} = p_{333} = \frac{1}{10}, \quad (10.4.37)$$

$$p_{112} = \frac{1}{A^r}\iint \zeta_1^2\zeta_2\, dR = \frac{2!\,2!}{5!} = \frac{1}{30}, \quad (10.4.38)$$

$$p_{211} = p_{121} = p_{223} = p_{322} = p_{232} = p_{332} = p_{233} = p_{323} =$$
$$p_{133} = p_{313} = p_{331} = p_{122} = p_{221} = p_{212} = p_{113} = p_{311} = p_{131} = \tfrac{1}{30}, \quad (10.4.39)$$

$$p_{123} = \frac{1}{A^r}\iint \zeta_1\zeta_2\zeta_3\, dR = \frac{2!}{5!} = \frac{1}{60}, \quad (10.4.40)$$

$$p_{231} = p_{321} = p_{132} = p_{213} = p_{312} = \frac{1}{60}. \quad (10.4.41)$$

These coefficients may be stored in three-dimensional arrays and used in computing $\mathbf{M}_f$. Similarly, the integral involving $g$ may be computed into a numeric symmetric matrix $\mathbf{M}_g$. Eqn. (10.4.32) becomes

$$[\mathbf{P} + \mathbf{M}_f]\, \underset{\sim}{\boldsymbol{\Delta\phi}} = \mathbf{M}_g - \mathbf{P}\underset{\sim}{\boldsymbol{\phi}}. \qquad (10.4.42)$$

This symmetric matrix equation may be assembled and solved repeatedly until $\underset{\sim}{\boldsymbol{\Delta\phi}}$ becomes small enough for us. $f$ and $g$ being exponential functions, it is the evaluation of these at the interpolation nodes that makes the solution of electron devices a relatively time-consuming operation on the computer.

The silicon-on-sapphire (SOS) device of Figure 10.4.2A is increasingly used for spacecraft applications. It was reduced to the

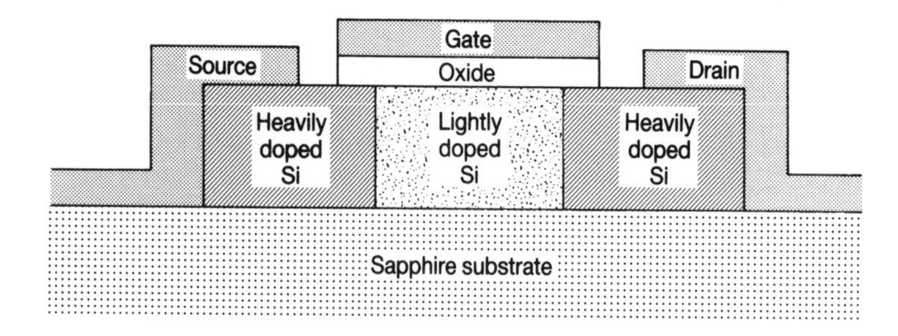


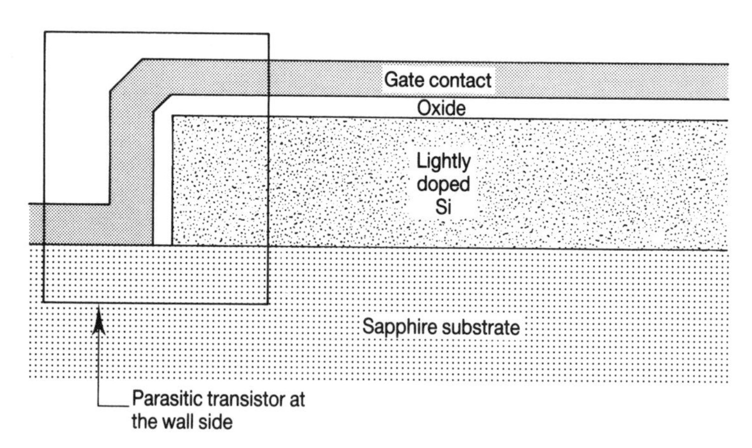


**Figure 10.4.2.** Silicon-on-Sapphire. **A.** The device. **B.** The boundary value problem. **C.** The solution.

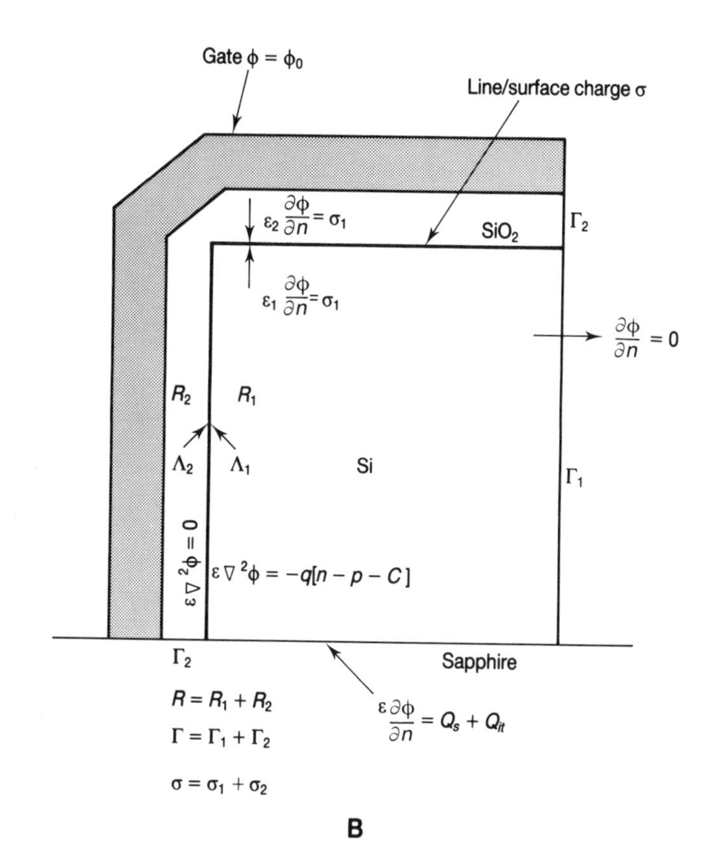

Gate $\phi = \phi_0$
Line/surface charge $\sigma$
$\varepsilon_2 \frac{\partial \phi}{\partial n} = \sigma_1$
$SiO_2$
$\Gamma_2$
$\varepsilon_1 \frac{\partial \phi}{\partial n} = \sigma_1$
$\frac{\partial \phi}{\partial n} = 0$
$R_2$
$R_1$
$\Lambda_2$
$\Lambda_1$
Si
$\Gamma_1$
$\varepsilon \nabla^2 \phi = 0$
$\varepsilon \nabla^2 \phi = -q[n - p - C]$
$\Gamma_2$
Sapphire
$R = R_1 + R_2$
$\Gamma = \Gamma_1 + \Gamma_2$
$\sigma = \sigma_1 + \sigma_2$
$\varepsilon \frac{\partial \phi}{\partial n} = Q_s + Q_{it}$


**B**

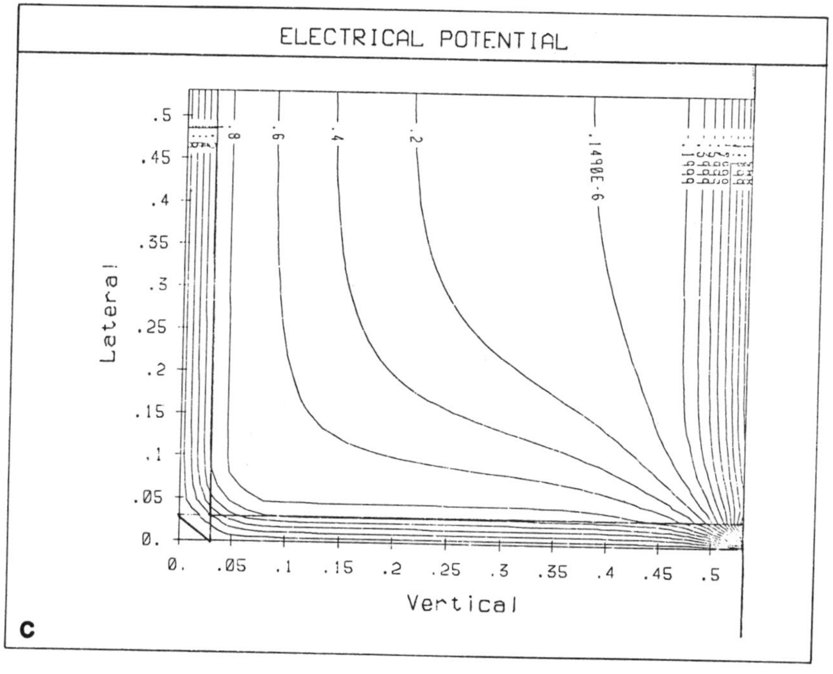

ELECTRICAL POTENTIAL
Lateral
Vertical
.8
.6
.4
.2
.1490E-6


**C**

two-dimensional boundary value problem of Figure 10.4.2B and solved by this scheme. A plot of the solution is shown in Figure 10.4.2C.

## 10.5 THE HELMHOLTZ EQUATION IN WAVEGUIDE PROBLEMS

Propagating waves, we have seen, may be described by the Helmholtz wave equation:

$$-\nabla^2\phi = s + k^2\phi, \tag{10.5.1}$$

where $k$ is the cutoff wave number. Comparing this equation with eqn. (10.3.4), the functional corresponding to the wave equation must be

$$\mathcal{L} = \iint [\tfrac{1}{2}\nabla\phi]^2 + \tfrac{1}{2}k^2\phi^2 - \phi s]\, dR. \tag{10.5.2}$$

The extremization of this functional is just as in eqn. (10.3.7) and may easily be implemented. In waveguides, the source term $s$ usually does not exist and we obtain the eigenvalue problem

$$\sum \mathbf{P}\underset{\sim}{\boldsymbol{\phi}} = \sum k^2 A^r \mathbf{T}^{n,n}\underset{\sim}{\boldsymbol{\phi}}, \tag{10.5.3}$$

or after assembly into global form,

$$\mathbf{P}^g\underset{\sim}{\boldsymbol{\phi}}^g = k^2\mathbf{T}^g\underset{\sim}{\boldsymbol{\phi}}^g, \tag{10.5.4}$$

which needs to be linearized. To this end, we split the nonsingular matrix $T^g$ into lower and upper Cholesky factors $\mathbf{LL}^t$ and rewrite the equation as

$$\mathbf{L}^{-1}\mathbf{P}^g\mathbf{L}^{-t}[\mathbf{L}^t\underset{\sim}{\boldsymbol{\phi}}^g] = k^2[\mathbf{L}^t\underset{\sim}{\boldsymbol{\phi}}^g], \tag{10.5.5}$$

for which the eigenvalues $k^2$ may be computed. Alternatively, when possible, $\mathbf{P}^g$ may be factorized as described in section 6.8.4 It has been shown by Silvester (1968, 1969) that, for the same number of nodes, a few high-order triangles give far superior results compared with several low-order triangles.

## 10.6 BOUNDARY INTEGRAL EQUATIONS AND FINITE ELEMENTS

We have worked several problems using integral equations, such as in working out the charge on a capacitor, the current in a conductor, and the magnetization in a permeable structure. Although we did not quite realize it at the time, the methods come under a Galerkin finite element scheme. We used "pulse" functions to generate the unknowns; i.e., in taking the unknown to be a constant in each element, we effectively said that, taking the charge problem as example:

$$q(x, y) = \sum_k q_k \alpha_k(x, y), \tag{10.6.1}$$

where the pulse function $\alpha_k$ is one on element $k$ and zero elsewhere. The residual of eqn. (2.1.6) we were solving becomes:

$$\mathcal{R} = \iiint \frac{q}{4\pi\varepsilon r} dV - \phi = \sum_j \iiint \left[\frac{q(x_j, y_j)}{4\pi\varepsilon r_j}\right] dR - \phi, \tag{10.6.2}$$

where $x_j$, $y_j$ is a point on element $j$, and $r_j$ is the distance from that point to the field point. Simplifying using eqn. (10.6.1),

$$\begin{aligned} \mathcal{R} &= \sum_j \iiint \sum_k \left[\frac{q_k \alpha_k(x_j, y_j)}{4\pi\varepsilon r_j}\right] dV - \phi \\ &= \sum_j q_j \iiint \left(\frac{1}{4\pi\varepsilon r_j}\right) dV - \phi. \end{aligned} \tag{10.6.3}$$

Applying the Galerkin principle with respect to the basis pulse function $\alpha_i$:

$$\iiint_V \mathcal{R}\alpha_i \, dV = \iiint_{V_i} \mathcal{R} \, dV = 0, \tag{10.6.4}$$

which gives us the familiar approximation:

$$\sum_j q_j \iiint \left(\frac{1}{4\pi\varepsilon r_{ij}}\right) dV - \phi_i = 0 \tag{10.6.5}$$

at every element $i$ that we have employed.

Clearly then, we may with greater accuracy use first and higher-order approximations on the unknowns (Lindholm 1981). The only difficulty with integral methods lies in systematically performing the required integrals of $\alpha_i/r$, $\alpha_i ln_e r$, etc., where $\alpha_i$ are the interpolation polynomials.

It is worth remarking that, in the differential formulations, the approximation usually must be at least of the same order as the differential operator of the equation, lest we have no terms when the polynomial is differentiated. An exception to this is the operator $\nabla^2$, which normally requires at least a second-order polynomial interpolation so that the weighted term $\alpha_j \nabla^2 \alpha_i$ in the Galerkin method will not vanish. But by using Gauss's theorem and converting $\alpha_j \nabla^2 \alpha_i$ to $\nabla \alpha_j \cdot \nabla \alpha_i$ we manage not only to work with first-order polynomials, but also to deal with a symmetric matrix. With integral operators, on the other hand, no such restriction on the approximation applies, since even a pulse function may be integrated. For the integrals over higher-order elements, the more interested reader is referred to McDonald et al (1973) and Collie (1976).

## 10.7 OPEN BOUNDARY PROBLEMS

### 10.7.1 Far Field Boundary Conditions

Many field problems are unbounded; i.e., they cannot be encased within a finite boundary along which either the normal or tangential component of the field may be specified. As a consequence, the actual boundary is at infinity, with both $\phi$ and $\nabla \phi \cdot \mathbf{u}_n$ zero thereupon. When dealing with integral methods this boundary condition is automatically taken care of, but then it is difficult to account for inhomogeneities with differential methods. On the other hand, differential methods easily account for inhomogeneities, but require the boundary to be nearby so that the matrix for solution is not infinitely large; but then, bringing the boundary close makes the potential fall (or rise) more rapidly and distorts our solution.

Overcoming this difficulty with exterior field problems has spurred a whole new area of research activity; as a result, several schemes are now available for tackling this. One regret of this writer is his inability, on account of pressures of schedule, to summarize a very recent, elegant, and most promising way of tackling open boundary problems presented by Lee and Cendes (1987), Chari (1987, 1988), and Chari and Bedrosian (1987). They use a circle as a boundary. The solution in the exterior region is obtained analytically and matched to the interior finite element solution.

## 10.7.2 Ballooning

The simplest scheme for accounting for open boundaries involves using a graded mesh that spaces out in steps of 1.5 or a similar factor, as shown in Figure 3.8.1. Caution is called for in dealing with the external triangles, which have large area and therefore can cause arithmetic overflow in a computer. The disadvantage in this scheme is that, although by grading we have reduced the number of nodes in the exterior region compared with a uniform mesh, the number of additional nodes is nonetheless high. For example, if we have a 10 × 10 mesh that must be extended by five extra graded nodes in all directions, the mesh of 100 nodes will take 264 nodes to model the exterior region—a situation clearly without balance.

Silvester et al. (1977) have introduced the concept of node condensation using a star point, as shown in Figure 10.7.1. The method is based in part on the fact that two similar triangles have the same finite element matrix **P** obtained by integrating $(\nabla\tilde{\alpha})^2$ in eqn. (7.6.9). Observe that the terms of $\tilde{\alpha}$ are dimensionless, and that

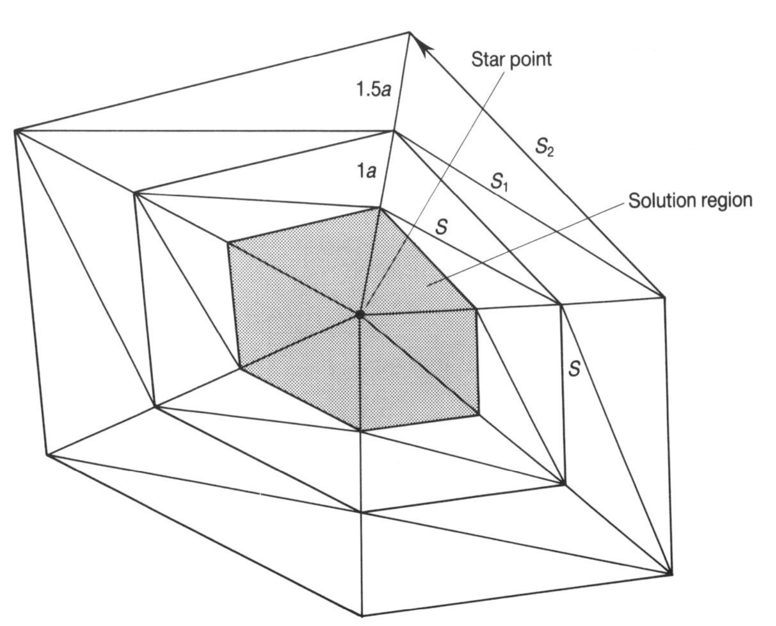


**Figure 10.7.1.** Star point for ballooning.

the square of the differential introduces the square of length in the denominator, which is taken away upon integration over an area. All the extensions to the interior mesh $R$ with boundary $S$ containing $n$ nodes are made with respect to a star point inside $R$, and the exterior regions are added in annuli, bounded by $S$ and $S^1$, $S^1$ and $S^2$, etc., spaced in ever increasing factors $k$, $k^2$, $k^3$ etc., where $k$ is 1.5 as before or any factor slightly greater than one. Each surface $S^i$ contains the same number of nodes $n$ as $S$. The nodes on the boundaries $S^i$ are numbered in the same order as their corresponding nodes on $S$. The discretization of all the annuli is in like manner, so that each triangle in an annulus has a similar counterpart in the first. Because of the similarity of triangles, the finite element matrices **P** from a triangle in any annulus will be the same as the corresponding matrix from the similar triangle in the first annulus.

Consider the finite element matrix equation for an annulus, bounded by **S** and $\mathbf{S}^1$, for which we may compute the coefficients of the block matrix equation:

$$\begin{bmatrix} \mathbf{A}_{11} & \mathbf{A}_{12} \\ \mathbf{A}_{21} & \mathbf{A}_{22} \end{bmatrix} \begin{bmatrix} \boldsymbol{\phi} \\ \boldsymbol{\phi}^1 \end{bmatrix} = \begin{bmatrix} \mathbf{0} \\ \mathbf{0} \end{bmatrix}, \tag{10.7.1}$$

where the elements of $\boldsymbol{\phi}$ are the potentials on $S$ and those of $\phi^1$ are the potentials on $S^1$. Similarly for the second annulus, we have:

$$\begin{bmatrix} \mathbf{S}_{11} & \mathbf{S}_{12} \\ \mathbf{S}_{21} & \mathbf{S}_{22} \end{bmatrix} \begin{bmatrix} \boldsymbol{\phi}^1 \\ \boldsymbol{\phi}^2 \end{bmatrix} = \begin{bmatrix} \mathbf{0} \\ \mathbf{0} \end{bmatrix}, \tag{10.7.2}$$

where the coefficients **A** and **S** are the same, but are written with different symbols for the sake of clarity as well as for a reason that will soon be apparent. Now for the combined region, by the usual rules of adding matrix contributions,

$$\begin{bmatrix} \mathbf{A}_{11} & \mathbf{A}_{12} & 0 \\ \mathbf{A}_{21} & \mathbf{A}_{22} + \mathbf{S}_{11} & \mathbf{S}_{12} \\ 0 & \mathbf{S}_{21} & \mathbf{S}_{22} \end{bmatrix} \begin{bmatrix} \boldsymbol{\phi} \\ \boldsymbol{\phi}^1 \\ \boldsymbol{\phi}^2 \end{bmatrix} = \begin{bmatrix} 0 \\ 0 \\ 0 \end{bmatrix}. \tag{10.7.3}$$

Provided the potentials $\boldsymbol{\phi}$ and $\boldsymbol{\phi}^2$, each of length $n$, are specified, the annulus between surfaces $S$ and $S^2$ is a closed Laplacian region with Dirichlet boundaries, so eqn. (10.7.3) presents a uniquely solvable boundary value problem for the "unknown" $\boldsymbol{\phi}^1$. Solving for $\boldsymbol{\phi}^1$ from row 2 of eqn. (10.7.3), we have

$$\boldsymbol{\phi}^1 = -[\mathbf{A}_{22} + \mathbf{S}_{11}]^{-1}[\mathbf{A}_{21}\boldsymbol{\phi} + \mathbf{S}_{12}\boldsymbol{\phi}^2]. \tag{10.7.4}$$

Substituting in the first and last rows, the nodes of the surface $S^1$ are, as it were, condensed:

$$\begin{bmatrix} \mathbf{A}_{11} - \mathbf{A}_{12}[\mathbf{A}_{22} + \mathbf{S}_{11}]^{-1}\mathbf{A}_{21} & -\mathbf{A}_{12}[\mathbf{A}_{22} + \mathbf{S}_{11}]^{-1}\mathbf{S}_{12} \\ -\mathbf{S}_{21}[\mathbf{A}_{22} + \mathbf{S}_{11}]^{-1}\mathbf{A}_{21} & \mathbf{S}_{22} - \mathbf{S}_{21}[\mathbf{A}_{22} + \mathbf{S}_{11}]^{-1}\mathbf{S}_{12} \end{bmatrix} \begin{bmatrix} \boldsymbol{\phi} \\ \boldsymbol{\phi}^2 \end{bmatrix} = \begin{bmatrix} \mathbf{0} \\ \mathbf{0} \end{bmatrix}. \tag{10.7.5}$$

Therefore, with the recursion

$$\mathbf{A}^1 = \begin{bmatrix} \mathbf{S}_{11} & \mathbf{S}_{12} \\ \mathbf{S}_{21} & \mathbf{S}_{22} \end{bmatrix}, \tag{10.7.6}$$

$$\mathbf{A}^{k+1} = \begin{bmatrix} \mathbf{A}_{11}^k - \mathbf{A}_{12}^k[\mathbf{A}_{22}^k + \mathbf{S}_{11}]^{-1}\mathbf{A}_{21}^k & -\mathbf{A}_{12}^k[\mathbf{A}_{22}^k + \mathbf{S}_{11}]^{-1}\mathbf{S}_{12} \\ -\mathbf{S}_{21}[\mathbf{A}_{22}^k + \mathbf{S}_{11}]^{-1}\mathbf{A}_{21}^k & \mathbf{S}_{22} - \mathbf{S}_{21}[\mathbf{A}_{22}^k + \mathbf{S}_{11}]^{-1}\mathbf{S}_{12} \end{bmatrix}, \tag{10.7.7}$$

we will, after $m$ applications, arrive at the equation

$$\begin{bmatrix} \mathbf{A}_{11}^m & \mathbf{A}_{12}^m \\ \mathbf{A}_{21}^m & \mathbf{A}_{22}^m \end{bmatrix} \begin{bmatrix} \boldsymbol{\phi} \\ \boldsymbol{\phi}^m \end{bmatrix} = \begin{matrix} 0 \\ 0 \end{matrix}. \tag{10.7.8}$$

This contribution may then be added to the contributions from the interior region, and after setting $\boldsymbol{\phi}^m$, practically at infinity for large $m$, to zero, we may solve for all the potentials. Two remarks are worth making here. First, the main work in condensation is in computing the inverse $[\mathbf{A}_{22}^k + \mathbf{S}_{11}]^{-1}$ and is proportional to the number of nodes $n$ on the boundary $S$. Second, the recursion results in dense matrices—since the multiple of two sparse matrices is not as sparse as the matrix factors. As a result, the matrix $\mathbf{A}_{11}^m$ will be fully populated for all practical purposes.

Figures 10.7.2A and 10.7.2B give the solution, by this ballooning scheme, of the open boundary value problem posed by a parallel plate capacitor. They represent a factor of 1.2 for stepping out and are for $m = 2$ and $m = 40$, respectively. Beyond $m = 40$ no changes in the field were observed. Figure 10.7.2C demonstrates the distortion that occurs when we artificially close the boundary.

### 10.7.3 Hybrid FE–BIM Schemes

Hybrid finite element–boundary integral (FE–BIM) schemes exploit the fact that the former are best in closed inhomogeneous regions and the latter in open homogeneous regions. That is, the BIM equations are used in the exterior and the FE equations in the interior of a field problem, with some kind of matching at the common boundary (Salon 1985).

The earliest of these was by Silvester and Hsieh (1971), who employed an "infinite element" to model the exterior region. They employed the functional

$$\mathscr{L} = \iint_I \varepsilon[\nabla\phi]^2 - 2\rho\phi]\, dR + \iint_E \varepsilon_0[\nabla\phi]^2\, dR, \qquad (10.7.9)$$

using the fact that there is no source in the exterior Laplacian region $E$. This requires the energy of the exterior region to be finite, which condition is readily met by physical problems. The energy from the exterior element is converted using Gauss's theorem (A13)

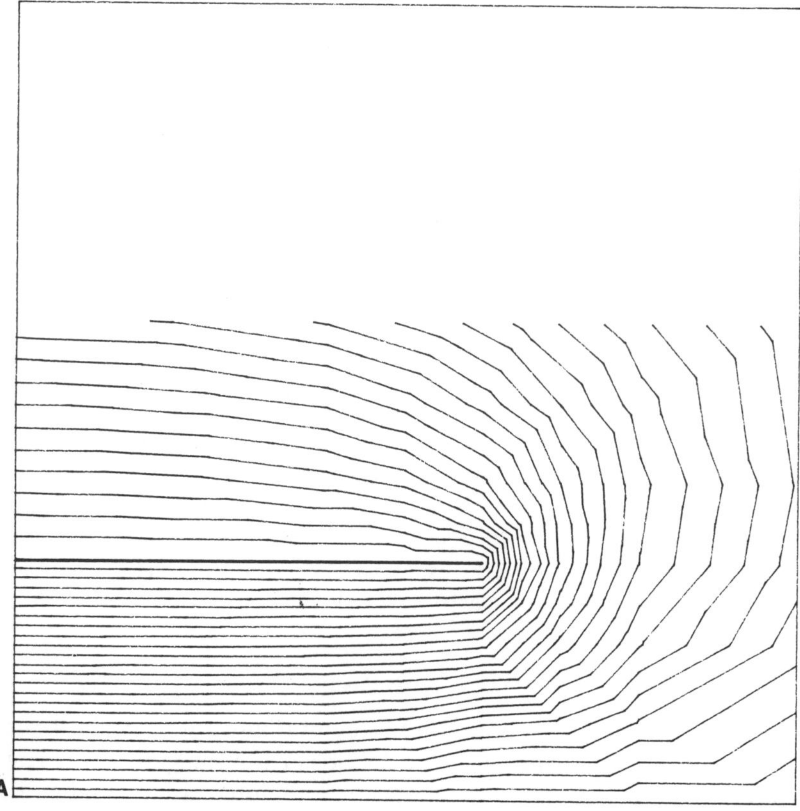


**Figure 10.7.2.** A capacitor. **A.** Solution with two steps of ballooning. **B.** Truly open: solution with 40 steps of ballooning. **C.** Closed artificial boundary.

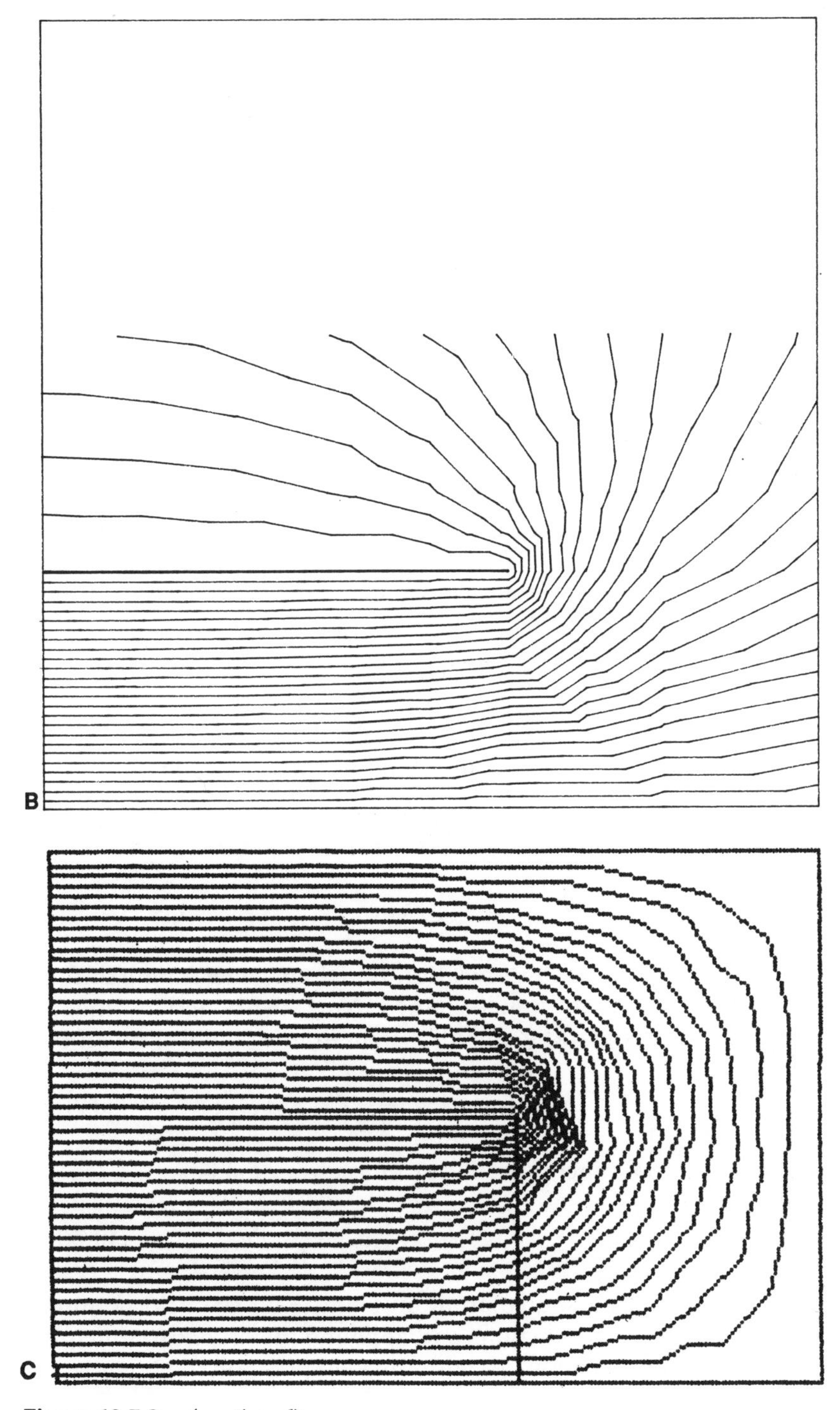


**Figure 10.7.2.** (*continued*)

to:

$$\mathscr{L}_E = \iint_E \varepsilon_0[\nabla\phi]^2\, dR = -\iint_E \varepsilon_0\phi[\nabla^2\phi]\, dR - \varepsilon_0\int \phi\nabla\phi \,.\, \mathbf{u}_n\, dS$$
$$= -\varepsilon_0\int \phi\nabla\phi \,.\, \mathbf{u}_n\, dS = -\varepsilon_0\int \phi\frac{\partial\phi}{\partial n}\, dS, \tag{10.7.10}$$

where the negative sign represents the fact that for the exterior region, the outward normal points inwards; and we have used the fact that in the exterior region $\nabla^2\phi$ is zero. The evaluation of the contribution of this functional requires us to express $\partial\phi/\partial n$ in terms of the potential in the interior. To this end, the boundary integral scheme is pressed into service.

Assume that $g = \partial\phi/\partial n$ and $\phi$ both vary as the same trial function along the boundary between the interior and exterior regions. Since $\phi$ must match the trial function of the mesh, for first-order triangles, for example, we may take first-order variations along the edges:

$$\phi = \zeta_1\phi_1 + \zeta_2\phi_2, \tag{10.7.11}$$

$$g = \zeta_1 g_1 + \zeta_2 g_2. \tag{10.7.12}$$

More generally, in the notation of Silvester and Hsieh (1971):

$$g = \sum b_i\beta_i, \tag{10.7.13}$$

$$\phi = \sum a_i\beta_i, \tag{10.7.14}$$

where $\beta_i$ are the set of interpolation functions generating $g$ and $\phi$, and $a_i$ and $b_i$ are as yet undetermined. Now, the boundary integral method stipulates that the potential at an exterior point is given by:

$$\phi = -\int_S g\frac{1}{4\pi\varepsilon r}\, dS, \tag{10.7.15}$$

or, when the trial functions are put in, the residual is

$$\mathscr{R} = \sum a_i\beta_i(\xi) + \sum\int_S b_i\frac{\beta_i(\eta)}{4\pi r(\eta)}\, d\eta, \tag{10.7.16}$$

where $\eta$ stands for a source point on $S$ and $\xi$ for the field point on $S$. Putting in the trial functions and applying the Galerkin principle, for the weighting functions $\beta_k$ with the field point on the boundary:

$$\int_S \sum \beta_k(\xi)a_i\beta_i(\xi)\, d\xi + \int_S\int_S \sum b_i\frac{\beta_k(\xi)\beta_i(\eta)}{4\pi r}\, d\eta\, d\xi, \tag{10.7.17}$$

which gives us the matrix equation

$$\mathbf{Ra} = \mathbf{Qb}, \tag{10.7.18}$$

where

$$R_{ij} = \int \beta_i(\xi)\beta_j(\xi)\, d\xi \tag{10.7.19}$$

and

$$Q_{ij} = \iint \frac{\beta_i(\xi)\beta_j(\eta)}{4\pi r}\, d\eta\, d\xi. \tag{10.7.20}$$

Since, given $\phi$, we may compute $g$ and vice versa, both $\mathbf{R}$ and $\mathbf{Q}$ are invertible nonsingular matrices. Observe also that they are both symmetric matrices.

Now, we are ready to put the trial functions into the exterior term of the functional (10.7.10):

$$\begin{aligned}\mathcal{L}_E &= -\varepsilon_0 \int_S \phi g\, dS = -\varepsilon_0 \sum_i \sum_j a_i b_j \iint \beta_i \beta_j\, dS \\ &= -\varepsilon_0 \mathbf{a}^t \mathbf{R} \mathbf{b} = -\varepsilon_0 \mathbf{a}^t \mathbf{R} \mathbf{Q}^{-1} \mathbf{R} \mathbf{a},\end{aligned} \tag{10.7.21}$$

from eqn. (10.7.18). Now, since we took the same interpolation for $\phi$ along the boundary as we did for $\phi$ inside the boundary, the coefficients $a$ are nodal potentials $\phi$. It is therefore straightforward to add the contribution of eqn. (10.7.21) to the interior functional by incorporation into the global matrix and solve the problem.

McDonald and Wexler (1972) offer a philosophical nuance in their approach. They regard the interior finite element method as a closed boundary value problem with $\phi$ known along the boundary. This value of $\phi$ is constrained by the boundary integral equations. The potentials $\phi$ are separated into two parts, $\phi_I$ in the interior and $\phi_B$ on the boundary, so that the interior finite element functional takes the form

$$\mathcal{L} = \boldsymbol{\phi}_I^t \mathbf{A}_{II} \boldsymbol{\phi}_I + \boldsymbol{\phi}_I^t \mathbf{A}_{IB} \boldsymbol{\phi}_B + \boldsymbol{\phi}_B^t \mathbf{A}_{BI} \boldsymbol{\phi}_I + \boldsymbol{\phi}_B^t \mathbf{A}_{BB} \boldsymbol{\phi}_B - 2\boldsymbol{\phi}_I^t \mathbf{b}_I - 2\boldsymbol{\phi}_B^t \mathbf{b}_B. \tag{10.7.22}$$

The boundary integral equations are applied to the Laplacian region bounded between infinity and a contour running just inside the boundary of the finite element mesh to yield

$$\boldsymbol{\phi}_B = \mathbf{M} \boldsymbol{\phi}_I. \tag{10.7.23}$$

The latter equation is used to eliminate $\boldsymbol{\phi}_B$, and the solution to the problem is obtained by extremizing the functional with respect to $\boldsymbol{\phi}_I$.

## 10.8 DIRECT VECTOR SOLUTION

Up to this point in this text, we have been dealing with potentials. The advantage in so doing is that, although the electric and magnetic fields are multivalued at material discontinuities, the potentials from which they are derived by differentiation are not. The disadvantage, however, is that in differentiating the potential to obtain the field densities, some loss of accuracy is incurred. But usually, the disadvantages of potential solution are considered to be far outweighed by the convenience of their formulations and amenability to being programmed into general purpose software.

Be that as it may, at least in homogeneous problems where the field intensities are continuous, a strong case exists for solving directly for the field density (Hoole and Cendes 1985). Consider a two-dimensional magnetic field problem, with the flux density on the $xy$ plane. Let us approximate the components $B_x$ and $B_y$ by the usual finite element trial functions over a triangular mesh and represent (Konrad 1974, 1976):

$$\mathbf{B} = \begin{bmatrix} \underset{\sim}{\mathbf{B}}_x \\ \underset{\sim}{\mathbf{B}}_y \end{bmatrix} = \begin{bmatrix} \tilde{\boldsymbol{\alpha}}^n \underset{\sim}{\mathbf{B}}_x \\ \tilde{\boldsymbol{\alpha}}^n \underset{\sim}{B}_y \end{bmatrix} = \begin{bmatrix} \tilde{\boldsymbol{\alpha}}^n & \mathbf{0} \\ \mathbf{0} & \tilde{\boldsymbol{\alpha}}^n \end{bmatrix} \begin{bmatrix} \underset{\sim}{\mathbf{B}}_x \\ \underset{\sim}{\mathbf{B}}_y \end{bmatrix}, \tag{10.8.1}$$

or

$$\mathbf{B} = [\boldsymbol{\alpha}]^n \underset{\sim}{\mathbf{B}}, \tag{10.8.2}$$

with corresponding notation. We therefore have, for the curl eqn. (1.2.14),

$$\begin{aligned} \nabla \times \mathbf{B} &= \mathbf{u}_z \left[ \frac{\partial B_y}{\partial x} - \frac{\partial B_x}{\partial y} \right] = \mathbf{u}_z [\tilde{\boldsymbol{\alpha}}^{n-1} \mathbf{D}_x \underset{\sim}{\mathbf{B}}_y - \tilde{\boldsymbol{\alpha}}^{n-1} \mathbf{D}_y \underset{\sim}{\mathbf{B}}_x] \\ &= \mathbf{u}_z \tilde{\boldsymbol{\alpha}}^{n-1} [-\mathbf{D}_y \quad \mathbf{D}_x] \underset{\sim}{\mathbf{B}} = \mathbf{u}_z \tilde{\boldsymbol{\alpha}}^{n-1} \mathbf{M}_c \underset{\sim}{\mathbf{B}} = \mathbf{u}_z \mu J, \end{aligned} \tag{10.8.3}$$

where the current density $J$ has been assumed to be a constant. Similarly for the divergence eqn. (1.2.10),

$$\begin{aligned} \nabla . \mathbf{B} &= \left[ \frac{\partial B_x}{\partial x} + \frac{\partial B_y}{\partial y} \right] = \mathbf{u}_z [\tilde{\boldsymbol{\alpha}}^{n-1} \mathbf{D}_x \underset{\sim}{\mathbf{B}}_x + \tilde{\boldsymbol{\alpha}}^{n-1} \mathbf{D}_y \underset{\sim}{\mathbf{B}}_y] \\ &= \tilde{\boldsymbol{\alpha}}^{n-1} [\mathbf{D}_x \quad \mathbf{D}_y] \underset{\sim}{\mathbf{B}} = \tilde{\boldsymbol{\alpha}}^{n-1} \mathbf{M}_d \underset{\sim}{\mathbf{B}} = 0, \end{aligned} \tag{10.8.4}$$

where the curl and divergence matrices, $\mathbf{M}_c$ and $\mathbf{M}_d$, have been defined. To solve the equation pair (10.8.3)–(10.8.4), let us use a least square error functional:

$$\begin{aligned}\mathscr{L} &= \iint \{[\nabla \times \mathbf{B} - \mu \mathbf{J}]^2 + [\nabla . \mathbf{B} - 0]^2\}\, dR \\ &= \iint \{[\tilde{\boldsymbol{\alpha}}^{n-1}\mathbf{M}_c\underset{\sim}{\mathbf{B}} - \mu J]^2 + [\tilde{\boldsymbol{\alpha}}^{n-1}\mathbf{M}_d\underset{\sim}{\mathbf{B}}]^2\}\, dR \\ &= \sum \{A^r\underset{\sim}{\mathbf{B}}^t\mathbf{M}_c^t\mathbf{T}^{n-1,n-1}\mathbf{M}_c\underset{\sim}{\mathbf{B}} - 2\mu J A^r\underset{\sim}{\mathbf{B}}^t\mathbf{M}_c^t\mathbf{T}^{n-1} + \mu^2 J^2 \\ &\quad + A^r\underset{\sim}{\mathbf{B}}^t M_d^t[\mathbf{T}]^{n-1,n-1}M_d\underset{\sim}{\mathbf{B}}\},\end{aligned} \tag{10.8.5}$$

where the matrix $\mathbf{T}$ is defined in eqn. (8.5.24). Minimizing with

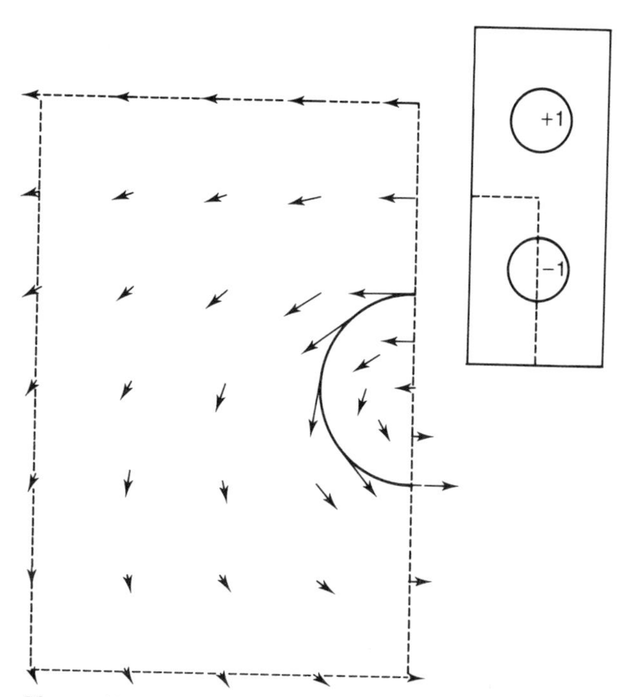


**Figure 10.8.1.** Direct solution in two-conductor line.

respect to the unknowns $\underset{\sim}{\mathbf{B}}$ and dividing by two, we have:

$$\sum \{A^r[\mathbf{M}_c^t\mathbf{T}^{n-1,n-1}\mathbf{M}_c + \mathbf{M}_d^t\mathbf{T}^{n-1,n-1}\mathbf{M}_d]\underset{\sim}{\mathbf{B}} = \sum \mu J A^r \underset{\sim}{\mathbf{B}} M_c^t \mathbf{T}^{n-1}. \tag{10.8.6}$$

The simple problem of the magnetic field around two conductors was solved by this method (Figure 10.8.1), as well as the matching second-order finite element vector potential formulation. Comparing with the well-known analytical solution, the errors are seen plotted in Figure 10.8.2. The accuracy in solving for the field directly is obvious. Here, however, the programming was simple in view of the homogeneity of the configuration, unlike more complex engineering problems.

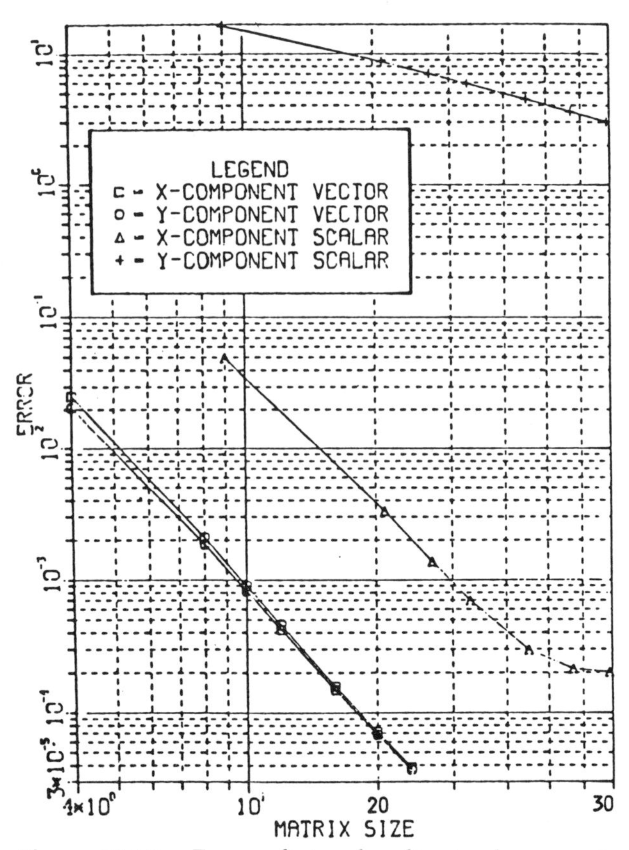


**Figure 10.8.2.** Errors: first-order direct scheme and second-order vector potential scheme.

# 10.9 THREE-DIMENSIONAL PROBLEMS

## 10.9.1 The Complexities in Three Dimensions

While several field problems may be conveniently reduced to two dimensions and, indeed, even one dimension, many problems do not lend themselves to such a reduction in dimensionality. If our object is accuracy in design, then we have no choice but to solve in three dimensions. The computational complexity alone is horrendous and mind-boggling to contemplate. A crude mesh with 10 nodes in each coordinate direction gives us 1000 nodes, and the matrix equation becomes correspondingly denser because each node is connected to about three times as many nodes as in two dimensions. Three dimensional vector potential formulations for magnetostatics present further problems because, unlike the scalar potential in electrostatics, three components are associated with each node. Also, the vector potential does not reduce to a single component, as with two dimensions. It is in fact possible to work with the magnetic scalar potential $\Omega$ of eqn. (4.2.8) after identifying a particular solution $\mathbf{H}_c$ (Simkin and Trowbridge 1979). In working with the scalar potential $\Omega$, automation of the differential formulation as a general program is difficult because of the interface conditions at inhomogeneities. In fact, once we reformulate the problem in terms of $\Omega$, it is easier to apply the boundary integral scheme (Trowbridge 1980).

Theoretical development of the boundary integral method in three dimensions, whether in electricity or magnetism, hardly differs from the exposition of sections 2.5.1 and 4.2.2 in two dimensions. The reader should have no difficulties in extending the methodology to three dimensions. We shall therefore restrict our discussion here to differential formulations.

Before switching to differential formulations, however, let us try to make a very rough comparison of the computational magnitudes of the boundary integral method and differential schemes employing first-order tetrahedral elements, as in Table 10.9.1. Let us take a system with roughly $n$ nodes in each direction. The BIM scheme with $n^2$ nodes on each of the six surfaces of a square region has approximately $6n^2$ unknowns. The scalar differential formulation will have $n^3$ unknowns and the vector differential formulation, with three unknowns per node, $3n^3$ unknowns. The BIM scheme results in a full matrix, so that storage will take a half of the $6n^2 \times 6n^2$

**Table 10.9.1.** Computing Requirements of Integral and Differential Schemes

| | *BIM* | *Differential Methods* | |
|---|---|---|---|
| | | *Scalar* | *Vector* |
| Nodes | $6n^2$ | $n^3$ | $n^3$ |
| Unknowns | $6n^2$ | $n^3$ | $3n^3$ |
| Connectivity | $6n^2$ | 20 | 60 |
| Storage | $18n^4$ | $25n^3$ | $225n^3$ |
| Solution Time | $36n^6$ | $n^{4.5}$ | $5.2n^{4.5}$ |

numbers of the matrix, if symmetry is exploited. Typically, for first-order tetrahedral elements, each node will be connected to 20 others (three times the six or seven for optimal first-order triangles). With symmetry, this number comes down to 10. Now assuming the Cholesky preconditioned conjugate gradients solver and the terminology of chapter 6, section 4, for a sparse and symmetric matrix of size $m$ and connectivity $2c$, we will need to store the $mc$ real numbers of the matrix, another $mc$ real numbers of the Cholesky factor, and the $mc$ integers that are the column numbers in **Col**. The vector **Diag**, of lower order $m$, may be neglected. The total requirements, then, are roughly $2.5\,mc$ real numbers. The scalar potential formulation, then, with one unknown at each node and connectivity 20, asks us to provide storage for $25n^3$ real numbers. The vector potential formulation, on the other hand, has three times the connectivity and three times as many unknowns, so that the storage requirements are $225n^3$. For the fully populated matrix yielded by the BIM scheme, the solution time is of the order of a sixth of the cube of the matrix size, of $36n^6$ (Cendes 1982). For sparse matrices by the conjugate gradients algorithm, this writer has achieved solution times of the order of matrix size raised by a power of 1.7, although a power of 1.2 has been reported elsewhere (Cendes 1982). The difference, of course, comes from the efficiencies of the algorithms used and in the convergence criteria used to stop the conjugate gradients iterations. Taking a power 1.5, we may complete Table 10.9.1.

Caution is urged in interpreting the solution times. As already pointed out, the actual answers depend on how efficiently the programs have been written. Moreover, the numbers relate to orders of magnitude rather than to actual numbers. For instance, when we say that the conjugate gradients algorithm takes time of the order of $m^{1.5}$, where $m$ is the matrix size, the actual time is $km^{1.5}$

where $k$ is a constant which will depend on the connectivity, among other factors. Therefore, the constant associated with the order $n^{4.5}$ for the scalar differential formulation is smaller than that associated with the $5.2n^{4.5}$ of the more connected matrix arising from the vector potential formulation. In other words, the time difference is bigger than indicated by the ratio 5.2:1. Similarly, the constant associated with the solution time $36n^6$ by the direct solution of the integral equations is much smaller than that associated with the conjugate gradients solutions. What happens then is that despite the solution time of the order of $36n^6$ with the direct scheme, for lower matrix sizes the solution time is lower. But, as matrix size goes up, the term varying as $n^6$ grows faster than that varying as $n^{4.5}$. That is, for lower matrix sizes, the prefixing constant dominates the comparison, whereas for larger matrix sizes, it is the order of the number (or the power of $n$) that dominates the behavior of solution time. It is difficult to obtain the prefixing constants, which depend on many factors such as the computing environment, without actually programming and running. Even then, the results would be valid only for that environment. The best use of the table we have constructed, therefore, is in making comparison for very large matrices where it is the order that dominates the solution time.

Another note of caution is that this comparison tells us little of the qualitative differences. For instance, an open boundary problem is better solved by integral schemes, whereas differential schemes are elegantly formulated and easy to program. These considerations also do not hold in three-dimensional problems where the configuration is thin in one direction, such as when we analyze the changes in a constant magnetic field in the presence of a slab.

Observe that the comparison of the integral scheme, such as where we solve for the magnetization **M** (as in section 4.2.1), is not and cannot be exact. The reason is that the number of unknowns in the integral scheme bears no strict relationship to $n$, the number of nodes in each direction. The number of unknowns is in fact a function of the quantity of permeable parts within the solution region. When the whole solution region is filled with different magnetic materials, the formulation will give a large matrix size, whereas if such material were to fill a tiny portion of the solution region, the scheme would require the inversion of a small matrix.

From Table 10.9.1 it is seen that three-dimensional problems make heavy demands on memory. It is, therefore, conventional wisdom to solve three-dimensional problems on mainframe computers (in the sense of the older use of the term, since the distinction between large and small machines is vanishing fast). However, most of the preprocessing and postprocessing may be done on personal computers and only the matrix solution passed on to a larger machine. Even here, difficulties exist in viewing three-dimensional objects in the two dimensions of a graphics screen or paper (Hoole 1985; Rajanathan, Bryant, and Freeman 1988).

### 10.9.2 Tetrahedral Elements and Trial Functions

Corresponding to the triangle in two dimensions, the tetrahedron is a convenient element to use in three dimensions. However, it is very difficult to visualize how a set of tetrahedrons[1] merge into a three-dimensional object and there is a real danger of creating meshes with a void in them. This is circumvented by resorting to building a mesh with octahedrons as the basic building blocks. These octahedrons may then be cut into five tetrahedrons each in one of two alternative ways as shown in Figures 10.9.1A and 10.9.1B. To be mathematically precise the cut along the common face between two octahedrons must be the same, so that two adjacent octahedrons must be cut in different ways. Failing to meet this criterion will result in nonconforming elements, which, as we have seen in section 8.8, somehow give us reasonable looking, but not the best results. Equations relating the number of tetrahedrons to the numbers of edges, vertices, and boundary nodes have been developed and may be employed to verify the integrity of the constructed mesh (Hoole, Jayakumaran, and Yoganathan 1986).

For the tetrahedron, we will have four tetrahedral (or homogeneous) coordinates $\zeta_i$:

$$\zeta_i = \frac{h_i}{H_i}, \tag{10.9.1}$$

where now $h_i$ is the height of a point from a triangular face of a tetrahedron and $H_i$ is the height of the vertex opposite that face. Going along the lines of section 7.5, we have for first-order

[1] The plural *tetrahedra* is commonly encountered in the engineering literature, but is not supported in any dictionary this writer has consulted. The root of the word *tetrahedron* is in the Greek words *tetra* (four) and *hedra* (base). The latter word casts further doubt on the correctness of *tetrahedra* as a plural. It is probably a rather fanciful invention by electrical engineers and we shall, like ordinary men, use the natural English plural *tetrahedrons.*

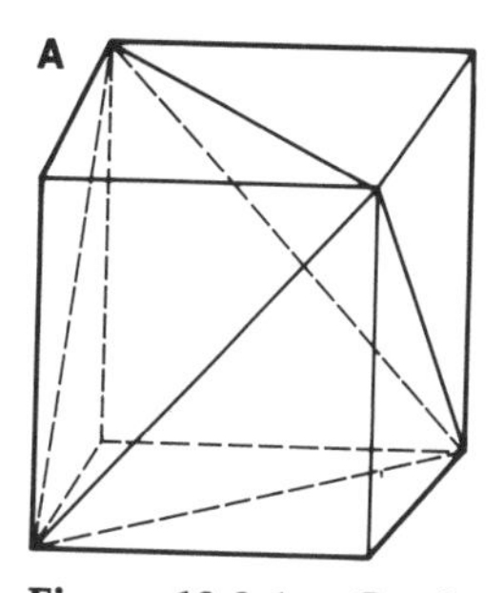


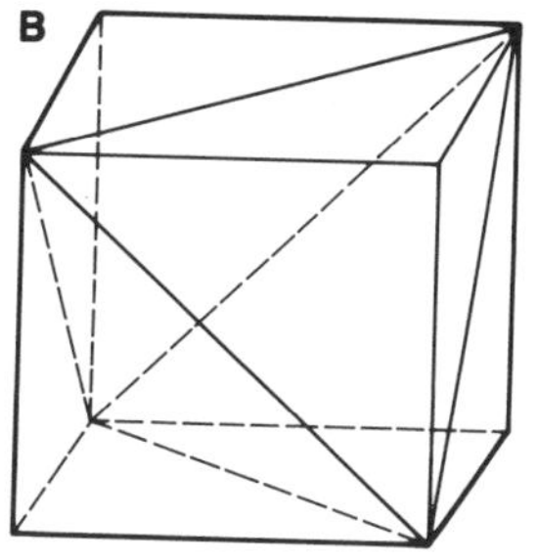


**Figure 10.9.1.** Cutting an octahedron into five tetrahedrons.

interpolation

$$\phi = \zeta_1\phi_1 + \zeta_2\phi_2 + \zeta_3\phi_3 + \zeta_4\phi_4 = \tilde{\boldsymbol{\alpha}}\underset{\sim}{\boldsymbol{\phi}}, \tag{10.9.2}$$

$$\zeta_1 + \zeta_2 + \zeta_3 + \zeta_4 = 1, \tag{10.9.3}$$

$$x_1\zeta_1 + x_2\zeta_2 + x_3\zeta_3 + x_4\zeta_4 = x, \tag{10.9.4}$$

$$y_1\zeta_1 + y_2\zeta_2 + y_3\zeta_3 + y_4\zeta_4 = y, \tag{10.9.5}$$

$$z_1\zeta_1 + z_2\zeta_2 + z_3\zeta_3 + z_4\zeta_4 = z. \tag{10.9.6}$$

Solving the preceding four equations for the four $\zeta$'s, we have:

$$\zeta_i = \frac{1}{6V}[a_i + b_i x + c_i y + d_i z], \tag{10.9.7}$$

where:

$$a_i = (-1)^{i1}[x_{i1}(y_{i2}z_{i3} - y_{i3}z_{i2}) + x_{i2}(y_{i3}z_{i1} - y_{i1}z_{i3}) + x_{i3}(y_{i1}z_{i2} - y_{i2}z_{i1})] \tag{10.9.8}$$

$$b_i = (-1)^{i}[(y_{i2}z_{i3} - y_{i3}z_{i2}) + (y_{i3}z_{i1} - y_{i1}z_{i3}) + (y_{i1}z_{i2} - y_{i2}z_{i1})] \tag{10.9.9}$$

$$c_i = (-1)^{i1}[(x_{i2}z_{i3} - x_{i3}z_{i2}) + (x_{i3}z_{i1} - x_{i1}z_{i3}) + (x_{i1}z_{i2} - x_{i2}z_{i1})] \tag{10.9.10}$$

$$d_i = (-1)^{i}[(x_{i2}y_{i3} - x_{i3}y_{i2}) + (x_{i3}y_{i1} - x_{i1}y_{i3}) + (x_{i1}y_{i2} - x_{i2}y_{i1})] \tag{10.9.11}$$

$$6V = a_1 + a_2 + a_3 + a_4, \tag{10.9.12}$$

where $i$, $i1$, $i2$, $i3$ are 1, 2, 3, and 4 or a cyclic permutation of them. The computation of these coefficients is presented in Alg. 10.9.1. This will give us the differentiation matrix

$$\frac{\partial\phi}{\partial x} = \sum_{i=1}^{4}\frac{\partial\phi}{\partial\zeta_i}\frac{\partial\zeta_i}{\partial x} = \sum_{i=1}^{4}\frac{b_i\phi_i}{6V} = \tilde{\mathbf{b}}\underset{\sim}{\boldsymbol{\phi}}, \tag{10.9.13}$$

**Algorithm 10.9.1. Computing First-Order Differentiation Matrices for a Tetrahedron**

```
Procedure Tetrahedron (b, c, d, V, x, y, z)
{Function: To compute the first order differentiation matrices
Outputs: b,c,d: The 1 × 4 first order differentiation matrices
defined in eqns.(7.6.3) and (7.6.4); V = volume of tetrahedron
Inputs: x,y,z = 4 × 1 vectors containing the coordinates of the
four vertices}
Begin
  Sign ← 1
  For i ← 1 To 4 Do
    i1 ← Mod(i, 4) + 1{Or i Mod 4 + 1}
    i2 ← Mod(i1, 4) + 1
    i3 ← Mod(i2, 4) + 1
    Sign ← −Sign
    a[i] ← −Sign*(x[i1]*(y[i2]*z[i3] − y[i3]*z[i2])
        + x[i2]*(y[i3]*z[i1] − y[i1]*z[i3]) + x[i3]*(y[i1]*z[i2] −
      y[i2]*z[i1]))
    b[i] ← Sign*(y[i2]*z[i3] − y[i3]*z[i2] + y[i3]*z[i1] −
      y[i1]*z[i3] + y[i1]*z[i2] − y[i2]*z[i1])
    c[i] ← −Sign*(x[i2]*z[i3] − x[i3]*z[i2] + x[i3]*z[i1] −
      x[i1]*z[i3] + x[i1]*z[i2] − x[i2]*z[i1])
    d[i] ← Sign*(x[i2]*y[i3] − x[i3]*y[i2] + x[i3]*y[i1] −
      x[i1]*y[i3] + x[i1]*y[i2] − x[i2]*y[i1])
  V ← a[1] + a[2] + a[3] + a[4]
  For i ← 1 To 4 Do
    b[i] ← b[i]/V
    c[i] ← c[i]/V
    d[i] ← d[i]/V
  V ← Abs(V/6)
End
```

and, similarly, the differentiation matrices with respect to $y$ and $z$ may be computed. Analogous to eqn. (D1), we have:

$$\iiint \zeta_1^i \zeta_2^j \zeta_3^k \zeta_4^l \, dV = \frac{i!\, j!\, k!\, l!\, 3!\, V}{(i + j + k + l + 3)!}, \tag{10.9.14}$$

so that the $\mathbf{T}$ matrix of eqn. (7.6.12) becomes in three dimensions

$$\mathbf{T}^{11} = V^{-1} \iiint \tilde{\alpha}^t \tilde{\alpha} \, dV = \frac{1}{20} \begin{bmatrix} 2 & 1 & 1 & 1 \\ 1 & 2 & 1 & 1 \\ 1 & 1 & 2 & 1 \\ 1 & 1 & 1 & 2 \end{bmatrix} \tag{10.9.15}$$

Section 8.5 for higher-order interpolation is also analogously modified (Silvester 1982a). In place of eqn. (8.5.4) we have, for order $n$ with an additional suffix $l$ for the fourth coordinate $\zeta_4$,

$$\alpha^n_{ijkl} = P^n_i(\zeta_i)P^n_j(\zeta_2)P^n_k(\zeta_3)P^n_l(\zeta_4), \tag{10.9.16}$$

while the polynomials $P^n_i(\zeta)$ are generated in the same way as in eqn. (8.5.5). The number of interpolation nodes for an $n$th-order tetrahedron is, analogous to eqn. (8.5.1) for triangles:

$$N(n) = \frac{1}{3!}(n + 1)(n + 2)(n + 3), \tag{10.9.17}$$

so that for first order we have four interpolation nodes and for second order, 10 interpolation nodes.

### 10.9.3 Electric Field Problems

Electric field problems are the least of our worries in three-dimensional finite elements. The functional eqn. (7.5.18) simply extends in three dimensions to:

$$\mathscr{L}[\phi] = \iiint \{0.5\varepsilon[\nabla\phi]^2 - \rho\phi\}\, dV. \tag{10.9.18}$$

Putting in the first-order approximation eqn. (10.9.2), noting that $[\nabla\phi]^2$ is $(\partial\phi/\partial x)^2 + (\partial\phi/\partial y)^2 + (\partial\phi/\partial z)^2$, and integrating as we did in two dimensions:

$$\mathscr{L}[\phi] = \sum_{\Delta} \{0.5V\varepsilon\underset{\sim}{\boldsymbol{\phi}}^t[\tilde{\mathbf{b}}^t\tilde{\mathbf{b}} + \tilde{\mathbf{c}}^t\tilde{\mathbf{c}} + \tilde{\mathbf{d}}^t\tilde{\mathbf{d}}]\underset{\sim}{\boldsymbol{\phi}} - \underset{\sim}{\boldsymbol{\phi}}^t V\mathbf{T}^{11}\underset{\sim}{\boldsymbol{\rho}}\}, \tag{10.9.19}$$

where $\rho$ has been approximated by a first-order interpolation. Extremizing this with respect to the unknowns $\underset{\sim}{\boldsymbol{\phi}}$:

$$\sum_{\Delta} \{V\varepsilon[\tilde{\mathbf{b}}^t\tilde{\mathbf{b}} + \tilde{\mathbf{c}}^t\tilde{\mathbf{c}} + \tilde{\mathbf{d}}^t\tilde{\mathbf{d}}]\underset{\sim}{\boldsymbol{\phi}} - V\mathbf{T}^{11}\underset{\sim}{\boldsymbol{\rho}}\} = 0. \tag{10.9.20}$$

The building of the local matrices is slightly different from Alg.

7.6.2 and should present no difficulty to the reader. The difference is in accounting for the additional term $\tilde{\mathbf{d}}^t\tilde{\mathbf{d}}$ and the change in the size of the matrix from $3 \times 3$ to $4 \times 4$. The placing of the local matrix in the global matrix is exactly as in Alg. 7.6.3, except that now the local matrix is $4 \times 4$ instead of $3 \times 3$ so that the Do loops on the variables Row and Col will run from 1 to 4 instead of 1 to 3.

One of the earliest papers on the solution of electrostatic fields using tetrahedral elements is by Zienkiewicz, Bahrani, and Arlett (1967). Even today the paper has much to offer in that it takes the reader through the basics of the method in three dimensions, including the chopping of octahedrons into tetrahedrons. Another interesting feature of the paper is that the authors, in considering a thin conductor in soil, model the conductor using one-dimensional elements only along whose lengths the potential varies, and model the soil in which it is buried in three dimensions. Thus, the input data will consist of one-dimensional and three-dimensional elements. The local matrices for the one-dimensional elements will be $2 \times 2$, whereas for the three-dimensional elements they will be $4 \times 4$. By modifying the algorithm GlobalPlace of Alg. 7.6.3 to take the matrix size $n$ as input, the same algorithm may be used for all dimensions.

### 10.9.4 Magnetic Field Problems

In three-dimensional magnetic field problems, the difficulty in dealing with the vector potential $\mathbf{A}$ arises on account of our having to satisfy two equations:

$$\mu^{-1}\nabla \times \nabla \times \mathbf{A} = \mathbf{J} \tag{1.2.25}$$

and

$$\nabla . \mathbf{A} = 0. \tag{1.2.26}$$

True, even in two dimensions we did satisfy both, but because $\mathbf{A}$ had only one component $A_z$ along the direction $z$ in which no variations occur, eqn. (1.2.26) reduced to $\partial A_z/\partial z$ and was automatically satisfied without our having to worry about it.

A functional satisfying eqn. (1.2.25) at its minimum may be easily identified from eqn. (7.5.19) by analogy, noting that the term $\frac{1}{2}\nu[\nabla A]^2$ arises from the stored energy density $\frac{1}{2}\mathbf{B} . \mathbf{H}$ and the term $JA$ from the work done in bringing the currents to where they are. Writing $\nabla \times \mathbf{A}$ for $\mathbf{B}$ and accounting for the currents in the $x$ and $y$ directions, which are also now present:

$$\mathscr{L}[A] = \iiint \{\tfrac{1}{2}\mu^{-1}[\nabla \times \mathbf{A}]^2 - \mathbf{J} . \mathbf{A}\}\, dV. \tag{10.9.21}$$

What we have intuitively derived may be mathematically justified by allowing the vector potential to take a small excursion $\boldsymbol{\delta}\mathbf{A}$:

$$\delta\mathcal{L} = \iiint \{\tfrac{1}{2}\mu^{-1}[\nabla \times (\mathbf{A} + \boldsymbol{\delta}\mathbf{A})]^2 - \mathbf{J} \,.\, (\mathbf{A} + \boldsymbol{\delta}\mathbf{A})\}\, dV$$

$$- \iiint \{\tfrac{1}{2}\mu^{-1}[\nabla \times \mathbf{A}]^2 - \mathbf{J} \,.\, \mathbf{A}\}\, dV$$

$$= \iiint \{\tfrac{1}{2}\mu^{-1}[(\nabla \times \mathbf{A})^2 + 2(\nabla \times \mathbf{A}) \,.\, (\nabla \times \boldsymbol{\delta}\mathbf{A}) + (\nabla \times \boldsymbol{\delta}\mathbf{A})^2]$$

$$- \mathbf{J} \,.\, (\mathbf{A} + \boldsymbol{\delta}\mathbf{A}) - \tfrac{1}{2}\mu^{-1}[\nabla \times \mathbf{A}]^2 + \mathbf{J} \,.\, \mathbf{A}\}\, dV$$

$$= \iiint \{\mu^{-1}(\nabla \times \mathbf{A}) \,.\, (\nabla \times \boldsymbol{\delta}\mathbf{A}) - \mathbf{J} \,.\, \boldsymbol{\delta}\mathbf{A} + \tfrac{1}{2}\mu^{-1}(\nabla \times \boldsymbol{\delta}\mathbf{A})^2\}\, dV. \tag{10.9.22}$$

Now, according to the identity eqn. (A8) with $\boldsymbol{\delta}\mathbf{A}$ for $\mathbf{A}$ and $\nabla \times \mathbf{A}$ for $\mathbf{B}$:

$$\nabla \,.\, [\boldsymbol{\delta}\mathbf{A} \times (\nabla \times \mathbf{A})] = (\nabla \times \mathbf{A}) \,.\, (\nabla \times \boldsymbol{\delta}\mathbf{A}) - \boldsymbol{\delta}\mathbf{A} \,.\, (\nabla \times \nabla \times \mathbf{A}). \tag{10.9.23}$$

Integrating this over the solution region $V$ bounded by the surface $S$ and applying Gauss's theorem (A13) to the left-hand-side term:

$$\iint [\boldsymbol{\delta}\mathbf{A} \times (\nabla \times \mathbf{A})] \,.\, d\mathbf{S} = \iiint [(\nabla \times \mathbf{A}) \,.\, (\nabla \times \boldsymbol{\delta}\mathbf{A}) - \boldsymbol{\delta}\mathbf{A}(\nabla \times \nabla \times \mathbf{A})]\, dV. \tag{10.9.24}$$

Using this in eqn. (10.9.22),

$$\delta\mathcal{L} = \iint \mu^{-1}[\boldsymbol{\delta}\mathbf{A} \times (\nabla \times \mathbf{A})] \,.\, d\mathbf{S} + \iiint [\mu^{-1}\, \boldsymbol{\delta}\mathbf{A} \,.\, (\nabla \times \nabla \times \mathbf{A} - \mathbf{J})]\, dV + \iiint \tfrac{1}{2}\mu^{-1}(\nabla \times \boldsymbol{\delta}\mathbf{A})^2\, dV. \tag{10.9.25}$$

If the surface integral should vanish, then the last term being positive, it is clear that when eqn. (1.2.25) is satisfied, the change in the functional $\delta\mathcal{L}$ is positive so that eqn. (1.2.25) is satisfied at the minimum of $\mathcal{L}$.

It is easy to verify from the identities eqns. (A2) and (A3) that for three vectors $\mathbf{P}$, $\mathbf{Q}$, and $\mathbf{R}$ $(\mathbf{P} \times \mathbf{Q}) \,.\, \mathbf{R}$ is the same as $(\mathbf{Q} \times \mathbf{R}) \,.\, \mathbf{P}$

and $(\mathbf{R} \times \mathbf{P}) \cdot \mathbf{Q}$; they are each equal to a $3 \times 3$ determinant with the components of each vector constituting a row. Now the surface integral has the term $[\boldsymbol{\delta}\mathbf{A} \times (\nabla \times \mathbf{A})] \cdot \mathbf{dS}$, where $\mathbf{dS}$ is along the normal to $S$. The surface term therefore vanishes if $\boldsymbol{\delta}\mathbf{A} \times \mathbf{dS}$ or $(\nabla \times \mathbf{A}) \times \mathbf{dS}$ vanishes on $S$. This will happen if $\boldsymbol{\delta}\mathbf{A}$ or $\mathbf{B} = \nabla \times \mathbf{A}$ is along the normal to $S$, since the cross product of two vectors is proportional to the sine of the angle between the vectors, as shown in eqn. (4.1.20). That is, if the tangential component of $\mathbf{A}$ is specified on $S$ (making the tangential component of $\boldsymbol{\delta}\mathbf{A}$ zero) or the tangential component of the flux density $\mathbf{B}$ is set to zero, then the surface integral will vanish. Considerations along the lines of section 8.3.1 will show that if we use this functional, then along those sections of the boundary where we do not specify the tangential component of $\mathbf{A}$, the tangential component of the flux density $\mathbf{B}$ will naturally turn out to be zero.

Demerdash et al (1981) have simply used this functional without any reference to the divergence equation. While typically no unique solution is possible without specifying the curl and divergence of a vector, it has been shown by Hoole, Rios, and Yoganathan (1988) that the finite element approximations provide sufficient restrictions to make the vector potential unique and that the curl, divergence, and trial function specifications together constitute an overdetermined set of equations. Hoole, Rios, and Yoganathan (1988) further show that since the divergence equation plays no physical role (except in flux plotting), throwing away the divergence equation allows for better satisfaction of the physically more important eqn. (1.2.25). Therefore, using this functional provides better values of flux density $\mathbf{B}$, when the solution has its curl taken. They also show that although the zero divergence specification is ignored, it is weakly imposed through the functional of eqn. (10.9.21). In fairness, it is added that some people of eminence are not comfortable with this writer's position on this subject and have privately expressed their reservations on throwing away the divergence equation.

To implement the functional in finite element analysis, we make the usual approximation on each component of $\mathbf{A}$ so that:

$$\mathbf{A} = [\boldsymbol{\alpha}]^n \underset{\sim}{\mathbf{A}}, \qquad (10.9.26)$$

as we did in eqn. (10.8.2) where the matrix $[\boldsymbol{\alpha}]^n$ is of size $3 \times N(n)$. This time, in three dimensions, the finite element vector $\underset{\sim}{\mathbf{A}}$ is of length $3N(n)$—or 12 for first-order tetrahedrons. We have from eqns. (A1) and (A3):

$$\nabla \times \mathbf{A} = \begin{bmatrix} \dfrac{\partial A_y}{\partial z} - \dfrac{\partial A_z}{\partial y} \\ \dfrac{\partial A_z}{\partial x} - \dfrac{\partial A_x}{\partial z} \\ \dfrac{\partial A_x}{\partial y} - \dfrac{\partial A_y}{\partial x} \end{bmatrix} = [\boldsymbol{\alpha}]^{n-1} \begin{bmatrix} 0 & \mathbf{D}_z & -\mathbf{D}_y \\ -\mathbf{D}_z & 0 & \mathbf{D}_x \\ \mathbf{D}_y & -\mathbf{D}_x & 0 \end{bmatrix} \begin{bmatrix} \underset{\sim}{\mathbf{A}}_x \\ \underset{\sim}{\mathbf{A}}_y \\ \underset{\sim}{\mathbf{A}}_z \end{bmatrix}$$

$$= [\boldsymbol{\alpha}]^{n-1}\mathbf{M}_c\underset{\sim}{\mathbf{A}}, \tag{10.9.27}$$

where $\mathbf{M}_c$ is a matrix curl operator. The size of this matrix is clearly $3N(n-1) \times 3N(n)$ where the function $N$ is defined in eqn. (10.9.17). Substituting in the functional eqn. (10.9.21) with a similar approximation on the components of $\mathbf{J}$:

$$\mathscr{L} = \sum_{\Delta} \tfrac{1}{2}\mu^{-1}V\underset{\sim}{\mathbf{A}}^t\mathbf{M}_c^t[\mathbf{T}]^{n-1}\mathbf{M}_c\underset{\sim}{\mathbf{A}} - V\underset{\sim}{\mathbf{A}}^t[\mathbf{T}]^n\underset{\sim}{\mathbf{J}}, \tag{10.9.28}$$

which upon extremization gives us:

$$\frac{\partial \mathscr{L}}{\partial \underset{\sim}{\mathbf{A}}} = \sum_{\Delta} \mu^{-1}V\mathbf{M}_c^t[\mathbf{T}]^{n-1}\mathbf{M}_c\underset{\sim}{\mathbf{A}} - V[\mathbf{T}]^n\underset{\sim}{\mathbf{J}} = \underset{\sim}{0}, \tag{10.9.29}$$

where the local matrix $\mu^{-1}V\mathbf{M}_c^t[\mathbf{T}]^{n-1}\mathbf{M}_c$ is $3N \times 3N$. For the simple first-order approximation, this will be $12 \times 12$, $\mathbf{M}_c$ being $3 \times 12$ and the matrix $[\mathbf{T}]^{n-1}$ reducing to a $3 \times 3$ unit matrix. Once the differentiation matrices $\mathbf{D}_x$, $\mathbf{D}_y$, and $\mathbf{D}_z$ (the $1 \times 4$ matrices $\tilde{\mathbf{b}}$, $\tilde{\mathbf{c}}$, and $\tilde{\mathbf{d}}$ for first order) are formed, the matrix $\mathbf{M}_c$ and the local matrix may easily be formed. Assembling the global matrix requires a minor elaboration of the principles expounded so far and is left as an exercise.

Coulomb (1981) imposes the divergence condition eqn. (1.2.26) along with eqn. (1.2.25) in a least-square sense and uses the functional:

$$\mathscr{L}[A] = \iiint \{\tfrac{1}{2}\mu^{-1}[\nabla \times \mathbf{A}]^2 - \mathbf{J}\,.\,\mathbf{A}\}\, dV + \iiint \tfrac{1}{2}\mu^{-1}[\nabla\,.\,\mathbf{A}]^2\, dV, \tag{10.9.30}$$

where the first part satisfies eqn. (1.2.25) and the second part satisfies eqn. (1.2.26). The term $\mu^{-1}$ weights the divergence condition for a numerical balance between the two equations. Chari et al (1985a, b) also feel that it is more appropriate to use this functional, and provide an extensive discussion of this method. Since for our approximations:

$$\nabla \cdot \mathbf{A} = \frac{\partial A_x}{\partial x} + \frac{\partial A_y}{\partial y} + \frac{\partial A_z}{\partial z} \tag{10.9.31}$$
$$= \tilde{\boldsymbol{\alpha}}^{n-1}[\mathbf{D}_x \quad \mathbf{D}_y \quad \mathbf{D}_z]\underset{\sim}{\mathbf{A}} = \tilde{\boldsymbol{\alpha}}^{n-1}\mathbf{M}_d\underset{\sim}{\mathbf{A}},$$

where $\mathbf{M}_d$ is a divergence matrix of size $N(n-1) \times 3N(n)$ or $1 \times 12$ for first order, it is easy to verify that the functional is now modified to:

$$\mathscr{L} = \sum_{\Delta} \tfrac{1}{2}\mu^{-1}V\underset{\sim}{\mathbf{A}}^t\{\mathbf{M}_c^t[\mathbf{T}]^{n-1}\mathbf{M}_c + \mathbf{M}_d^t\mathbf{T}^{n-1}\mathbf{M}_d\}\underset{\sim}{\mathbf{A}} - V\underset{\sim}{\mathbf{A}}^t[\mathbf{T}]^n\underset{\sim}{\mathbf{J}}, \tag{10.9.32}$$

which upon extremization gives us:

$$\frac{\partial \mathscr{L}}{\partial A} = \sum_{\Delta} \mu^{-1}V\{\mathbf{M}_c^t[\mathbf{T}]^{n-1}\mathbf{M}_c + \mathbf{M}_d^t\mathbf{T}^{n-1}\mathbf{M}_d\}\underset{\sim}{\mathbf{A}} - V[\mathbf{T}]^n\underset{\sim}{\mathbf{J}} = \underset{\sim}{\mathbf{0}}. \tag{10.9.33}$$

A third functional is given by Chari et al (1981), which requires certain material interface conditions to be explicitly imposed when inhomogeneities are involved. It is thus not widely used.

The competing, important three-dimensional scalar potential formulations should also be mentioned. We saw in section 4.2 that the total magnetic field $\mathbf{H}$ may be split into two components $-\nabla\Omega$ and $\mathbf{H}_c$, where $\Omega$ is the partial scalar potential and $\mathbf{H}_c$ is the field caused by the currents present in the system in the absence of magnetic materials (Simkin and Trowbridge 1979; Trowbridge 1980, 1981). $\mathbf{H}_c$ can be computed by the Biot–Savart law. The balance of the solution is obtained by solving for $\Omega$:

$$\nabla \cdot \mu\nabla\Omega = \nabla \cdot \mu\mathbf{H}_c. \tag{4.2.8}$$

The method of solution is the same as for the electric scalar potential of the previous section. In fact, this is probably the most widely used formulation today in view of the solution being reduced to that of a scalar potential. One of the problems of the method is that in permeable materials $-\nabla\Omega$ and $\mathbf{H}_c$ are oppositely directed and large in relation to the total field, so that numerical error tends to build up. This has been largely overcome whenever the magnetizable parts do not carry currents. When this is so, the total field $\mathbf{H}$ in the magnetizable parts is represented by $-\nabla\Psi$, where $\Psi$ is called the *total scalar potential.* Naturally, matching of $\Omega$ and $\Psi$ at partial–total potential interfaces is required. In a very recent development Mayergoyz, Chari, and D'Angelo (1987) have provided an alternative formulation that overcomes this problem entirely.

We should have perceived by now that three-dimensional problems are easy to formulate but costly to implement by the finite element method. On the other hand, by the competing boundary element method, inhomogeneities are clumsy to handle while the resulting matrix equation is small, albeit dense. In the next few years, therefore, we may expect to see more of these two methods being combined to exploit their advantages (Chari, d'Angelo, and Neumann 1985; Mayergoyz and Doong 1985).

# 11 AUTOMATION AND INTERACTIVE GRAPHICS

## 11.1 GENERAL REQUIREMENTS

By now we are familiar with the mathematics and techniques of field analysis. The basic process of field analysis may be said to be the following:

1. Formulation; e.g., finite elements, finite difference.
2. Data preparation; e.g., dividing the solution region into elements.
3. Assembling the matrix equation.
4. Solving the matrix equation.
5. Result interpretation; e.g., pictorially presenting flux lines.

In a real design environment, several operations (such as getting flux densities from various parts of a device) are repetitive and tedious. This is best avoided through automation, using the powerful graphics features now commonly available in most computing environments. For example, it is far simpler to point to a part of a device presented on a graphics screen, using a graphics pen, than it is to type on the computer keyboard the exact coordinates of the part we are referring to. Indeed, the difficulty is not only in the typing, but also in first obtaining the coordinates from a drawing. Moreover, the above procedures 1–5 become a cycle. We may proceed from 1 to 5 and then decide, for example, that our solution is crude and therefore choose to go back to 2 to refine the mesh. Or else, in an optimization/synthesis effort concerning a device, we may even wish to modify the geometry slightly at step 2 to improve some parameter such as force. It sometimes even happens, although far less commonly, that from step 5 we return to step 1—as, when we try a boundary integral formulation and, upon being dissatisfied with the solution, wish to try solving the same problem by another formulation.

Be it noted that in the above operations the principal problem-dependent operation is the assembly of the matrix at step 3. Another possible difference may be in the way vectors for the plots, such as flux density and velocity, are derived—for example, they may be derived from the gradient of a scalar potential or from the curl of a vector potential as seen in eqns. (1.2.33) and (1.2.34). But for these minor differences, the computer programs for the solution of differential equations by a given formulation are similar.

In view of these similarities it is easy to construct a graphics library that may be commonly and conveniently applied to various finite element problems and formulations. In general, the desirable attributes of the library are that it should

1. allow us, the users, to make decisions, by being interactive and menu driven;
2. allow automatic mesh generation;
3. be as problem independent as possible;
4. allow as many choices as possible to be made through the graphics system;
5. permit transport to another computer system with little change.

In this chapter, using the fact that most operations are common and independent of the equation, we present ways to form a graphics library that will allow us to pose and solve difficult field problems rapidly (Hoole, Hoole, and Jayakumaran 1986). Programming is a very personal matter involving style. The algorithms given here are not meant to dictate style, but rather to give the analyst a starting point. From this start, it is hoped that thought will be provoked so as to result in improvements in software methodology.

## 11.2 GRAPHICS PRIMITIVES

So as to be able to use the proposed library on any machine easily, we shall rely only on a few indispensable graphics primitives in implementing it. Thus, while the elements we use may not be available on every work station, their counterparts will surely exist and the appropriate routines may be employed by redefining these primitive routines as procedures.

The primitive elements used are:

1. **Move($x, y$):** Moves the graphics cursor to the location $x,y$ on the screen.
2. **Line($x, y$):** Draws a line from where the cursor is to $x,y$.
3. **Set_Color($i$):** $i = 1$ is for drawing in white and $i = 0$ for erasing, $2, \ldots, 9$ for the colors violet, indigo, blue-green, yellow, orange, and red.
4. **Graphics_Text('String'):** Writes a prescribed string on the screen as graphics text, starting from where the cursor is.
5. **Locator($i, x, y$):** Reads the digitized coordinate $x,y$ from a graphics tablet. The index $i$ determines the kind of cursor echo, such as one for a simple cross-hair, two for rubber-band (a line anchored at the previous cursor position).
6. **Char_Size($H, W$):** Sets the height and width of graphics text.
7. **Rotate_Text($\alpha$):** Sets the direction, $\alpha$ degrees with respect to the $x$ axis, of text output.
8. **Graphics_Initialize:** Sets up the graphics system by defining the input device (joystick, mouse, tablet, screen) and the output device, which is usually the screen in interactive graphics (since several drawings are usually made) and possibly a plotter/printer (if we want a hard copy of the solution).
9. **Clear_Display:** Wipes the graphics screen clean.

We write the library assuming that the above routines are system routines. To use these on a particular machine, we need to make sure that these terms are understood. To this end, all we need to do is write a small secondary library to make these nine terms, assumed in our graphics library, sensible to the computer. If we need to take it to a machine with equivalent primitives, we construct a few new procedures (subroutines) defining the nine primitives described above. For example, if we are using the commercially available package called Template to drive our graphics and we are working with FORTRAN, then the primitive corresponding to our primitive Move($x, y$) is the Template command UMove($x, y$). Thus, the subroutine required to define the command Move($x, y$) extensively by the library is shown in Alg. 11.2.1A.

It is also possible that the graphics environment to which we are moving cannot support some of the commands that we have assumed. For example, it may not support color and may allow only drawing and erasing respectively, by, say, the commands Draw(1)

**Algorithm 11.2.1. Redefining Assumed Primitives**

**a.**

```
Subroutine Move(x, y)
Real x, y
UMove(x, y)
Return
End
```

**b.**

```
Procedure Set_Color(i: Integer);
Begin{Procedure}
  If i = 0
    Then Draw(0)
    Else Draw(1)
End; {Procedure}
```

**c.**

```
Procedure Locator(i: Integer; Var x, y:Real);
Var p,q: Integer;
Begin{Procedure}
  Pen(p, q);
  x:=p;
  y:=q;
End;{Procedure}
```

and Draw(0). If we are using PASCAL, then Alg. 11.2.1B will allow the command Set_Color($i$) to be understood by the computer. In this we have made all colors appear as simple drawings in the color of the screen (usually a bluish white, or green or brown) and erasing appears as erasing. Another possibility is that in the command Locator($i, x, y$), we may not be allowed several types of cursor display given by the integer $i$, so that we may have to write extensive routines to make $i$ meaningful, or, in defining the procedure Locator, we may just ignore the term $i$, as shown in Alg. 11.2.1C in PASCAL, where the command Pen($p, q$) with a simple cursor cross-hair display is assumed where $p$ and $q$ are integers.

Thus, on the new computing system, the primitive variable Locator will now be defined and the library itself will not have to be meddled with.

## 11.3 DATA STRUCTURES

The data for finite element field analysis of electric and magnetic fields require triangles covering the whole solution region; each

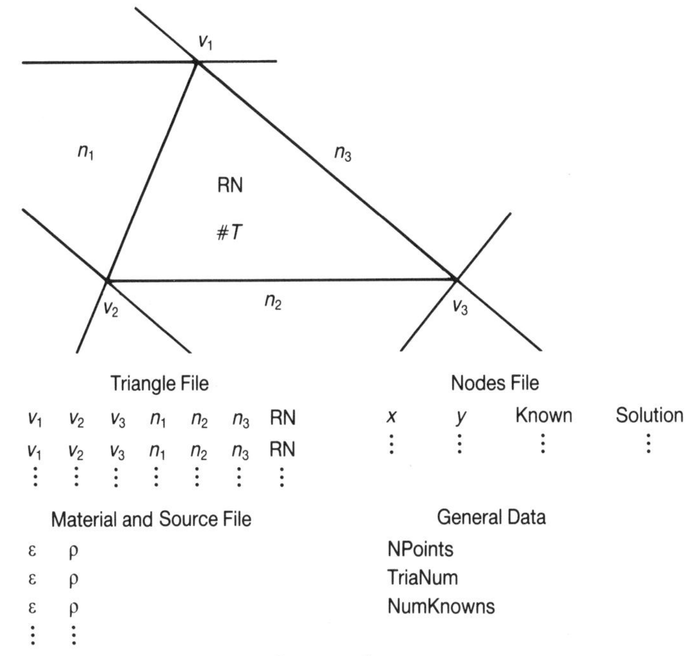


**Figure 11.3.1.** Finite element data structures.

triangle is defined by the nodes on it, the material value $\varepsilon$ of the Poisson equation, and the source value $\rho$. For first-order triangles we will have three nodes, and more for higher orders. Now in field problems, many triangles will have the same pair of values $\varepsilon$ and $\rho$, and therefore some economy in storage may be achieved by associating a region with a unique pair $(\varepsilon, \rho)$ and we will need to store only the region number with the triangle. The region numbers will be pointers to the locations of $\rho$ and $\varepsilon$ in another file, as shown in Figure 11.3.1.

Moreover, economy in computation may be achieved by storing the three triangle neighbors with each triangle (Keinstreuer and Hadenan 1980). As we shall see, among other things, having the neighbors permits us to identify speedily a triangle pointed at by a user and compute the optimality of nodal connections (Hoole 1986).

We shall also require another file containing the nodal data. This will consist of the nodal coordinates and the solutions, the $i$th entry standing for the $x$ and $y$ coordinates of the $i$th node and the solution thereat, which we shall compute and write. Up to now we have assumed that the known nodes are numbered last. This is fine so long as we have one shot at the solution. But in practical situations, we iteratively refine the mesh until we achieve an acceptably accurate solution. In the process of refinement, new nodes will be introduced, some known, others not. And rigidly adhering to numbering the knowns last requires our renumbering the nodes, which also means redefining the triangles—a difficult task that is quickly avoided by assigning an additional field to the nodal data consisting of just an integer, saying whether it is known or not. Calling this item **Known**($i$) for triangle $i$, we may set it to 0 if it is unknown and to some negative integer, say $-1$, when it is known. In the latter event, the known value is written in the field of the node reserved for the solution. At the time of solution, we may search through the field **Known** and perform a simple mapping of the variable number in the matrix equation associated with each node. For example, if we had six nodes with the third and fifth known, then the field **Known** will be [0, 0, $-1$, 0, $-1$, 0]. The mapping is done by searching through this and overwriting [1, 2, $-1$, 3, $-1$, 4]; so that, for example, in the matrix equation, the four unknowns to be computed (1, 2, 3, and 4) correspond to the field values at nodes 1, 2, 4, and 6. The difference this makes to our Alg. 7.6.3. GlobalPlace is that whether a node $i$ is known or not is identified, not by the question whether $i \leq$ NUnk, the number of unknowns, but by asking if **Known**($i$) is less than 0; if not, then it is an unknown node and row Row of the local matrix associated with the node **v**[Row] should go into row **Known**(**v**[Row]) of the global matrix. Previously, an unknown node **v**[Row] was the variable **v**[Row] of the global matrix, and therefore we simply added row Row of the local matrix to row **v**[Row]. A similar modification should be made to the column entries.

We shall then at this point stipulate the storage of the following:

1. a file containing the triangle data, with our storing for each triangle the integers $v_1, v_2, v_3$, which are the three vertices (for first order, say), $n_1$, $n_2$, $n_3$, the three neighbors, and RN, the region number;
2. a file containing the nodal parameters, with our storing for each node the $x$ coordinate, the $y$ coordinate, and entry of **Known**, the latter being set to zero for unknown nodes and to $-1$ otherwise. In addition we shall have a real number field for the solution to be written; or if the solution is already known through boundary conditions, it merely forms a part of the data;
3. a short file consisting of the material $\varepsilon$ and source $\rho$ corresponding to each region number RN. Thus each entry will be a pair $\varepsilon$ and $\rho$.

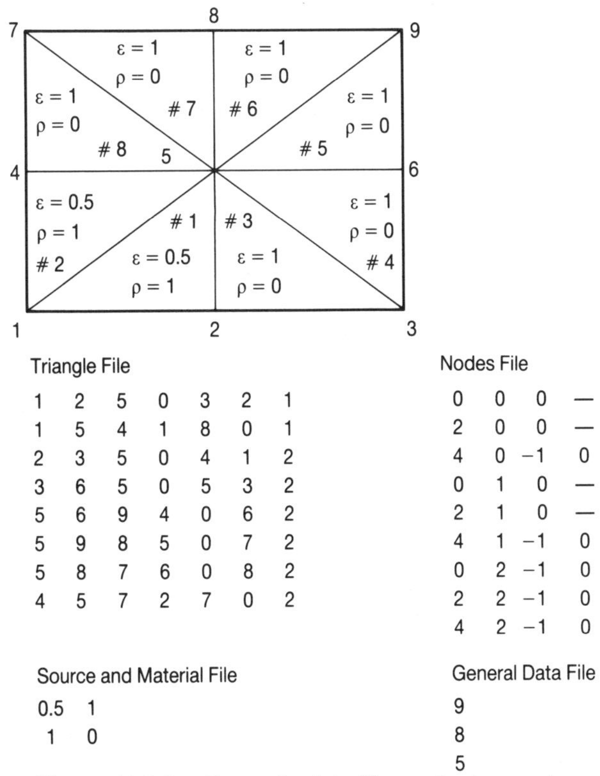


Triangle File

| | | | | | | |
|---|---|---|---|---|---|---|
| 1 | 2 | 5 | 0 | 3 | 2 | 1 |
| 1 | 5 | 4 | 1 | 8 | 0 | 1 |
| 2 | 3 | 5 | 0 | 4 | 1 | 2 |
| 3 | 6 | 5 | 0 | 5 | 3 | 2 |
| 5 | 6 | 9 | 4 | 0 | 6 | 2 |
| 5 | 9 | 8 | 5 | 0 | 7 | 2 |
| 5 | 8 | 7 | 6 | 0 | 8 | 2 |
| 4 | 5 | 7 | 2 | 7 | 0 | 2 |

Nodes File

| | | | |
|---|---|---|---|
| 0 | 0 | 0 | — |
| 2 | 0 | 0 | — |
| 4 | 0 | −1 | 0 |
| 0 | 1 | 0 | — |
| 2 | 1 | 0 | — |
| 4 | 1 | −1 | 0 |
| 0 | 2 | −1 | 0 |
| 2 | 2 | −1 | 0 |
| 4 | 2 | −1 | 0 |

Source and Material File

| | |
|---|---|
| 0.5 | 1 |
| 1 | 0 |

General Data File

9
8
5

**Figure 11.3.2.** Example data file contents.

The number of nodes, number of triangles, and number of knowns in the mesh may be extracted from the files of 1 and 2 by reading them until the ends of the files are reached and incrementing a counter as each entry is read. Alternatively and conveniently, we may have another file containing these three numbers. This modified data structure for the problem of Figure 7.8.1, with a permeability of 2 ($\varepsilon = 0.5$), is given in Figure 11.3.2.

With this economic data structure, it is even possible to stop a menu-driven program halfway through, write all the values on file, and resume processing at a convenient time later by first reading in the data.

## 11.4 OUR OWN LIBRARY

### 11.4.1 Windows

In this section we describe the elements of the graphics library built from the graphics primitives described above in section 11.2. Before writing the library, we ought to define the use of the computer screen. Some computers permit division of a screen into several windows, so that we may assign one window for alphanumeric output, another for menus, yet another for prompts to the user, and so on. Some other computers such as the Hewlett Packard series 200 machine even have both alphanumerics and graphics on the same screen, but with provisions for a toggle switch that permits us to switch off the one or the other; in effect, therefore, the computer has two windows. Unfortunately, however, not all computers have this feature of multiple windows and as a result the screen will become cluttered with the alphanumerics and the graphics overlaid on each other. We shall therefore follow here the strategy of writing the program without assuming these multiwindow features, so that we may use our program without too much modification when we at some future time need to change over to another computer; i.e., we shall divide the screen into three parts, say the top right corner for menu-driven choices, the lower part (5%) for messages to the user describing the menu, and the rest for drawing the object under analysis.

## 11.4.2 Draw_Box

Drawing a box is such a common operation in the interactive operations envisaged that we shall make a separate procedure out of it. The algorithm is simple, as seen in Alg. 11.4.1, and is accomplished through Move and Line Commands. It takes the coordinates of the four vertices as input.

**Algorithm 11.4.1. Drawing a Box**

> **Procedure Draw_Box**($x1, y1, x2, y2, x3, y3, x4, y4$)
> {Function: Draws a box with coordinates $(x_1, y_1), \ldots, (x_4, y_4)$; inputs $x1$,$y1$,$x2$,$y2$,$x3$,$y3$,$x4$,$y4$ in screen coordinates}
> **Begin**
>   **Move**($x1, y1$)
>   **Line**($x2, y2$)
>   **Line**($x3, y3$)
>   **Line**($x4, y4$)
>   **Line**($x1, y1$)
> **End**

## 11.4.2 Screen—Menu for Compulsory Decisions

This procedure is for drawing a large screen-sized menu as in Figure 11.4.1, and may be used to drive an interactive program at main command level, where, by choosing one of the options provided, the user directs the program. The use of the menu is compulsory in that until the user selects what he or she wants, nothing may happen. It takes as input the number of options and a string associated with each option. The procedure returns the option selected. Thus, for the menu of Figure 11.4.1A, the number of options is five and the associated strings "PreProcess," "Solve-Once," "Solve-Adaptively," "PostProcess," and "Quit." This is algorithmically described in Alg. 11.4.1, which allows a maximum of 10 options. Thus, to produce the menu of Figure 11.4.1A on a Hewlett Packard screen with coordinates going from $(-1, -1)$ to

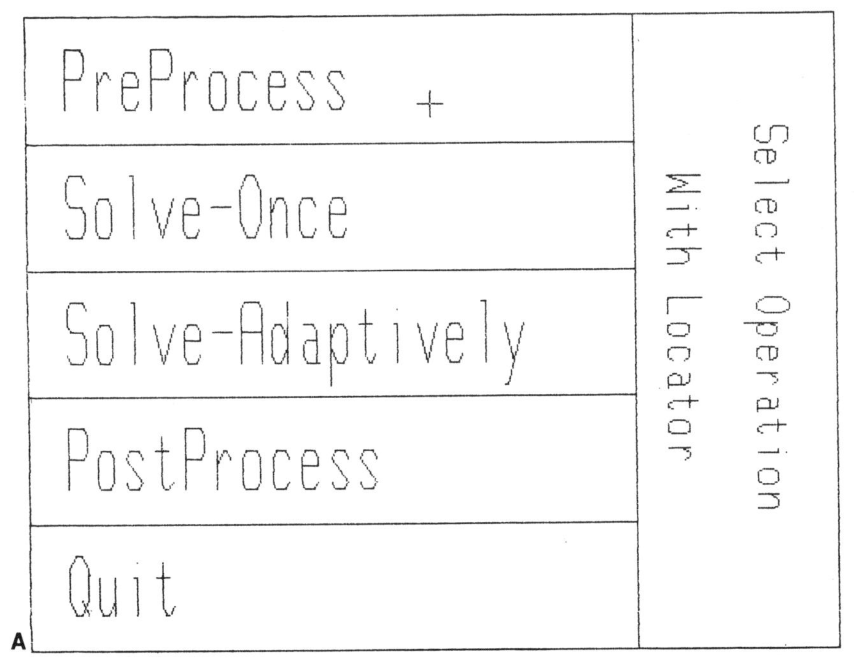


**Figure 11.4.1.** Principal graphics menus. **A.** Menu at main command level. **B.** Menu for preprocessing. **C.** Menu for postprocessing.

Start Mesh
Define Line Elements
Manual Refine
Auto Refine
Refine Boundary
Boundary Conditions
Define Boundary Inputs
Change Materials
Change Mesh
Quit
Select Operation
With Locator
B

Plot
Force
Potential
Intensity
Inductance
Draw Graphs
Quit
Select Operation
With Locator
C

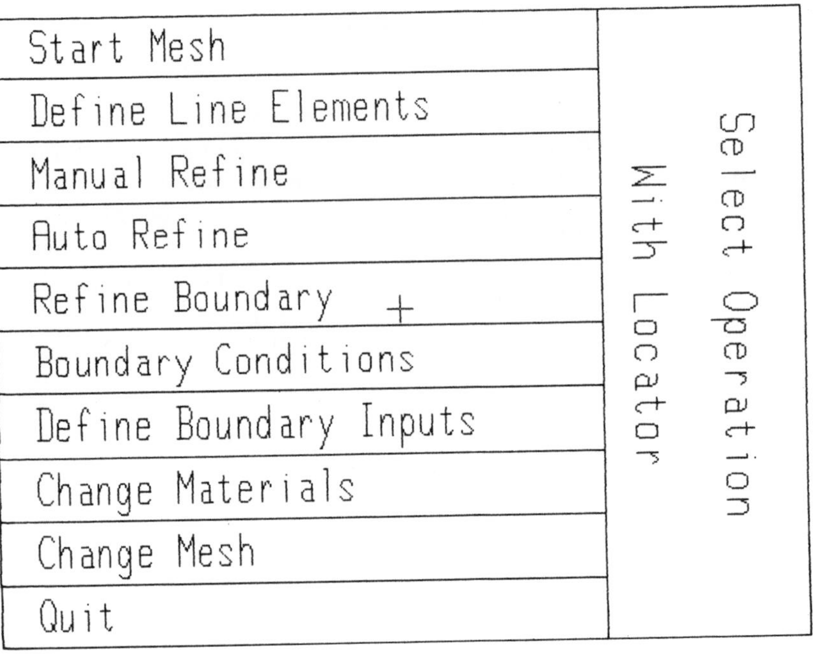

**Figure 11.4.1.** *(continued)*

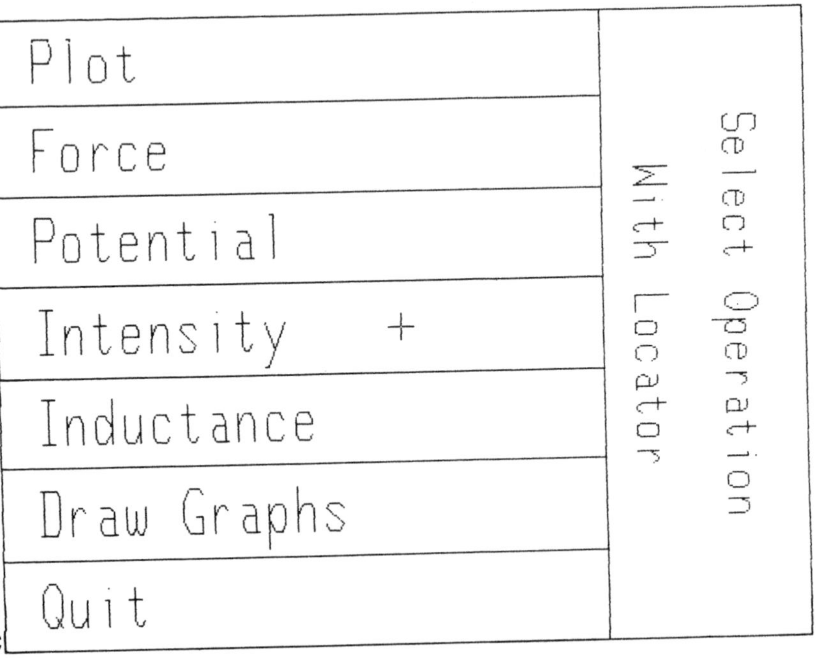

**Algorithm 11.4.2. Drawing a Screen-Sized Graphics Menu**

```
Procedure Screen_Menu (Choice, Options, S1, S2, S3, S4, S5,
S6, S7, S8, S9, S10, x_L_S, x_U_S, y_L_S, y_U_S)
{Function: Presents a maximum of 10 choices to the user and
returns the choice made by the user
Inputs: n = number of options being offered to the user, ≤10;
S1, . . . , S10 = strings of text describing each option; x_L_S,
y_L_S = lower left screen coordinates; x_U_S, y_U_S =
upper right screen coordinates
Output: The integer choice made by the user}
Begin
  AS[1] ← S1      {AS is an array of strings of size 10}
  AS[2] ← S2
  · ← · {etc.}
  · ← · {etc.}
  AS[10] ← S10
  BoxHeight ← (x_U_S − x_L_S)/Options
  Char_Size(0.3*BoxHeight, 0.9*BoxHeight)
  yBoxTop ← y_U_S
  xBoxTop ← x_L_S + 0.75*(x_U_S—x_L_S)
  For i ← 1 To Options Do
    yBoxBot ← yBoxTop − BoxHeight
    Draw_Box(x_L_S,yBoxBot,xBoxTop,yBox_Bot,
      xBoxTop,yBoxTop, x_L_S,yBoxTop)
    Move(x_L_S,BoxBot + 0.1*BoxHeight)
    Graphics_Text(AS[i]);
    yBoxTop ← yBoxBot
  Draw_Box(xBoxTop, y_L_S, x_U_S, y_L_S, x_U_S,
    y_U_S, xBoxTop, y_U_S)
  Rotate_Text(90)
  Move(xBoxTop + 0.125*(x_U_S − x_L_S), y_L_S)
  Graphics_Text('Select Operation')
  Move(x_U_S,y_L_S);
  Graphics_Text('With Locator');
  Locator(1, x, y) {simple cross-hair used}
  Choice ← 0
  yTop ← y_U_S
  While y < yTop Do
    yTop ← yTop − BoxHeight
    Choice ← Choice + 1
End
```

(1, 1), the call is:

Screen_Menu(Choice, 5, "PreProcess";
"Solve-Once," "Solve-Adaptively,"
"PostProcess," "Quit," ," " ," " ," " ," " ," " −1, 1, −1, 1);

Note the five dummy strings used for compatibility between the call and the procedure. To implement this, the height of the screen is divided into five character sizes accordingly chosen using Char_Size, lines drawn using Move and Line, and the strings output using Graphics_Text and Rotate_Text. Once the user digitizes in the appropriate box and the Locator command has read the $x$ and $y$ coordinates, the $y$ coordinate is used to evaluate the box in which digitization took place and this determines the selection.

## 11.4.3 Color_Box for Filled Boxes

This procedure, Color_Box, draws several straight lines closely adjacent and parallel to each other, using the Move and Line commands, inside a box with edges parallel to the principal axes and predesignated lower vertex, width, and height (at the time of calling the procedure). It takes an integer, Color, for choice of color

**Algorithm 11.4.3. Filling a Box with Color**

**Procedure Color_Box**($x0$,$y0$,Width,Height,Color)
{Function: Fills a Box with lower left vertex at $x_0$, $y_0$ and size Width × Height
Inputs: $(x_0, y_0)$ = lower left coordinates of box;
Width = width of box; Height = height of box; Color = color to be filled}
**Begin**
  **Set_Color**(Color)
  $n \leftarrow 50$
  Del ← Height/$n$
  $xu \leftarrow x0$ + Width
  $y \leftarrow y0$
  **For** $i \leftarrow 1$ **To** $n$ **Do**
    **Move**($x0$,$y$)
    **Line**($xu$, $y$)
    $y \leftarrow y$ + Del
  **Set_Color**(1)
**End**

so that, for example, when we call it with Color = 1, the region inside the box is made white, and with Color 0, anything in the box is erased. This algorithm is described in Alg. 11.4.3.

The use of this, as we shall see in the next section, is principally in erasing portions of a screen without disturbing the other portions.

### 11.4.4 Small_Menu for Optional Decisions

The procedure Small_Menu works very much like Screen_Menu, except for the menu being restricted to the top right part of the screen as in Figures 11.4.2 and 11.4.5A. Under Screen_Menu, as seen in Figure 11.4.1, the user must make a decision. But sometimes it happens that a user is performing a repetitive task, from which he may call a special feature. This procedure, Small_Menu, is therefore useful when some operation is required in the main part of the screen and we may wish to have access to other suboperations. Thus, as in Figure 11.4.2A, we may, while adding points to a

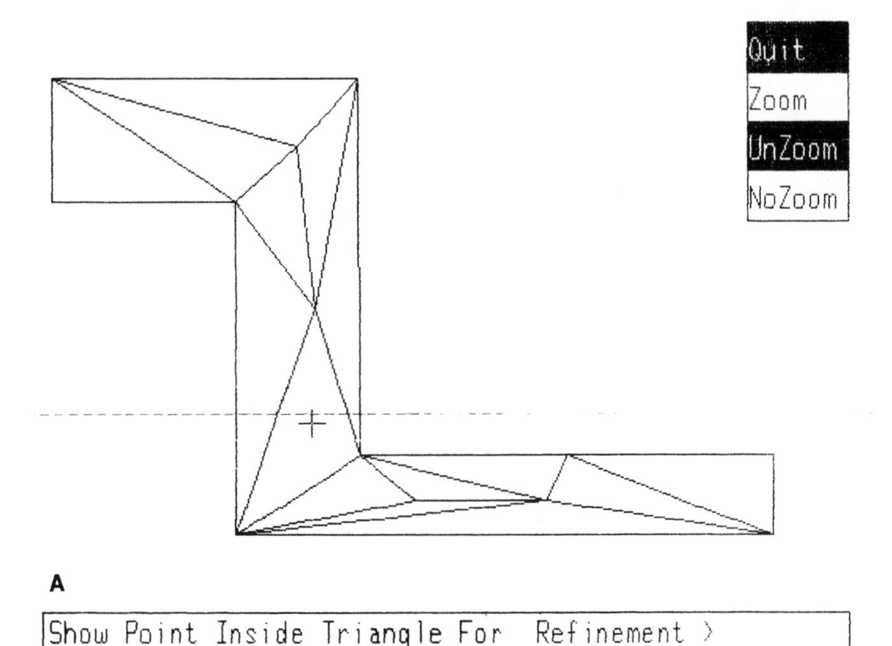


**Figure 11.4.2.** Internal triangle refinement. **A**. Adding a point. **B**. After adding a point.

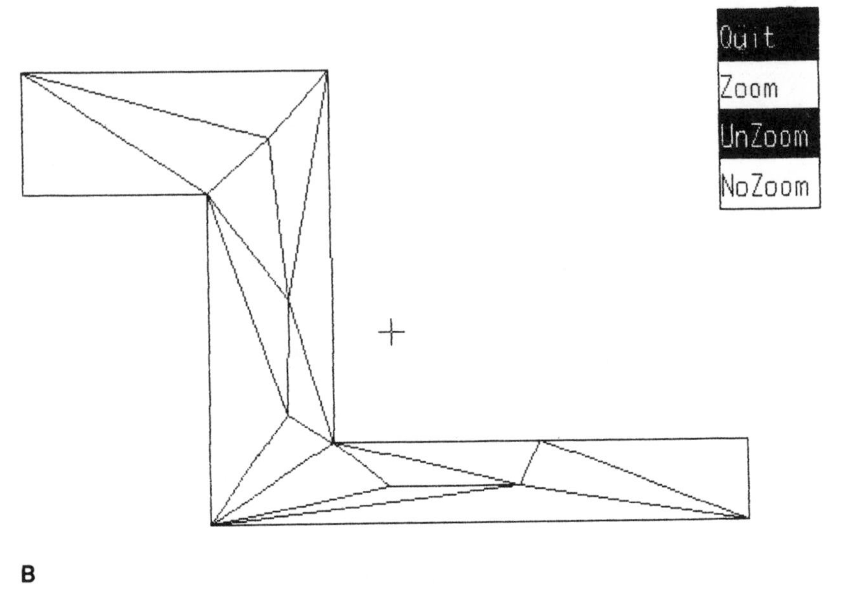


**Figure 11.4.2.** (*continued*)

finite element mesh, suddenly want to call a zooming feature. As a point $x$, $y$ is digitized by the user, we first determine if it is in the domain of the menu; if so, we determine the option selected and accordingly allow the execution of that option; otherwise, we take the coordinate prescribed as indicative of the point to be added to the mesh in refinement. We see the same menu in use in Figure 11.4.5A, where we are about to operate on the boundary conditions of a field problem.

The white background (which comes out black in the figures) is achieved by calling Color_Box with color 1 and the position of the box as input. To get the string "Quit," for example, in the white background of Figure 11.4.2, before outputting the text using Graphics_Text, we set the color to 0 and effectively perform an erasure. The development of this algorithm is very similar to Screen_Menu and is left as an exercise to the more interested reader.

## 11.4.5 Quit_Box for Quitting Region

Quit_Box operates similarly to Small_Menu, but with only one alternative option—to Quit—as shown in Figures 11.4.3, 11.4.4, and

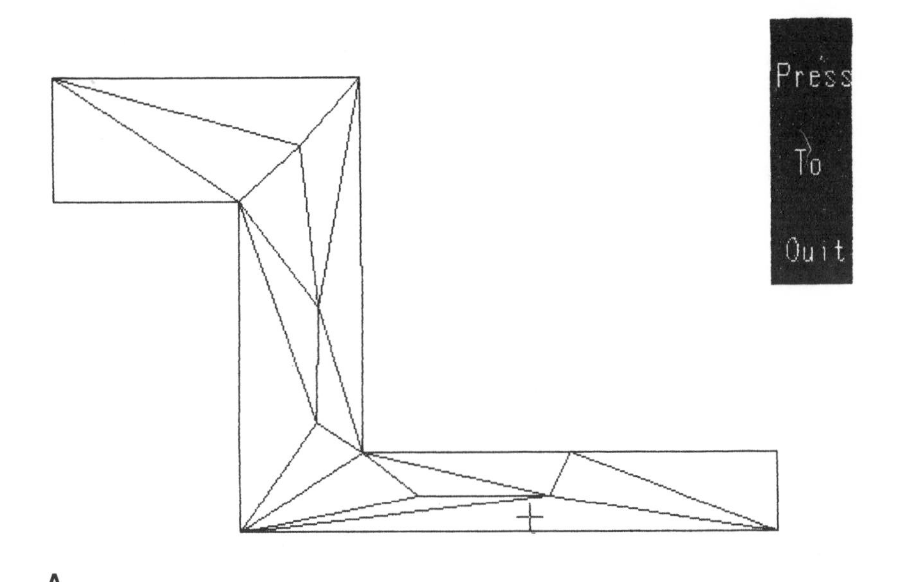


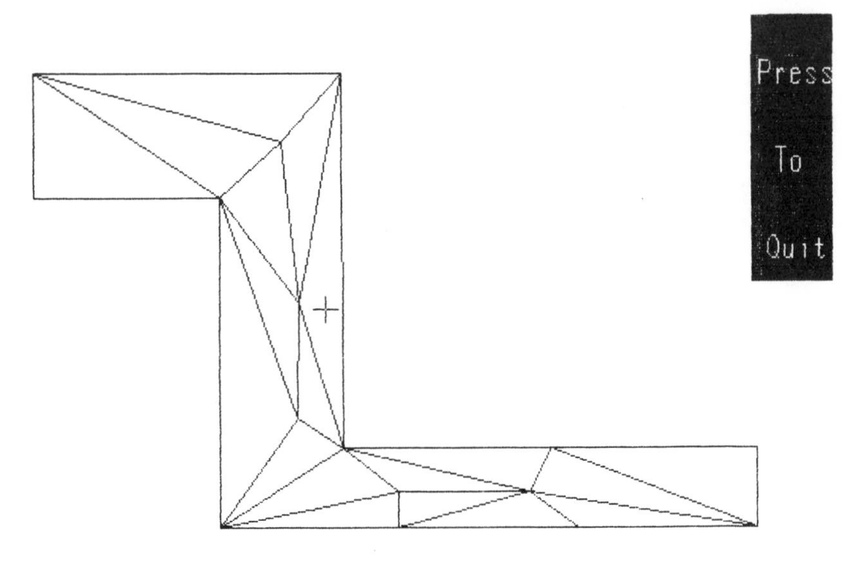


**Figure 11.4.3.** Edge refinements. **A**. Adding a point. **B**. After adding a point.

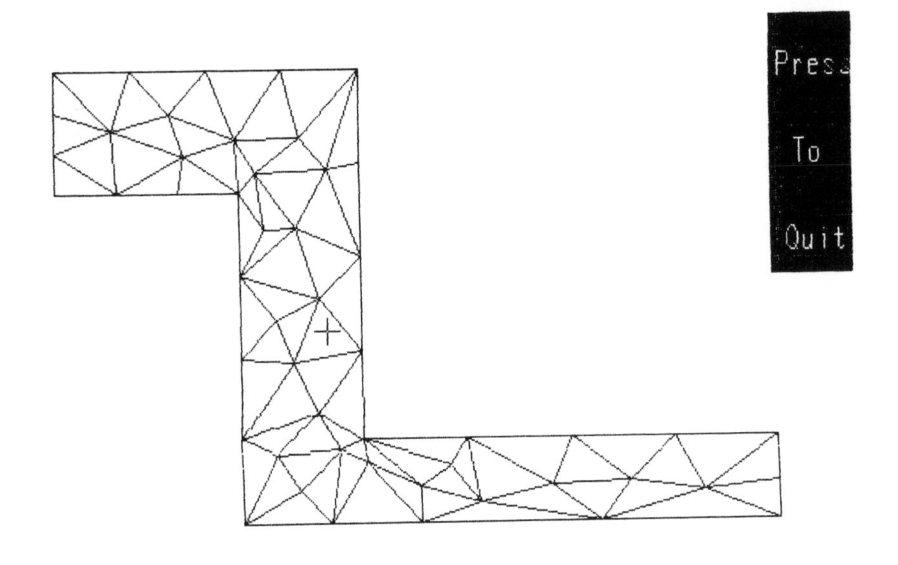


**Figure 11.4.4.** Extracting flux density. Interactive imposition at boundary conditions.

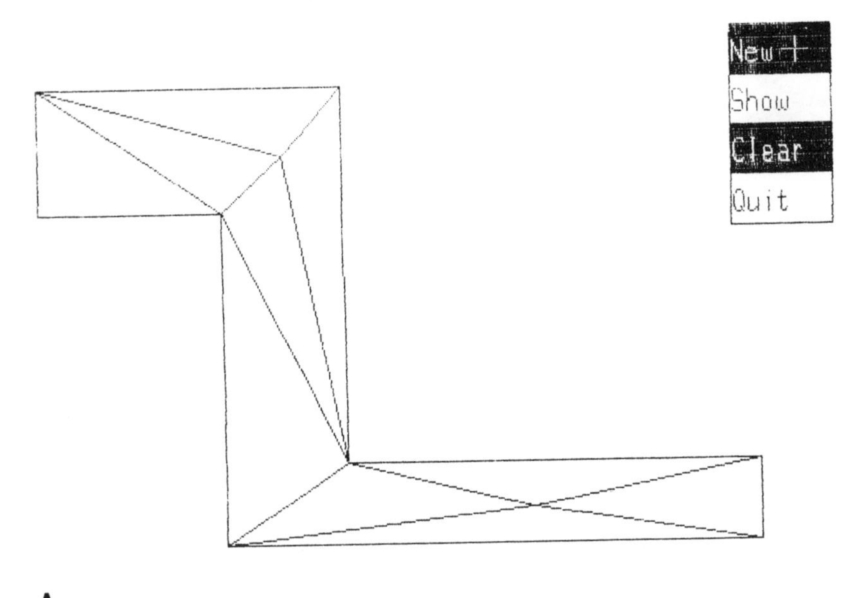


**Figure 11.4.5.** Interactive imposition at boundary conditions. **A.** Options for boundary conditions. **B.** Setting boundary conditions.

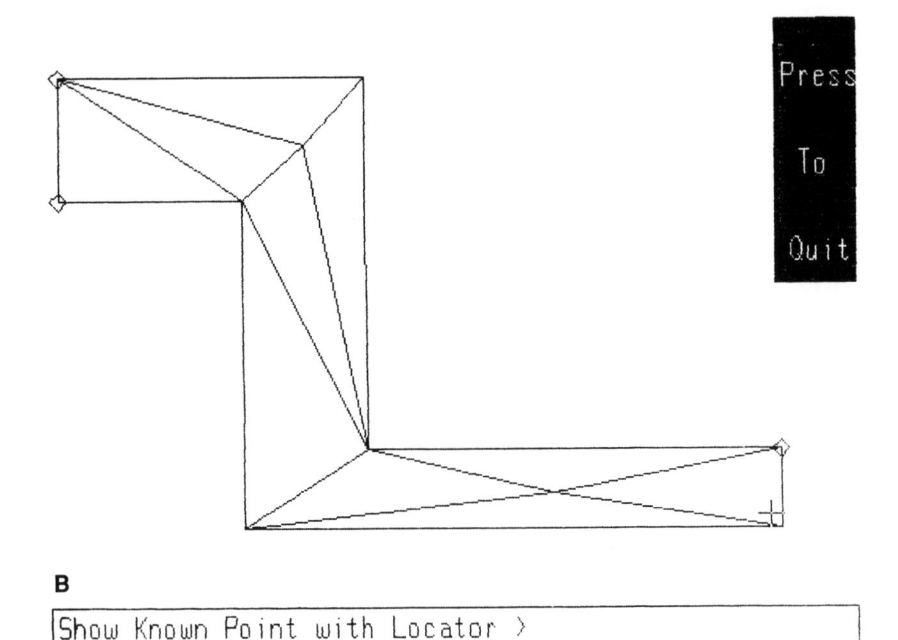


**Figure 11.4.5.** (*continued*)

11.4.5B. Thus, as seen in Figure 11.4.3, while refining an edge we may either digitize inside the boundary triangle to refine the edge or digitize in the Quit_Box region to terminate the operation of edge refinement. As before, therefore, every time we digitize we need to check if the coordinates are inside or outside the box.

## 11.4.6 Write_Message and Erase_Message for User Prompts

When a computer system does not permit multiple windows, the screen can become an unintelligible clutter of alphanumerics and graphics, one overlaid upon the other. Thus, when the program is to prompt the user with some string, we optimally would like to avoid alphanumerics, which cannot usually be erased. One alternative is to use another device such as a printer to print the prompt. Not only is this wasteful of paper but it also diverts the attention of the user from the screen on which he or she is concentrating. This problem may be avoided by using graphics prompts.

The procedure Write_Message takes a string as input, draws a long and narrow box along the lower 5% of the screen, moves the cursor to the left edge of the box, just inside, sets the text size, and outputs the text. This is useful in writing such messages as "Show Point Inside Triangle For Refinement > " at the bottom of Figure 11.4.2A, and similar messages in Figures 11.4.3–11.4.5, to prompt the user. Once the user has responded to the prompt, the procedure Erase_Message is called to erase the message that was written earlier. This procedure does exactly the same thing as Write_Message, but only after setting the color to zero. Thus, this scheme is useful in erasing and replacing the prompt at the bottom of the screen, without having to regenerate the whole screen, as, for example, in going from Figure 11.4.5A to 11.4.5B.

An alternative scheme involves writing the prompt as alphanumerics, and then, to get rid of the prompt on a screen that allows, say, 22 lines of text, writing dummy lines 22 times in place of Erase_Message, so that with each dummy line the previously written alphanumeric text moves up one line until it finally disappears from the screen. The only disadvantage in this is that it is relatively time consuming vis à vis the approach of writing prompts in graphics and then erasing them. Moreover, on systems such as the IBM personal computer series that distinguish between graphics and alphanumerics only in the manner of output, even the graphics part of the screen will scroll with the alphanumerics. As a result, when the picture of the device scrolls up and the user points to a part of the device, the coordinate read will be above the point actually desired.

## 11.4.7 IDTriangle for Identification of Triangles

The procedure IDTriangle identifies that triangle of the mesh containing the coordinates $x$, $y$ given as input. This is done by searching through the list of triangles making up the mesh and computing the triangular coordinates of $x$, $y$ in that triangle, until the triangle containing it is encountered. We recognize the triangle, because it is the only one having all three triangular coordinates positive. This may be useful in refinement as in Figures 11.4.2A and 11.4.2B, or in asking for the flux density at a particular point as in Figure 11.4.4. The algorithm is presented in Alg. 11.4.4, which uses eqns. (7.5.13)–(7.5.17) to compute the three triangular coordinates of $(x, y)$ in each triangle using the procedure ReturnZeta. A faster algorithm for identifying a triangle pointed out by the user has been given by Hoole (1986), and may be referred to by the more interested reader.

**Algorithm 11.4.4. Identification of a Triangle**

```
Procedure ReturnZeta(Zeta, xP, yP, xT, yT)
{Function: Computes the triangular coordinates zeta of a point
xP,yP in a triangle whose vertex coordinates are stored in 3 × 1
arrays xT and yT}
Begin
  Delta ← xT[2]*yT[3] − xT[3]*yT[2] + xT[3]*yT[1] −
    xT[1]*yT[3] + xT[1]*yT[2] − xT[2]*yT[1]      {Eq. (7.5.17)}
  For i ← 1 To 3 Do
    i1 ← i Mod 3 + 1
    i2 ← i1 Mod 3 + 1
    Ai ← (xT[i1]*yT[i2] − xT[i2]*yT[i1])/Delta   {Eqn. (7.5.14)}
    Dxi ← (yT[i1] − yT[i2])/Delta                 {Eqn. (7.5.15)}
    Dyi ← (xT[i2] − xT[i1])/Delta                 {Eqn. (7.5.16)}
    Zeta[i] ← Ai + Dxi*xP + Dyi*yP                {Eqn. (7.5.13)}
End
```

```
Procedure IDTriangle(Tria, Found, TriaNum, v,xG,yG,x,y)
{Function: Identifies the triangle containing the point x,y
Inputs: TriaNum = the number of triangles; v = TriaNum × 3
array containing the vertices of each triangle; xG,yG = arrays
containing the coordinates of all nodes, x,y = the coordinates
of the point we are searching for
Outputs: Found = a Boolean set to True if the triangle is found;
Tria = the found triangle}
Begin
  Found ← False
  i ← 0
  Repeat
    i ← i + 1
    For j ← 1 To 3 Do
      xTriangle[j] ← xG[v[i,j]]
      yTriangle[j] ← yG[v[i,j]]
    ReturnZeta(Zeta, x,y,xTriangle,yTriangle)
    If ((Zeta[1] > 0) And (Zeta[2] > 0) And (Zeta[3] > 0))
      Then Found ← True
  Until Found Or (i = TriaNum)
  Tria ← i
End
```

When using a rectangular mesh in finite element or finite difference analysis, a similar but simpler algorithm may be devised to permit the identification—by checking if $x$ and $y$ lie within the lower and upper limits of the $x$ and $y$ coordinates of the rectangle. The program may run through the nodes of the finite difference mesh, establish using Alg. 9.2.4 whether a rectangle lies in the first quadrant about the node, and, if so, check whether that rectangle contains the point $(x, y)$. Devising the precise details of the algorithm is very straightforward.

## 11.4.8 IDPoint for Identifying a Point

The procedure IDPoint takes coordinates $x$, $y$ as input and identifies the closest node of the mesh. This is useful in setting boundary conditions, as in Figure 11.4.5B, where we may set the value of a node by pointing to it on the screen with the graphics tablet. This may also be used in moving nodes under "Change Mesh" of Figure 11.4.1B, while preprocessing.

When a user points out a node on the screen, the locator command will read the $x$ and $y$ coordinates—which necessarily will not exactly be those of the node. The algorithm will run through each node of the mesh of coordinates $(x^p, y^p)$ and compute the distance $\surd\{(x - x^p)^2 + (y - y^p)^2\}$ and select the point having the shortest distance to the digitized point.

## 11.4.9 Equi_Plot for Plotting Equipotentials

The procedure Equi_Plot, given values of some scalar quantity $\phi$ at the vertices of the triangular mesh, draws 50 contours along which that quantity is equal. We have already seen the use of this in section 9.2, and the procedure has been described in Alg. 9.2.3.

In this general library algorithm, however, we first search through the nodes for the maximum and minimum $\phi$Max and $\phi$Min so that the value difference between two contours $d = (\phi\text{Max} - \phi\text{Min})/49$ and the contour values $\phi$Min, $\phi\text{Min} + d, \ldots, \phi\text{Max}$ may be kept general for whatever field solution turns up. In Figures 9.2.3A and 11.4.6A, we see the equipotential lines being drawn, obviously triangle by triangle, but the overall result becomes continuous as seen in Figures 9.2.3B and 11.4.6B.

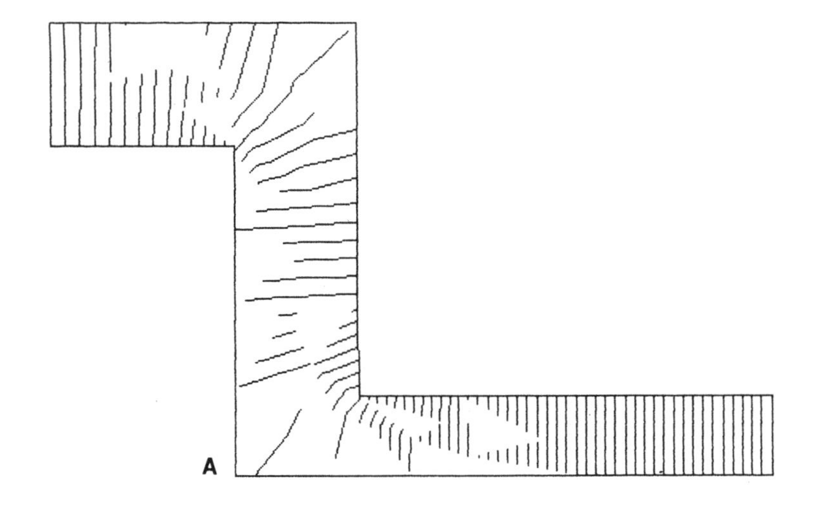


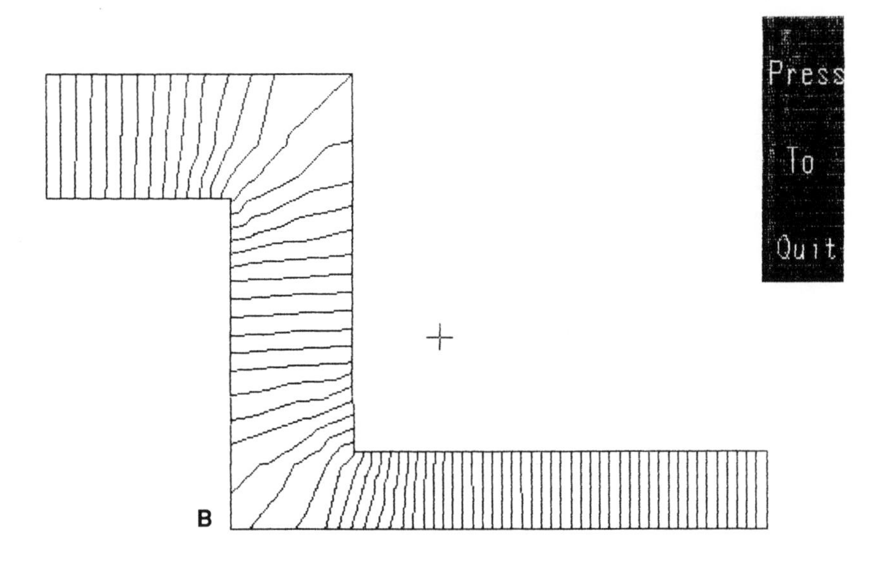


Show Point For Field Density >

**Figure 11.4.6.** Graphical display of fields. **A**. Halfway through plotting equipotentials. **B**. Final equipotential plot. **C**. Corresponding field strength plot.

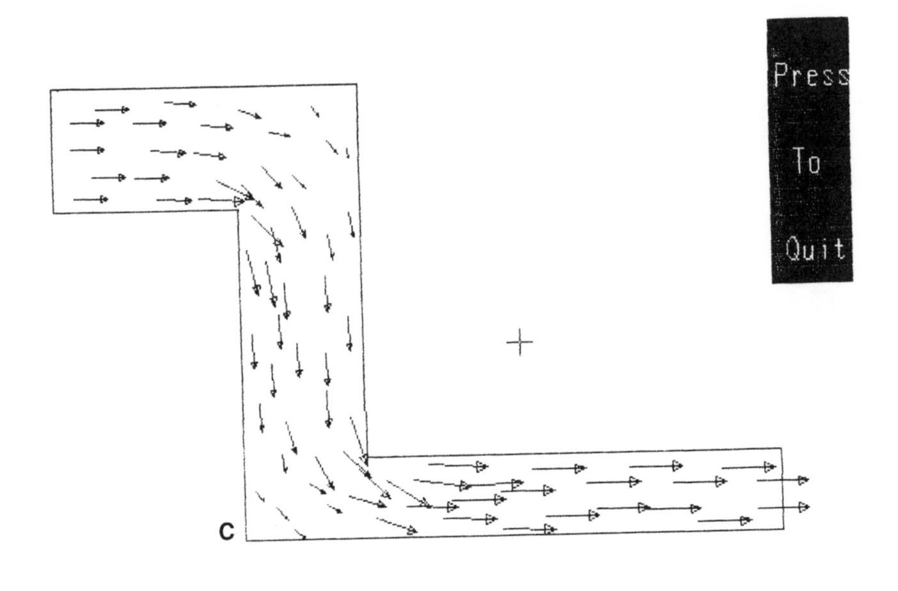


**Figure 11.4.6.** *(continued)*

## 11.4.10 Mesh_Draw for Drawing the Mesh

This procedure, Mesh_Draw, draws the finite element mesh on the screen. The use is obviously in refinement, where we may want to have a picture of the mesh on the screen to see where we may wish to add points to the mesh. This is accomplished by taking every triangle, and, using the Move and Line commands, drawing straight lines along the three edges of every triangle. Obviously, we will be drawing each interior edge twice, but this is less time-consuming than searching for those we have already drawn.

In drawing the device, we are placing a device in world coordinates on the screen, so the procedure OurMove of Alg. 9.2.3 and a similarly defined OurLine will need to be used. It is the procedure Mesh_Draw that produced the meshes of Figures 11.4.2–11.4.5.

### 11.4.11 Draw—Device for Drawing the Device Outline

In plotting equipotentials or drawing the mesh as described above, it is also necessary to show where in relation to the device these pictures are; i.e., for example, as in Figure 11.4.6, the outline of the device also should be provided for a meaningful flux plot. And this procedure should be kept as general as possible so that once we write our program, we may use it for all shapes of device we may choose to analyze in the future.

Drawing a device in finite element analysis is quite easily done because of the data structure used in section 11.3, where for each triangular element we store the three vertices, the three neighbors, and a fourth integer indicating a region number with a unique material and source value pair. Thus, as we search through the mesh we take a triangle and examine the three neighbors along each side. If no neighbor exists, then it is on the boundary, and therefore that edge needs to be drawn as a part of the device outline. If a neighbor exists, then we examine the region numbers (the seventh indices) of the two neighboring triangles to see if they are the same, and, if they are not, then again the edge is a boundary (such as between air and steel) and we need to draw the edge.

Once we establish the need to draw an edge, knowing the node numbers, we may get their coordinates and draw them. It is desirable to have the device outline in a thicker line style, which may be permitted in many graphics systems. Alternatively, we may use Alg. 11.4.4 to draw thick lines, which uses a slight offset (a hundredth of device dimensions) to draw a second line to give a thick appearance.

In dealing with a finite difference mesh, a similar algorithm may be devised using the type statement for a node and by checking for Neumann and Dirichlet boundaries.

### 11.4.12 Arrow for Drawing Arrows

This procedure draws an arrow beginning at a given point $x$, $y$ and of a specified $x$ and $y$ component. This may be used to produce the flux density plot of Figure 11.4.6C, which is for the same configuration as that of Figure 11.4.6B. A procedure to accomplish this, which we shall call "Arrow," may be easily written using Move and Line commands as shown in Figure 11.4.7. Arrow merely takes the

**Algorithm 11.4.5. Drawing a Thick Line**

```
Procedure Thick_Line(x1,y1,x2,y2,x_L_W,x_U_W,y_L_
W,y_U_W)
{Function: Draw a thick line from point x1,y1 to point x2,y2 in
world coordinates with limits x_L_W,y_L_W and x_U_
W,y_U_W}
Begin
  OurMove(x1,y1)
  OurLine(x2,y2)
  Off_Set ← Sqrt((x_U_W − x_L_W)² + (y_U_W −
  y_L_W)²)/100
  Den ← y2 − y1
  If Den < 1.e^-8
    Then {Line is horizontal}
      p1 ← x1
      q1 ← y1 + Off_Set
      p2 ← x2
      q2 ← y2 + Off_Set
    Else
      Slope ← −(x2 − x1)/Den
      Costh ← 1/Sqrt(1 + Slope²)
      Sinth ← Slope*Costh
      xShift ← Off_Set*Costh
      yShift ← Off_Set*Sinth
      p1 ← x1 + xShift
      p2 ← x2 + xShift
      q1 ← q1 + yShift
      q2 ← q2 + yShift
  OurMove(p1, q1)
  OurLine(p2, q2)
End
```

starting and finishing points of the arrow as input. The dimensions of Figure 11.4.7, suitably modified by our own aesthetic predilections, may be used to personalize the procedure.

## 11.5 USING THE LIBRARY

The typical flow chart in finite element analysis is shown in Figure 11.5.1. We preprocess and define a problem; solve; postprocess and examine the solution; and if we are satisfied with the solution we

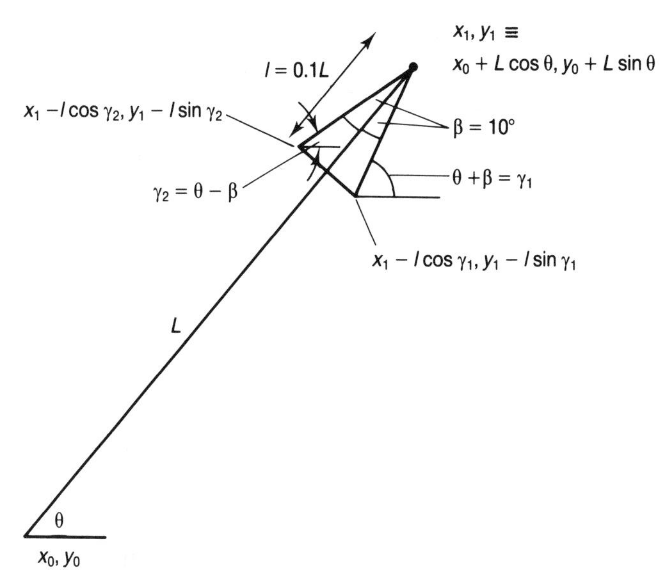


**Figure 11.4.7.** The arrow.

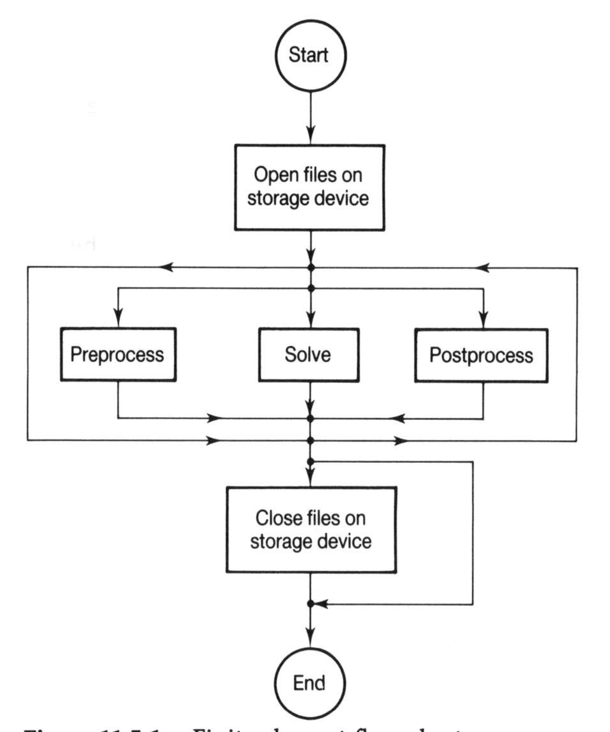


**Figure 11.5.1.** Finite element flow chart.

quit or go back to preprocessing and redefine the problem. To implement this, among the many possibilities in our scheme, we call Screen_Menu with five options and the appropriate strings to get Figure 11.4.1A. If the option returned is 1 for Preprocessing or 4 for Postprocessing, we initialize the screen and accordingly call Screen_Menu again, with 10 or 7 options, respectively, and the respective strings to get Figure 11.4.1B or 11.4.1C. If option 2 is returned from the menu of Figure 11.4.1A, we then call the solver and we quit if the option selected is 5.

Under Preprocessing, as another example of the use of the graphics elements, consider our wishing to set boundary conditions for the mesh by selecting the appropriate option in Figure 11.4.1B. As seen in Figure 11.4.5, we would, after initializing the display, call the routines Write_Message (with "New, Show, or Clear Boundary Conditions?" as string input), Small_Menu (with four options corresponding to the strings "New," "Show," "Clear," and "Quit"), and Mesh_Draw to produce the picture of Figure 11.4.5A. Thereafter, we read the Locator address given by the user and determine the option. For the selection of "New" boundary conditions, to avoid erasing the whole screen and producing Figure 11.4.5B as before, we first call Erase_Message with the same string ("New, Show, or Clear Boundary Conditions?") as input, call Write_Message with the new prompt "Show Known Point with Locator > " as input, erase the menu section of the screen in Figure 11.4.5A by calling Color_Box with color 0 for erasure, and then call Quit_Box to replace the menu to get Figure 11.4.5B. Now we read the locator position and quit if it is in the Quit_Box region; otherwise we call IDPoint, identify the node pointed at, and set the value of the node according to what the user feeds in.

A similar example is found in Figure 11.4.4, where after solving the electric current flow problem in a z shaped conductor, we have, under postprocessing, upon selecting the option "Intensity" under the Screen_Menu of Figure 11.4.1C, drawn the mesh using Mesh_Draw, written "Show Point For Field Density > " using Write_Message, and allowed the option to quit using the Quit_Box. Every time the user digitizes, we first determine if the coordinates are in the Quit_Box and if so we quit; otherwise, we call IDTriangle and identify the triangle in which the user digitized, and through the interpolation of eqn. (7.5.3) identify the previously unknown (but now computed) potential $\phi$ and compute its gradient according to eqns. (1.2.33), (1.2.34), (7.6.3), and (7.6.4), and display it on the screen. In Figure 11.4.2 again, we can see the procedure IDTriangle in use, this time when we choose "Manual Refine" in

Figure 11.4.1B, under "PreProcess" of Figure 11.4.1A. Here in Figure 11.4.2, we are manually adding points to the mesh; as the user digitizes in a triangle of Figure 11.4.2A, we identify that in which the point lies, connect the three vertices of the triangle to the new point (display the new connections by Move and Line commands), optimize the mesh by the methods of section 11.6, and, possibly, change some of the connections to get the mesh of Figure 11.4.2B, to which we may further add points or quit.

## 11.6 AUTOMATIC MESH GENERATION

### 11.6.1 Defining a Good Triangular Finite Element Mesh

One elementary rule for a simple mesh is that we want more points where the field variation is high. The logic of this is clear, especially in one dimension—to fit a graph by straight lines, we need several short lines when the graph changes rapidly, whereas one long straight line will do when the graph varies linearly. And this is what a first-order finite element solution does—it tries to fit the field as best it may, with straight lines whose endpoints along the $x$ axis are fixed by the interpolation nodes.

Reverting to two dimensions, once we decide on where we want our points, they should be connected together into a triangular mesh. A good finite element mesh, in general, avoids having obtuse angled triangles (Babuska and Aziz 1976). While the truth of this statement may be rigorously proved, it may also be grasped intuitively. For a triangle with an obtuse angle will have the side opposite the large angle much larger than the others. Consequently, thinking in terms of first-order interpolation, a linear variation being assumed along that edge will introduce a large error in the trial function. This large error along the long side is accompanied by a small error along the other two correspondingly smaller sides. However, the error of the whole does not average out, but rather corresponds to the largest error. As a result, we must use triangles that are as close to equilateral as possible.

While maintaining close-to-equilateral triangles is a general guideline, other simple rules may be obtained by specific considerations of the problem. At a Dirichlet corner, for example, as shown in Figure 11.6.1A, nodal connections that do not cut the corner region into two or more parts will give us a triangle 123, all

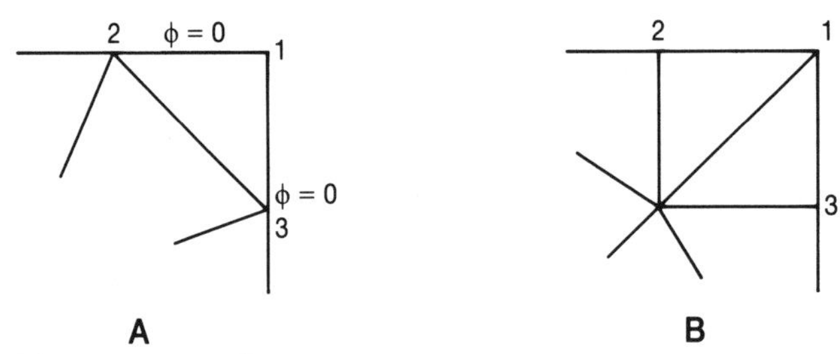


**Figure 11.6.1.** Discretization at corners.

three of whose vertices are at specified potential. The finite element interpolation therefore makes the potential known everywhere in that triangle and effectively eliminates that corner triangle from the solution region. Connections as shown in Figure 11.6.1B will properly reflect the boundary conditions. The same considerations hold if node 1 is a Neumann corner—the normal gradient cannot be prescribed along both edges 12 and 13 of Figure 11.6.1A.

A procedure commonly found convenient in setting up a mesh is to start with a crude mesh, which is then refined to obtain the desired level of accuracy. As shown in Figure 11.5.1, we would obtain the solution of a given mesh, view the solution graphically to gauge the error, and, if necessary, refine the mesh and repeat this cycle. The details of how this is done are given below.

## 11.6.2 Automatically Setting Up a Starting Mesh

A starting mesh may easily be defined by giving as input several polygons, each with a region number and an internal point. The assembly of a lifting magnet system is as shown in Figure 11.6.2.

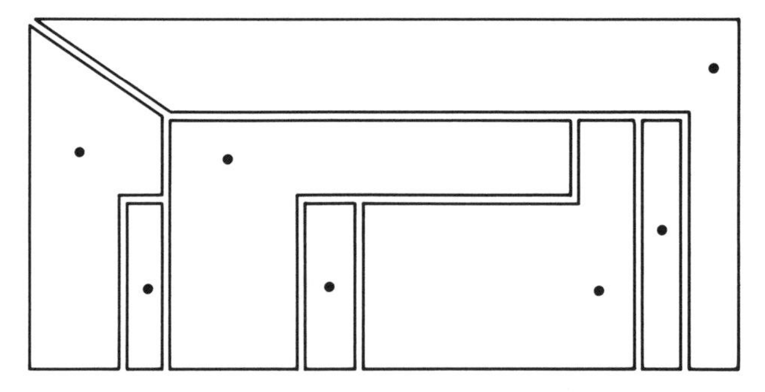

**Figure 11.6.2.** Assembling polygons with interior points into a starting mesh for lifting magnet.

The polygon with $n$ vertices $\mathbf{p}[1], \mathbf{p}[2], \ldots, \mathbf{p}[n]$, with interior point $q$, forms the triangles $\mathbf{p}[i]$, $\mathbf{p}[i+1]$, $q$ for $i$ going from 1 to $n-1$, and the last triangle is $\mathbf{p}[n]$, $\mathbf{p}[1]$, $q$. The same region number will then apply to all the triangles.

The polygon themselves may be defined by feeding the nodes on a graphics screen (perhaps using a tablet and drawing). A counterclockwise definition of the polygon should be adopted so as to preserve the counterclockwise numbering of the triangle nodes. As each point is added to the first polygon (as demonstrated for the lifting magnet of Figures 11.6.3A–11.6.3D), the counter NPoints giving the number of nodes of the mesh is incremented, the digitized coordinates written in the data file for the new point NPoints, and the number NPoints put into the array $p$. At the end of the definition of the polygon, we merely digitize in the Quit_Box and, upon our giving the interior point through the

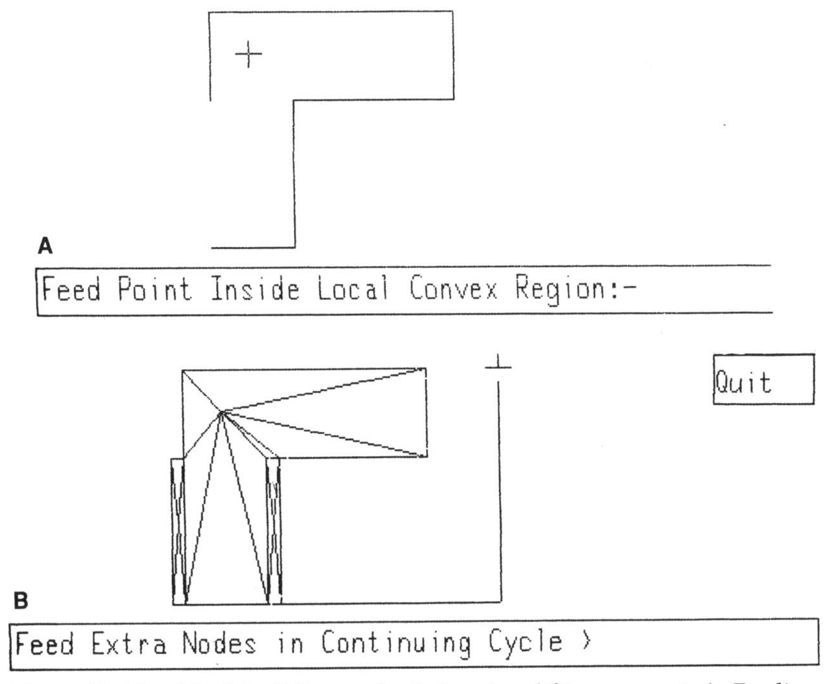


**Figure 11.6.3.** Mesh building and solution in a lifting magnet. **A.** Feeding point in first polygon. **B.** Extending the mesh. **C.** Further extensions. **D.** Arbitrarily connected initial mesh. **E.** Optimized initial mesh—before triangle refinement. **F.** Mesh after triangle refinement and before edge refinement. **G.** Mesh after edge refinement. **H.** Final mesh after refinement.

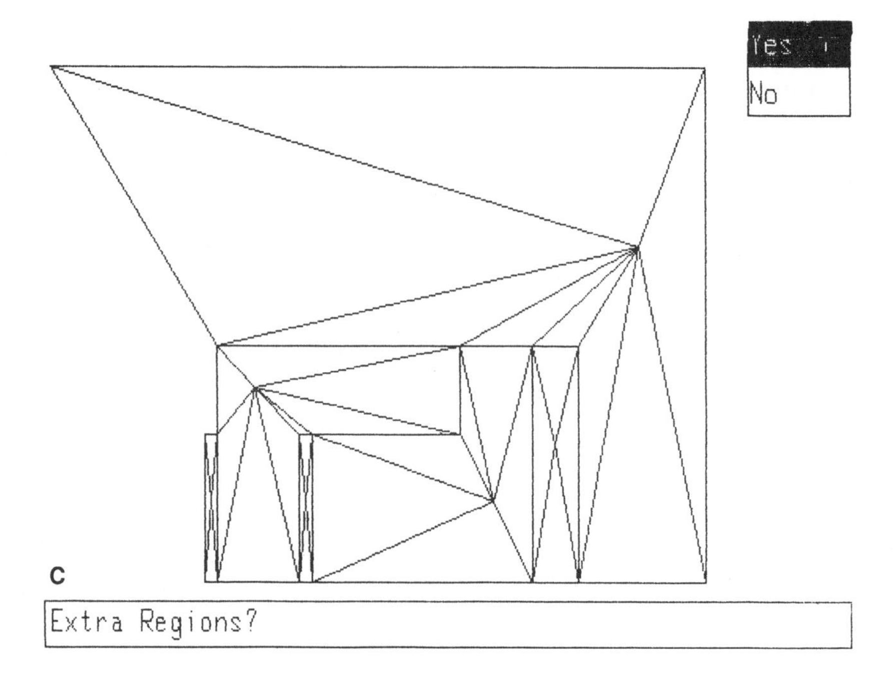


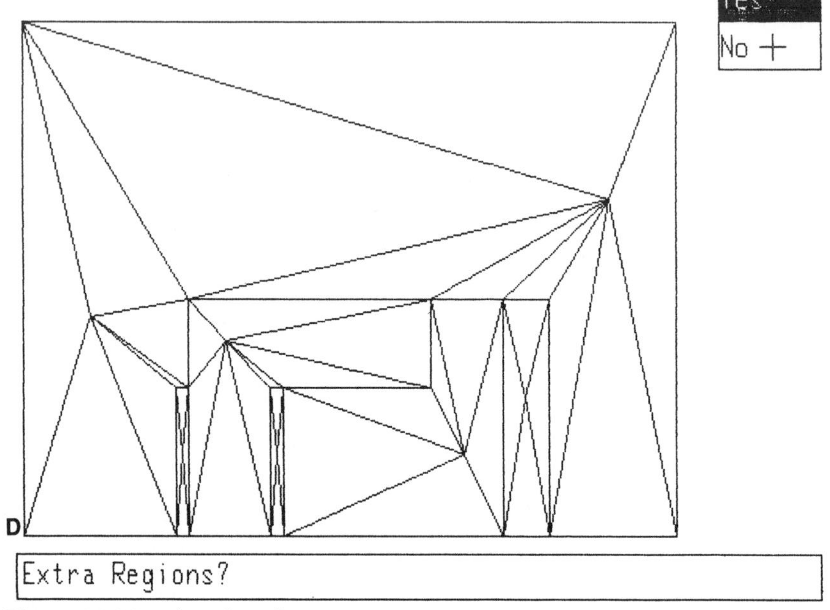


**Figure 11.6.3.** (*continued*)

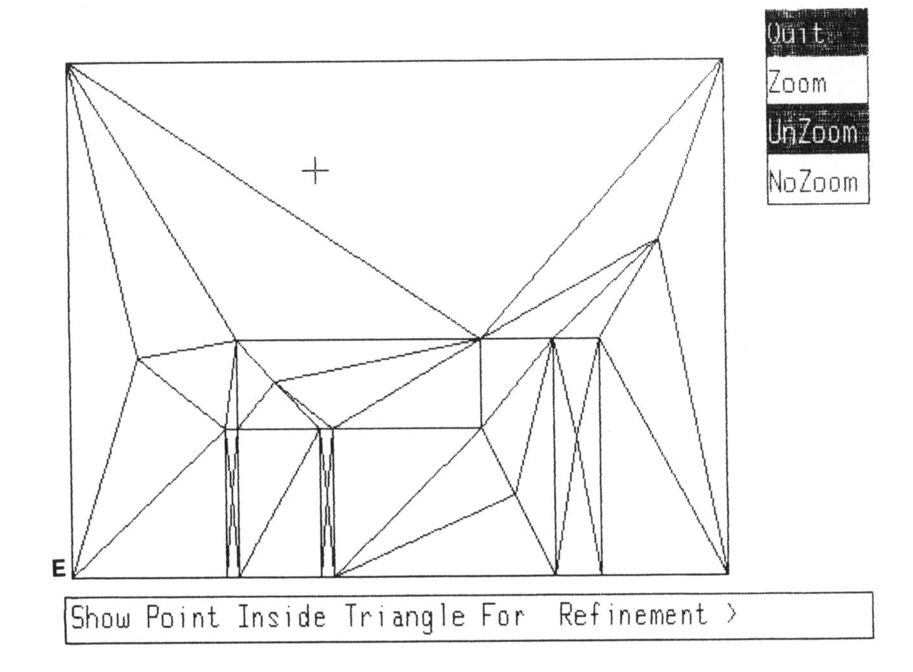
Quit
Zoom
UnZoom
NoZoom
E
Show Point Inside Triangle For Refinement >

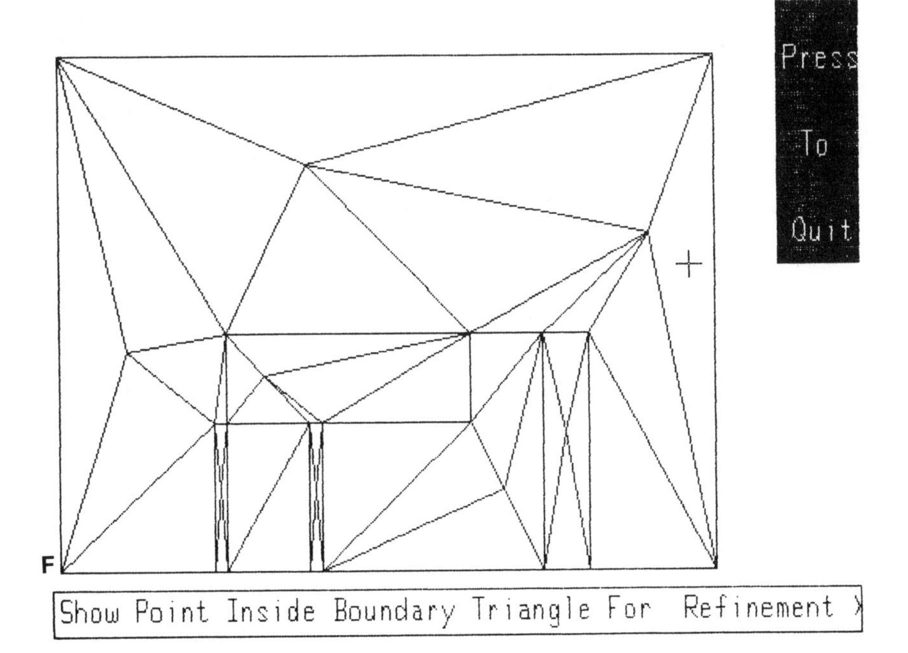
Press
To
Quit
F
Show Point Inside Boundary Triangle For Refinement >

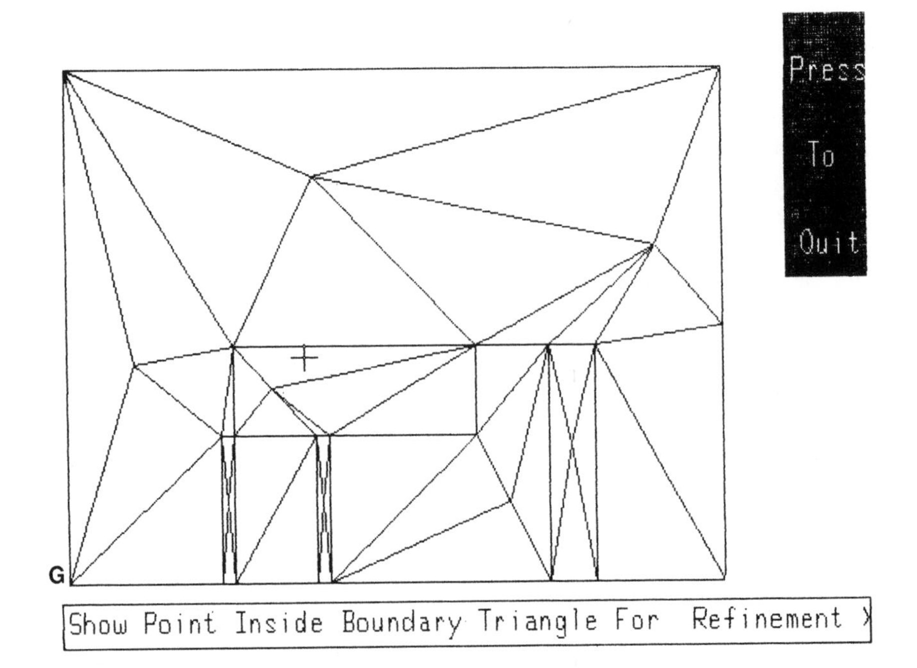


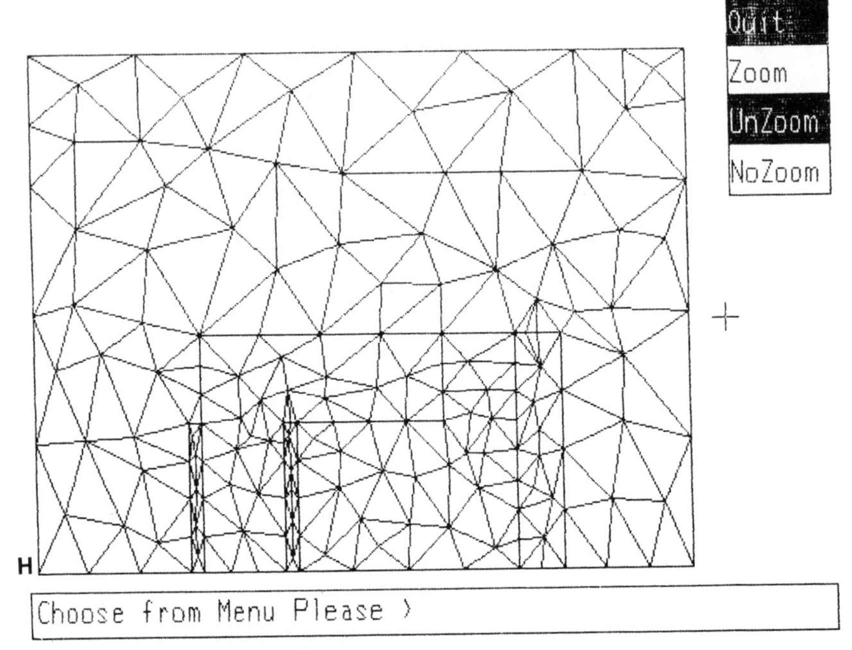


**Figure 11.6.3.** (*continued*)

digitizer, the triangulation is done. If a second polygon is required, we first check if the digitized point is "very close" to an already defined point, by computing the distance. If so, the digitized point is already defined and instead of incrementing NPoints, we take the point as that already defined. Otherwise, the procedure will be similar. This way we would have defined all the triangles of the mesh.

But the neighbors of triangles, which we stipulated as a part of our data structure are, as yet, undefined. These may, however, be computed. Taking each triangle, we read all others and compare if two nodes are common to the pair. If so, they are neighbors. If no common points may be identified with any other triangle along an edge, then that edge is along the exterior boundary and so the neighbor is zero. This computation is shown in Alg. 11.6.1.

### 11.6.3 The Delaunay Criterion for Optimal Meshes

At this point, we have defined the geometry through a crude mesh of triangles, which do not necessarily connect the nodes optimally in the sense of avoiding obtuse angles. Many schemes exist for optimizing the connections. All these schemes try to make connections yielding triangles as close to equilateral as possible within the constraints imposed by the given locations of the nodes and essentially give the same result (Lawson 1977). The Delaunay criterion (Keinstreuer and Hadenan 1980, Hermeline 1982; Cendes, Shenton, and Shahnasser 1983) is one such scheme and is easy to implement with generality.

The Delaunay criterion for optimizing triangular meshes tells us how to connect a given set of points into a triangular mesh so as to minimize the occurrence of obtuse angles. Referring to Figure 11.6.4, in essence the method looks at four neighboring points 1, 2, 3, and 4 and determines whether the region encompassed by the quadrilateral 1234 is best covered by triangles 123 and 243 or triangles 124 and 143. The criterion may be implemented by examining the circumcircle of triangle 123 and checking whether node 4 is in or out of that circle. If node 4 is outside the circle then the triangle pair 123, 243 serves us best, and if node 4 is inside the circumscribing circle then the region 1234 is best represented by triangles 124 and 143.

The theorem from plane geometry that opposite angles of a circular quadrilateral sum to $\pi$ is useful in understanding and implementing the Delaunay criterion. As node 4 moves in from the

**Algorithm 11.6.1. Making the Neighbors of a Triangle**

```
Procedure MakeNeighbors(T, TriaNum)
{Function: Computes the three neighbors N[1], N[2], N[3] of
triangle T
Input: TriaNum = number of triangles}
Begin
  Get vertex numbers V[1], V[2], V[3] of triangle T
  For j ← 1 To 3 Do
    N[j] ← 0
  For i ← 1 To TriaNum Do
    If i < > T
      Then
        Get vertex numbers Vn[1], Vn[2], Vn[3] of triangle i
        Common ← 0
        For j ← 1 To 3 Do
          If((Vn[j] = V[1])Or(Vn[j] = V[2])Or(Vn[j] = V[3]))
            Then Common ← Common + 1
        If Common = 2
          Then
            j ← 1
            While ((V[j] < > Vn[1]) And (V[j] < > Vn[2])
            And (V[j] < > Vn[3])) Do
              j ← j + 1 {first matching vertex j = 1 or 2}
            k ← j + 1
            While ((V[k] < > Vn[i]) And (V[k] < > Vn[2])
                And (V[k] < > Vn[3])) Do
                  k ← k + 1 {second matching vertex k = 2
                    or 3}
            If j = 1
              Then
                If k = 2
                  Then N[1] ← i
                  Else N[3] ← i
              Else N[2] ← i
  Put N in the neighbor data for triangle T
End
```

perimeter of the circumcircle of triangle 123, the sum of the angles $\angle 312$ and $\angle 342$ exceeds $\pi$ and keeps increasing. On the other hand, as point 4 moves into the circle, the angles $\angle 134$ and $\angle 124$ keep decreasing. As a result, the diagonal 41 is drawn in place of 23 so as to cut the pair of angles $\angle 312$ and $\angle 342$ into smaller angles in the mesh in representing the quadrilateral 1234. When 4 moves out of

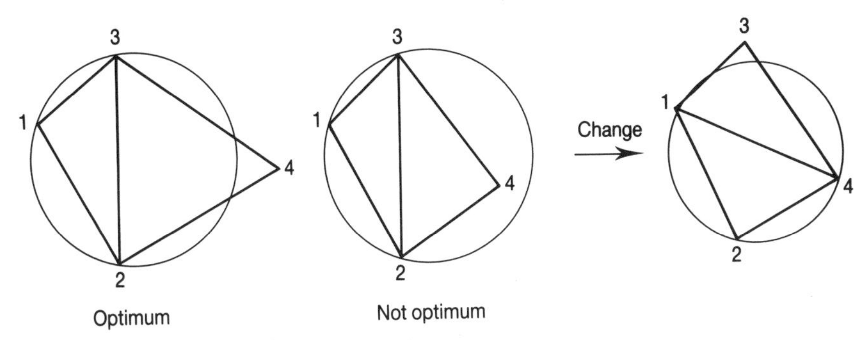


**Figure 11.6.4.** The Delaunay criterion.

the circumcircle, the opposite happens and we keep the diagonal 23.

## 11.6.4 Recursive Application of the Delaunay Criterion

The Delaunay criterion tells us how a given set of points are to be connected to form an optimal mesh. The process uses what are known as *Voronoi polygons.* Using the criterion, we have in section 11.6.3 the rule for determining how best to connect four points. It is, however, algorithmically simpler, compared with using the Voronoi polygons, to connect our points in any way we please as we did the starting mesh of Figure 11.6.3 and, then, to check for the optimality of all pairs of neighbors. Figure 11.6.3E gives the optimized version of the arbitrarily connected mesh of Figure 11.6.3D. Thus mesh generation may be easily accomplished automatically. Alg. 11.6.2 gives the procedure for checking the optimality of the two neighbors of Figure 11.6.5A. Figure 11.6.5B gives the new configuration after the connections between the nodes are changed, if so dictated by the Delaunay rule.

To optimize the starting mesh, therefore, we take all the triangles one by one looking for neighbors, and if a triangle has a neighbor, we check for optimality using Alg. 11.6.2. In the process the mesh will become optimized. However, a complication arises as a result of the changes. For example, say that we have already looked at triangle $N[i]$ of Figure 11.6.5A with its neighbors so that triangles $T$ and $N[i]$ are optimally connected. Now while examining the triangles $T$ and $Tn$ we find that they are not optimally connected and have had to change the common edge around. The new triangle

**Algorithm 11.6.2. Delaunay Optimization of Two Neighboring Triangles**

```
Procedure ModifyNeighbor(T,NOld,NNew)
{Function: Change neighbor NOld to NNew for triangle data
T}
Begin
  Get three neighbors N of triangle T
  For i ← 1 To 3 Do
    If N[i] = NOld
      Then N[i] ← NNew
    Put three neighbors N of triangle T
End

Procedure Delaunay(T, Tn)
{Function: Delaunay—optimizes the triangle neighbors T and
Tn, which are integers corresponding to the triangle numbers
Input: The triangle numbers T and Tn of Fig. 11.6.2
Output: Possible new values of vertices and neighbor in
triangle data file
Assumes the real functions X(i) and Y(i), which fetch the
coordinates of node i, either through file reading or from arrays
in which they are stored. Also uses the procedure
UncommonAngle. Has nested function L2. Assumes the system
functions Sqr and Sqrt and ArcCos for the square, square root,
and inverse cosine. Put and Get are data file operations if files
are used. If arrays are used, they are appropriately modified}

Function L2(i, j)
{Finds the real square of the length between nodes i and j using
the real functions x and y}
Begin
  L2 ← Sqr(x(i) − x(j)) + Sqr(y(i) − y(j))
End

Procedure UnCommonAngle(A,V1,V2,k,k1,k2)
{Finds angle A of triangle V1 where A is not a vertex of
neighboring triangle V2. k is the position of vertex A in triangle
V1}
Begin
  For i ← 1 To 3 Do
    If V1[i] < > (V2[1] or V2[2] or V2[3])
      Then k ← i
  k1 ← kMod3 + 1
  k2 ← k1Mod3 + 1
```

```
    a2 ← L2(V1[k1],V1[k2])
    b2 ← L2(V1[k],V1[k1])
    c2 ← L2(V1[k],V1[k2])
    cosA ← (b2 + c2 − a2)/(2*Sqrt(b2*c2))
    A ← ArcCos(CosA)
End


Begin
  Get three vertices and three neighbors of triangle T in arrays
  V and N and region number RN
  Get three vertices and three neighbors of triangle Tn in arrays
  Vn and Nn and region number RNn
  If RN = RNn
    Then {can apply Delaunay scheme}
      UncommonAngle(A,V,Vn,i,i1,i2)
      UnCommonAngle(B,Vn,V,j,j1,j2)
      If A + b > π
        Then {we need to change the triangles T and Tn}
          {Redefine changed vertices of new triangles T and Tn}
            V[i2] ← Vn[j]
            Vn[j2] ← V[i]
            {Now if the triangles T and Tn have neighbors
            Nn[j2] and N[i2], redefine the changed neighbors
            of the neighbors}
            If Nn[j2] > 0
              Then ModifyNeighbor(Nn[j2], Tn, T)
            If N[i2] > 0
              Then ModifyNeighbor(N[i2], T, Tn)
            {Define changed neighbors of new triangles T and
            Tn}
            N[i1] ← Nn[j2]
            Nn[j1] ← N[i2]
            N[i2] ← Tn
            Nn[j2] ← T
            Put V and N in triangle data T
            Put Vn and Nn in triangle data Tn
            {Optimize neighbors of neighbors only if recur-
            sion is allowed by programming language}
            Delaunay(T,N[i])
            Delaunay(T,N[i1])
            Delaunay(Tn, Nn[j])
            Delaunay(Tn, Nn[j1])
 End
```

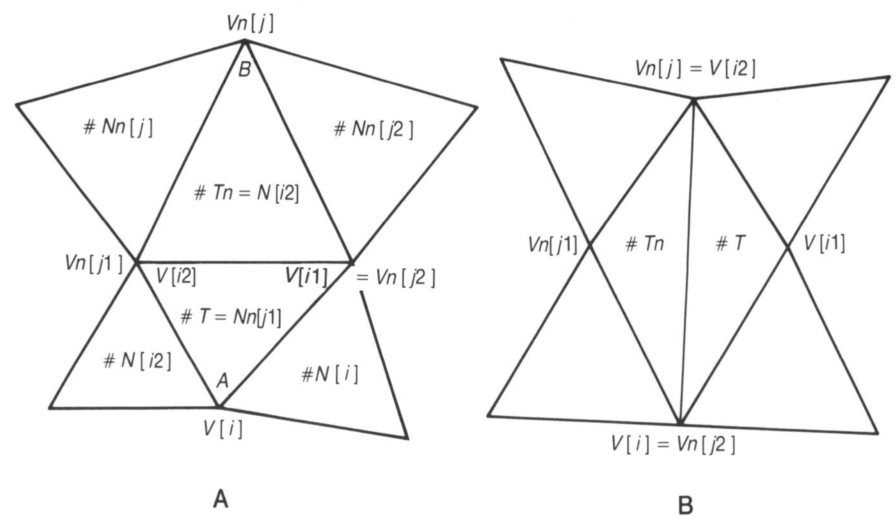


**Figure 11.6.5.** Definitions for Alg. 11.6.2. **A**. Before switch. **B**. After switch.

*T* of Figure 11.6.5B is no longer guaranteed to be optimal with respect to triangle **N**[*i*]. Therefore the Delaunay rule has to be applied to the new pair *T* and **N**[*i*]. If these require swapping of the diagonals, then triangle **N**[*i*] needs to be checked again with its other neighbors. Fortunately the process converges so as to stop swapping with a few neighbors of neighbors (Lee and Schachter 1980), and this repeated checking may be done by recursively calling the Delaunay procedure within itself as shown in Alg. 11.6.2. Delaunay is also called to check along the neighbors **Nn**[*j*2], **Nn**[*j*], and **N**[*i*2]. Figure 11.6.3D gives the starting mesh for a lifting magnet, and Figure 11.6.3E its optimized version.

## 11.6.5 Interactive Mesh Refinement

Once we set up an initial mesh and use it to solve a problem, we may find that the flux plots are crude in certain parts of the device, and therefore wish to refine the mesh in those parts. Certainly we would wish to avoid setting up a fresh but finer mesh. We may more simply pursue constructing a fine mesh by adding points to the existing mesh.

Such addition of points is easily accomplished by flashing the existing mesh on the screen using the Mesh_Draw command of section 11.4.10. Once this is done, using the Write_Message command of section 11.4.6, the user is asked "Show Point Inside Triangle for Refinement > " as seen in Figures 11.4.2A and 11.6.3E. As the user digitizes and the point is identified as not being in the Quit_Box region, using the IDTriangle command of section 11.4.7, that triangle $T$ in which this new point lies is computed. This triangle then is to be cut into three new triangles as shown in Figure 11.6.6. We also delete the triangle $T$ from the data records since it no longer exists. The number of triangles therefore increases by two. We increment NPoints, the number of points, by one and have to redefine the neighbors. How this is done is shown in Alg. 11.6.3.

After the addition of the point in the middle of the triangle and the creation of two new triangles, the optimality of the mesh will have been destroyed locally in the vicinity of triangle $T$. The angles subtended at the new point will be large, and the interpolations along the old edges from **V**[1] to **V**[2], **V**[2] to **V**[3], and **V**[3] to **V**[1] remain unchanged. If we add another point inside the triangle, the angle subtended there will be even larger. If we merely solve after refinement, the solution will be found virtually unchanged except slightly in the triangle $T$. Therefore we call the Delaunay algorithm and check for the optimality of the connections, and change them if necessary. Figure 11.6.7 shows the solution before optimization and after optimization in a conductor, and the advantages of optimizing are self-evident. The meshes that result from the refinement of the meshes of Figures 11.4.2A and 11.6.3E are given, respectively, in Figures 11.4.2B and 11.6.3F.

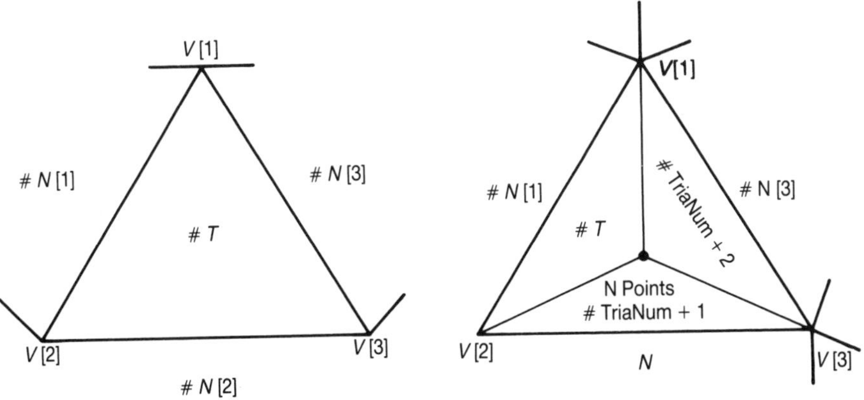


**Figure 11.6.6.** Interior refinement of a triangle.

**Algorithm 11.6.3. Refine Triangle for Mesh Refinement**

**Procedure WriteTria** (*T*, *v*1, *v*2, *v*3, *n*1, *n*2, *n*3, *RN*)
{Function: Puts in data files the three vertices, three neighbors, and region number for triangle number *T*}
**Begin**
  **Put** *v*1, *v*2, *v*3, *n*1, *n*2, *n*3, *RN* in data file for triangle *T*
**End**

**Procedure RefineTriangle**(*T*,*x*,*y*);
{Function: Refines triangle *T* using the point $(x, y)$. Uses the procedure WriteTria}
**Begin**
  **Get** three vertices, three neighbors, and *RN* of triangle *T* in arrays **V**, **N**
  NPoints ← NPoints + 1
  **Put** Coords *x*, *y* in nodal record for new node NPoints
  **WriteTria**(*T*,**V**[1],**V**[2],NPoints,**N**[1],TriaNum + 1,TriaNum + 2,*RN*)
  **WriteTria**(*T*,**V**[1],**V**[2],NPoints,**N**[1],TriaNum + 1,TriaNum + 2,*RN*)
  **WriteTria**(TriaNum + 1,**V**[2],**V**[3],NPoints,**N**[2],TriaNum + 2,*T*,*RN*)
  **WriteTria**(TriaNum + 2,*V*[3],*V*[1], NPoints,*N*[3],*T*,TriaNum + 1,*RN*)
  **If** *N*[2] > 0
  **Then ModifyNeighbor** (**N**[2],*T*,TriaNum + 1)
  **If N**[3] > 0
  **Then ModifyNeighbor**(**N**[3],T,TriaNum + 2)
  TriaNum ← TriaNum + 2
  **Delaunay**(*T*,**N**[1])
  **Delaunay**(TriaNum − 1,**N**[2])
  **Delaunay**(TriaNum,**N**[3])
**End**

## 11.6.6 Edge Refinement

We know how to refine a mesh and get rid of obtuse angles by applying the Delaunay criterion. However, the Delaunay criterion is applicable only when the diagonal of two triangles forming a quadrilateral may be switched. For example, when the edge that needs to be switched is along a material interface, if we change the connections between nodes, the validity of the mesh is destroyed.

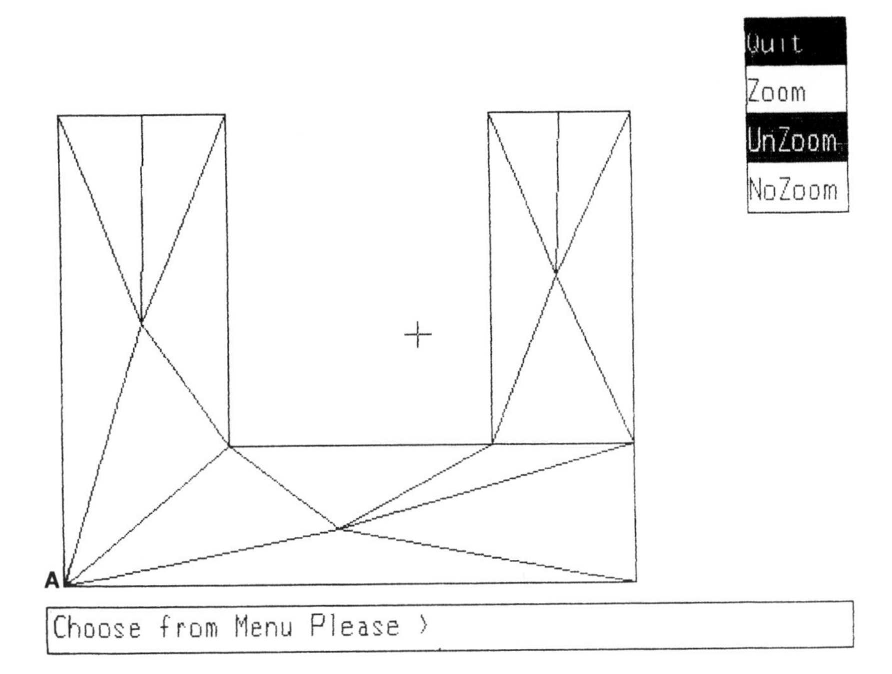


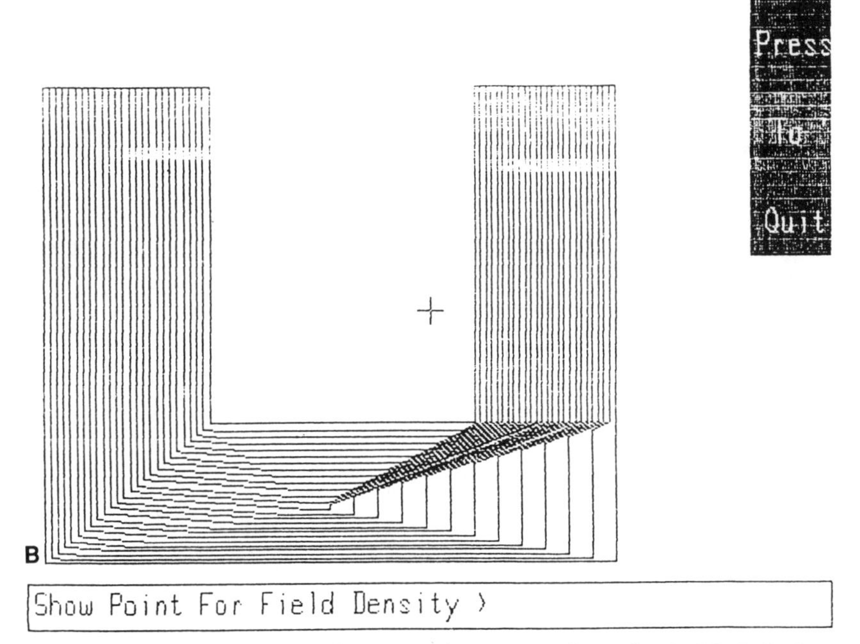


**Figure 11.6.7.** Improved solution with optimized mesh. **A**. Mesh before optimization. **B**. Solution before optimization. **C**. Mesh after optimization. **D**. Solution after optimization.

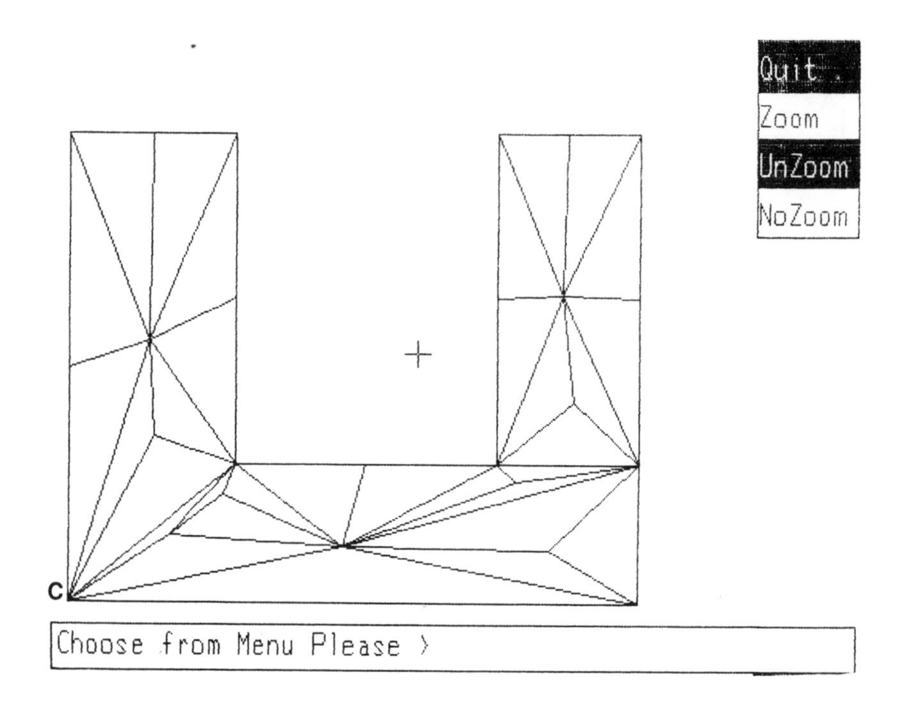


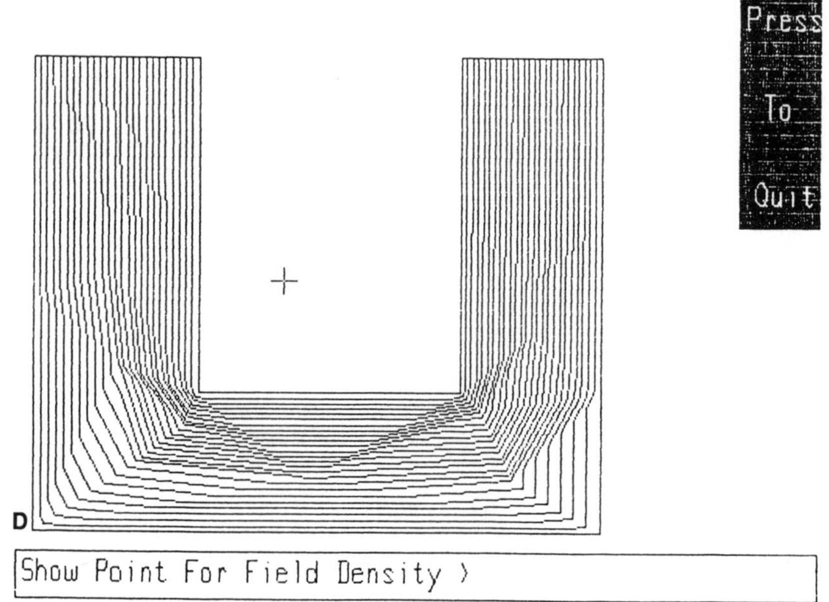


**Figure 11.6.7.** (*continued*)

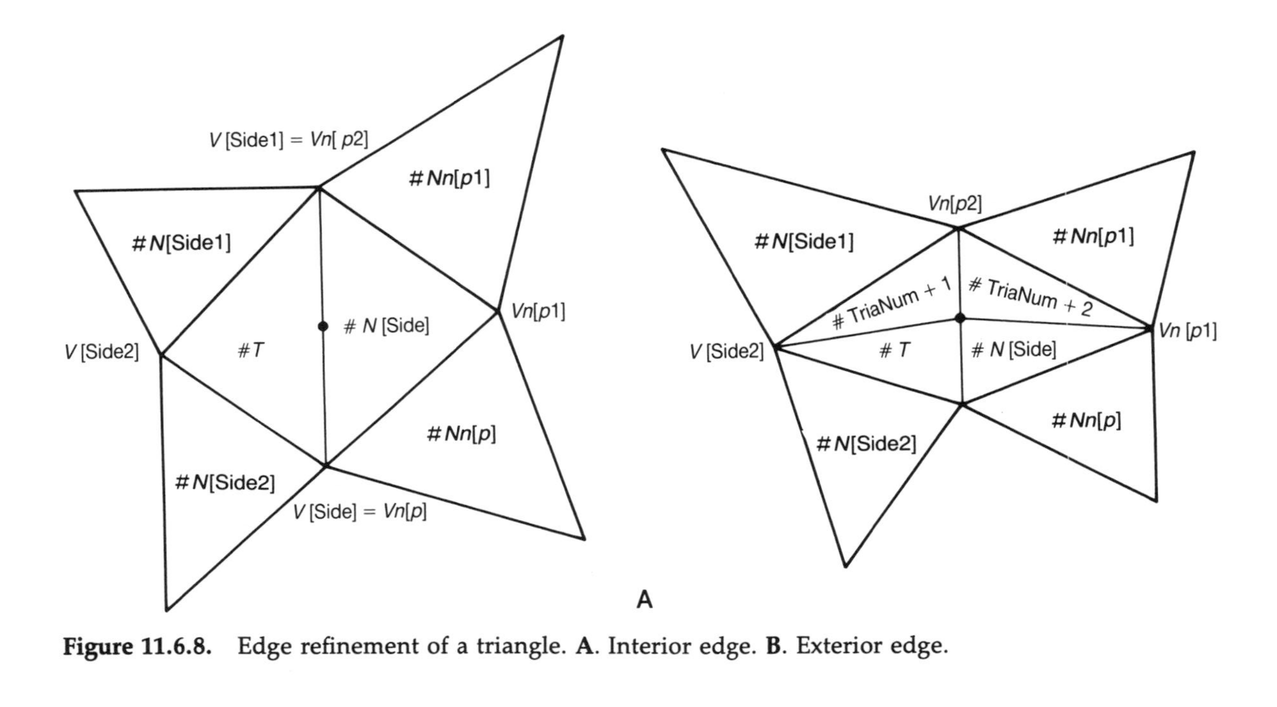


**Figure 11.6.8.** Edge refinement of a triangle. **A**. Interior edge. **B**. Exterior edge.

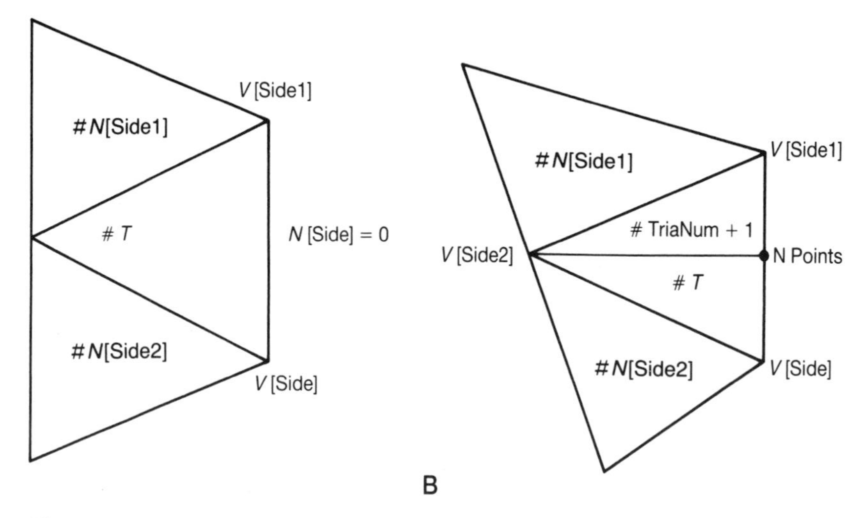


**Figure 11.6.8.** *(continued)*

Also when the edge is along a boundary, there is no neighbor with which switching may be contemplated. For this reason, we need to allow for edge refinement as shown in the menu of Figure 11.4.1C. The resulting changes in data are depicted in Figure 11.6.8. An application of this algorithm is seen in Figures 11.4.3A and 11.6.3F, and the meshes resulting from edge refinement in Figures 11.4.3B and 11.6.3G.

Under boundary refinement, the mesh is drawn on the screen and the user is asked to show the boundary triangle for refinement as in Figure 11.4.3. As the user digitizes, if the digitized point is not in the Quit_Box, the triangle $T$ in which the point was digitized is computed. Then we call the procedure RefineBoundary of Alg. 11.6.4 to refine the boundary. Two cases are considered: first for the edge's being along the exterior boundary of the solution region, and second for its being along a material interface.

## 11.7 ADAPTIVE MESHES AND ERROR ESTIMATION

### 11.7.1 Expert Workstations

Usually in finite element analysis, it is we, the users, who solve the field problem, look at flux plots, refine the mesh where the flux

**Algorithm 11.6.4. Refining a Finite Element Boundary**

```
Procedure RefineBoundary(T)
{Function: Refines the boundary edge of a triangle T}
Begin
  Get three vertices, three neighbors, and RN of triangle T in
    arrays V, N
  j ← 0
  Side ← 0
  While Side = 0 And j < 0 Do
    j ← j + 1
    If N[j] = 0
      Then Side ← j
      Else
        Get three vertices, three neighbors, in arrays Vn, Nn
        and RNn of triangle N[j]
          If RN < > RNn
            Then Side ← j
  If Side < > 0
    Then
      Side 1 ← Side Mod 3 + 1
      Side 2 ← Side1 Mod 3 + 1
      NPoints ← NPoints + 1
      Get Coords x1,y1 and x2,y2 of nodes V[Side] and
        V[Side1]
      x ← 0.5*(x1 + x2)
      y ← 0.5*(y1 + y2)
      Put Coords x,y for node NPoints in data file
      If N[Side] = 0
        Then {refining a boundary edge—Figure 11.6.8B}
          WriteTria(T,V[Side],NPoints,V[Side2], 0,
            TriaNum + 1, N[Side2], RN)
          WriteTria(TriaNum + 1,NPoints,V[Side1],
            V[Side2],0,N[Side1], T, RN)
          ModifyNeighbor(N[Side1],T,TriaNum + 1)
          Delaunay(T,N[Side2])
          Delaunay(TriaNum + 1,N[Side1])
          TriaNum ← TriaNum + 1
        Else {refining material interface—Figure 11.6.8A}
            For j ← 1 To 3 Do
              If Vn[j] = V[Side]
                Then p ← j
```

```
            p1 ← p Mod 3 + 1
            p2 ← p1 Mod 3 + 1
            WriteTria(T,V[Side],NPoints,V[Side2],N[Side],
              TriaNum + 1, N[Side2], RN)
            WriteTria(TriaNum + 1,NPoints,V[Side1],V[Side2],
              TriaNum + 2,N[Side1], T, RN)
            WriteTria(N[Side],V[Side],Vn[p1],NPoints, Nn[p],
            TriaNum + 2,T,RNn)
            WriteTria(TriaNum + 2,NPoints,Vn[p1],Vn[p2],
              NSide, Nn[p1],TriaNum + 1,RNn)
            ModifyNeighbor(N[Side1],T,TriaNum + 1)
            ModifyNeighbor(Nn[p1],N[Side],TriaNum + 2)
            Delaunay(T,N[Side2])
            Delaunay(TriaNum + 1,N[Side1])
            Delaunay(NSide,Nn[p])
            Delaunay(TriaNum + 2,Nn[p1])
            TriaNum ← TriaNum + 2
       Else Write ("Not a boundary triangle")
   End
```

lines are not smooth enough for us, solve again, and so on, until an acceptable solution is arrived at. This procedure is prone to much error because we estimate error by sight. Moreover, it calls upon the user to be an expert who knows how the solution is expected to behave, so that he or she might be in a position to determine where refinements are required.

However, if we can impart to the finite element workstation the intelligence to determine where the solution is incorrect, then this process of mesh generation through refinement may be completely automated. The elements having relatively high error may be refined for the next iteration on the mesh. Hopefully then, design tasks may even be carried out by the uninitiated. Doing so requires our determining in a scientific and systematic manner the error in the finite element solution. And since we do not know the exact solution (if we knew it there would be no need for finite elements), we are called upon to estimate the error from the finite element solution. For this reason, the term *a priori* error estimation (reasoning from what is prior, tracing from cause to effect) is used.

## 11.7.2 Error Estimates

To estimate the error in a finite element solution, several schemes have been put forward. Zienkiewicz and Morgan (1983) suggest using the residual of the equation

$$\mathcal{R} = -\varepsilon\nabla^2\phi - \rho. \tag{11.7.1}$$

Since the operator $\nabla^2$ requires differentiation twice, the potential $\phi$ has to be at least of order 2. However, Zienkiewicz and Morgan overcome this problem by considering a hypothetical hierarchic element of one higher order and computing the addition to the finite element solution if the higher-order term had been present; in this the residual is computed from the hypothetical higher-order element. Babuska and Rheinboldt (1978) look at a local finite-element problem surrounding a node, recompute its value, and compare the answer with that yielded by the finite element solution.

In another scheme (Penman and Fraser 1982; Thatcher 1982) the same problem is solved by the two complementary formulations of section 8.3.2, which have energy converging from opposite sides. For example, Figure 8.3.2 gives the solution in a fuse. This is a current flow problem as in section 3.3.2. Since the current density has no divergence or curl, it may be defined through a scalar potential as $-\nabla\phi$ or the complementary vector potential as $\nabla \times \mathbf{u}_z T$. The convergence of energy may be shown to be from opposite directions. Thus convergence is measured by the difference in energy by the two methods, and those elements with large differences may be discerned as having large error and reserved for refinement. The disadvantages, however, are (1) having to solve the field problem twice, (2) our inability to identify exactly the corresponding boundary conditions of the complementary formulation, and (3) the difficulties in implementation when the problem is not Laplacian, as here, but Poissonian. In the latter event, in magnetics for example, the scalar potential formulation requires our finding a particular solution as in eqn. (4.2.4) before we can set up the Poisson eqn. (4.2.8) for $\Omega$; even then, we need to ensure that continuity conditions on the normal flux density and tangential field intensity are imposed. Despite these weaknesses, it is one of few adaptive schemes with a rigidly mathematical proof on convergence, whereas most schemes are intuitively justified. On the basis of the complementary energies' convergence from opposite sides, an interesting application is in computing energy from a crude mesh—the average of the complementary energies from a crude mesh turns out to be very close to the exact solution for energy. Indeed, another scheme of refinement (Hoole, Yoganathan, and Jayakumaran 1985) measures the angle of bending of the electric and magnetic fields at finite element edges and compares with the bending required by the continuity conditions. All edges on which the actual and predicted deflections deviate from each other by a prescribed angle, or more, are refined. The disadvantage here is particularly at the corners of boundaries, where a discontinuity in the field vectors is inherent to the finite element method, and this is overcome by searching for the convergence of energy to stop further refinement.

A simple idea for error estimation—the nodal perturbation scheme—has been presented by Hoole (1987). Here we take a starting mesh, solve the field problem posed by it, and, on the first iteration on the mesh, refine every triangle. But before adding the new points to the mesh, we determine through the finite element interpolations the solution at the location of each node. After the addition we solve for the mesh, and at every node we now have the solution from the previous mesh and that from the present mesh. The perturbation in the nodal potential then is a measure of error. Hereafter, we refine only those triangles in which the solution at any of the three nodes has changed by more than a prescribed amount, such as 0.01% for example. The advantage of the scheme is its simplicity. Jayakumaran and Hoole (1988) have applied this scheme to the adaptive refinement of boundary element meshes.

Other equally elegant schemes for adaptive refinement have been provided by Carey and Humphrey (1979), Girdino et al (1983), Penman and Grieve (1985), Pinchuk and Silvester (1985), and Biddlecomb, Simkin, and Trowbridge (1986). The more interested reader is referred to the original works referenced above.

### 11.7.3 The Solution of a Magnetic Recording Head by an Expert Workstation

We have seen that a finite element workstation may be made an expert by imparting to it the ability to discern those regions contributing the most to the solution error. Such expertise carries the design of electromagnetic devices to its apogee.

We shall therefore conclude this text with an example of an analysis of a magnetic recording head by an expert workstation, based on the simple nodal perturbation scheme. Figure 11.7.1A

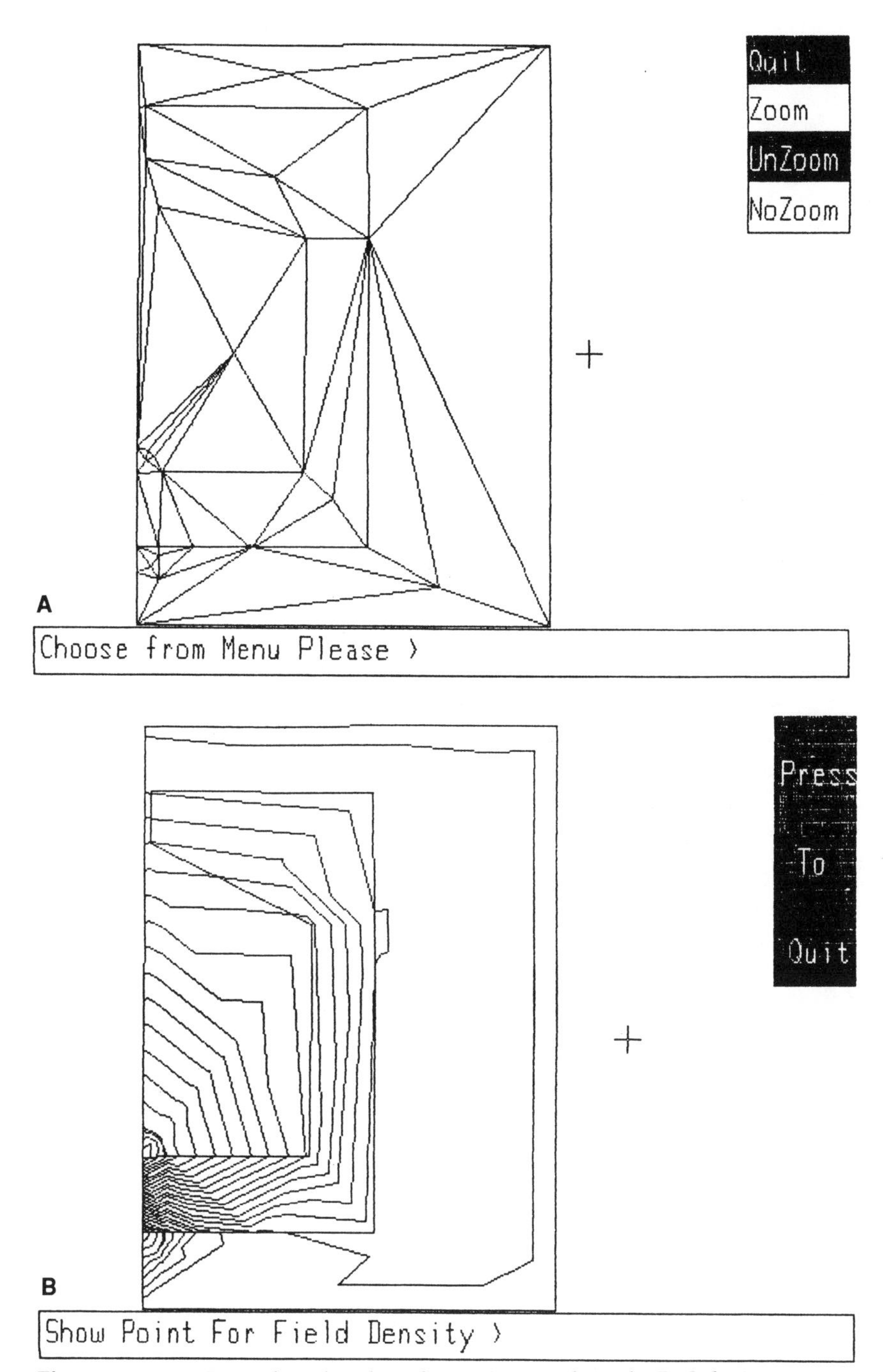


**Figure 11.7.1.** Recording head analysis. **A**. Initial mesh. **B**. Solution.

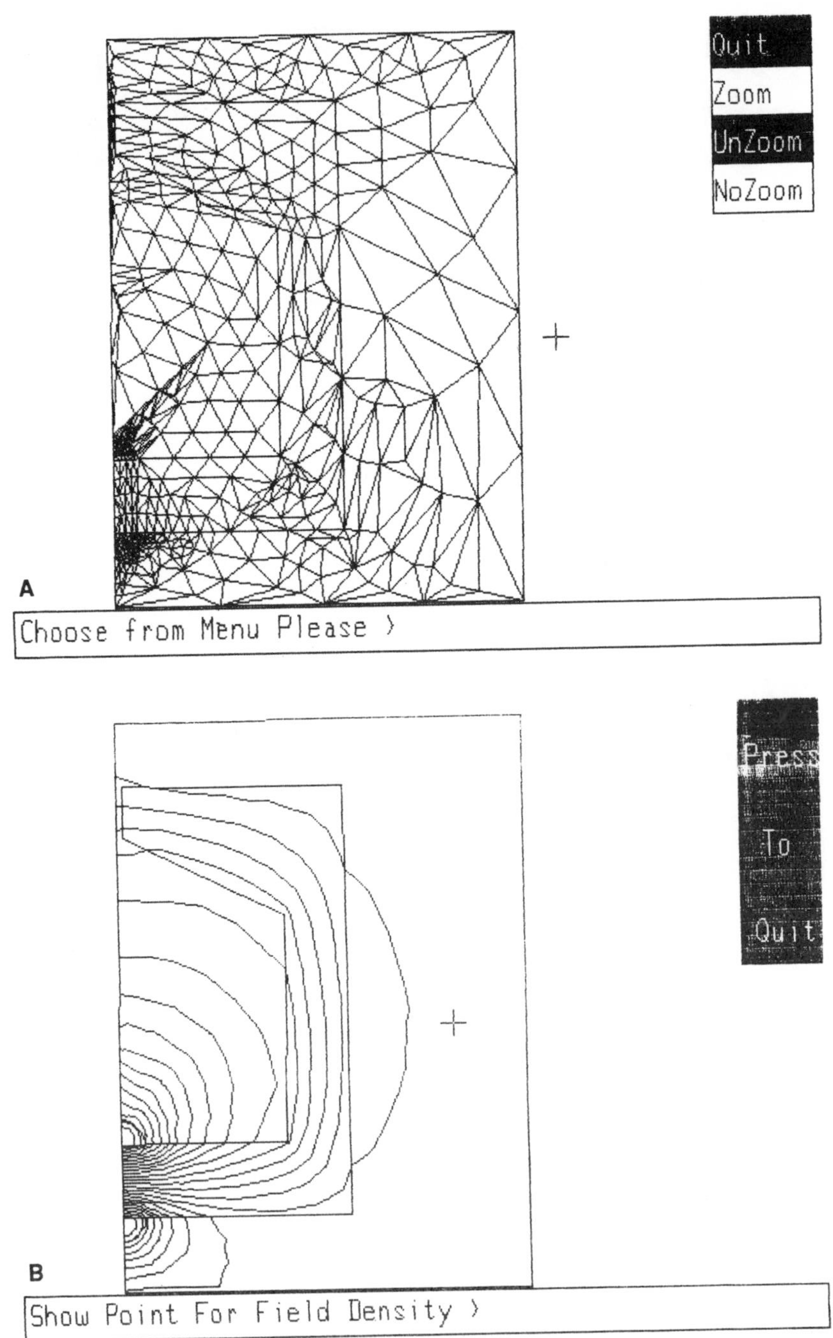


**Figure 11.7.2.** Recording head analysis. **A**. Final mesh. **B**. Solution.

gives a simple, crude mesh for the right half of the magnetic recording head sourced by the coil along the horizontal limb. The corresponding solution, hardly acceptable, is seen in Figure 11.7.1B. The expert workstation takes over and alternately solves and refines until the nodal perturbations are below 0.01%. The corresponding mesh and smooth solution are presented in Figures 11.7.2A and B. By making the convergence criterion more stringent, a superior and smoother solution could have been obtained.

Observe further that we may also impose the more stringent convergence criterion only in the region of the gap if we want a higher accuracy there. To do this, we would need to modify the data structures of Figure 11.3.1. The source and material file will be given an extra field, the percentage accuracy desired. In defining the mesh, then, we will make the region of the gap a different region with a higher accuracy (with a lower percentage change for convergence). When examining the nodal perturbations to determine refinement, we will then only need to use the predefined tolerance of the region.

# Appendix A
# SOME USEFUL VECTOR IDENTITIES

The purpose of this appendix is to give useful vector relationships repeatedly used in the book. For a proof of the relationships, the more interested reader is referred to Ferraro (1970) and Panofsky and Phillips (1969).

$$\nabla = \mathbf{u}_x \frac{\partial}{\partial x} + \mathbf{u}_y \frac{\partial}{\partial y} + \mathbf{u}_z \frac{\partial}{\partial z} \tag{A1}$$

$$\mathbf{A} \cdot \mathbf{B} = A_x B_x + A_y B_y + A_z B_z \tag{A2}$$

$$\begin{aligned} \mathbf{A} \times \mathbf{B} = {} & \mathbf{u}_x(A_y B_z - A_z B_y) + \mathbf{u}_y(A_z B_x - A_x B_z) \\ & + \mathbf{u}_z(A_x B_y - A_y B_x) \end{aligned} \tag{A3}$$

$$\nabla \times \nabla = 0 \tag{A4}$$

$$\nabla \cdot \nabla \times = 0 \tag{A5}$$

$$\nabla^2 \phi = \nabla \cdot \nabla \phi \tag{A6}$$

$$\nabla^2 \mathbf{A} = \mathbf{u}_x \nabla^2 A_x + \mathbf{u}_y \nabla^2 A_y + \mathbf{u}_z \nabla^2 A_z \tag{A7}$$

$$\nabla \cdot (\mathbf{A} \times \mathbf{B}) = \mathbf{B} \cdot (\nabla \times \mathbf{A}) - \mathbf{A} \cdot (\nabla \times \mathbf{B}) \tag{A8}$$

$$\nabla \times \nabla \times \mathbf{A} = \nabla \nabla \cdot \mathbf{A} - \nabla^2 \mathbf{A} \tag{A9}$$

$$\nabla(\phi \psi) = \phi \nabla \psi + \psi \nabla \phi \tag{A10}$$

$$\nabla \cdot (\phi \mathbf{A}) = \phi \nabla \cdot \mathbf{A} + \mathbf{A} \cdot \nabla \phi \tag{A11}$$

$$\nabla \times (\phi \mathbf{A}) = \phi \nabla \times \mathbf{A} - \mathbf{A} \times \nabla \phi \tag{A12}$$

$$\iint \mathbf{A} \cdot \mathbf{dS} = \iiint (\nabla \cdot \mathbf{A}) \, dC \qquad \text{(Gauss's Theorem)} \tag{A13}$$

$$\int \mathbf{A} \cdot \mathbf{dL} = \iint (\nabla \times \mathbf{A}) \cdot \mathbf{dS} \qquad \text{(Stokes's Theorem)} \tag{A14}$$

# Appendix B UNIQUENESS OF THE SCALAR POTENTIAL

In this appendix, we establish the conditions for the uniqueness of the solution of the Poisson equation

$$\varepsilon\nabla^2\phi = \varepsilon\frac{\partial^2}{\partial x^2}\phi + \varepsilon\frac{\partial^2}{\partial y^2}\phi = -\rho. \tag{B1}$$

The conditions are that $\phi$ should be specified at one point on the boundary of the solution region, and at all other points either $\phi$ or its derivative in the normal direction should be specified.

We shall prove this by deriving the conditions under which any two alternative solutions to eqn. (B1), $\phi_1$ and $\phi_2$, differ by 0; that is, we determine the conditions for

$$\psi = \phi_1 - \phi_2 = 0. \tag{B2}$$

By putting $\phi_1$ and $\phi_2$ into eqn. (B1), which they both, being alternative solutions, satisfy, and subtracting one from the other, we obtain

$$\varepsilon\nabla^2\psi = 0. \tag{B3}$$

Now, consider the vector identity of eqn. (A10) with $\psi = \phi$ and $\nabla\psi = \mathbf{A}$:

$$\nabla\,.\,(\psi\nabla\psi) = [\nabla\psi]\,.\,[\nabla\psi] + \psi\nabla^2\psi. \tag{B4}$$

The last term is zero in view of eqn. (B3). Integrating the rest of eqn. (B4) over the region of solution $R$, and using Gauss's divergence theorem (A12) on the left-hand-side terms of eqn. (B4), we obtain

$$\iint [\nabla\psi]^2\,dR = \int \psi[\nabla\psi\,.\,\mathbf{dS}] = \int \psi\frac{\partial}{\partial n}\psi\,dS, \tag{B5}$$

where $n$ is the direction of the normal to the boundary $S$. Since the integrand on the left, being a square, is nonnegative, we may conclude that $\nabla\psi$ is zero everywhere in $R$, provided that either $\psi$ or $\partial\psi/\partial n$ is zero. The latter conditions apply whenever $\phi$ or its normal derivative on the boundary of solution is specified; for then, $\phi_1$ and $\phi_2$ will both have either the same value or the same normal derivative and therefore $\psi$, their difference, will necessarily be either zero or have zero normal derivative.

The specification of $\phi$ along $R$ is known as a *Dirichlet boundary condition*, and that of its normal derivative as a *Neumann boundary condition*. What we have shown is that if a Neumann or Dirichlet condition applies on every part of the boundary, then $\nabla\phi$ is unique. However, that does not mean that $\phi$ itself is unique; but we may assert that

$$\phi_1 = \phi_2 + c, \tag{B6}$$

since when taking the gradients on both sides, we would have $\nabla\phi_1 = \nabla\phi_2$ only if $c$ is a constant; for any inconstant function $c(x, y)$ would have a gradient and invalidate the uniqueness of the gradient. Therefore, with Neumann and Dirichlet conditions everywhere on $R$, $\phi$ is unique within a constant $c$. But, so long as the Dirichlet condition applies at least at one point on the boundary, then $\phi_1 = \phi_2$ at that point, making $c$ zero, so that $\phi$ is unique everywhere in $R$.

In conclusion, then, the Poisson eqn. (B1) has a unique solution so long as the Dirichlet condition applies at least at one point and the Neumann condition applies wherever the Dirichlet condition is not at play.

# Appendix C
# UNIQUENESS OF VECTOR FIELDS

When the curl and divergence of a vector field $\mathbf{V}$ are defined everywhere in a region $R$ bounded by a surface $S$, we shall here determine sufficient boundary conditions to make that vector uniquely determinable. We prove in this section that it is sufficient for the normal or tangential component of that vector to be specified along the boundary to make the vector unique everywhere in the solution region.

Consider two alternative solutions $\mathbf{V}_1$ and $\mathbf{V}_2$ to the pair of vector equations:

$$\nabla \times \mathbf{V} = \mathbf{c}, \tag{C1}$$

$$\nabla \cdot \mathbf{V} = s. \tag{C2}$$

Proceeding as in the previous section, since both the alternative solutions satisfy eqns. (C1) and (C2), their difference

$$\mathbf{U} = \mathbf{V}_1 - \mathbf{V}_2, \tag{C3}$$

satisfies

$$\nabla \times \mathbf{U} = 0 \tag{C4}$$

and

$$\nabla \cdot \mathbf{U} = 0. \tag{C5}$$

Now, comparing eqn. (C4) with the vector identity (A4), we may say that

$$\mathbf{U} = \nabla\phi, \tag{C6}$$

which, when put into eqn. (C5), yields

$$\nabla^2\phi = 0. \tag{C7}$$

We know from Appendix B that this has a unique $\nabla\phi$ (and therefore unique $U$ from eqn. (C6)) if $\phi$ or its normal gradient is specified on the boundary $S$. Clearly from eqn. (C6), the specification of $\phi$ along $S$ implies that $\partial\phi/\partial t$, which is the tangential component of the vector field, is specified on the boundary. Similarly, the Neumann condition implies that the normal component of the vector field is specified. Of course, when the vector field is defined by a vector potential (or stream function), then the specification of $\phi$ is the specification of the normal component of the vector field, and the imposition of a Neumann condition is the same as specifying the tangential component of the vector field.

# Appendix D INTEGRATION OF TRIANGULAR COORDINATES

The integration of triangular coordinates repeatedly occurs in finite element analysis, during formation of the element matrices as well as during the integration of our approximation functions, such as when we need to compute by numerical integration the stored energy of the system. In this section we give a formal proof of the often-required formula used in such integrations:

$$\iint_{\Delta} \zeta_1^i \zeta_2^j \zeta_3^k \, dx\, dy = \frac{i!\, j!\, k!\, 2!\, A}{(i + j + k + 2)!}, \tag{D1}$$

where $A$ is the area of the triangle. To prove this, we first require a second formula for translating from the integrations with respect to $x$ and $y$ to any two of the $\zeta$'s:

$$dx\, dy = 2A\, d\zeta_1\, d\zeta_2 = 2A\, d\zeta_2\, d\zeta_3 = 2A\, d\zeta_3\, d\zeta_1. \tag{D2}$$

To prove eqn. (D2) consider the triangle $ABC$, with node numbering in the same order, shown in Figure D.1, where for convenience we have assumed that the side $BC$ is along the $x$ axis. As a result $y$ corresponds to the altitude $h_1$ of a point within the triangle, with respect to $BC$ (which is our side 1). It is also seen from Figure D.1 that

$$dh_3 = dx \sin\beta. \tag{D3}$$

From the definition of the triangular coordinates we have, therefore,

$$H_1\, d\zeta_1 = dh_1 = dy, \tag{D4}$$

$$H_3\, d\zeta_3 = dh_3 = dx \sin\beta, \tag{D5}$$

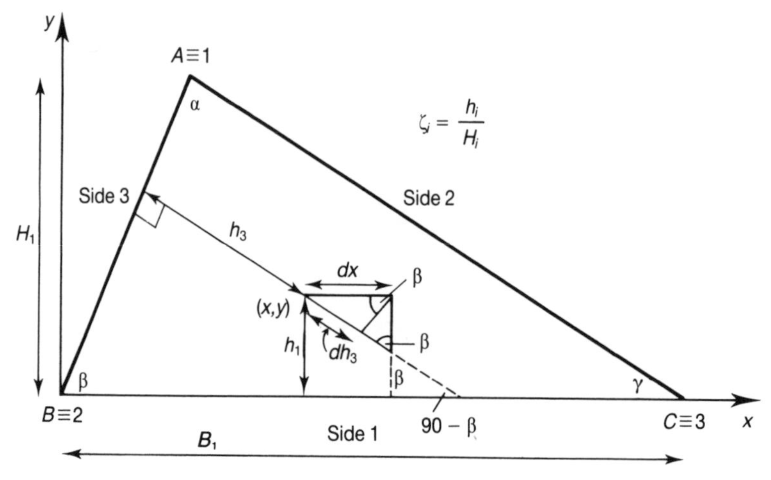


**Figure D.1.** Elemental lengths and triangular coordinates.

where $dh_3$ corresponds to incrementing $x$ while holding $y$ constant. From eqns. (D4) and (D5) we therefore have

$$
\begin{aligned}
dx\,dy &= \frac{H_3\,d\zeta_3}{\sin\beta}\,.\,H_1\,d\zeta_1 \\
&= B_1 H_1\,d\zeta_1\,d\zeta_3 \\
&= 2A\,d\zeta_1\,d\zeta_3,
\end{aligned}
\tag{D6}
$$

where $B_1$, the base or length of side 1, has been substituted for $H_3/\sin\beta$, and the geometric definition of the area of the triangle as a half of the base times the height has been used. By symmetry about the coordinates, eqn. (D2) follows. Using eqn. (D2) to evaluate the integral of interest to us, we may write:

$$
I = \iint_\Delta \zeta_1^i \zeta_2^j \zeta_3^k\,dx\,dy = 2A \iint_\Delta \zeta_1^i \zeta_2^j \zeta_3^k\,d\zeta_2\,d\zeta_3. \tag{D7}
$$

We have already seen that only two of the three $\zeta$ coordinates are independent, with the third fixed by:

$$
\zeta_1 + \zeta_2 + \zeta_3 = 1. \tag{6.5.9}
$$

Using this, we have

$$
I = \iint_\Delta \zeta_1^i \zeta_2^j \zeta_3^k\,dx\,dy = 2A \int_{\zeta_3=0}^{1} \int_{\zeta_2=0}^{1-\zeta_3} (1 - \zeta_2 - \zeta_3)^i \zeta_2^j \zeta_3^k\,d\zeta_2\,d\zeta_3, \tag{D7}
$$

where the limits of integration have been obtained by considera-

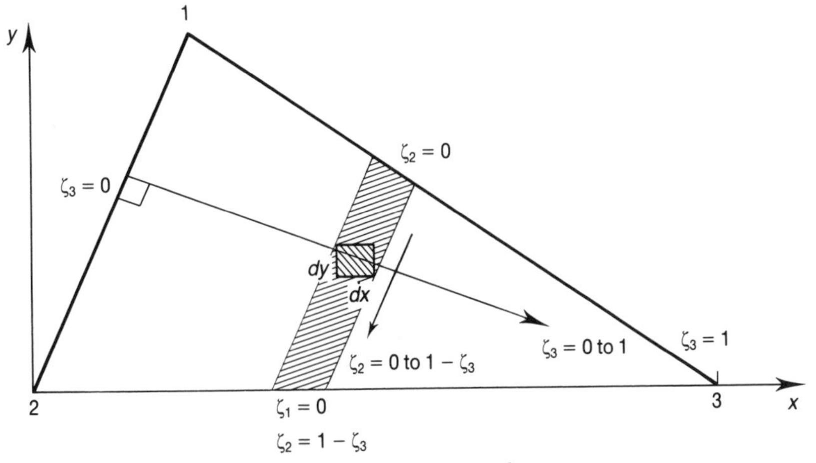


**Figure D.2.** Limits integrating over a triangle.

tions from Figure D.2. We shall attempt to evaluate this difficult integral in a piecemeal fashion. Let us consider the inner integral:

$$I_{i,j} = \int_0^p (p - \zeta_2)^i \zeta_2^j \, d\zeta_2 \qquad \text{where } p = 1 - \zeta_3,$$

$$= \frac{\int_0^p (p - \zeta_2)^i d[\zeta_2^{j+1}]}{j + 1}$$

$$= \left\{ \frac{(p - \zeta_2)^i \zeta_2^{j+1}]}{j + 1} \right\}_0^p + \int_0^p i(p - \zeta_2)^{i-1} \frac{\zeta_2^{j+1}}{j + 1} d\zeta_2 \qquad \text{by parts} \tag{D8}$$

$$= \int_0^p \frac{i(p - \zeta_2)^{i-1} \zeta_2^{j+1}}{j + 1} d\zeta_2$$

$$= \frac{i}{j + 1} I_{i-1,j+1}.$$

Repeatedly employing this recursive relationship,

$$I_{i,j} = \int_0^p (p - \zeta_2)^i \zeta_2^j \, d\zeta_2 \qquad \text{where } p = 1 - \zeta_3$$

$$= \frac{i}{j + 1} I_{i-1,j+1}$$

$$= \frac{i(i-1)}{(j+1)(j+2)} I_{i-2,j+2} \tag{D9}$$
$$= \frac{i(i-1)(i-2)\ldots 2 \times 1}{(j+1)(j+2)\ldots(j+i)} I_{0,i+j}$$
$$= \left\{\frac{i!\,j!}{(i+j)!}\right\} I_{0,i+j}.$$

Evaluating $I_{0,i+j}$ from the definition of eqn. (D8) and substituting $1 - \zeta_3$ for $p$,

$$\begin{aligned} I_{i,j} &= \frac{i!\,j!}{(i+j)!} \int_0^{1-\zeta_3} \zeta_2^{i+j}\, d\zeta_2 \\ &= \frac{i!\,j!\,(1-\zeta_3)^{i+j+1}}{(i+j+1)!}. \end{aligned} \tag{D10}$$

Before trying to evaluate $I$, we shall get an additional relationship from eqn. (D9) by substituting $\zeta_3$ for $\zeta_2$ and 1 for $p$, which we shall need recourse to in evaluating $I$. The resulting integral $J_{i,j}$:

$$\begin{aligned} J_{i,j} &= \int_0^1 (1-\zeta_3)^i \zeta_3^j\, d\zeta_3 \\ &= \left\{\frac{i!\,j!}{(i+j)!}\right\} J_{0,i+j}. \end{aligned} \tag{D11}$$

Now putting eqn. (D10) into the double integral for $I$, in eqn. (D7), we have

$$\begin{aligned} I &= \left\{\frac{i!\,j!\,2A}{(i+j+1)!}\right\} \int_0^1 (1-\zeta_3)^{i+j+1} \zeta_3^k\, d\zeta_3 \\ &= \left\{\frac{i!\,j!\,2A}{(i+j+1)!}\right\} J_{i+j+1,k} \\ &= \left\{\frac{i!j!2A}{(i+j+1)!}\right\}\left\{\frac{(i+j+1)!\,k!}{(i+j+1+k)!}\right\} J_{0,i+j+k+1} \qquad \text{from eqn. (D11)} \\ &= 2A\left\{\frac{i!\,j!\,k!}{(i+j+k+1)!}\right\} \int_0^1 \zeta_3^{i+j+k+1}\, d\zeta_3 \\ &= 2A\left\{\frac{i!\,j!\,k!}{(i+j+k+1)!}\right\}\left\{\frac{\zeta_3^{i+j+k+2}}{(i+j+k+2)}\right\}_0^1 \\ &= \left\{\frac{i!\,j!\,k!\,2A}{(i+j+k+2)!}\right\}, \end{aligned} \tag{D12}$$

which is the result that we seek.

# Appendix E
# INTEGRAL OF $r^{-1}$ OVER A RECTANGLE

In solving field problems by integral methods, we are often required to integrate $r^{-1}$ over an element. In this appendix to the text, we shall prove that, over a rectangular element of size $2a \times 2b$,

$$\iint r^{-1}\, dR = \int_{-a}^{a} \int_{-b}^{b} \frac{1}{\sqrt{(x^2 + y^2)}}\, dx\, dy$$

$$= 2a ln_e \frac{\sqrt{(a^2+b^2)}+b}{\sqrt{(a^2+b^2)}-b} + 2b ln_e \frac{\sqrt{(a^2+b^2)}+a}{\sqrt{(a^2+b^2)}-a}. \tag{E1}$$

Before trying to prove this, we shall first derive the integral of $(x^2 + a^2)^{-1/2}$ and $ln_e[\sqrt{(x^2 + a^2)} + a]$ with respect to $x$, which are required in the proof.

Consider first the integral of $(x^2 + a^2)^{-1/2}$. Let us make the substitution

$$x = a \tan \theta, \tag{E2}$$

so that

$$dx = a \sec^2 \theta\, d\theta, \tag{E3}$$

and we have:

$$\int (x^2 + a^2)^{-1/2}\, dx = \int a^{-1}(1 + \tan^2 \theta)^{-1/2} a \sec^2 \theta\, d\theta$$

$$= \int \sec \theta\, d\theta = \int \sec \theta \frac{\sec \theta + \tan \theta}{\sec \theta + \tan \theta} d\theta$$

$$= \int \frac{\sec^2 \theta + \sec \theta \tan \theta}{\sec \theta + \tan \theta} d\theta$$

$$= \int \frac{1}{\sec\theta + \tan\theta} d(\sec\theta + \tan\theta) \tag{E4}$$

$$= ln_e(\sec\theta + \tan\theta) = ln_e\left[\sqrt{1 + \frac{x^2}{a^2}} + \frac{x}{a}\right] \quad \text{from eqn. (E2)}$$

$$= ln_e \frac{\surd(x^2 + a^2) + x}{a}$$

$$\equiv ln_e[\surd(x^2 + a^2) + x],$$

where the constant $ln_e\, a$ has been dropped on the understanding that it is lumped with the arbitrary constant of integration, which vanishes when limits are imposed.

To evaluate the second integral of $ln_e[\surd(x^2 + a^2) + a]$, first observe that this results as one of the terms when $x ln_e[\surd(x^2 + a^2) + a]$ is differentiated with respect to $x$:

$$\frac{d}{dx} x ln_e[\surd(x^2 + a^2) + a] = ln_e[\surd(x^2 + a^2) + a] + x\frac{d}{dx} ln_e[\surd(x^2 + a^2) + a]$$

$$= ln_e[\surd(x^2 + a^2) + a] + x . \frac{\frac{1}{2} . 2x/\surd(x^2 + a^2)}{\surd(x^2 + a^2) + a} \tag{E5}$$

$$= ln_e[\surd(x^2 + a^2) + a] + \frac{x^2}{(x^2 + a^2) + a\surd(x^2 + a^2)}$$

Also observe that, making the substitutions of eqns. (E2) and (E3), we have:

$$\int \frac{x^2}{(x^2 + a^2) + a\surd(x^2 + a^2)} dx = \int \frac{a^2 \tan^2\theta}{a^2 \sec^2\theta + a^2 \sec\theta} a \sec^2\theta\, d\theta$$

$$= \int \frac{a \sec\theta(\sec^2\theta - 1)}{1 + \sec\theta} d\theta \tag{E6}$$

$$= \int a \sec\theta[\sec\theta - 1]\, d\theta = \int a \sec^2\theta\, d\theta - \int a \sec\theta\, d\theta$$

$$= a \tan\theta - a ln_e(\sec\theta + \tan\theta) \quad \text{from eqn. (E4)}$$

$$\equiv x - a ln_e[\surd(x^2 + a^2) + x]$$

We now obtain the desired integral by integrating eqn. (E5) and using the result, eqn. (E6):

$$\int ln_e[\sqrt{(x^2 + a^2)} + a]\, dx$$

$$= \int \frac{d}{dx} x ln_e[\sqrt{(x^2 + a^2)} + a]\, dx - \int \frac{x^2}{(x^2 + a^2) + a\sqrt{(x^2 + a^2)}}\, dx$$

$$= x ln_e[\sqrt{(x^2 + a^2)} + a] - x + a ln_e[\sqrt{(x^2 + a^2)} + x]. \tag{E7}$$

With these results ready, we are now prepared for the task at hand:

$$\int_{-a}^{a}\int_{-b}^{b} \frac{dx\, dy}{\sqrt{[x^2 + y^2]}} = \int_{-a}^{a} \left\{ \int_{-b}^{b} \frac{dy}{\sqrt{[x^2 + y^2]}} \right\} dx$$

$$= \int_{-a}^{a} \{ln_e[y + \sqrt{(x^2 + y^2)}]\}_{-b}^{b}\, dx \qquad \text{from eqn. (E4)}$$

$$= \int_{-a}^{a} \{ln_e[\sqrt{(x^2 + b^2)} + b] - ln_e[\sqrt{(x^2 + b^2)} - b]\}\, dx \tag{E8}$$

$$= \{x ln_e[\sqrt{(x^2 + b^2)} + b] - x + b ln_e[\sqrt{(x^2 + b^2)} + x]\}_{-a}^{a}$$
$$-\{x ln_e[\sqrt{(x^2 + b^2)} - b] - x - b ln_e[\sqrt{(x^2 + b^2)} + x]_{-a}^{a}\}$$

from eqn. (E7) with $a = b$ and $a = -b$

$$= \left\{ x ln_e \frac{\sqrt{(x^2 + b^2)} + b}{\sqrt{(x^2 + b^2)} - b} + 2b ln_e[\sqrt{(x^2 + b^2)} + x] \right\}_{-a}^{a}$$

$$= 2a ln_e \frac{\sqrt{(a^2 + b^2)} + b}{\sqrt{(a^2 + b^2)} - b} + 2b ln_e \frac{\sqrt{(a^2 + b^2)} + a}{\sqrt{(a^2 + b^2)} - a}.$$

Observe that for the special case $a = b$ of a square plate, this result reduces to:

$$\int_{-a}^{a}\int_{-a}^{a} \frac{dx\, dy}{\sqrt{[x^2 + y^2]}} = 4a ln_e \frac{\sqrt{2} + 1}{\sqrt{2} - 1}$$

$$= 8a ln_e[\sqrt{2} + 1] = 4\sqrt{(4a^2)} ln_e 2.4142 \tag{E9}$$

$$= 3.525\sqrt{A},$$

where $A$ is $4a^2$, the area of the rectangle.

# REFERENCES

Abramovitz, M., and Stegun, I., eds., 1964, *Handbook of Mathematical Functions with Formulas, Graphs, and Mathematical Tables*, U.S. Dept. of Commerce, Natl. Bureau of Standards, Applied Mathematics Series 55, Washington, D.C.

Adachi, T., Yoshii, A., and Sudo, T., 1979, "Two-Dimensional Semiconductor Analysis Using Finite Element Method," IEEE Trans. Electron Devices, Vol. ED-26, No. 7, pp. 1026–1031, July.

Adkins, B., 1962, *The General Theory of Electrical Machines*, Pitman, London.

Ali, K. F., Ahmed, M. T., and Burke, P. E., 1987, "Surface Impedance–BEM Technique for Nonlinear TM–Eddy Current Problems," J. Appl. Phys., Vol. 61, No. 8, pp. 3925–3927.

Armor, A. F., and Chari, M. V. K., 1976, "Heatflow in the Stator Core of Large Turbine Generators by the Method of Three-Dimensional Finite Elements," IEEE Trans. Pow. App. Sys., Vol. PAS-95, pp. 1648–1656.

Babuska, I., and Aziz, A. K., 1976, "On the Angle Condition in the Finite Element Method," SIAM J. Numer. Anal., Vol. 13, No. 2, pp. 214–216.

Babuska, I., and Miller, A., 1984, "The Postprocessing Approach in the Finite Element Method—Part 3: A Posteriori Error Estimates and Adaptive Mesh Selection," Int. J. Num. Meth. in Eng., Vol. 20, pp. 2311–2324.

Babuska, I., and Rheinboldt, W. C., 1978, "A Posteriori Error Estimates for the Finite Element Method," Int. J. Num. Meth. in Eng., Vol. 12, pp. 1579–1615.

Bank, R. E., Rose, D. J., and Fichtner, W., 1983, "Numerical Methods for Semiconductor Device Simulation," IEEE Trans. Electron Devices, Vol. ED-30, No. 9, pp. 1031–1041, Sept.

Bank, R. E., and Sherman, A. H., 1981, "An Adaptive Multi-level Method for Elliptical Boundary Value Problems," Computing, Vol. 26, pp. 91–105.

Bank, R. E., and Weiser, A., "Some A Posteriori Error Estimates for Elliptic Partial Differential Equations," Mathematics of Computation, to appear.

Barnes, J. J., and Lomax, R. J., 1974, "Two-dimensional Finite Element Simulation of Semiconductor Devices," Electron. Lett., Vol. 10, pp. 341–343, Aug.

Barnes, J. J., and Lomax, R. J., 1977, "Finite Element Methods in Semiconductor Device Simulation," IEEE Trans. Electron Devices, Vol. ED-24, pp. 1082–1089, Aug.

Benedek, P., and Silvester, P., 1972, "Capacitance of Parallel Rectangular Plates Separated by Dielectric Sheets," IEEE Trans. Microwave Theory Tech., Vol. MTT-20, No. 8, pp. 504–510.

Berk, A. D., 1956, "Variational Principles for Electromagnetic Resonators and Waveguides," IRE Trans. Antenna and Propagation, Vol. AP-4, pp. 104–111, April.

Bewley, L. V., 1948, "Two-Dimensional Fields in Electrical Engineering," Macmillan, New York; reprinted 1963, Dover, New York.

Biddlecomb, C. S., Simkin, J., and Trowbridge, C. W., 1986, "Error Analysis in Finite Element Models of Electromagnetic Fields," IEEE Trans. Magn., Vol. MAG-22, pp. 811–813.

Binns, K. J., and Lawrenson, P. J., 1963, *Analysis and Computation of Electric and Magnetic Field Problems,* Oxford, Pergamon Press.

Brauer, John R., 1975, "Simple Equations for the Magnetization and Reluctivity Curves of Steel," IEEE Trans. Magn., Vol. MAG-11, 1, p. 81, Jan.

Brebbia, C. A., ed., 1978a, *Recent Advances in Boundary Element Analysis,* Pentech, London.

Brebbia, C. A., 1978b, *The Boundary Element Method for Engineers,* John Wiley, New York.

Campbell, P., 1986, "Application of Modern Permanent Magnets to Consumer Products," *Proc. Symp. on Soft and Hard Magnetic Materials,* Amer. Soc. Metals, Lake Buena Vista, Florida, pp. 119–130.

Campbell, P., Hoole, S. R. H., and Tsals, I., 1984, "Finite Element Field Analysis in 2-D and 3-D on a Personal Computer," IEEE Trans. Magn., Vol. MAG-20, No. 5, pp. 1903–1905, Sept.

Carey, G. F., and Humphrey, D. L., 1979, "Residuals, Adaptive Refinement and Nonlinear Iterations for Finite Element Computations," in Shepard, M. S. and Gallagher, R. H., eds., *Finite Element Grid Optimization,* Amer. Soc. of Mech. Eng, New York, pp. 437–448.

Carpenter, C. J., 1959, "Surface-Integral Methods of Calculating Forces on Magnetized Iron Parts," Monograph No. 342, IEE, London (Also 1960, Proc. IEE, Vol. C107).

Carpenter, C. J., 1967, "Theory and Application of Magnetic Shells," Proc. IEE, Vol. 114, No. 7, pp. 995–1000.

Carpenter, C. J., 1968, "Magnetic Equivalent Circuits," Proc. IEE, Vol. 115, 1503–1511.

Carpenter, C. J., 1975, "Finite Element Network Models and Their Application to Eddy-Current Problems," Proc. IEE, Vol. 122, pp. 455–462.

Carpenter, C. J., 1977, "Comparison of Alternative Formulation of Three-Dimensional Magnetic Field and Eddy Current Problems," Proc. IEE, Vol. 124, pp. 66–74.

Carpenter, C. J., and Djurovic, M., 1975, "Three-Dimensional Numerical Solution of Eddy Currents in Thin Plates," Proc. IEE, Vol. 122, pp. 681–688.

Carpenter, C. J., and Wyatt, E. A., 1976, "Efficiency of Numerical Techniques for Computing Eddy Currents in Two and Three Dimensions," *Proc. Compumag 76,* Rutherford Laboratory, Oxford, pp. 242–250.

Carter, G. W., 1961, *The Electromagnetic Field in Its Engineering Aspects,* Longmans, London.

Cendes, Z. J., 1982, "Notes on the Preconditioned Conjugate Gradient Method," Computational Electromagnetics Seminar Series, Carnegie-Mellon University, Nov. 29, Pittsburgh, Pennsylvania.

Cendes, Z. J., and Shenton, D., 1985, "Adaptive Mesh Refinement in the Finite Element Computation of Magnetic Fields," IEEE Trans. Magn., Vol. MAG-21, pp. 1811–1816, Sept.

Cendes, Z. J., Shenton, D. N., and Shahnasser, H., 1983, "Magnetic Field Computation Using Delaunay Triangulation and Complementary Finite Element Methods," IEEE Trans. Magn., Vol. MAG-19, pp. 2551–2554.

Chari, M. V. K., 1970, "Finite Element Analysis of Nonlinear Magnetic Fields in Electric Machines," Ph.D. thesis, McGill University, Montreal, March.

Chari, M. V. K., 1974, "Finite Element Solution of the Eddy Current Problem in Magnetic Structures," IEEE Trans. Power Apparatus Syst., Vol. PAS-93, pp. 62–72.

Chari, M. V. K., 1987, "Electromagnetic Field Computation of Open Boundary Problems by a Semi-Analytical Approach," IEEE Trans. Magn., Vol. MAG-23, pp. 3566–3568.

Chari, M. V. K., and Bedrosian, G., 1988, "A Hybrid Method for Eddy Current Open Boundary Field Computation," J. of Appl. Phys., Vol. 63, No. 8, pp. 3019–3021, April 15.

Chari, M. V. K., Csendes, Z. J., Silvester, P., Konrad, A., and Palmo, M. A., 1981, "Three-Dimensional Magnetostatic Field Analysis of Electrical Machinery by the Finite Element Method," IEEE Trans. Power Apparatus Syst., Vol. PAS-100, pp. 4007–4019.

Chari, M. V. K., and Bedrosian, G., 1987, "Hybrid Harmonic/Finite Element Method for Two-Dimensional Open Boundary Problems," IEEE Trans. Magn., Vol. MAG-23, pp. 3572–3574.

Chari, M. V. K., d'Angelo, J., and Neumann, T. W., 1985, "Three-Dimensional Nonlinear Analysis of a Permanent Magnet Machine by a Hybrid Finite-Element Method," J. Appl. Phys., Vol. 57, No. 8, pp. 3866–3868.

Chari, M. V. K., d'Angelo, J., Palmo, M. A., and Sharma, D. K., 1985a, "Application of Three-Dimensional Electromagnetic Analysis Methods to Electrical Machinery and Devices—Part I: Theory," IEEE Power Eng. Soc. Winter Meeting, Vancouver, July.

Chari, M. V. K., d'Angelo, J., Palmo, M. A., and Sharma, D. K., 1985b, "Application of Three-Dimensional Electromagnetic Analysis Methods to Electrical Machinery and Devices—Part II: Applications," IEEE Power Eng. Society Winter Meeting, Vancouver, July.

Chari, M. V. K., and Silvester, P., 1971, "Analysis of Turboalternator Magnetic Fields by Finite Elements," IEEE Trans. Power Apparatus Syst., Vol. PAS-90, No. 2, pp. 454–464.

Chari, M. V. K., and Silvester, P. P., eds., 1980, *Finite Elements in Electric and Magnetic Field Problems*, New York, John Wiley.

Cherry, E. C., 1949, "The Analogies between the Vibrations of Elastic Membranes and the Electromagnetic Fields in Guides and Cavities," Proc. IEE, Vol. 96, No. 3, pp. 346–360.

Christoffel, E. B., 1870, "Über die Abbildung einer einfach einblattrigen zusammenhangenden Flache auf den Kreis," Göttinger Nachrichten.

Chung, K. C., 1981, "A Generalized Finite-Difference Method for Heat Transfer Problems of Irregular Geometries," Numer. Heat Transfer, Vol. 4, pp. 345–357.

Collie, C. J., 1976, "Magnetic Fields and Potentials of Linearly Varying Current or Magnetisation in a Plane Bounded Region," *Proc. Compumag 76*, Rutherford Laboratory, Oxford, pp. 86–95.

Coulomb, J.-L., 1981, "Finite Elements 3-D Magnetic Field Computation," IEEE Trans. Magn., Vol. MAG-17, pp. 3241–3246.

Coulomb, J.-L., 1983, "A Methodology for the Determination of Global Electromechanical Quantities from a Finite Element Analysis and Its Application to the Evaluation of Magnetic Forces, Torques and Stiffness," IEEE Trans. Magn., Vol. MAG-19, pp. 2514–2519.

Coulomb, J.-L., and Meunier, G., 1984, "Finite Element Implementation of Virtual Work Principle for Magnetic or Electric Force and Torque Computation," IEEE Trans. Magn., Vol. MAG-20, pp. 1894–1896.

Crank, J., and Nicholson, P., 1947, "A Practical Method for Numerical Solutions of Partial Differential Equations of Heat Conduction Type," Proc. Camb. Phil. Soc., Vol. 43, pp. 50–67.

Csendes, Z. J., 1980, "The High-Order Polynomial Finite Element Method in Electromagnetic Field Computation," in Chari and Silvester, (1980), pp. 125–143.

Csendes, Z. J., and Silvester, P., 1970, "Numerical Solution of Dielectric Loaded Waveguides: I—Finite Element Analysis," IEEE Trans. Microwave Theory Tech., Vol. MTT-17, pp. 684–690.

Csendes, Z. J., Weiss, J., and Hoole, S. R. H., 1982, "Alternative Vector Potential Formulations for 3-D Magnetostatics," IEEE Trans. Magn., Vol. MAG-18, pp. 367–372.

Cuthill, E., and McKee, J., 1969, "Reducing the Bandwidth of Sparse Symmetric Matrices," *ACM Proc. of 24th Natl. Conf.*, New York.

Davis, J. L., and Hoburg, J. F., 1985, "Enhanced Capabilities for a Student-Oriented Finite Element Electrostatic Potential Program," IEEE Trans. Educ., Vol. E-28, pp. 25–28.

Demerdash, N. A., and Nehl, T. W., 1976, "Flexibility and Economics of the Finite Element and Difference Techniques in Nonlinear Magnetic Fields of Power Devices," IEEE Trans. Magn., Vol. MAG-12, pp. 1036–1038.

Demerdash, N. A., Nehl, T. W., Fouad, F. A., and Mohammed, O. A., 1981, "3-D Finite Element Vector Potential Formulation for Magnetic Fields in Electrical Apparatus," IEEE Trans. Power Apparatus Syst., Vol. PAS-100, pp. 4104, 4111.

Draper, A., 1964, *Electrical Circuits including Machines*, Longmans, London.

Duffin, R. J., 1959, "Distributed and Lumped Networks," J. Math. Mech., Vol. 8, pp. 793–826.

Edwards, T. W., and Van Bladel, J., 1961, "Electrostatic Dipole Moment of a Dielectric Cube," Applied Science Research Section B., Vol. 9, pp. 151–155.

Ehrlich, L. W., 1959, "Monte Carlo Solutions of Boundary-value Problems," J. Ass. Comp. Mach., Vol. 6, No. 2, p. 204.

Ellison, A. J., 1965, *Electromechanical Energy Conversion*, Harrap, London.

English, W. J., 1971, "Variational Solutions of the Maxwell Equations with Applications to Wave Propagation in Inhomogeneously Filled Waveguides," IEEE Trans. Microwave Theory Tech., Vol. MTT-19, pp. 9–18, Jan.

Fawzi, T. H., Ahmed, M. T., and Burke, P. E., 1985, "On the Use of the Impedance Boundary Conditions in Eddy Current Problems (Invited)," IEEE Trans. Magn., Vol. MAG-21, No. 5, pp. 1834–1840.

Ferrari, R. L., and Maile, G. L., 1978, "Three-Dimensional Finite Element Method for Solving Electromagnetic Problems," Electron. Lett., Vol. 14, No. 15, pp. 467–468.

Ferraro, V. C. A., 1970, *Electromagnetic Theory*, ELBS, London.

Fichtner, W., Rose, D. J., and Bank, R. E., 1983, "Semiconductor Device Simulation," IEEE Trans. Electron Devices, Vol. ED-30, No. 9, pp. 1018–1030, Sept.

Forsythe, G. E., and Wasow, W. R., 1960, *Finite Difference Methods for Partial Equations*, John Wiley, New York.

Fouad, F. A., Nehl, T. W., and Demerdash, N. A., 1981, "Permanent Magnet Modeling for Use in Vector Potential Finite Element Analysis in Electrical Machinery," IEEE Trans. Magn., Vol. MAG-17, pp. 3002–3004.

Fritts, M. J., 1979, "Numerical Approximation on Distorted Lagrangian Grids," *Advances in Computer Methods for Partial Differential Equations*, Vol. III, IMACS, pp. 137–142.

Gelfand, I. M., and Fomin, S. V., 1963, *Calculus of Variations*, Prentice Hall, Englewood Cliffs, New Jersey.

George, A., 1973, "Nested Dissection of Regular Finite Element Mesh," SIAM J. Numer. Anal., Vol. 10, pp. 345–363.

Girdino, P., Molfino, P., Molinari, G., Puglisi, L., and Viviani, A., 1983, "Finite Difference and Finite Element Grid Optimization by the Grid Iteration Method," IEEE Trans. Magn., Vol. MAG-19, pp. 2543–2546.

Gould, S. H., 1957, *Variational Methods for Eigenvalue Problems*, University of Toronto Press, Canada.

Gummel, H. K., 1964, "A Self-Consistent Iterative Scheme for One-Dimensional Steady-State Transistor Calculations," IEEE Trans. Electron Devices, Vol. ED-11, pp. 455–465, Oct.

Harrington, R. F., 1968, *Field Computation by the Moment Method*, Krieger, Florida.

Hassan, M. A., and Silvester, P., 1977, "Radiation and Scattering by Wire Antenna Structures Near a Rectangular Plate Reflector," Proc. IEE, Vol. 124, pp. 429–435.

Heimer, H. H., 1973, "A Two-Dimensional Numerical Analysis of a Silicon N–P–N Transistor," IEEE Trans. Electron Devices, Vol. ED-20, pp. 708–714.

Hermeline, F., 1982, "Automatic Triangulation of a Polyhedron in Dimension N," RAIRO Numer. Anal., Vol. 16, No. 3, pp. 211–242.

Heubner, K. H., and Thornton, E. A., 1982, *The Finite Element Method for Engineers,* 2d ed., John Wiley, New York.

Hoburg, J. F., and Davis, J. L., 1983, "A Student Oriented Finite Element Program for Electrostatic Potential Problems," IEEE Trans. Educ., Vol. E-26, No. 4, pp. 138–142, Nov.

Hoole, S. R. H., 1982, "Direct Finite Element Solution of the Magnetic Field Vector", Ph.D. Dissertation, Carnegie–Mellon University, December.

Hoole, S. R. H., 1984, "A Memory Economic 3-D Finite Element Mesh Generator for a Microcomputer," IEEE/CS Proc. of the Small Computer (R)evolution, pp. 111–115, Sept.

Hoole, S. R. H., 1985, "The Fictitious Potential for 3-D Magnetic Flux Plots," Electron. Lett., Vol. 21, No. 1, pp. 15–16, Jan.

Hoole, S. R. H., 1986, "Enhancing Interactivity and Automation through Finite Element Neighbours," Comm. Appl. Numer. Meth., Vol. 2, pp. 509–517.

Hoole, S. R. H., 1987, "Nodal Perturbations in Adaptive Expert Finite Element Mesh Generation," IEEE Trans. Magn., Vol. MAG-23, Sept., pp. 2635–2637.

Hoole, S. R. H., 1988, "A Novel Proof of Natural Boundary Conditions for the Poisson Equation," IEEE Trans. Educ., Vol. E-31, Feb, pp. 4–8.

Hoole, S. R. H., Anandaraj, A. W., and Bhatt, J. J., 1987, "Choice of Matrix Solution Algorithms in Adaptive Finite Elements—Subjective and Objective Considerations," under review.

Hoole, S. R. H., and Carpenter, C. J., 1985, "Surface Impedance Models for Corners and Slots," IEEE Trans. Magn., Vol. MAG-21, No. 5, pp. 1841–1843.

Hoole, S. R. H., and Cendes, Z. J., 1985, "Direct Vector Solution of Three-Dimensional Magnetic Field Problems," J. Appl. Phys., Vol. 57, pp. 3835–3837, April.

Hoole, S. R. H., Cendes, Z. J., and Hoburg, J. F., 1985, "Three-Dimensional Analysis of a Slot Motor Solving Directly for Flux Density," J. Appl. Phys., Vol. 57, No. 1, pp. 3875–3877, April.

Hoole, S. R. H., Cendes, Z. J., and Hoole, P. R. P., 1986, "Preconditioning and Renumbering in the Conjugate Gradients Algorithm," in Cendes, Z. J., ed., *Computational Electromagnetics,* North–Holland, New York, pp. 91–99.

Hoole, S. R. H., and Hoole, P. R. P., 1984, "Building an Expert System for Interactive Finite Element Design," Proc. of the IEEE/CS Third Annual Workshop on Interactive Computing: CAE: Electrical Engineering Education, IEEE Computer Society Press, Pittsburgh, Pennsylvania, pp. 90–93.

Hoole, S. R. H., and Hoole, P. R. P., 1985, "On Finite Element Force

Computation from Two- and Three-Dimensional Magnetostatic Fields," J. Appl. Phys., Vol. 57, No. 8, pp. 3825–3875, April.

Hoole, S. R. H., and Hoole, P. R. P., 1986a, "Finite Element Programs for Teaching Electromagnetics," IEEE Trans. Educ., Vol. E-29, No. 1, pp. 21–26, Feb.

Hoole, S. R. H., and Hoole, N. R. G., 1986b, "Reluctivity Characteristics of Steel for Nonlinear Finite Element Analysis," IEEE Trans. Magn., Vol. MAG-22, No. 5, pp. 1352–1353.

Hoole, P. R. P., and Hoole, S. R. H., 1986c, "Finite Element Computation of Magnetic Fields from Lightning Return Strokes," In Cendes, Z. J., ed., *Computational Electromagnetics*, North-Holland, New York, pp. 229–237.

Hoole, P. R. P., and Hoole, S. R. H., 1987, "Computing Transient Electromagnetic Fields Radiated from Lightning," J. Appl. Phys., Vol. 61, No. 8, pp. 3473–3475.

Hoole, S. R. H., Hoole, P. R. P., and Jayakumaran, S., 1986, "A Graphics Library for the Finite Element Modelling of Differential Equations," in Koval, D. O., ed., *Applied Simulation and Modelling*, Acta Press, Calgary, pp. 207–210.

Hoole, S. R. H., Hoole, P. R. P., Jayakumaran, S., and Hoole, N. R. G., 1988, "Teaching Electromagnetics through Finite Elements. Part II: The Instructional Program," Int. J. for Elect. Eng. Educ., Vol. 25, pp. 151–161.

Hoole, S. R. H., Jayakumaran, S., Anandaraj, A. W., and Hoole, P. R. P., 1987, "Relevant Purpose Based Error Criteria for Expert Finite Element Mesh Generation," J. of Electromagnetic Waves and Applications, in press.

Hoole, S. R. H., Jayakumaran, S., and Yoganathan, S., 1986, "Tetrahedrons, Edges and Nodes in 3-D Finite Element Analysis," Electron. Lett., Vol. 22, No. 14, pp. 735–737.

Hoole, S. R. H., Rios, R., and Yoganathan, S., 1988, "Vector Potential Formulations and Finite Element Trial Functions," Int. J. Numer. Meth. in Eng., Vol. 26, pp. 95–108.

Hoole, S. R. H., and Shin, Y., 1987, "Three Dimensional Solution of Sinusoidally Excited Saturated Magnetic Fields," J. Appl. Phys., Vol. 61, No. 8, pp. 3928–3930.

Hoole, S. R. H., Weeber, K., and Hoole, N. R. G., 1988, "The Natural Finite Element Formulation of the Impedance Boundary Condition in Shielding Structures," J. Appl. Phys., Vol. 63, No. 8, pp. 3022–3024.

Hoole, S. R. H., Yoganathan, S., and Jayakumaran, S., 1986, "Implementing the Smoothness Criterion in Adaptive Meshes," IEEE Trans. Magn., Vol. MAG-22, Sept., pp. 808–810.

Hrenikoff, J., 1941, "Solutions of Problems in Elasticity by the Framework Method," J. Appl. Mech., Vol. 8, pp. 169–175.

Irons, B. M., 1970, "A Frontal Solution Program for Finite Element Analysis," Int. J. Num. Meth. in Eng., Vol. 2, pp. 5–32.

Irons, B. M., and Draper, K. J., 1965, "Inadequacy of Nodal Connections in a Stiffness Solution for Plate Bending," J. Amer. Inst. Aero. and Astronautics, Vol. 3, pp. 5–8.

Jayakumaran, S., and Hoole, S. R. H., 1988, "Perturbation in Adaptive Refinement of Boundary Elements," J. Appl. Phys., Vol. 63, No. 8, pp. 3013–3015.

Jeng, G., and Wexler, A., 1977, "Isoparametric, Finite Element, Variational Solution of Integral Equations for Three-Dimensional Fields," Int. J. Num. Meth. in Eng., Vol. 11, pp. 1455–1471.

Jennings, A., 1977, *Matrix Computation for Engineers and Scientists*, John Wiley, London.

Jensen, P. S., 1972, "Finite Difference Techniques for Variable Grids," Comput. Struct., Vol. 2, pp. 17–29.

Johnson, E. E., and Green, C. H., 1927, "Graphical Determination of Magnetic Fields—Comparison of Calculation and Tests," AIEE Trans., Vol. XLVI, pp. 136–140.

Kalaichelvan, S., and Lavers, J. D., 1987, "Singularity Evaluation in Boundary Integral Equations of Minimum Order for 3-D Eddy Currents," IEEE Trans. Magn., Vol. MAG-23, pp. 3053–3055.

Kamminga, W., 1975, "Finite Elements Solutions for Devices with Permanent Magnets," J. Appl. Phys. (U.K.), Vol. 8.

Karplus, W. J., 1958, *Analog Simulation*, McGraw–Hill, New York.

Keinstreuer, C., and Hadenan, J. T., 1980, "A Triangular Finite Element Mesh Generation for Fluid Dynamic Systems of Arbitrary Geometry," Int. J. Num. Meth. in Eng., Vol. 15, pp. 1325–1334.

Keran, S., and Lavers, J. D., 1986, "A Skin-Depth Independent Finite Element Method for Eddy Current Problems," IEEE Trans. Magn., Vol. MAG-22, No. 5, Sept., pp. 1248–1250.

Kershaw, D. S., 1978, "The Incomplete Cholesky Conjugate Gradient Method for Iterative Solution of Systems of Linear Equations," J. Comp. Phys., Vol. 26, pp. 43–65.

Konrad, A., 1974, "Triangular Finite Elements for Vector Fields in Electromagnetics," Ph.D. Thesis, McGill University, Montreal.

Konrad, A., 1976, "Vector Variational Formulations of Electromagnetic

Fields in Anisotropic Media," IEEE Trans. Microwave Theory Tech., Vol. MTT-24, No. 9, pp. 553–559.

Konrad, A., 1985a, "Eddy Currents in Modelling (Invited)," IEEE Trans. Magn., Vol. MAG-21, No. 5, pp. 1805–1810.

Konrad, A., 1985b, "Electromagnetic Devices and the Application of Computational Techniques in their Design (Invited)," IEEE Trans. Magn., Vol. MAG-21, No. 6, pp. 2382–2387.

Kornhauser, E. T., and Stakgold, I., 1952, "A Variational Theorem for $\nabla^2 u + \lambda u = 0$ and its Application," J. Math. Phys., Vol. 31, pp. 45–54.

Kotiuga, P. R. and Silvester, P. P., 1982, "Vector Potential Formulations for Three-Dimensional Magnetostatics," J. Appl. Phys., Vol. 54, No. 11, pp. 8399–8401.

Kron, G., 1959, *Tensors for Circuits*, Dover, New York.

Laithwaite, E. R., 1967, "Magnetic Equivalent Circuits for Electrical Machines," Proc. IEE, Vol. 114, No. 11, pp. 1805–1809.

Lawson, C. L., 1977, "Software for $C^1$ Surface Interpolation," in Rice et al., Eds., *Mathematical Software III*. New York, Academic Press.

Lean, M. H., and Wexler, A., 1985, "Accurate Numerical Integration of Singular Boundary Element Kernels over Boundaries with Curvature," Int. J. for Numer. Meth. in Eng., Vol. 21, pp. 211–228.

Lee, D.-T., and Schachter, B. J., 1980, "Two Algorithms for Constructing a Delaunay Triangulation," Int. J. Comp. Inf. Sci., Vol. 9, No. 3, pp. 219–241.

Lee, J.-F., and Cendes, Z. J., 1987, "Transfinite Elements: A Highly Efficient Procedure for Modeling Open Field Problems," J. Appl. Phys., Vol. 16, No. 8, pp. 3913–3915.

Liebmann, G., 1949, "Precise Solution of Partial Differential Equations by Resistance Networks," Nature, Vol. 164, p. 149.

Liebmann, G., 1952, "The Solution of Waveguide and Cavity-Resonator Problems with the Resistance-Network Analogue," Proc. IEE, Vol. 99, No. 4, Radio Section, pp. 260–272.

Lindholm, D. A., 1980, "Notes on Boundary Integral Equations for Three Dimensional Magnetostatics," IEEE Trans. Magn., Vol. MAG-16, pp. 1409–1413.

Lindholm, D. A., 1981, "Application of Higher Order Boundary Integral Equations to Two-Dimensional Magnetic Head Problems," IEEE Trans. Magn., Vol. MAG-17, pp. 2445–2452.

Lowther, D. A., Saldhana, C. M., and Choy, G., 1985, "The Application of Expert Systems to CAD in Electromagnetics," IEEE Trans. Magn., Vol. MAG-21, No. 6, pp. 2559–2562.

MacLean, William, 1954, "Theory of Strong Electromagnetic Waves in Massive Iron," J. Appl. Phys., Vol. 25, No. 10, October.

MacNeal, R. H., ed., 1981, *MSC/NASTRAN Handbook,* The MacNeal–Schwendler Corporation, December.

Maergoiz, I. D., 1970, "Pulse Magnetization of a Ferromagnetic Lamination," Automation and Remote Control (translated from Avtomatika i Telemekhanika), 1969, No. 10, pp. 137–146.

Mari, A. de, 1968, "An Accurate Numerical Steady-State One-dimensional Solution of the *p–n* Junction," Solid-State Electron., Vol. 11, pp. 33–58.

Maxwell, C., 1904, *A Treatise on Electricity and Magnetism,* Vols. I and II, Clarendon Press, England.

Mayergoyz, I., 1981, "Theoretical Investigation of the Nonlinear Skin Effect," Archiv für Elecktrotechnik, Vol. 64, pp. 153–162.

Mayergoyz, I. D., 1983, "A New Approach to the Calculation of Three-Dimensional Skin Effect Problems," IEEE Trans. Magn., Vol. MAG-19, No. 5, pp. 2198–2200.

Mayergoyz, I. D., 1986a, "Solution of the Nonlinear Poisson Equation of Semiconductor Device Theory," J. Appl. Phys., Vol. 59, No. 1, pp. 195–199.

Mayergoyz, I. D., 1986b, "Mathematical Models of Hysteresis," IEEE Trans. Magn., Vol. MAG-22, No. 5, pp. 603–608.

Mayergoyz, I. D., Abdel-Kader, F. M., and Emad, F. P., 1984, "On Penetration of Electromagnetic Fields in Nonlinear Conducting Ferromagnetic Media," J. Appl. Phys., Vol. 55, No. 3, pp. 618–629.

Mayergoyz, I. D., Chari, M. V. K., and D'Angelo, J., 1987, IEEE Trans. on Magnetics, Vol. MAG-23, pp. 3889–3894.

Mayergoyz, I. D., Chari, M. V. K., and Konrad, A., 1983, "Boundary Galerkin's Method for Three-Dimensional Finite Element Electromagnetic Field Computation," IEEE Trans. Magn., Vol. MAG-19, No. 6, pp. 2333–2336.

Mayergoyz, I. D., and Doong, T., 1985, "Hybrid Boundary-Volume Galerkin's Method for Nonlinear Magnetostatic Problems," J. Appl. Phys., Vol. 57, No. 8, pp. 3838–3840.

Mayergoyz, I. D., and Korman, C., 1987, "A Parallel and Globally Convergent Iterative Method for the Solution of the Nonlinear Poisson Equation for Semiconductor Device Theory," Proc. IEEE Workshop on Electromagnetic Field Computation, pp. C-01-C-05, Schenectady, New York, Oct.

McDonald, B. H., Friedman, M., Decreton, M., and Wexler, A., 1973, "Integral Finite Element Approach for Solving the Laplace Equation," Electron. Lett., Vol. 9, No. 11, pp. 242–244, May.

McDonald, B. H., Friedman, M., and Wexler, A., 1974, IEEE Trans. on Micro. Theory and Tech., Vol. MTT-22, No. 3, pp. 237–248.

McDonald, B. H., and Wexler, A., 1972, "Finite Element Solution of Unbounded Field Problems," IEEE Trans. Microwave Theory Tech., Vol. MTT-20, No. 12, pp. 841–847, Dec.

McHenry, D., 1943, "A Lattice Analogy for the Solution of Plane Stress Problems," J. Inst. Civil Eng., London, Vol. 21, pp. 59–85.

McPherson, G., 1981, *An Introduction to Electric Machines and Transformers,* John Wiley, New York.

Meijerink, J. A., and van der Vost, H. A., 1977, "An Iterative Solution Method for Linear Systems of which the Coefficient Matrix is a Symmetric *M*-Matrix," Math. Comp., Vol. 31, No. 137, pp. 148–162.

Melosh, R. J., 1961, "A Stiffness Matrix for the Analysis of Thin Plates in Bending," J. Aero Sci., Vol. 28, pp. 34–42.

Mikhlin, S. G., and Smolitsky, K. L., 1967, *Approximate Methods for the Solution of Differential Equations,* Elsevier, New York.

Mishra, M., Forghani, B., Lowther, D. A., and Silvester, P. P., 1983, "Solution of Three Dimensional Electromagnetic Field Problems on a Mini-Computer," IEEE Trans. Magn., Vol. MAG-19, No. 6, pp. 2663–2666.

Molinari, G., and Viviani, 1979, "Grid Iteration Method for Finite Element Grid Optimization," in Shephard, M. S., and Gallagher, R. H., eds., *Finite Element Grid Optimization,* Amer. Soc. Mech. Eng., New York, pp. 49–59.

Mohammed, O. A., Davis, W. A., Popovic, B. D., Nehl, T. W., and Demerdash, N. A., 1982, "On Uniqueness of Solution of Magnetostatic Vector-Potential Problems by Three-Dimensional Finite Element Methods," J. Appl. Phys., Vol. 54, No. 11, pp. 8402–8404.

Nakata, T., and Takahashi, N., 1983, "New Design Method of Permanent Magnets by Using the Finite Element Method," IEEE Trans. Magn., Vol. MAG-19, No. 6, pp. 2494–2497.

Nakata, T., Takahashi, N., and Kawase, Y., 1985, "New Approximate Method for Calculating Three-Dimensional Magnetic Fields in Laminated Cores," IEEE Trans. Magn., Vol. MAG-21, pp. 2374–2377.

Namjoshi, H. V., and Biringer, P. P., 1985, "Influence of the Leakage Field in Crossfield Heating Systems," IEEE Trans. Magn., Vol. MAG-21, No. 5, pp. 1844–1846.

Newman, M. J., Trowbridge, C. W., and Turner, L. R., 1972, "GFUN: An Interactive Program as an Aid to Magnet Design," 4th Int. Conf. on Magnet Technology, Brookhaven. Also Rutherford Laboratory, Oxford, RPP/A94.

Nguyen-Van-Phai, 1982, "Automatic Mesh Generation with Tetrahedral Elements," Int. J. Num. Meth. Eng., Vol. 18.

Okon, E. E., 1982, "Some Vector Potential Expansion Functions for Quadrilateral and Triangular Domains," Int. J. Num. Meth. Eng., Vol. 18, pp. 727–735.

Orloff, J., and Swanson, L. W., 1979, "An Asymmetric Electrostatic Lens for Field Emission Microprobe Applications," J. Appl. Phys. Vol. 50, No. 4, April, pp. 2494–2501.

Panofsky, W. K. H., and Phillips, M., 1969, *Classical Electricity and Magnetism*, Addison–Wesley.

Park, R. H., 1929, "Two-Reaction Theory of Synchronous Machines—Generalized Method of Analysis—Part I," AIEE Trans., Vol. 48, July, pp. 716–727.

Penman, J., and Fraser, J. R., 1982, "Complementary and Dual Energy Finite Element Principles in Magnetostatics," IEEE Trans. Magn., Vol. MAG-18, No. 2, March, pp. 319–324.

Penman, J., and Grieve, M. D., 1985, "An Approach to Self Adaptive Mesh Generation," IEEE Trans. Magn., Vol. MAG-21, pp. 2567–2570.

Phillips, H. B., 1934, "Effect of Surface Discontinuity on the Distribution of Potential," J. Math. Phys., Vol. 13, pp. 261–267.

Pichler, P., Jungling, W., Selberherr, S., Guerrero, E., and Potzl, H. W., 1985, "Simulation of Critical IC-Fabrication Steps," IEEE Trans. Electron Devices, Vol. ED-32, No. 10, pp. 1940–1953.

Pinchuk, A. R., and Silvester, P. P., 1985, "Error Estimation for Automatic Adaptive Finite Element Mesh Generation," IEEE Trans. Magn., Vol. MAG-21, pp. 2551–2554.

Prenter, P. M., 1975, *Splines and Variational Methods*, John Wiley, New York.

Rafferty, C. S., Pinto, M. R., and Dutton, R. W., 1985, "Iterative Methods in Semiconductor Device Simulation," IEEE Trans. Electron Devices, Vol. ED-32, No. 10, Oct., pp. 2018–2027.

Rajanathan, C. B., Bryant, C. F., and Freeman, E. M., 1988, "A Post-Viewer for 3-D Finite Element Solutions," IEEE Trans. Magn., Vol. MAG-24, No. 1, pp. 393–396.

Rall, L. B., 1969, *Computational Solution of Nonlinear Operator Equations*, John Wiley, New York.

Reddy, J. N. 1984, *An Introduction to the Finite Element Method*, McGraw–Hill, New York.

Saito, Y., Saotome, H., Hayano, S., and Yamamura, T., 1983, "Modelling of

Hysteretic and Anisotropic Magnetic Field Problems," IEEE Trans. Magn., Vol. MAG-19, No. 6, pp. 2510–2513.

Salon, S. J., 1982, "Solution of Poisson's Equation in Two Dimensions Using the Boundary Integral Method," in *Computer Workshop in Finite Element Methods of Analysis of Electric and Magnetic Field Problems*, ch. 7, July 26–30, Union College, Schenectady, New York.

Salon, S. J., 1985, "The Hybrid Finite Element–Boundary Element Method in Electromagnetics" (invited), IEEE Trans. Magn., Vol. MAG-21, pp. 1829–1834.

Salon, S. J., and Istfan, B., 1986, "Inverse Non-Linear Finite Element Problem," IEEE Trans. Magn., Vol. MAG-22, pp. 817–818.

Salon, S. J., and Schneider, J. M., "A Hybrid Finite Element-Boundary Integral Formulation of the Eddy Current Problem," IEEE Trans. on Magn., Vol. MAG-18, pp. 461–466.

Sathiaseelan, V., Iskander, M. F., Howard, G. C. W., and Bleehen, 1986, "Theoretical Analysis and Clinical Demonstration of the Effect of Power Pattern Control Using the Annular Phased Array Hyperthermia System," IEEE Trans. on Microwave Theory and Tech., Vol. MTT-34, pp. 514–519.

Schneider, J. M., and Salon, S. J., 1980, "A Boundary Integral Formulation of the Eddy Current Problem," IEEE Trans. Magn., Vol. MAG-16, pp. 1086–1088.

Schwarz, H. A., 1869, "Über einige Abbildungsaufgaben," Crelle's J. Mathematik, Vol. 70, p. 105.

Seeger, K., 1982, *Semiconductor Physics*, 2d ed., Springer-Verlag, New York.

Seshan, C., and Cendes, Z., 1985, "Computing Magnetic Domain Patterns in Thin Film Soft Magnetic Materials," IEEE Trans. Magn., Vol. MAG-21, No. 6, pp. 2378–2381.

Sheingold, L. S., 1953, "The Susceptance of a Circular Obstacle to an Incident Dominant Circular-Electric Wave," J. Appl. Phys., Vol. 24, pp. 414–422.

Silvester, P., 1967, "Network Analogue Solution of Skin and Proximity Effect Problems," Trans. Am. Inst. Elect. Eng., Vol. PAS-86, pp. 241–247.

Silvester, P., 1968, *Modern Electromagnetic Fields*, Prentice Hall, Englewood Cliffs, New Jersey.

Silvester, P., 1969a, "Finite Element Solution of Homogeneous Wave Guide Problems," Alta Frequenza, Vol. 38, pp. 313–317.

Silvester, P., 1969b, "A General High-Order Finite-Element Waveguide Analysis Program," IEEE Trans. Microwave Theory Tech., Vol. MTT-17, No. 4, April, pp. 204–210.

Silvester, P., 1969c, "High Order Polynomial Triangular Finite Elements for Potential Problems," Int. J. Eng. Sci., Vol. 7, pp. 849–861.

Silvester, P., 1970, "Symmetric Quadrature Formulae for Simplexes," Math. Comp., Vol. 24, pp. 95–100, Jan.

Silvester, P., 1978, "Construction of Triangular Finite Element Universal Matrices," Int. J. Num. Meth. in Eng., Vol. 12, pp. 237–244.

Silvester, P., 1982a, "Universal Finite Element Matrices for Tetrahedra," Int. J. Num. Meth. in Eng., Vol. 18, No. 7, pp. 1055–1061.

Silvester, P., 1982b, "Permutation Rules for Simplex Finite Elements," Int. J. Num. Meth. in Eng., Vol. 18, No. 7, pp. 1245–1259.

Silvester, P., Cabayan, H. S., and Browne, B. T., 1973, "Efficient Techniques for Finite Element Analysis of Electric Machines," IEEE Trans. Power Apparatus Syst., Vol. PAS-92, No. 4, pp. 1274–1281, July/Aug.

Silvester, P., and Chan, K. K., 1972, "Bubnov–Galerkin Solutions to Wire Antenna Problems," Proc. IEE, Vol. 119, pp. 1095–1099.

Silvester, P., and Chari, M. V. K., 1970, "Finite Element Solution of Saturable Magnetic Field Problems," IEEE Trans. Power Apparatus Syst., Vol. PAS-89, No. 7, pp. 1642–1650, Sept./Oct.

Silvester, P. P., and Ferrari, R. L., 1983, *Finite Elements for Electrical Engineers*, Cambridge University Press, Cambridge, England.

Silvester, P., and Haslam, C. R. S., 1972, "Magnetoelluric Modelling by the Finite Element Method," Geophysical Prospecting, Vol. 20, pp. 872–891.

Silvester, P., and Hsieh, M. S., 1971, "Finite Element Solution of 2-Dimensional Exterior Field Problems," Proc. IEE, Vol. 118, No. 10, pp. 1743–1747.

Silvester, P., and Konrad, A., 1973, "Axisymmetric Triangular Finite Elements for the Scalar Helmholtz Equation," Int. J. Num. Meth. in Eng., Vol. 5, pp. 481–497.

Silvester, P. P., Lowther, D. A., Carpenter, C. J., and Wyatt, E. A., 1977, "Exterior Finite Elements for 2-Dimensional Field Problems with Open Boundaries," Proc. IEE, Vol. 124, No. 12, pp. 1267–1270.

Simkin, J., 1986, "Eddy Current Modelling in Three Dimensions (Invited)," IEEE Trans. Magn., Vol. MAG-22, pp. 609–613.

Simkin, J., and Trowbridge, C. W., 1979, "On the Use of the Total Scalar Potential in the Numerical Solution of Field Problems in Electromagnetics," Int. J. Num. Meth. in Eng., Vol. 14, pp. 423–440.

Skitek, G. G., and Marshall, S. V., 1982, *Electromagnetic Concepts and Applications*, Prentice Hall, Englewood Cliffs, New Jersey.

Southwell, R. V., 1940, *Relaxation Methods in Engineering Science*, Clarendon Press, Oxford.

Stevenson, A. R., and Park, R. H., 1927, "Graphical Determination of Magnetic Fields—Theoretical Considerations," Trans. AIEE., Vol. XLVI, pp. 112–135.

Stewart, G. W., 1973, *Introduction to Matrix Computations*, Academic Press, New York.

Stoll, R. L., 1974, *The Analysis of Eddy Currents*, Clarendon Press, Oxford.

Stone, H. L., 1968, "Iterative Solution of Implicit Approximation of Multi-Dimensional Partial Differential Equations," SIAM J. Numer. Anal., Vol. 5, pp. 530–558.

Streetman, B. G., 1980, *Solid State Electronic Devices*, Prentice Hall, Englewood Cliffs, New Jersey.

Sze, A. M., 1981, *Physics of Semiconductor Devices*, John Wiley, New York.

Thatcher, R. W., 1982, "Assessing the Error in a Finite Element Solution," IEEE Trans. Microwave Theory Tech., Vol. MTT-30, No. 6, pp. 911–915.

Tiberghien, J., 1982, *The Pascal Handbook*, Hewlett Packard, Palo Alto.

Timoshenko, S. P., and Woinowsky-Krieger, S., 1959, *Theory of Plates and Shells*, McGraw–Hill, New York.

Toyabe, T., Masuda, H., Aoki, Y., Kawaguchi, H., and Hagiwara, T., 1985, "Three-Dimensional Device Simulator Caddeth with Highly Convergent Matrix Solution Algorithms," IEEE Trans. CAD, Vol. CAD-4, No. 4, pp. 482–488, October.

Trowbridge, C. W., 1980, "Application of Integral Equation Methods to the Numerical Solution of Magnetostatic and Eddy-Current Problems," in Chari and Silvester (1980), pp. 191–213.

Trowbridge, C. W., 1981, "Three-Dimensional Field Computation," IEEE Trans. Magn., Vol. MAG-18, pp. 293–297.

Tseng, A., 1984, "A Generalized Finite Difference Scheme for Convection-Dominated Metal Forming Problems," Int. J. Num. Meth. in Eng., Vol. 20, pp. 1885–1900.

Turner, L. R., 1973, "Direct Calculation of Magnetic Fields in the Presence of Iron as Applied to the Computer Program GFUN," Rutherford Laboratory, Oxford, Report RL-73-102.

Turner, M. J., Clough, R. W., Martin, H. C., and Topp, L. J., 1956, "Stiffness and Deflection Analysis of Complex Structures," J. Aero Sci., Vol. 23, pp. 805–823.

van Bladel, J., 1964, *Electromagnetic Fields*, McGraw-Hill, New York.

Van de Vooren, A. I., and Vliegenthart, A. C., 1967, "On the 9-Point Difference Formula for Laplace's Equation," J. of Eng. Math., Vol. 1, pp. 187–202.

Varga, R. S., 1962, *Matrix Iterative Analysis*, Prentice Hall, Englewood Cliffs, New Jersey.

Visser, W., 1965, "A Finite-Element Method for the Determination of Non-Stationary Temperature Distribution and Thermal Deformations," *Proc. Conf. on Matrix Methods in Struct. Mech.*, Air Force Inst. of Tech., Wright Patterson Air Force Base, pp. 925–943.

Weeber, K., Vidyasagar, S., and Hoole, S. R. H., 1988, "Linear-Exponential Functions for Eddy Current Analysis," J. Appl. Phys., Vol. 62, No. 8, pp. 3010–3012.

Weiss, J., Konrad, A., and Silvester, P. P., "Scalar Finite Element Solution of Magnetic Fields in Axisymmetric Boundaries," IEEE Trans. Magn., Vol. MAG-18, pp. 270–274.

Wexler, A., 1969, "Computation of Electromagnetic Fields," IEEE Trans. Microwave Theory Tech., Vol. MTT-17, pp. 416–439.

Williams, C. G., and Cambrell, G. K., 1972, "Efficient Numerical Solution of Unbounded Field Problems," Electron. Lett., Vol. 8, No. 9, pp. 247–248, May.

Winslow, A. M., 1967, "Numerical Solution of the Quasilinear Poisson Equation in a Nonuniform Triangle Mesh," J. Comp. Phys., Vol. 2, pp. 149–172.

Zienkiewicz, O. C., 1977, *The Finite Element Method in Engineering Science*, 2d ed., McGraw-Hill, London.

Zienkiewicz, O. C., Bahrani, A. K., and Arlett, P. L., 1967, "Solution of Three-Dimensional Field Problems by the Finite Element Method," Engineer, Oct., pp. 547–550.

Zienkiewicz, O. C., and Cheung, Y. K., 1965, "Finite Elements in the Solution of Field Problems," Engineer, Sept., pp. 507–510.

Zienkiewicz, O. C., and Löhner, R., 1985, "Accelerated 'Relaxation' or Direct Solution? Future Prospects for FEM," Int. J. Num. Meth. in Eng., Vol. 21, pp. 1–11.

Zienkiewicz, O. C., Lyness, J., and Owen, D. R. J., 1977, "Three Dimensional Magnetic Field Determination Using Scalar Potential—A Finite Element Solution," IEEE Trans., Vol. MAG-13, pp. 1649–1656.

Zienkiewicz, O. C., and Morgan, K., 1983, *Finite Elements and Approximation*, John Wiley, New York.

# INDEX

## U



## V



## W